Study Guide & Solutions Manual

Volume 1

to accompany

ORGANIC CHEMISTRY

Third Edition

Study Guide & Solutions Manual

Volume 1

to accompany

ORGANIC CHEMISTRY

Third Edition

G. Marc Loudon
Joseph G. Stowell
Purdue University

The Benjamin/Cummings Publishing Company, Inc.

Menlo Park, California • Reading, Massachusetts
New York • Don Mills, Ontario • Wokingham, U.K. • Amsterdam
Bonn • Paris • Milan • Madrid • Sydney • Singapore • Tokyo
Seoul • Taipei • Mexico City • San Juan, Puerto Rico

Executive Editor: Sally Elliot
Sponsoring Editor: Anne Scanlan-Rohrer
Associate Editor: Leslie With
Editorial Assistant: Sharon Sforza
Accuracy Reviewers: Morton Golub; Steven Hardinger, California State University
    at Fullerton; Martin Newcomb, Wayne State University
Production Editor: Larry Olsen
Manufacturing Supervisor: Merry Free Osborn
Cover Designer: Yvo Riezebos
Cover Illustration: Joseph Maas

The table on pages 294 and 295 is reprinted by permission from the Ph.D. dissertation
of Dr. T.J. Curphey, Harvard University, 1961, as modified by H.A. Morrison.

ISBN 0-8053-6651-2

1 2 3 4 5 6 7 8 9 10—CRS— 99 98 97 96 95

The Benjamin/Cummings Publishing Company, Inc.
2725 Sand Hill Road
Menlo Park, CA 94025

Preface

The purpose of this preface is to give you an overview of the organization and features of this Study Guide and Solutions Manual so that you might use it more effectively.

Organization

Each chapter corresponds to the chapter of the text with the same number and title. The following sections are found within each chapter.

1. **Terms**

    This is a list of new or important terms encountered within the corresponding chapter of the text along with the location of each term. The lists of terms in this manual comprise the vocabulary of organic chemistry that you will need to master the subject. These lists will probably be most useful if you can *write* a definition of each term *in your own words* and give an example, if appropriate. Don't forget that if you need a more general index of terms, the text itself has a very detailed index that can be used to locate the definition of any term of interest.

2. **Concepts**

    This is a concise summary of the concepts within the chapter in outline form. In some cases, these are presented from a somewhat different perspective, and organized somewhat differently, than they are in the text. These outlines can be used in various ways. For example, they can provide a quick review after you have read the chapter; they can be used for ready reference while you are working problems; or they can serve as a refresher prior to class periods.

3. **Reactions**

    This summary consolidates the reactions of the chapter in one place and presents not only the reactions themselves but also the essential features of their mechanisms and stereochemistry for cases in which these issues were discussed in the text. Studying and learning reactions is a skill that should be cultivated, and Study Guide Link 5.1 on page 88 of this manual will help you to use these reaction summaries to best advantage.

4. **Study Guide Links**

These are short extensions of text material that are called out by margin icons within the text. Study Guide Links are of two types. A Study Guide Link of the first type, flagged with a checked (✓) icon, provides additional explanations of topics that typically cause difficulty. For example, Study Guide Link 5.1, "How to Study Organic Reactions," (mentioned in item 3 above) is flagged with a check at a point in the text at which some students begin to have difficulty learning reactions. This Study Guide Link provides detailed hints on how to study and learn organic reactions efficiently. A Study Guide Link of the second type, called out in the text by an unchecked icon, provides more in-depth information about the particular topic or a different way of looking at the same topic. Although it may at first seem that the material in the Study Guide Links should have been included in the text, doing so would have made the text unacceptably long, and, in some cases, would have interrupted the logical flow of material.

5. **Solutions**

We have endeavored not only to provide answers to problems but also to show how these answers are deduced. Solutions to the problems within the text come first, and these are followed by solutions to the additional problems. Only solutions to the *asterisked* problems are provided. (See the discussion of paired problems, below.) The best use of solutions is to try to work problems with the solutions closed or covered, and consult a solution only when you have answered the problem or have given it a reasonable effort.

Features

This manual contains certain features that are designed to help you use it more effectively.

1. **Typesetting Conventions**

The text and typesetting conventions are the same ones used in the text.

2. **Index Tabs**

You will find on the first right-hand page of each Solutions section a black rectangle at the edge of the page. (For example, the first of these appears on page 11.) The edges of these rectangles can be seen when the manual is viewed end-on and the pages are bent back slightly. These serve as index tabs that can be used to locate the solutions sections rapidly. In Volume 1, they are arranged in two rows of nine, and in Volume 2, they are arranged in one row of nine. Thus, to locate the solutions for Chapter 12, bend back the pages of Volume 1 and open to the third index mark in the second row.

3. **Icon Comments**

Within the solutions to the problems, you will occasionally find comments marked with icons of two different types.

 This icon indicates a comment that provides additional information about the topic covered in the problem, an alternate correct answer, or another way of looking at the problem.

 This icon indicates a special caution about some aspect of the problem or the material on which the problem is based—something you should be careful about to avoid confusion.

4. Paired Problems

Many of the problems within the text are *paired*. This means that some problems (or problem parts) marked with an asterisk are followed by problems (or problems parts) of a similar type, in most cases of equal or lesser difficulty, which are not asterisked. Only the answers to the asterisked problems appear in the Solutions sections of this manual. If you understand the solution to an asterisked problem (or part), you should be able to work the unasterisked problem (or part) that follows using similar reasoning. (In one or two cases, an unasterisked problem follows a worked-out Study Problem in the text rather than an asterisked problem.) Note that not every problem is paired in this way; some asterisked problems have no unasterisked partner. You may want to use the unasterisked problems as a "test bank" that can be used for a review prior to examinations.

This manual has undergone three separate accuracy checks by the authors and a critical reading by three accuracy checkers. We have tried hard to make this manual as free of errors as possible. Nevertheless, if errors are found, we would like to know about them so that we can correct them on reprint if possible. You can send these or any comments to us by electronic mail at *loudonm@sage.cc.purdue.edu,* by FAX at 317-494-7880, or by mail at RHPH Building, Purdue University, West Lafayette IN 47907-1330.

We would like to gratefully acknowledge the many comments of the users of the previous edition of this manual, especially Mark Cushman, John Schwab, Ron Magid, and Phil Fuchs, as well as their teaching assistants and students. We particularly appreciate the conscientious work of the accuracy checkers, Professor Martin E. Newcomb, Professor Steve Hardinger, and Dr. Morton Golub.

We sincerely hope that you find this manual useful in your study of organic chemistry.

June 26, 1995
G. Marc Loudon and Joseph G. Stowell

Contents

Introduction to Stereochemistry 110

Cyclic Compounds and Stereochemistry of Reactions 132

Introduction to Alkyl Halides, Alcohols, Ethers, Thiols, and Sulfides 163

Chemistry of
Alkyl Halides   185

Chemistry of Alcohols,
Glycols, and Thiols   216

Chemistry of Ethers,
Epoxides, and Sulfides   245

Infrared Spectroscopy and Mass Spectrometry 270

Nuclear Magnetic Resonance Spectroscopy 285

Chemistry of Alkynes 312

Dienes, Resonance,
and Aromaticity 331

Chemistry of Benzene
and Its Derivatives 360

Allylic and
Benzylic Reactivity 390

Chemistry of Aryl Halides,
Vinylic Halides, and Phenols   414

1

Chemical Bonding
and Chemical Structure

Terms

The glossary in this and subsequent chapters is a list of the key terms and concepts contained in the chapter. These terms and concepts will be used throughout the text. It probably will not help you to memorize the exact definitions given in the text; rather, define each of these terms and concepts in your own words and give an example if appropriate.

Concepts

I. Atoms

A. PERIODIC TABLE

1. The elements in the first two rows of the periodic table along with K, Ca, Br, and I are of greatest importance in organic chemistry.
2. The arrangement of electrons in atoms is suggested by the periodic table.
3. A neutral atom of each element contains a number of both protons and electrons equal to its atomic number.
4. The ease with which neutral atoms lose electrons to form positive ions increases to the left and towards the bottom of the periodic table.
5. The ease with which neutral atoms gain electrons to form negative ions increases to the right and towards the top of the periodic table.

B. VALENCE ELECTRONS

1. The electrons in the outermost shell of an atom are called valence electrons.
2. The number of valence electrons for any neutral atom in an A group of the periodic table (except He) equals its group number.
3. Stable ions are formed when atoms gain or lose valence electrons in order to have the same number of electrons as the noble gas of closest atomic number.

II. Chemical Bonds

A. IONIC BONDS

1. The ionic bond—a force that hold atoms together within molecules—is an electrostatic attraction between ions of opposite charge.
2. The ionic bond is most likely to form between atoms at opposite ends of the periodic table.
3. The formation of ions tends to follow the octet rule—each atom is surrounded by eight valence electrons (two electrons for hydrogen).
4. The ionic bond is the same in all directions (it has no preferred orientation in space).
5. When an ionic compound dissolves in water:
   a. ionic bonds are broken;
   b. the ionic compound dissociate into free ions.

$$Na^+Cl^- \xrightarrow{\ H_2O\ } Na^+ + Cl^-$$

B. COVALENT BONDS

1. Covalent bonds are formed when the orbitals of different atoms overlap and share electrons.
2. Both electrons in a covalent bond are shared equally between the bonding atoms.
3. Covalent bonding can be understood to arise from the filling of bonding molecular orbitals by electron pairs.
4. The bonding in covalent compounds tends to follow the octet rule—each atom is surrounded by eight valence electrons (two electrons for hydrogen).
5. The covalent bond has a definite direction in space.

C. POLAR COVALENT BONDS

1. In polar covalent compounds, electrons are shared unequally between bonded atoms, and a bond dipole results.
   a. The tendency of an atom to attract electrons to itself in a covalent bond is indicated by its electronegativity.
   b. A polar bond is a bond between atoms of significantly different electronegativities.
2. Partial charge in a polar covalent bond is indicated by a delta (δ), which reads "partially" or "somewhat."

3. The uneven electron distribution in a compound containing covalent bonds is measured by a quantity called the dipole moment μ in units of debyes D.
    a. Dipole moments of typical polar organic molecules are in the 1–3 D range.
    b. Each polar bond has associated with it a dipole moment contribution, called a bond dipole.
4. A bond dipole is a vector quantity; the dipole moment is oriented from the partial positive charge to the partial negative charge.
    a. The vector sum of all bond dipoles in a molecule is its dipole moment.
    b. Two vectors of equal magnitude oriented in opposite directions always cancel.

5. Molecules that have permanent dipole moments are called polar molecules.
6. The polarity of a molecule can significantly affect its chemical and physical properties.

D. OCTET RULE

1. An atomic species tends to be especially stable when its valence shell contains eight electrons.
2. The tendency of atoms to gain or lose valence electrons to form ions with the noble-gas configuration is known as the octet rule.
    a. The octet rule is often obeyed in covalent bonding.
    b. The sum of all bonding electrons and unshared pairs for atoms in the second period of the periodic table will under no circumstances be greater than eight (octet rule). These atoms may, however, have fewer than eight electrons.
    c. In some cases, atoms below the second period of the periodic table may have more than eight electrons. However, these cases occur so infrequently that Rule b should be followed until exceptions are discussed.
3. To determine if an atom has a complete octet, count all unshared valence electrons and all bonding electrons.

E. LEWIS STRUCTURES

1. Lewis strucutres are molecular structures that use lines and/or dot symbols between atoms to denote bonds and dot symbols on atoms to denote unshared electrons.
    a. Nonvalence electrons are not shown in Lewis structures.
    b. A hydrogen can share no more than two electrons.
2. A bond consisting of one electron pair is called a single bond (bond order = 1).
3. A bond consisting of two electron pairs is called a double bond (bond order = 2).
4. A bond consisting of three electron pairs is called a triple bond (bond order = 3).
5. A charge on each atom assigned by electronic bookkeeping is called its formal charge.
    a. Formal charge = $U + \frac{1}{2}S - V$, where U is the number of unshared electrons, S is the number of shared electrons, and V are the number of valence electrons in the neutral atom.
    b. Formal charge, if not equal to zero, is indicated with a plus or minus sign.

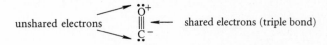

unshared electrons — shared electrons (triple bond)

III. Molecular Structure

A. GENERAL

1. The structure of a molecule is determined by its bond lengths, bond angles, and conformation.
2. Sources of molecular structure are
    a. X-ray crystallography
    b. electron diffraction
    c. microwave spectroscopy
3. How a molecule reacts chemically is closely linked to its structure.

B. BOND LENGTHS

1. The bond length is the distance between the centers of the bonded nuclei (in Å; Å = 10^{-10} m).
2. Bond lengths are governed, in order of importance, by:
   a. the period (row) of the periodic table from which the bonded atoms are derived—bond lengths increase significantly down a column of the periodic table;
   b. the bond order (the number of covalent bonds shared by two atoms)—bond length decreases with increasing bond order;
   c. the column (group) of the periodic table from which the atoms in the bond are derived—bond length decreases to the right along a given row (period) of the periodic table.
3. An unshared electron pair can be considered as a bond without a nucleus at one end.

C. BOND ANGLES

1. The bond angle is the angle between each pair of bonds to the same atom.
2. The bond angles within a molecule determine its shape.
3. Approximate bond angles can be predicted by assuming that the groups bound to a central atom are as far apart as possible.
   a. An atom bonded to four groups has tetrahedral geometry (bond angle = 109.5°).
   b. An atom bonded to three groups has trigonal planar geometry (bond angle = 120°).
   c. An atom bonded to two groups has linear geometry (bond angle = 180°).

D. CONFORMATIONS

1. Conformation is the spatial relationship of the bonds on adjacent atoms and is important for molecules that contain many atoms.
2. The angular relationship of planes is called the dihedral angle.
3. The conformations of a molecule are described by specifying the dihedral angles between bonds on adjacent atoms.
4. Newman projections are representations of molecules drawn by sighting along the bond of interest.

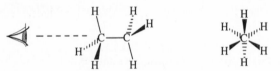

| view down the carbon-carbon bond | end-on view | Newman projection |

E. RESONANCE STRUCTURES

1. A molecule that is not adequately described by a single Lewis structure is represented as a resonance hybrid.
   a. A resonance hybrid is a weighted average of two or more fictitious Lewis structures.
   b. When resonance structures are identical, they are equally important descriptions of the molecule.
   c. Resonance between two Lewis structures is shown by a double-headed arrow.
   d. Resonance structures are not in equilibrium; the molecule is one structure that is a hybrid of the fictitious Lewis structures.

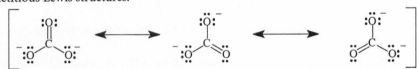

2. A molecule represented by resonance structures is more stable than its fictional resonance contributors and is said to be resonance-stabilized.

IV. *Nature of the Electron*

A. WAVELIKE PROPERTIES

1. In the submicroscopic world of the electron and other small particles, there is no real distinction between particles and waves.
   a. The behavior of very small objects simultaneously as waves and particles is called the wave-particle duality.

a. Orbitals are descriptions of the wave properties of electrons in atoms and molecules, including their spatial distribution.

b. As a consequence of their wave properties, electrons in atoms and molecules can exist only in certain allowed energy states (orbitals).

2. The theory that describes the behavior of electrons in molecules is called quantum mechanics.

a. Heisenberg uncertainty principle: the exact position and velocity of an electron cannot be specified; only the probability that it occupies a certain region of space can be specified.

b. Pauli exclusion principle: no two electrons may have all four quantum numbers the same. A maximum of two electrons may be placed in any one atomic orbital and these electrons must have opposite spins.

c. Hund's rules: electrons are distributed among identical orbitals of equal energy, single electrons are placed into separate orbitals before the orbitals are filled, and the spins of these unpaired electrons are the same.

d. Aufbau principle: to determine the electronic configuration of an atom, electrons are placed one by one into orbitals of the lowest possible energy in a manner consistent with the Pauli exclusion principle and Hund's rules.

3. Electron density within an orbital is a matter of probability (Heisenberg uncertainty principle); think of an orbital as a "smear" of electron density.

B.   QUANTUM NUMBERS

1. Electrons in orbitals are characterized by quantum numbers which, for atoms, are designated n, l, m, and s.

a. The principal quantum number n can assume any integral value greater than zero ($n = 1, 2, 3, \ldots$).

b. The angular momentum quantum number l depends on the value of n and can assume any integral value from zero through $n - 1$ ($l = 0, 1, 2, \ldots, n - 1$) where $l = 0 \Rightarrow s$; $l = 1 \Rightarrow p$, $l = 2 \Rightarrow d$, and $l = 3 \Rightarrow f$.

c. The magnetic quantum number m depends on the value of l and can assume integral values between $-l$ and $+l$.

d. The spin quantum number s can assume a value of $+\frac{1}{2}$ or $-\frac{1}{2}$.

2. Each orbital is described by a series of three quantum numbers:

a. The principal quantum number n governs the energy of an orbital; orbitals of higher n have higher energy.

b. The angular momentum quantum number l governs the shape of an orbital: $l = 0$ (s orbitals) are spheres; orbitals with $l = 1$ (p orbitals) are dumbbells.

c. The magnetic quantum number m governs the orientation of an orbital.

C.   ENERGIES

1. The energy of an electron is said to be quantized; that is, limited to certain values.

a. The higher the principal quantum number n of an electron, the higher the energy.

i. In atoms other than hydrogen, the energy is also a function of the l quantum number.

ii. Electrons with the same principal quantum number n but different values of l have different energies.

b. Gaps between energy levels become progressively smaller as the principal quantum number increases.

c. The energy gap between orbitals that differ in principal quantum number is greater than the gap between two orbitals within the same principal quantum number.

V. *Orbitals*

A.   ATOMIC ORBITALS

1. Some orbitals contain nodes which separate the wave-peak parts of the orbitals from the wave-trough parts.

a. An orbital with principal quantum number n has $n - 1$ nodes.

b. The number of nodes in an atomic orbital increases with its principal quantum number.

c. The greater number of nodes in orbitals with higher n is a reflection of their higher energies.

2. Each atomic orbital is characterized by a three-dimensional region of space in which the electron is most likely to exist.
    a. The size of an atomic orbital is governed by its principal quantum number n; that is, the larger is n, the larger the region of space occupied by the orbital.
    b. The angular momentum quantum number l governs the shape of an atomic orbital.
    c. The magnetic quantum number m governs the directionality of an atomic orbital.
3. Atomic orbitals are populated with electrons according to the Aufbau principle.
4. The distribution of electron density in a given type of orbital has a characteristic arrangement in space:
    a. An electron in an s atomic orbital is a smear of electron density in the shape of a sphere.
    b. An electron in a p orbital has a smear of electron density in the shape of a dumbbell consisting of two lobes that are part of the same orbital and are directed in space.
    c. The three p orbitals are mutually perpendicular.

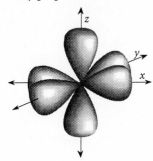

5. The electronic configuration of an atom or ion describes the distribution of electrons among the atomic orbitals.

$$\text{chlorine atom} \quad 1s^2 2s^2 2p^6 3s^2 3p^5$$

B. HYBRID ORBITALS

1. Hybrid orbitals arise from the mixing of pure orbitals.
2. The hybridization of an atom and the geometry of the attached atoms are closely related.
    a. Hybridization affords bonds that are as far apart as possible.
    b. Hybridization of an atom provides orbitals that have the bulk of their electron density directed toward the nuclei of the bonded atoms.
    c. Hybridization gives stronger, more stable bonds.
3. One $2s$ orbital and three $2p$ orbitals can be mixed to give four equivalent orbitals which are one part s character and three parts p character, called sp^3 orbitals.
    a. The electron density of an sp^3 orbital is highly directed in space.
    b. One of the lobes is very large and is directed towards the bonded atom; the other lobe is very small.
    c. All sp^3-hybridized atoms have tetrahedral geometry.

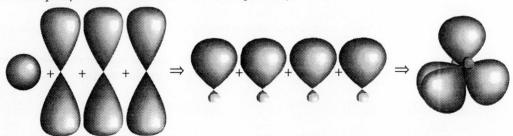

one $2s$ and three $2p$ orbitals   four sp^3 orbitals   tetrahedral arrangement

C. MOLECULAR ORBITALS

1. Chemical bonding is an energetically favorable process.
2. When atoms combine into a molecule, the electrons of each atom are no longer localized on individual atoms but "belong" to the entire molecule.

3. Combination of *x* atomic orbitals gives *x* molecular orbitals.
    a. One molecular orbital of the hydrogen molecule is formed by the addition (additive overlap) of the individual atomic orbitals (a bonding molecular orbital).
        i. Bonding molecular orbitals arise through wave peak-wave peak or wave trough-wave trough interactions.
        ii. The energy of an electron in the bonding molecular orbital is lower than that of an electron in an isolated atom.
    b. The second molecular orbital of the hydrogen molecule is generated by the subtraction (subtractive overlap) of the individual atomic orbitals (an antibonding molecular orbital).
        i. Antibonding molecular orbitals arise through wave peak-wave trough interactions.
        ii. The energy of an electron in an antibonding molecular orbital is higher than that of an electron in an isolated atom.
4. Molecular orbitals are populated with electrons according to the Aufbau principle.
    a. The electrons occupy the bonding molecular orbitals before occupying the antibonding orbitals.
    b. The most common bonding situations occur when bonding molecular orbitals contain electron pairs and antibonding molecular orbitals are empty.
5. The Lewis view of the electron-pair bond is approximately equivalent to the quantum-mechanical idea of a bonding molecular orbital occupied by a pair of electrons.
    a. Bonds that have cylindrical symmetry about the internuclear axis are called σ bonds (sigma bonds).
    b. The carbon in methane has an arrangement of orbitals that is different from the orbitals in atomic carbon.

Study Guide Links

✓ 1.1   Formal Charge

After you have a small amount of practice assigning formal charges, you'll begin to notice certain patterns that emerge. For example, an oxygen with three single bonds and one unshared electron pair *always* has a +1 formal charge:

$$ —\overset{\overset{\textstyle +}{\cdot\cdot}}{\underset{|}{O}}— $$

Some other common patterns are shown below.

carbon:	*nitrogen:*	*oxygen:*

(Verify these patterns by calculating the formal charge in each one!) Recognition of these patterns can be a great time-saver in applying formal-charge rules. Conversely, when you calculate a formal charge, always double-check your calculation if you appear to be violating one of these common patterns.

1.2   Dipole Moments

The mathematical definition of the dipole moment μ is as follows:

$$ \mu = q\mathbf{r} \tag{SG1.1} $$

where q is the magnitude of the separated charge and $\mathbf{r}$ is a vector from the site of positive charge to the site of negative charge. When μ is a bond dipole, as in H—Cl, the magnitude of $\mathbf{r}$ is simply the bond length; in the case of H—Cl, it is the length of the H—Cl bond. In the HCl molecule the dipole moment vector μ lies along the H—Cl bond, and points from the H (the site of positive charge) to the Cl (the site of negative charge), as shown in the text. That is, μ and $\mathbf{r}$ *have the same direction.*

The dimensions of the dipole moment, as suggested by Eq. SG 1.1 above, correspond to charge × length. As the text indicates, the units of the dipole moment are called *debyes*. A dipole moment of one debye results when opposite charges, each with a magnitude of 1×10^{-10} electrostatic unit (esu), are separated by one Ångstrom. That is,

$$ 1 \text{ debye} = 1 \times 10^{-10} \text{ esu Å} $$

These units were established prior to the current trend toward the use of standard international units. To put these units on more familiar ground, the charge on an electron is 4.8×10^{-10} esu, and one Ångstrom = 10^{-10} meter = 100 picometers. (In standard international (SI) units, in which the unit of length is the meter and the unit of charge is the coulomb (C), 1 D = 3.34×10^{-30} C-m, or coulomb-meter.)

An actual calculation of the charge separation in H—Cl can provide a more quantitative sense of the meaning of the dipole moment. A table of experimentally measured dipole moments gives μ =1.08 D for HCl. A table of bond lengths gives the length of the H—Cl bond as 1.274 Å; this is r in Eq. SG 1.1. The only unknown remaining in Eq. SG 1.1 is q. Because μ and **r** have the same direction, q can be calculated by diving the magnitude of μ by that of **r**.

$$q = |\mu| \div |\mathbf{r}| = 1.08 \times 10^{-10} \text{ esu Å} \div 1.274 \text{ Å} = 0.848 \times 10^{-10} \text{ esu}$$

As noted above, the charge on an electron is 4.8×10^{-10} esu. Consequently, the charge separation in HCl in electronic charge units is

$$q \text{ (in electrons)} = (0.848 \times 10^{-10} \text{ esu}) \div (4.8 \times 10^{-10} \text{ esu per electron})$$
$$= 0.18 \text{ electron}$$

In other words, the dipole moment of HCl, 1.08 D, means that a partial negative charge of 0.18 units ("18% of an electron") is on the chlorine of HCl, and a partial positive charge of 0.18 units is on the hydrogen.

Notice that the definition of dipole moment in Eq. SG 1.1 contains two elements: the *amount* of charge separated (q) and the *distance* between the separated charges (**r**). A smaller amount of charge separated by a long bond can result in as significant a dipole moment as a larger amount of charge separated by a shorter bond. (See Problem 1.31 in Chapter 1.)

✓1.3   **Structure Drawing Conventions**

Don't be confused by typesetting conventions that seem to ignore what you've just learned about molecular geometry. For example, you might see methane (CH_4) written as follows:

$$\begin{array}{c} \text{H} \\ | \\ \text{H---C---H} \\ | \\ \text{H} \end{array}$$

You now know that methane is tetrahedral. But there is a good reason for using simpler but less accurate structures anyway. When geometrically accurate structures are not needed, there is no point in going to the additional trouble (and in the case of typesetting, the expense) of using them. In other words, we use the simplest structures that accomplish the purpose at hand, and so should you! When the shape of a molecule is an important issue, use line-and-wedge formulas or other types of structures that convey spatial information. Otherwise, use the simpler structures.

Solutions

Solutions to In-Text Problems

1.1   (a) Because sodium (Na) is in Group 1A, it has one valence electron.
(c) Oxygen has six valence electrons; hence, O^{2-} has eight.

1.2   (a) Because neon has ten valence electrons, the negative ion requested in the problem also has ten. The negative ion with ten valence electrons is F^-.
(c) The neon species isoelectronic with neutral fluorine has nine valence electrons, and is therefore Ne^+.

1.3   (a)   (c)   (e)

chloroform   ammonia   hydronium ion

1.4   One structure is that of ethanol, and the other is that of dimethyl ether.

ethanol   dimethyl ether

> In all but the simplest cases there are many structures that have the same atomic composition. For molecules of moderate size hundreds or even thousands of structures are possible. Compounds that have the same atomic composition, but different atomic connectivities, are called **constitutional isomers**. Thus, dimethyl ether and ethanol are constitutional isomers. You'll learn about isomers in Chapter 2.

1.5   (a)

allene

1.6   The formal charge on boron is –1; the formal charge on nitrogen is +1; the formal charge on the hydrogens is zero. The net charge on the entire structure is zero.

1.7   Of the atoms in the molecule, carbon and hydrogen differ least in electronegativity (Table 1.1). Consequently, the C—H bonds are the least polar bonds. The carbon with the most partial positive character is the one bound to the greatest number of electronegative atoms:

atom with most partial positive character

1.8   (a) Water has bent geometry; that is, the H—O—H bond angle is approximately tetrahedral. Repulsion between the lone pairs and the bonds reduces this bond angle somewhat. (The actual bond angle is about 105°.)

(c) The formaldehyde molecule has trigonal planar geometry. Thus, both the H—C—H bond angle and the H—C=O bond angle are about 120°.

1.10   In the following Newman projections, the near carbon is shown as a circle and the far carbon is hidden from view.

(a)   (b)

1.12   The resonance structures of benzene:

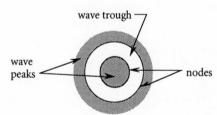

1.13   (a) A 3s orbital is three concentric spheres of electron density, each separated by a node, as shown in the following "cutaway" diagram:

1.14   (a) Nitrogen atom: $1s^2 2s^2 2p^3$. The three $2p$ electrons are unpaired; each one is in a different p orbital.
(c) Sodium atom: $1s^2 2s^2 2p^6 3s^1$
(d) Magnesium atom: $1s^2 2s^2 2p^6 3s^2$

1.15   (a) The pictures of the orbitals are exactly the same as they are in Fig. 1.14 of the text. However, the various species differ in their electron occupancies.

(a1)   The He_2^+ ion contains three electrons. By the Aufbau principle, two electrons occupy the bonding molecular orbital, and one occupies the antibonding molecular orbital. This is shown on the left side of the electron-occupancy diagram in Fig. SG1.1 at the top of the next page. Because more electrons occupy the bonding molecular orbital, this species is stable.

(a3)   The H_2^{2-} ion contains four electrons; two occupy the bonding molecular orbital, and two occupy the antibonding molecular orbital. This is shown on the right side of the occupancy diagram in Fig. SG1.1. This species has no energetic advantage over two dissociated hydride ($H:^-$) ions, and therefore readily dissociates.

(b) Because the H_2^+ ion has one electron in a bonding molecular orbital, it should have about half the stability of H_2 itself relative to dissociated fragments. Thus, 217 kJ/mol (52 kcal/mol) is an estimate of the bond dissociation energy of this species.

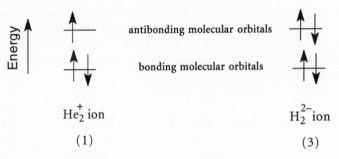

Figure SG1.1   *Electron-occupancy diagram for the solution to Problem 1.15.*

1.16   If the oxygen of water is sp^3-hybridized, then two of the sp^3 hybrid orbitals contain unshared electron pairs. The other two sp^3 hybrid orbitals contain one electron each; each of these overlaps with the $1s$ orbital of a hydrogen atom (which contains one electron) to give the two sp^3-$1s$ σ bonds (the O—H bonds) of water.

Solutions to Additional Problems

1.17   (a) Compound (3) would be most likely to exist as an ionic species, because Na and At (astatine, the halogen below iodine) come from opposite ends of the periodic table.

1.18   Species (b) and (c) have complete octets on every atom (duets for hydrogens); in (a), (d), and (e), carbon, boron, and iodine, respectively, have electronic sextets. The formal charges are:
   (a)   +1 on carbon   (b)   0 on all atoms
   (c)   −1 on carbon   (d)   0 on all atoms
   (e)   +1 on iodine

1.19   (a)   (c)

H₂C=CH—Cl   C₂H₃Cl   H₂C=C=Ö   C₂H₂O
                                              ketene

1.20   (a) Perchlorate: −1 on each oxygen, +3 on chlorine, −1 overall.
   (c) Methylene: zero on all atoms, zero overall.
   (e) Trimethylamine oxide: −1 on oxygen, +1 on nitrogen, zero overall.
   (g) Hypochlorite: zero on chlorine, −1 on oxygen, −1 overall.
   (h) Ethyl radical: zero on all atoms, zero overall.

1.21   (a) Chlorine atom: $1s^2 2s^2 2p^6 3s^2 3p^5$
   (c) Argon atom: $1s^2 2s^2 2p^6 3s^2 3p^6$

1.22   The $2d$ orbital is not permitted. The maximum l value for an orbital with principal quantum number $= 2$ is $l = 1$; but d would require that $l = 3$.

1.23   (a) The carbon is trigonal-planar; consequently, the H—C—H bond angle is about 120°.
   (c) Same as (a).

(e) The central oxygen is bound to three "groups," two oxygens and an electron pair. Hence, it is trigonal planar, and the O=O—O bond angle is 120°.

(g) The nitrogen is trigonal planar; the O=N—O bond angle is 120°, as are both C—N—O bond angles. The carbon is bound to four groups, and is tetrahedral; all bond angles centered on carbon are about 109.5°.

(i) Because the carbon is trigonal-planar, all bond angles are about 120°.

1.24   (a) The longest bond is the carbon-carbon bond.

(b) The shortest bonds are the carbon-hydrogen bonds.

(c) The two carbon-oxygen bonds are equivalent by resonance.

(d) The carbon-oxygen bonds are the most polar, because they connect the atoms (C and O) that differ most in electronegativity.

1.25   (a) Because two negative charges are shared equally by three oxygens, there is 2/3 of a negative charge per oxygen.

(b) The bond order is equal to the bond order in each structure times the weight of each structure. Because the structures are equivalent, their weights are equal (1/3 each). Thus, the bond order is (1/3)(2) + (2/3)(1) = 2/3 + 2/3 = 4/3.

1.26   (c) The resonance structure on the right violates the octet rule for nitrogen; that is, nitrogen is surrounded by ten electrons. The reason that neither the analogous structure for the allyl cation does not violate the octet rule is that the positively-charged carbon is surrounded by only six electrons and can therefore accomodate the additional electron pair that completes its octet.

    Notice from this example that a positive charge does not necessarily mean that an atom lacks an octet. This point will be considered further in Chapter 3.

1.27   (a) Figure 1.8 shows that a 2s orbital has a single spherical node; and Figure 1.10 shows that the 2p orbital has one planar node. Figure 1.11 shows that the 3p orbital has two nodes—one planar node and one spherical node. In all cases, the total number of nodes is one less than the principal quantum number. In the 2s orbital, the value of *l* is 0, and the number of planar nodes is zero. In the 2p and 3p orbitals, the value of *l* is 1, and each orbital has one planar node.

(b) From the generalizations in part (a) of the problem, a 5s orbital has four spherical nodes. (In fact, it is five concentric spheres of electron density.) A 3d orbital has two nodes; because the value of *l* is 2, both nodes are planar. (See the following problem.)

1.28   There are two planar nodes, one along the *x*-axis and one along the *y*-axis.

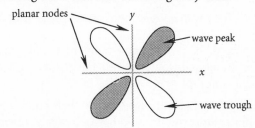

1.30   A 4p orbital is somewhat like a 3p orbital, except that it has one additional spherical node and two additional regions of electron density. (See Fig. SG1.2 at the top of page 14.)

1.31   Because fluorine is more electronegative than chlorine, a larger dipole moment would be expected for methyl fluoride than for methyl chloride, other things being equal. However, Study Guide Link 1.2 indicates that the magnitude of a dipole moment is the product of charge separation and the distance by which the charges are separated, that is, the bond length. The greater carbon-halogen bond length in

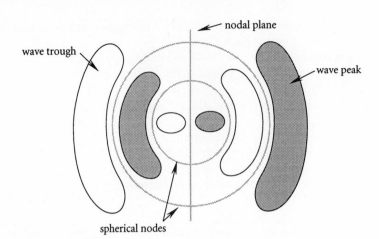

Figure SG1.2.   *Cutaway diagram of a 4p orbital for the solution to Problem 1.30.*

methyl chloride offsets the smaller electronegativity of chlorine; hence, the dipole moment of methyl chloride and methyl fluoride are nearly the same.

1.32   No matter how any CH_2 group is turned, the resultant bond dipole is the same:

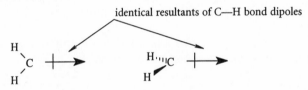

Hence, both conformations of ethylene should have zero dipole moment because the resultants of the two CH_2 groups cancel each other in either conformation. Thus, the observation of zero dipole moment does not permit a choice between these two conformations.

1.33   Quantum theory states that the energy E of light absorbed must exactly equal the difference between the orbitals, which, from the statement of the problem, is
1.635×10^{-18} J. Solving the given equation for λ and substituting,

$$\lambda \;=\; hc/E \;=\; (6.626 \times 10^{-34} \text{ J sec})(3 \times 10^8 \text{ m sec}^{-1})/(1.635 \times 10^{-18} \text{ J})$$
$$=\; (1.216 \times 10^{-7} \text{ m})(10^{10} \text{ Å m}^{-1}) = 1216 \text{ Å}$$

As noted in the problem, light with a wavelength of 1216 Å is in the far ultraviolet region of the spectrum.

1.34   A linear water molecule would have zero dipole moment, because the O—H bond dipoles would be oriented in opposite directions and would cancel. Hence, the nonzero dipole moment for water shows that it is a bent molecule. Out with Professor Szents!

1.35   When the O—H bonds lie at a dihedral angle of 180° their bond dipoles cancel; consequently, a hydrogen peroxide molecule in this conformation would have a dipole moment of zero. This conformation is *ruled out* by the observation of a significant dipole moment. (See Fig. SG2.2 at the top of p. 15.)

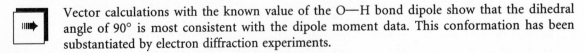

Vector calculations with the known value of the O—H bond dipole show that the dihedral angle of 90° is most consistent with the dipole moment data. This conformation has been substantiated by electron diffraction experiments.

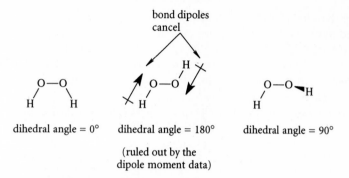

dihedral angle = 0°   dihedral angle = 180°   dihedral angle = 90°

(ruled out by the
dipole moment data)

Figure SG2.2.   *The conformations of hydrogen peroxide to accompany the solution to Problem 1.35.*

1.36   (a) and (b):   The overlap of the two $2p$ orbitals is side-to-side as shown in the following diagram:

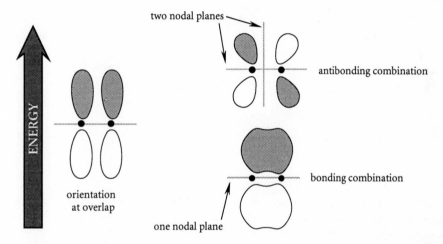

Notice that the molecular orbital of higher energy has more nodes than the one of lower energy.

(c) When two electrons occupy the bonding molecular orbital, the resulting bond is *not* a σ bond because the molecular orbital is not cylindrically symmetrical.

> This type of bond, called a pi (π) bond, is important in the carbon-carbon double bond, and is discussed in Chapter 4.

1.38   When a hydrogen molecule absorbs light, an electron jumps from a bonding molecular orbital into an antibonding molecular orbital:

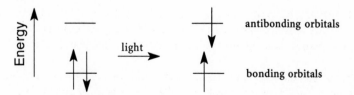

After light absorption, the hydrogen molecule has no excess of bonding electrons, and there is no energetic advantage to bonding. Hence, the molecule can readily dissociate into hydrogen atoms.

1.39   (a) The configuration of the atom with atomic number = 8: $1s^3 2s^3 2p^2$

(b) The second noble gas has a filled $2p$ orbital; but there are only two $2p$ orbitals ($m = 0, +1$); so the $2p$ level holds six electrons (three in each of the $2p$ orbitals). With filled $1s$, $2s$, and $2p$ levels, the second noble gas has $(3 + 3 + 6) = 12$ electrons; hence, its atomic number is 12.

(c) Because there are three values of spin, each orbital can hold three electrons. [You had to know this to get the right answer to parts (a) and (b)].

(d) Because a filled shell contains nine electrons (three electrons each in one s and two p orbitals), the corresponding rule might be called a "nonet rule."

2

Alkanes

Terms

Concepts

I. Hydrocarbons

A. GENERAL

1. Hydrocarbons are compounds that contain only the elements carbon and hydrogen.
2. Hydrocarbons are derived from petroleum and are important raw materials for the preparation of other organic compounds.
3. Some hydrocarbons are used as fuels.

B. Aliphatic hydrocarbons

1. Alkanes (paraffins) contain only single bonds.
   a. Normal alkanes (*n*-alkanes) contain unbranched carbon chains.
   b. Alkanes have sp^3-hybridized carbon atoms with tetrahedral geometry.
   c. Alkanes may contain branched chains, unbranched chains, or rings (cycloalkanes).

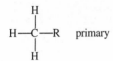 **propane**   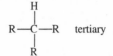 **cyclopropane**

   i. A primary carbon is bonded to one other carbon.
   ii. A secondary carbon is bonded to two other carbons.
   iii. A tertiary carbon is bonded to three other carbons.
   iv. A quaternary carbon is bonded to four other carbons.

2. Alkenes (olefins) contain carbon-carbon double bonds.
3. Alkynes (acetylenes) contain carbon-carbon triple bonds.

C. Aromatic Hydrocarbons

1. Aromatic hydrocarbons consist of benzene and its substituted derivatives.
2. Aromatic hydrocarbons are also called arenes.

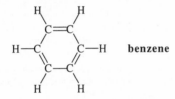 **benzene**   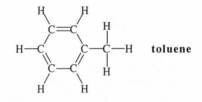 **toluene**

II. Molecular Structure

A. Empirical Formula

1. The empirical formula (obtained from elemental analysis) of a compound gives the smallest whole-number molar proportions of the elements present in the compound.
   a. Convert the relative masses of the elements into molar proportions by dividing the mass percent of each element by its atomic mass.
   b. Divide the molar proportion of each element by that of the element present in the smallest proportion.
   c. Multiply the resulting proportions by successive integers (2, 3, 4, …) until whole-number proportions for all elements are obtained.

$$\text{butane: } 82.66 \text{ \% C}; 17.34 \text{ \% H} \Rightarrow C_{\frac{82.66}{12.011}}H_{\frac{17.34}{1.008}} \Rightarrow C_{6.88}H_{17.2} \Rightarrow C_1H_{2.5} \Rightarrow C_2H_5$$

B. Molecular Formula

1. The molecular formula of a compound gives its atomic composition.
   a. All noncyclic alkanes have the general formula C_nH_{2n+2} in which *n* is the number of carbon atoms in the alkanes.
   b. The family of unbranched alkanes form a series in which successive members differ from one other by each —CH_2— (methylene) group.

$$\text{decane} \qquad C_{10}H_{22} \qquad (n = 10, H = 2n + 2 = 22)$$

2. The molecular formula may or may not differ from the empirical formula.

$$\text{butane} \qquad C_2H_5 \text{ (empirical formula)} \qquad C_4H_{10} \text{ (molecular formula)}$$

C. ISOMERS

1. Isomers are different compounds with the same molecular formula.
2. Isomers that differ in the connectivity of their atoms are termed constitutional isomers (structural isomers).

C_4H_{10}    butane   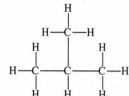   isobutane (2-methylpropane)

D. STRUCTURAL FORMULA

1. The structural formula of a molecule is its Lewis structure, which shows the connectivity of its atoms.
    a. Connectivity is the order in which the atoms are connected.
    b. A condensed structural formula conveys the same information as that of the structural formula.

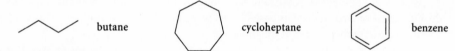   butane $\Rightarrow$ $CH_3CH_2CH_2CH_3$ condensed formula

E. SKELETAL STRUCTURES

1. Skeletal structures show only the carbon-carbon bonds.
2. Carbons are located at each vertex and at the ends of the structures.

butane   cycloheptane   benzene

F. FUNCTIONAL GROUP

1. A functional group is a characteristically bonded group of atoms that has about the same chemical reactivity whenever it occurs in a variety of compounds.
2. Alkyl groups are substituent groups derived from alkanes; R is the abbreviation of an alkyl group.
3. Aryl groups are substituent groups derived from benzene and its derivatives; the phenyl group is the simplest aryl group.
    a. Ph is the abbreviation for a phenyl group.
    b. Ar is the abbreviation for a general aryl (substituted phenyl) group.

III. *Molecular Conformation*

A. GENERAL

1. Alkanes exist in various staggered conformations that are rapidly interconverted at room temperature by internal rotation.
    a. Internal rotation is the rotation about a single bond.
    b. Staggered conformations about single bonds are favored.
2. The conformation that minimizes van der Waals repulsions has the lowest energy and is the predominant one.
    a. The van der Waals radius is a measure of an atom's size.
    b. Van der Waals repulsion is the extra energy required to force two nonbonded atoms within the sum of their van der Waals radii.
3. Conformations in which larger groups are brought closer together are less stable than conformations in which these groups are farther apart.

B. ETHANE CONFORMATIONS

1. The staggered conformation of ethane has a dihedral angle of 60°.

2. The eclipsed conformation of ethane has a dihedral angle of 0°.

C. BUTANE CONFORMATIONS

1. The *anti* conformation of butane has a dihedral angle of 180° and is the major conformation.
2. The *gauche* conformations of butane have dihedral angles of ±60° and exist to a lesser extent than the *anti* conformation.

IV. Nomenclature

A. GENERAL

1. Alkanes are named systematically according to the substitutive nomenclature rules of the IUPAC.
2. The name of a compound is based on its principal chain, which for an alkane is the longest continuous carbon chain in the molecule.
   a. Branching groups are in general termed substituents; substituents derived from alkanes are called alkyl groups.
   b. An alkyl group is named by substituting the *ane* suffix in the name of the parent alkane with a *yl* suffix: alkyl group name = alkane name – *ane* + *yl*.

B. UNBRANCHED ALKANES

1. The names of the unbranched alkanes are derived from the number of carbons.
2. The names of the unbranched alkanes are listed in Table 2.1 on page 47 of the text.

8 carbons

H—C—C—C—C—C—C—C—C—H   octane

C. BRANCHED ALKANES

1. Determine the principal chain: alkanes containing branched carbon chains are named according to the principal chain.
   a. The principal chain is the longest continuous chain in the molecule no matter how the molecule is drawn.
   b. In structures having two or more chains with the same length, the principal chain is the one containing the greater number of branches.

2. Number the carbons of the principal chain consecutively from one end to the other in the direction that gives the branches the lower numbers.
   a. When there are multiple substituent groups on the principal chain, each substituent receives its own number.

b. When there are substituent groups at more than one carbon of the principal chain, alternative numbering schemes are compared number by number, and the one is chosen that gives the lower number at the first point of difference.

c. When the numbering of different groups is not resolved by the other rules, the first-cited group receives the lowest number.

3. Name each branch.
   a. Identify the carbon number of the principal chain at which it occurs.
   b. The prefixes di-, tri-, tetra-, etc., are used to indicate the number of identical substituents.

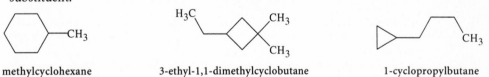

4. Alphabetize the substituent groups regardless of their location on the principal chain.
   a. The prefixes iso-, neo-, and cyclo- are considered in alphabetizing.
   b. The numerical prefixes di-, tri-, etc., are ignored in alphabetizing.
   c. The prefixes *tert-* and *sec-* are ignored in alphabetizing.

   tert-butyl, isopropyl, dimethyl

5. Construct the name by writing the carbon number of the principal chain at which the substituent occurs, a hyphen, the name of the branch, and the name of the alkane corresponding to the principal chain.

   8-*tert*-butyl-4-isopropyl-3,5-dimethylundecane

D. Cycloalkanes

1. Cycloalkanes are alkanes that contain carbon chains in closed loops, or rings.
2. The ring is treated as the principal chain if it has the greatest number of carbons.
   a. The prefix cyclo- is attached to the alkane name corresponding to the number of carbons in the ring.
   b. The numerical prefix 1- is not necessary for cycloalkanes having only one substituent.
3. When a noncyclic carbon chain contains more carbons than an attached ring, the ring is treated as a substituent.

methylcyclohexane   3-ethyl-1,1-dimethylcyclobutane   1-cyclopropylbutane

V. *Physical Properties*

A. General

1. The physical properties of an organic compound determine the conditions under which the compound is handled and used, and are also important in characterizing the compound.
2. When two molecules approach each other closely, as in a liquid, the electron clouds of one molecule interact with the electron clouds of the other.
   a. As a result, both molecules temporarily acquire small localized separations of charge called induced dipoles.
   b. The attraction between the dipoles of different molecules is an example of an attractive van der Waals force (dispersion force).

B. BOILING POINT

1. The boiling point is the temperature above which a substance is transformed spontaneously and completely from the liquid to the gaseous state.
   a. The boiling point is a crude measure of the attractive (van der Waals) forces among molecules in the liquid state compared to those in the gaseous state.
   b. Boiling points increase with increasing molecular weight within a homologous series—typically 20–30 °C per carbon atom.
2. The shape of a molecule is also important in determining its boiling point.
   a. Large molecules have relatively high boiling points.
   b. Highly branched molecules have relatively low boiling points because they have less molecular surface available for van der Waals attractions.

C. MELTING POINT

1. The melting point of a substance is the temperature at which it undergoes a reversible transition from the solid to the liquid state.
   a. Melting points tend to increase with increasing molecular weight within a series.
   b. Highly symmetrical molecules have particularly high melting points.
2. A sawtooth pattern of melting point behavior is observed within some series.

· ·

Reactions

· ·

I. *Reaction of Hydrocarbons*

A. COMBUSTION

1. Combustion is the most important reaction of alkanes.
2. Complete combustion affords carbon dioxide and water as the only combustion products.

$$C_nH_{2n+2} \;+\; \frac{3n+1}{2}O_2 \;\longrightarrow\; n\,CO_2 \;+\; (n+1)\,H_2O$$

3. Combustion can be used in elemental analysis, which is the quantitative determination of elemental compositions.
   a. The mass percents of the elements obtained from a combustion analysis in a compound can be used to determine directly the empirical formula.

· ·

Study Guide Links

✓2.1 Newman Projections

The discussion of butane in the text utilizes Newman projections about the central carbon-carbon bond. You should realize that a Newman projection can be drawn for *any* bond in a molecule. For example, a Newman projection about the bond between carbons 1 and 2 in butane is as follows:

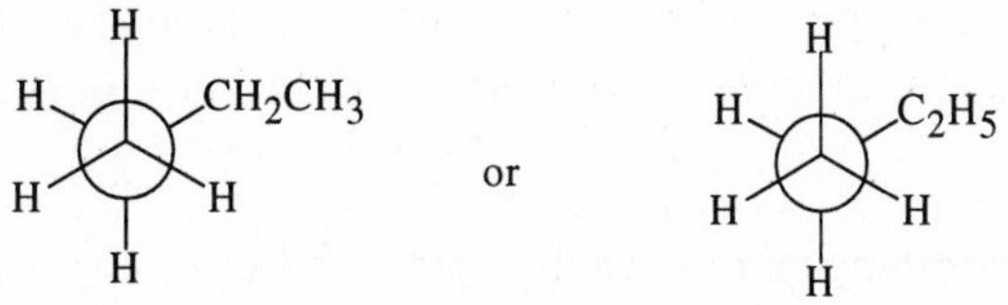

$$ \text{(Newman projections)} \quad \text{or} \quad \text{(Newman projections)} $$

Of course, this projection doesn't show the *gauche* or *anti* relationship between the CH_3 groups at each end of the molecule; this relationship is shown only by a Newman projection about the *central* carbon-carbon bond. When we use Newman projections, we project the bond that shows the relationships of interest.

✓2.2 Nomenclature of Simple Branched Compounds

The IUPAC recognizes as valid the following older names for branched alkane isomers of four, five, and six carbon atoms because of their common historical usage.

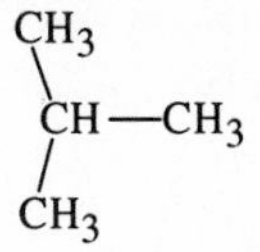

$$\begin{array}{c} CH_3 \\ | \\ CH-CH_3 \\ | \\ CH_3 \end{array} \quad \text{or} \quad (CH_3)_3CH$$

isobutane

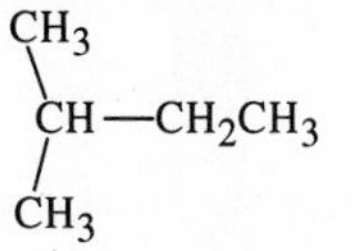

$$\begin{array}{c} CH_3 \\ | \\ CH-CH_2CH_3 \\ | \\ CH_3 \end{array} \quad \text{or} \quad (CH_3)_2CHCH_2CH_3$$

isopentane

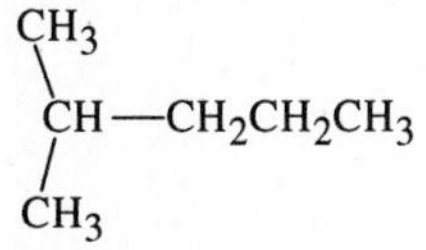

$$\begin{array}{c} CH_3 \\ | \\ CH-CH_2CH_2CH_3 \\ | \\ CH_3 \end{array} \quad \text{or} \quad (CH_3)_2CHCH_2CH_2CH_3$$

isohexane

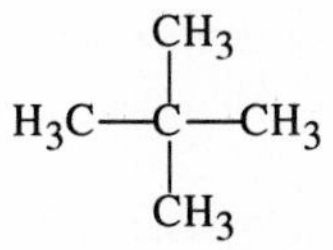

$$\begin{array}{c} CH_3 \\ | \\ H_3C-C-CH_3 \\ | \\ CH_3 \end{array} \quad \text{or} \quad (CH_3)_4C$$

neopentane

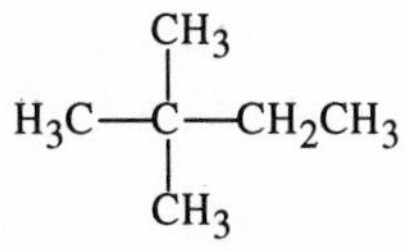

$$\begin{array}{c} CH_3 \\ | \\ H_3C-C-CH_2CH_3 \\ | \\ CH_3 \end{array} \quad \text{or} \quad (CH_3)_3CCH_2CH_3$$

neohexane

The distinguishing feature of an alkane named with the "iso" prefix is a pattern of two methyl branches at the end of a carbon chain. The distinguishing feature of an alkane with a "neo" prefix is a pattern of three methyl branches at the end of a carbon chain.

If you refer to Table 2.2 in the text, you'll see that the names of the branched alkyl groups in that table are derived from the names shown above for the corresponding alkanes.

Solutions

Solutions to In-Text Problems

2.1   (a) The number of hydrogens in an alkane (branched or unbranched) is $2n + 2$, where n is the number of carbons. Consequently, there are $2(18) + 2 = 38$ hydrogens in the alkane with eighteen carbons.
(b) No; all alkanes—indeed, all hydrocarbons—must have an even number of hydrogens.

2.2   (a) The difference in the boiling points of nonane, undecane, and undecane in Table 2.1 is about 20° per carbon. Consequently, the boiling point of tridecane is estimated by adding 20° to the boiling point of dodecane to obtain 236°. (The actual value is 235°.) The condensed structural formula of tridecane is $CH_3(CH_2)_{11}CH_3$.

2.3   (a) The staggered conformations of isopentane are *A*, *C*, and *E*; the eclipsed conformations are *B*, *D*, and *F*.

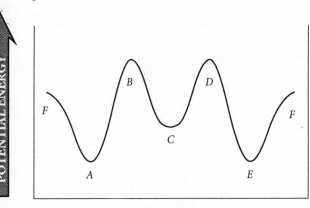

(b) Curve of potential energy vs. angle of rotation:

(c) Conformations *A* and *E* have the lowest energy because they have one fewer *gauche* interaction than conformation *C*; hence, either of these conformations is present in greater concentration than conformation *C*. As the problem is worded, either *A* or *E* is a correct answer.

2.5   (a) The principal chain has nine carbons; the name is 5-ethyl-4-methylnonane.
(c) It is probably easier to name the compound if you first draw the structure in a less condensed form.

$$CH_3CH_2CH_2\underset{\underset{CH_2CH_2CH_3}{|}}{\overset{\overset{CH_2CH_3}{|}}{C}}HCHCH_2CH_3$$

More than one chain contains seven carbons; the principal chain is the one that has the most branches. Consequently, the name is 3-ethyl-4-propylheptane.

2.6 This type of problem requires a systematic approach. First, draw the structure of heptane itself.

$$CH_3CH_2CH_2CH_2CH_2CH_2CH_3 \quad \text{heptane}$$

Next, draw the structures with a principal chain of six carbons and one methyl branch:

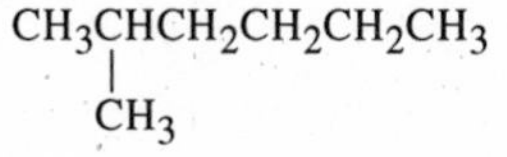

$$CH_3CHCH_2CH_2CH_2CH_3 \qquad CH_3CH_2CHCH_2CH_2CH_3$$
$$\quad\ |\qquad\qquad\qquad\qquad\qquad\quad |$$
$$\quad\ CH_3 \qquad\qquad\qquad\qquad\qquad CH_3$$

2-methylhexane 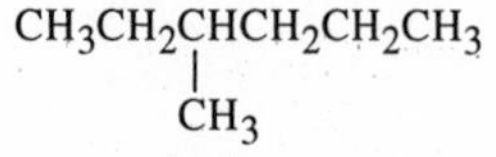 3-methylhexane

Notice that moving the methyl branch one more carbon to the right gives an identical structure. Next, draw the structures with five carbons in their principal chains and two methyl branches:

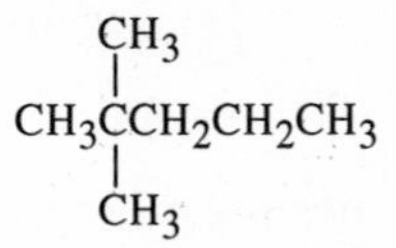

$$
\begin{array}{cccc}
CH_3 & CH_3 & CH_3 & CH_3 \\
| & | & | & | \\
CH_3CCH_2CH_2CH_3 & CH_3CH_2CCH_2CH_3 & CH_3CHCHCH_2CH_3 & CH_3CHCH_2CHCH_3 \\
| & | & | & | \\
CH_3 & CH_3 & CH_3 & CH_3
\end{array}
$$

2,2-dimethylpentane 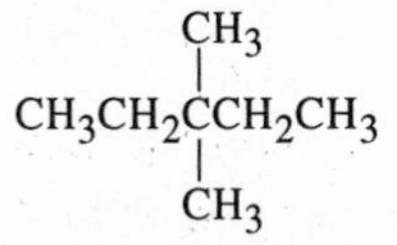 3,3-dimethylpentane 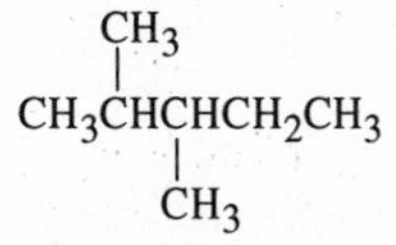 2,3-dimethylpentane 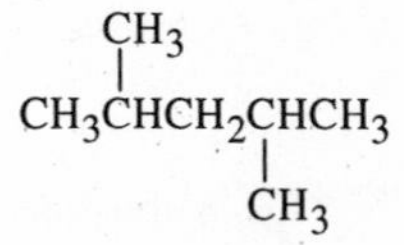 2,4-dimethylpentane

Then there is the structure with one ethyl branch:

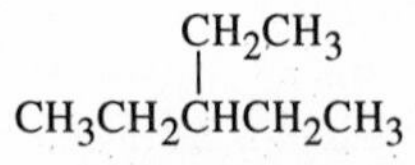

$$CH_2CH_3$$
$$|$$
$$CH_3CH_2CHCH_2CH_3$$

3-ethylpentane

Finally, there is one structure with a four-carbon principal chain:

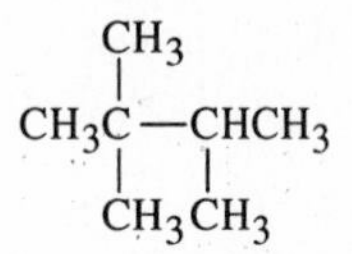

$$CH_3$$
$$|$$
$$CH_3C\!-\!CHCH_3$$
$$|\qquad |$$
$$CH_3\ CH_3$$

2,2,3-trimethylbutane

Notice that any other structures with substituents on a four-carbon chain are identical to structures already considered.

2.7 (a)

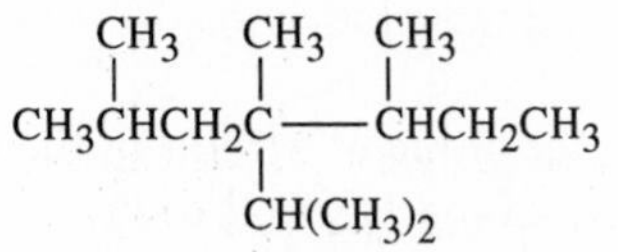

$$
\begin{array}{ccc}
CH_3 & CH_3 & CH_3 \\
| & | & | \\
CH_3CHCH_2C\!-\!\!-\!\!-\!CHCH_2CH_3 & & \text{4-isopropyl-2,4,5-trimethylheptane} \\
| & & \\
CH(CH_3)_2 & &
\end{array}
$$

2.8 (a) In the following structure, p = primary, s = secondary, t = tertiary, and q = quaternary.

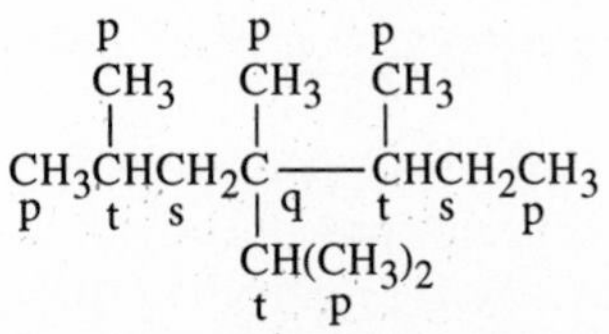

$$
\begin{array}{ccc}
p & p & p \\
CH_3 & CH_3 & CH_3 \\
| & | & | \\
CH_3CHCH_2C\!-\!\!-\!\!-\!CHCH_2CH_3 \\
p\ \ t\ \ s\ \ \ |\ q\quad t\ \ s\ \ p \\
CH(CH_3)_2 \\
t\ \ \ p
\end{array}
$$

(b) The methyl groups are the CH_3 groups in the structure of part (a) with carbons that are labeled with a "p". The other groups are circled in the structures below.

(solution continues)

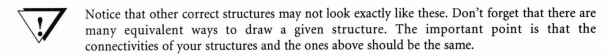

2.10   (a)   (c)

!   Notice that other correct structures may not look exactly like these. Don't forget that there are many equivalent ways to draw a given structure. The important point is that the connectivities of your structures and the ones above should be the same.

2.11   (a) The numbering scheme is 1,2,4-. The ethyl group receives the number 1 by Rule 10, text p. 61. The name is 1-ethyl-2,4-dimethylcyclopentane.

2.12   The more branched and compact molecule 2,2,3,3-tetramethylbutane should have the lower boiling point; therefore, its boiling point is 106.5°. Because it is very symmetrical, it should also have the higher melting point; its melting point is 100.7°. Note that this compound is a liquid over only about a six-degree range; as a solid, it has a very high vapor pressure, especially at temperatures approaching its melting point.

2.13   Water is not very effective in distinguishing a gasoline fire because the gasoline is not "smothered" by the water. Because it is less dense than water, gasoline always rises to the surface of water where the air supports further combustion.

2.14   The formula $C_{16}H_{36}$ would not be acceptable because the maximum number of hydrogens a sixteen-carbon hydrocarbon could have is 2(16) + 2, or 34.

2.15   (a) C, 48.64 mass %; H, 8.16 mass %; O, 43.20 mass %.

2.16   (a) The empirical formula of a compound with molecular formula $C_3H_6O_2$ is the same as the molecular formula, because this formula is the smallest one that contains whole-number ratios for the elements.

2.17   (a) Follow the procedure used in Study Problem 2.6. First, the number of mg of carbon in the compound is the fraction of CO_2 that is carbon (= 12.01/44.01) times the number of mg of CO_2 produced.

$$\text{mg C} = (12.01 \text{ mg C}/44.01 \text{ mg CO}_2)(31.95 \text{ mg CO}_2) = 8.72$$

The number of mg of hydrogen is given by

$$\text{mg H} = (2 \times 1.008 \text{ mg H}/18.02 \text{ mg H}_2\text{O})(11.44 \text{ mg H}_2\text{O}) = 1.28$$

Thus, the mass percent carbon in the sample is (8.72/10.00) × 100 = 87.2; the mass percent hydrogen is 12.8. Notice that the entire mass of *X* is accounted for by carbon and hydrogen; consequently, *X* is a hydrocarbon. The empirical formula is then $C_{(87.2/12.01)}H_{(12.8/1.008)}$ or $C_{7.26}H_{12.7}$. Dividing through by 7.26, the formula becomes $C_1H_{1.75}$ or $C_{4/4}H_{7/4}$. Multiplication by 4 gives the empirical formula of C_4H_7. This cannot be the molecular formula, because a hydrocarbon must have an even number of hydrogens. The molecular mass contribution of C_4H_7 is 55.1. Since the molecular mass of *X* is 110, then the molecular formula is twice C_4H_7, or C_8H_{14}.

2.18 (a) Many structures could be drawn that correspond to the molecular formula. One possible structure of an amine (back cover) has the form RNH_2. In such a case, $R = C_4H_9$, which could correspond to a butyl group, an isobutyl group, a *tert*-butyl group, or a *sec*-butyl group. All of these are possible, as are many others:

$$CH_3CH_2CH_2CH_2NH_2 \quad (CH_3)_2CHCH_2NH_2 \quad (CH_3)_3CNH_2 \quad CH_3CH_2\underset{\underset{NH_2}{|}}{C}HCH_3$$

(c) A carboxylic acid has the form RCO_2H. The mass percents result in an empirical formula $C_{3.33}H_{6.66}O_{3.33}$, or CH_2O. Because a carboxylic acid must have at least two oxygens, the smallest possible molecular formula is $C_2H_4O_2$. A structure corresponding to this formula is that of acetic acid:

$$H_3C\overset{\overset{O}{\underset{\|}{}}}{-}C-OH \quad \text{acetic acid}$$

2.19 Any structures containing nitrogen are ruled out; thus, (a) is excluded. Any structures that have double or triple bonds are ruled out, because a compound containing only carbon, hydrogen, and oxygen with five carbons and twelve hydrogens can have no double bonds. Thus, (c), (d), and (f) are also excluded. The compound could be either (b) an ether or (e) an alcohol or both. There are many examples of structures having the molecular formula $C_5H_{12}O_2$ that contain two ether groups, two alcohol groups, or one ether group and one alcohol group. Three examples are the following:

$$CH_3\underset{\underset{OH}{|}}{C}H CH_2 \underset{\underset{OH}{|}}{C}HCH_3 \qquad CH_3\underset{\underset{OCH_3}{|}}{C}HCH_2CH_2OH \qquad CH_3\underset{\underset{OCH_3}{|}}{C}HCH_2OCH_3$$

Solutions to Additional Problems

2.20 (a) Because the two compounds have the same functional groups and the same number of carbons, their boiling points should be very similar. An estimate is 152 °C.
(c) Use the rule of thumb that each additional carbon adds about 20–30° to the boiling point. Because the second compound has two more carbons than the first, its boiling point should be about 50° higher than the first compound. The estimated boiling point is 195–215°.

2.21 (a) The isomers of octane with five carbons in their principal chains:

2,3,4-trimethylpentane   2,2,3-trimethylpentane   2,2,4-trimethylpentane

2,3,3-trimethylpentane   3-ethyl-2-methylpentane   3-ethyl-3-methylpentane

(solution continues)

Notice the name 3-ethyl-2-methylpentane rather than 3-isopropylpentane. This choice is governed by Rule 3, text page 57: the principal chain is the one with the greater number of branches.

You may have drawn other structures, such as "2-ethyl-2-methylpentane;" notice that such a structure has *six* carbons in its principal chain, not five.

$$CH_3CCH_2CH_2CH_3 \qquad \text{3,3-dimethylhexane}$$

2.22   (a) In the following structures, p = primary, s = secondary, t = tertiary, and q = quaternary.

   Don't forget that in skeletal structures enough hydrogens are *understood* to be present at each vertex to give four bonds to each carbon.

2.23   (a) Neopentane has five carbons, and all hydrogens are primary:

$$H_3C-\overset{\overset{\displaystyle CH_3}{|}}{\underset{\underset{\displaystyle CH_3}{|}}{C}}-CH_3 \qquad \text{neopentane}$$

(c) This one is tricky. The structures that fit this description have several rings fused together. Among these are the following, along with their trivial names.

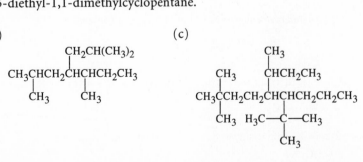

prismane   cubane   tetrahedrane

2.24   (a) 5-ethyl-4,4-dimethyloctane. Note that the numbering is governed by the "first point of difference" rule (Rule 8, text page 60).
(c) 2,3-dimethylhexane
(e) In this example, alternative numbering schemes are compared: (1,2,2,3), (1,1,2,5), and (1,4,5,5); notice that the first point of difference among these schemes occurs at the second number, and that the *second* of these schemes has the lowest number at the first point of difference. The name, then, is 2,5-diethyl-1,1-dimethylcyclopentane.

2.25   (a)   (c)

$$\underset{\underset{\displaystyle CH_3}{|}}{CH_3CHCH_2}\underset{\underset{\displaystyle CH_3}{|}}{\overset{\overset{\displaystyle CH_2CH(CH_3)_2}{|}}{CH}}CHCH_2CH_3$$

$$CH_3\overset{\overset{\displaystyle CH_3}{|}}{\underset{\underset{\displaystyle CH_3}{|}}{C}}CH_2CH_2\overset{\overset{\displaystyle \overset{\displaystyle CH_3}{|}}{CH}CH_2CH_3}{CH}CHCH_2CH_2CH_3}$$

2.26 (a) 5-neopentyldecane is a correct name.

$$CH_3CH_2CH_2CH_2\underset{\underset{\displaystyle CH_2C(CH_3)_3}{|}}{C}HCH_2CH_2CH_2CH_2CH_3$$

(c) Ima forgot about the "first point of difference" rule.

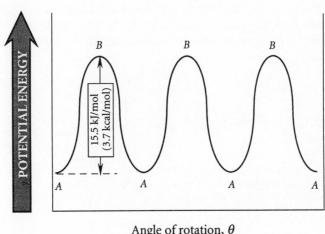

4-cyclopropyl-1,2-dimethylcyclohexane

2.27 (a) The connectivity of A, starting near the observer, is [CH$_3$, C connected to C$_2$H$_5$ and to CH$_3$, CH(CH$_3$)$_2$]. The connectivity of B is [(CH$_3$)$_2$CH, CH connected to C$_2$H$_5$, CH$_2$, CH$_3$], which is the same as [(CH$_3$)$_2$CH, CH(C$_2$H$_5$)$_2$], which is the same as C. Hence, structures B and C are Newman projections of the same compound, 3-ethyl-2-methylpentane. Compound A is 2,3,3-trimethylpentane.

(b) Structures A and C both have the connectivity [(CH$_3$)$_2$CH, CH$_2$, CH(CH$_3$), CH$_2$, CH$_2$, CH$_3$]. The connectivity of B is different. Consequently, structures A and C represent the same compound, 2,4-dimethylheptane. Compound B is 3-ethyl-2-methylheptane.

2.28 All staggered forms (A) are identical and thus have identical energies; all eclipsed forms (B) are identical, and also have identical energies. The energy barrier is the energy difference between forms B and A.

2.30 (a) The conformations of 1,2-dichloroethane:

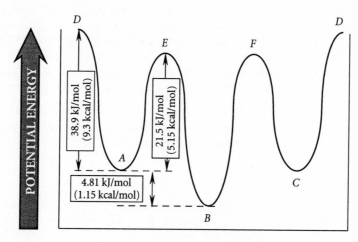

The potential-energy graph is shown below. Notice that the *C-F* barrier is identical to the *A-E* barrier; the *C-D* barrier is identical to the *A-D* barrier.

Angle of rotation, θ

(b) Conformation *B* is present in greatest amount, because it has the lowest energy. It has the lowest energy for two reasons. First, the chlorines have no van der Waals repulsions when they are *anti*. Second, the C—Cl bond dipoles are in a more repulsive arrangement in the *gauche* conformation than they are in the *anti* conformation:

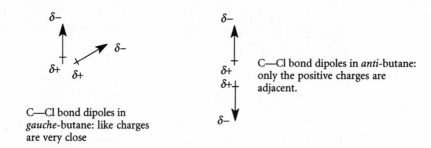

C—Cl bond dipoles in *gauche*-butane: like charges are very close

C—Cl bond dipoles in *anti*-butane: only the positive charges are adjacent.

2.31 There are severe van der Waals repulsions between the *tert*-butyl groups in compound *A*. The only way for the molecule to relieve these repulsions is to stretch the appropriate bonds and flatten itself by widening the C—C—C angles. This also costs energy, but presumably not as much as retaining the shorter bond lengths and narrower angles.

2.32 The conformational analysis of butane shows that the *anti* conformation is favored at each internal carbon-carbon bond. *Anti* conformations at each carbon-carbon bond of hexane lead to a zig-zag shape for the carbon backbone. The model and a line-wedge formula are as follows:

(solution continues)

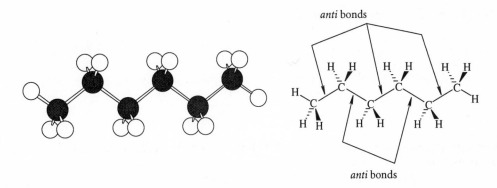

anti bonds

anti bonds

2.33   A major contributor to the barrier to internal rotation is the van der Waals repulsions between the methyl groups. In compound (b), these repulsions are smaller because the distance between methyl groups is greater. This distance is greater because the Si—Si bond is longer than the C—C bond.

2.35   (a) The expected dipole moment for the *anti* form is zero, because the polar C—Br bonds are oriented in opposite directions; hence their bond dipoles cancel. Likewise, for every C—H bond, there is another C—H bond pointing in the opposite direction.

(b) It follows from part (a) that the observed dipole moment is due solely to the relatively small amount of *gauche* form present. Hence,

$$\mu = \mu_{anti}N_{anti} + \mu_{gauche}N_{gauche}$$

or $$1.0 = (0)(0.82) + \mu_{gauche}(0.18) = \mu_{gauche}(0.18)$$

Solving for μ_{gauche}, this equals 5.56 D. This is a very large dipole moment. But this is what would be expected for a compound in which two C—Br bond dipoles (as well as the smaller C—H bond dipoles) are oriented in nearly the same direction.

2.36   A balanced equation for combustion of a cycloalkane C_nH_{2n}:

$$C_nH_{2n} + (3n/2)O_2 \longrightarrow nCO_2 + nH_2O$$

2.37   (a) The mass contributed by carbon is $3(12.01) = 36.03$. The mass contributed by hydrogen is $6(1.008) = 6.05$. The mass contributed by oxygen is 16.00. The total mass is 58.08. The mass percent carbon is the part of the mass that is carbon divided by the total mass expressed as a percentage: $(36.03/58.08) \times 100 = 62.04\%$.

2.38   (a) Follow the procedure in Study Problem 2.7 to obtain $C_{7.26}H_{12.72}$, or $CH_{1.75}$, or C_4H_7. The molecular formula of a hydrocarbon must have an even number of hydrogens; consequently, the smallest possible molecular formula is C_8H_{14}.
(c) An alkane consistent with this formula that has one quaternary carbon and all other carbons secondary (that is, no primary hydrogens and no tertiary carbons) :

2.39   (a) First, determine the fraction of the mass of CO_2 that is oxygen (O):

$$\text{fraction O in } CO_2 = (32.00 \text{ mg of O})/(44.01 \text{ mg of } CO_2)$$
$$= 0.7271 \text{ mg of O per mg of } CO_2$$

The mass of oxygen in the CO_2 formed is then

$$\text{mass of O} = (0.7271 \text{ mg of O per mg of } CO_2)(21.96 \text{ mg of } CO_2) = 15.97 \text{ mg}$$

Likewise, the mass of oxygen in the H_2O formed is

$$\text{mass of O} = (16.00 \text{ mg of O}/18.02 \text{ mg of } H_2O)(8.99 \text{ mg of } H_2O) = = 7.98 \text{ mg}$$

By conservation of mass, the total oxygen consumed is equal to the total oxygen in the products, and is therefore

$$15.97 \text{ mg} + 7.98 \text{ mg} = 23.95 \text{ mg}.$$

This weight of oxygen consumed corresponds to

$$(23.95 \text{ mg of O})/(32.00 \text{ mg of O per mmol of } O_2) = 0.748 \text{ mmol of } O_2$$

At 0 °C (273 K), this corresponds to

$$(0.748 \text{ mmol})(22.4 \text{ mL/mmol}) = 16.76 \text{ mL}$$

Applying a gas-law correction to this number, the volume at 25 °C (298 K) is

$$(16.76 \text{ mL})(298 \text{ K})/(273 \text{ K}) = 18.29 \text{ mL}$$

(b) To calculate the molecular formula, apply the techniques described in Study Problem 2.6. The mass of carbon in the sample is derived from the amount of carbon dioxide, and the mass of hydrogen is derived from the amount of water:

$$\text{Mass of C} = (21.96 \text{ mg of } CO_2)(12.01 \text{ mg of C}/44.01 \text{ mg of } CO_2) = 5.99 \text{ mg}$$
$$\text{Mass of H} = (8.99 \text{ mg of } H_2O)(2.016 \text{ mg of H}/18.02 \text{ mg of } H_2O) = 1.006 \text{ mg}$$

The empirical formula is $C_{(5.99/12.01)}H_{(1.006/1.008)} = C_{0.5}H_1 = CH_2$. Because the group mass of CH_2 is 14 mass units, the molecular formula of the hydrocarbon is $(CH_2)_{10}$, or $C_{10}H_{20}$.

2.40   (a) First, calculate the molecular formula by the process used in Study Problems 2.6 and 2.7. In doing so, assume (since we are told that the unknown compound is an amide) that the percentage unaccounted for is oxygen, which is therefore 18.37%. (Amides contain oxygen; see tables inside rear cover of text.) The molecular formula is

$$C_{(55.14/12.01)}H_{(10.41/1.008)}N_{(16.08/14.01)}O_{(18.37/16.00)}, \text{ or } C_{4.59}H_{10.33}N_{1.15}O_{1.15}, \text{ or } C_4H_9NO.$$

Since the compound contains an isopropyl group $(CH_3)_2CH-$, two possible structures for X are

2.41   Consult the inside rear cover of the text for a list of common functional groups. The functional groups in acebutolol:

3

Acids and Bases; the Curved-Arrow Formalism

Terms

Concepts

I. Acids and Bases

A. LEWIS ACIDS AND BASES

1. A Lewis acid is
   a. an electrophile (an "electron-loving" species).
   b. a compound that reacts by accepting an electron pair.
2. A Lewis base is
   a. a nucleophile (a "nucleus-loving" species).
   b. a compound that reacts by donating an electron pair to a Lewis acid, that is, by "attacking" a Lewis acid.

3. All Lewis acid-base reactions involve either the reactions of Lewis bases with electron-deficient compounds, or electron-pair displacements.

   a. A Lewis acid-base association reaction occurs when a Lewis acid and a Lewis base combine to give a single product.

   Lewis acid   Lewis base

   b. A Lewis acid-base dissociation reaction is the reverse of a Lewis acid-base association reaction.

   c. An electron-pair displacement reaction occurs when a Lewis base attacks an atom that is not electron-deficient.

   Lewis acid   Lewis base

B.   BRØNSTED-LOWRY ACIDS AND BASES

1. A Brønsted acid is a species that reacts by donating a proton.

   a. When a Brønsted acid loses a proton, its conjugate base is formed. For example, the conjugate base of a carboxylic acid is called a carboxylate ion.

   b. A Brønsted acid and its conjugate base comprise a conjugate acid-base pair.

   acetic acid   acetate ion
   (a carboxylic acid)   (a carboxylate ion)
   Brønsted acid   conjugate base

2. A Brønsted base is a species that reacts by accepting a proton.

   a. When a Brønsted base accepts a proton, its conjugate acid is formed.

   b. A Brønsted base and its conjugate acid comprise a conjugate acid-base pair.

   Brønsted base   conjugate acid

3. An amphoteric compound can act as either an acid or a base. For example, water is an amphoteric compound.

   Brønsted acid   conjugate base   Brønsted base   conjugate acid

4. Brønsted acid-base reactions are electron-pair displacement reactions that occur by attack of an electron pair on a proton.

   a. The equilibrium in a Brønsted acid-base reaction always favors the side with the weaker acid and weaker base.

   b. Two structural effects on Brønsted acidity are the element effect and the polar effect.

C.   DISSOCIATION CONSTANT

1. The dissociation constant K_a is a measure of the strength of an acid—how well a Brønsted acid donates a proton to water, a Brønsted base. The dissociation constant K_a for each Brønsted acid is unique.

2. A logarithmic pK_a scale is used, in which $pK_a = -\log K_a$; a stronger acid has a smaller pK_a value.

3. The basicity constant K_b is a measure of how well a Brønsted base accepts a proton from water, a Brønsted acid.
    a. The strength of a Brønsted base is inferred from the K_a (or pK_a) of its conjugate acid.
    b. The stronger base has the conjugate acid with the greater pK_a.

II. Free Energies of Reactions and Structural Effects on Acidity

A. STANDARD FREE ENERGY $\Delta G°$

1. The standard free energy of dissociation $\Delta G_a°$ is equal to the standard free energies of the ionization products minus the standard free energy of the un-ionized acid.
    a. When $\Delta G°$ is positive, the concentration of ionization products is less than the concentration of un-ionized acid—the acid is a weak acid.
    b. When $\Delta G°$ is negative, the concentration of ionization products is greater than the concentration of the un-ionized acid—the acid is a strong acid.
2. The equilibrium constant K_{eq} for a reaction is related to the standard free-energy difference $\Delta G°$ between products and reactants by the relationship $\Delta G° = -2.3RT \log K_{eq}$.
    a. Reactions with $K_{eq} < 1$ have positive $\Delta G°$ values and favor reactants at equilibrium.
    b. Reactions with $K_{eq} > 1$ have negative $\Delta G°$ values and favor products at equilibrium.
3. The equilibrium constant for a reaction indicates which species in a chemical equilibrium are present in highest concentration.
    a. A chemical equilibrium favors the species of lower free energy.
    b. The more two compounds differ in free energy, the greater is the difference in their concentrations at equilibrium.
    c. When the free energy of a conjugate base is lower (when a conjugate base is more stable), the pK_a of the acid is smaller (the acid is more acidic).

B. ELEMENT EFFECT

1. For an acid H—A, a smaller bond dissociation energy increases acidity.
2. For an acid H—A, a larger electron affinity of A• increases acidity.
3. Brønsted acidity increases as the atom to which the acidic proton is attached has a greater atomic number within a column (group) of the periodic table.
    a. This trend is dominated by the H—A bond dissociation energies of the acids.
    b. Stronger acids have weaker H—A bonds.
4. Brønsted acidity increases as the atom to which the acidic proton is attached is farther to the right within a row (period) of the periodic table.
    a. This trend is dominated by electron-attracting abilities (electron affinities) of the elements A in the acids H—A.
    b. Stronger acids have A groups that are more electron-attracting.
5. Electronegative substituent groups such as halogens increase the acidities of carboxylic acids by stabilizing their conjugate-base carboxylate ions.

C. POLAR EFFECT

1. The polar effect (inductive effect) on acidity is caused largely by the interaction of charges formed in an acid-base reaction with the bond dipoles of polar bonds or with other charged groups in an acid or base molecule.
    a. The electrostatic law (Eq 3.40, text p. 110) governs the interaction of charges and/or dipoles.
    b. Electronegative groups exert an electron-withdrawing effect by pulling electrons towards themselves and away from the carbon to which they are attached.
    c. Some atoms exhibit an electron-donating polar effect by pushing electrons away from themselves and towards the carbon to which they are attached.

III. The Curved-Arrow Formalism

1. The curved-arrow formalism is a symbolic device for depicting the flow of electron pairs in chemical reactions.

   a. The reaction of a Lewis base with an electron-deficient compound requires one curved arrow.

   Lewis acid   Lewis base

   b. An electron-pair displacement reaction requires two curved arrows

   c. The sum of the formal charges in the reactants and products must be the same.

2. The curved arrow originates at the source of the electrons and terminates at the destination of the electron pair.

   a. Redraw all atoms just as they were in the reactants.

   b. Draw the bonds and electron pairs that do not change.

   c. Draw the new bonds or electron pairs indicated by the curved-arrow formalism.

   d. Complete the formal charges to give the product.

3. The curved-arrow formalism can be used to derive resonance structures that are related by the movement of one or more electron pairs.

Study Guide Links

√3.1   The Curved-Arrow Formalism

The curved-arrow formalism is a symbolic notation that must be used precisely. However, some things matter and some things don't. In drawing curved arrows, the only thing that matters is *the starting point and the destination of the arrows.* It doesn't matter whether an arrow curves up or down, or whether it goes to the top or bottom of the atom. Your curved arrows might appear different from the ones in the text or solutions manual; what matters is where they start and where they end. For example, in each of the following examples the curved arrows mean the same thing: breaking of a C—F bond with the electron pair moving onto the fluorine.

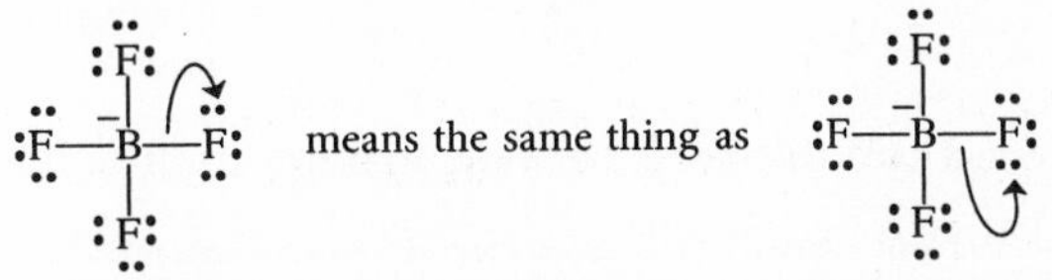

means the same thing as

It also doesn't matter where the arrowhead points on an atom. For example, in the above example, the arrow can point directly to the F, or to an electron pair on the F. In each case, the meaning is that fluorine is taking on another electron pair.

The following is an example of the curved-arrow formalism written in three different ways for an electron-pair displacement reaction. All of these are correct. They all refer to the same process: removal of a proton from the ammonium ion by hydroxide ion to give water and ammonia.

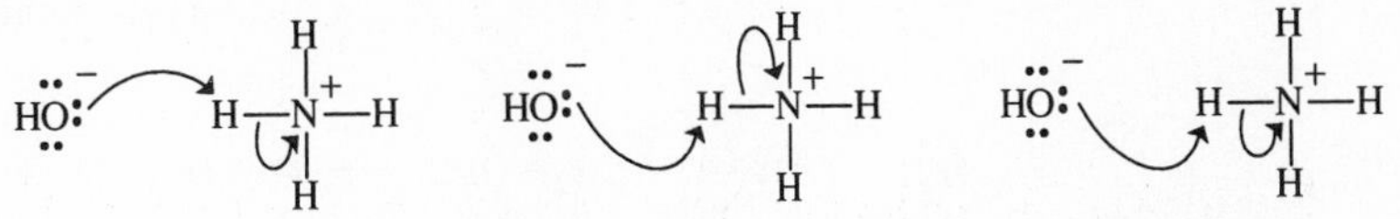

All describe the same process which gives the following as products:

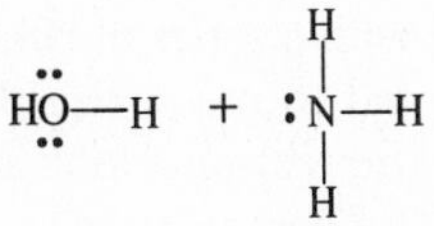

√3.2   Rules for Use of the Curved-Arrow Formalism

Now that you've seen several examples in which the curved-arrow formalism is used, let's summarize the use of the formalism with six simple rules. If you follow these you will always use the curved-arrow formalism correctly. The first two rules reiterate points that have been made repeatedly in the text. Examples of the last four rules are provided below.

Rules for use of the curved-arrow formalism:

The first two rules are very important and general:

1. Curved arrows indicate the flow or "movement" of electron pairs, not atoms or nuclei.
2. A curved arrow is always drawn with its tail at the source of electrons (a bond or unshared pair) and its head at the destination.

The second two rules deal with the manipulation of unshared electron pairs:

3.   When an atom donates an unshared electron pair, a curved arrow must indicate that the electron pair is used to form an additional bond to the same atom.

4.   When an atom accepts an unshared electron pair, a curved arrow must show that the electron pair originates from a bond to the same atom.

The following example illustrates both of these rules:

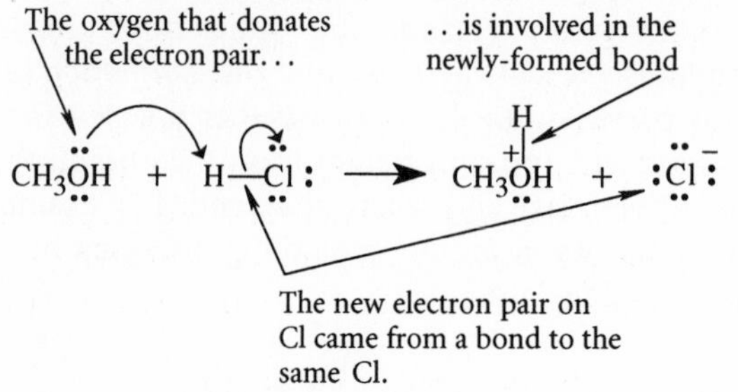

The oxygen that donates the electron pair. . .   . . .is involved in the newly-formed bond

$$CH_3\overset{..}{\underset{..}{O}}H \;+\; H\!-\!\overset{..}{\underset{..}{Cl}}: \;\longrightarrow\; CH_3\overset{+}{\underset{..}{O}}\overset{H}{\underset{|}{H}} \;+\; :\overset{..}{\underset{..}{Cl}}:^{-}$$

The new electron pair on Cl came from a bond to the same Cl.

The last two rules deal with the manipulation of bonds:

5.   When a bond is lost, a curved arrow must show its conversion either into an unshared electron pair on one of the two bonded atoms, or into a new bond involving one of the two original bonded atoms.

6.   When a bond is formed, a curved arrow must show that it originates from either an electron pair on one of the two bonded atoms, or from a bond involving one of the two bonded atoms.

The following example illustrates these rules:

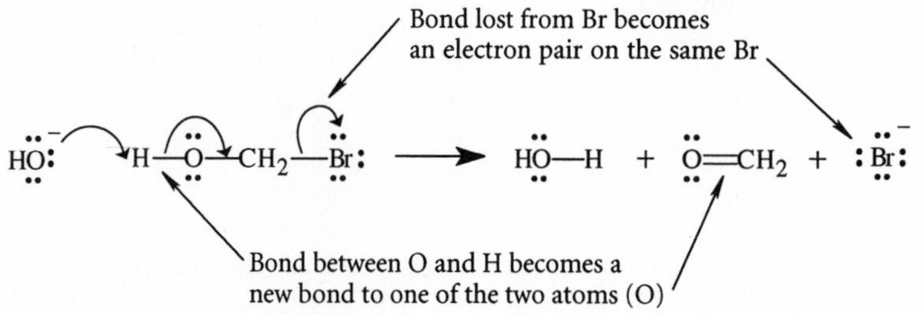

Bond lost from Br becomes an electron pair on the same Br

$$H\overset{..}{\underset{..}{O}}:^{-} \quad H\!-\!\overset{..}{\underset{..}{O}}\!-\!CH_2\!-\!\overset{..}{\underset{..}{Br}}: \;\longrightarrow\; H\overset{..}{\underset{..}{O}}\!-\!H \;+\; \overset{..}{\underset{..}{O}}\!=\!CH_2 \;+\; :\overset{..}{\underset{..}{Br}}:^{-}$$

Bond between O and H becomes a new bond to one of the two atoms (O)

When confronted with a series of rules like these for use of the curved-arrow formalism, some students are tempted to memorize them. *Don't do it!* What should you do with these rules? First, go back in the text and look at about three or four equations that show the curved-arrow formalism and demonstrate to yourself that these rules are applied correctly. Then, the next few times you use the formalism (as in Problem 3.3, for example) use your intuition to answer the problem, but then check your result against these rules before you look at the answer. If you do this three or four times, you're not likely to misuse the formalism.

✓3.3   Identification of Acids and Bases

This section stresses an important subtlety in dealing conceptually with acids and bases. An acid or a base—whether of the Lewis type or the Brønsted type—is defined by a *reaction,* and not by a *structure.* For example, some compounds react by donating a proton in certain reactions, in which case they are Brønsted acids; in other reactions, they react in other ways, in which case, they are *not* Brønsted acids.

You can sometimes tell from a structure whether a compound *might* react as an acid or a base. When someone looks at a structure and says, "That compound is an acid," what that person really means is "That compound is *capable of acting as an acid.*" It *is* an acid only in reactions in which it is acting like one. Thus, a compound with an unshared electron pair *might* react as a Lewis base; but if the same compound has an acidic hydrogen, it might also react as a Brønsted acid. The water molecule, with its amphoteric character, is an example of exactly this situation. Whether water is an acid or a base depends on the reaction. An electron-deficient compound certainly has the capability to react as a Lewis acid, and we might say, "That compound is a Lewis acid," but it is not a Lewis acid until it reacts like one. It might undergo other types of reactions in which it does not act as a Lewis acid. You have to see what happens to a compound in a reaction before you can classify the compound as an acid or base; and then the acid or base character is defined for that particular reaction.

Consider the following analogy. Suppose you have a friend who has just purchased a sleek, red convertible. He or she asks you, "Is that a fast car, or what?" Logically, you can't answer the question until you see the car perform, even though you suspect that it may have the capability of running at great speed. In the Indianapolis 500, your friend's car may *not* be fast at all. On the back roads of rural Indiana, it may be the fastest car anyone has ever seen. When someone shows you a structure and asks, "Is this a strong Brønsted acid or what?" your logical answer should be, "Show me the reaction and I'll tell you."

3.4   Factors That Affect Acidity

The bond dissociation energies and electron affinities used to explain the element effect are *gas phase* data. Yet the pK_a values presented in the text are solution data. Without the gas-phase data we could not be sure whether the trends in solution pK_a values come from some effect of the solvent (water) or from the molecules themselves. It is the very nature of the gas phase that the properties of gaseous molecules at low pressure are essentially unaffected by surrounding molecules. Consequently, because the element effect is observed in the gas phase, we know that this same effect in solution data is caused by the structures of the molecules themselves.

Solutions

Solutions to In-Text Problems

3.1   (a1)   The product results from attack of the oxygen of H_2O, which has unshared pairs, on the electron-deficient carbon of the cation:

attacking atom

$$CH_3-\overset{\overset{\displaystyle CH_3}{|}}{\underset{\underset{\displaystyle CH_3}{|}}{C}}+ \quad :\ddot{O}H_2 \quad \rightleftharpoons \quad CH_3-\overset{\overset{\displaystyle CH_3}{|}}{\underset{\underset{\displaystyle CH_3}{|}}{C}}-\underset{+}{\ddot{O}H_2}$$

Lewis acid   Lewis base

(b1)   The curved-arrow formalism for both forward and reverse reactions:

$$CH_3-\overset{\overset{\displaystyle CH_3}{|}}{\underset{\underset{\displaystyle CH_3}{|}}{C}}+ \quad :\ddot{O}H_2 \quad \rightleftharpoons \quad CH_3-\overset{\overset{\displaystyle CH_3}{|}}{\underset{\underset{\displaystyle CH_3}{|}}{C}}-\overset{+}{\ddot{O}}H_2$$

3.2   (a)   (b)

$$\begin{array}{c} H_2C \diagdown \\ \quad \quad \overset{\ddot{O}:}{\diagup} \\ \quad :\ddot{O}: \\ H_2C \diagup \\ \quad \quad \overset{\ddot{O}:}{\diagdown} \end{array}$$

$$\overset{CH_3}{\underset{CH_3}{\overset{+}{C}}}-CH_2 \overset{H}{} \qquad :\overset{..}{\underset{..}{Br}}:^{-}$$

3.3   (a)

$$CH_3\overset{..}{\underset{..}{O}}: \quad H-CH_2-\overset{\underset{\displaystyle CH_3}{|}}{CH}-\overset{..}{\underset{..}{Br}}:$$

3.4   (a) Deriving the resonance structure that shows sharing of the negative charge and double-bond character requires use of the curved-arrow formalism for electron-pair displacement reactions:

$$\left[:\overset{-}{CH_2}-CH=CH_2 \quad \longleftrightarrow \quad CH_2=CH-\overset{..}{CH_2} \right]$$

(c) The resonance structures for benzene are obtained by a series of electron-pair displacements in a cyclic fashion:

$$\left[\hexagon \quad \longleftrightarrow \quad \hexagon \right]$$

3.5   (a) In this reaction, one ammonia molecule acts as an acid and the other acts as a base:

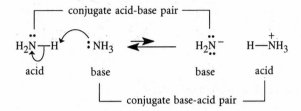

 Don't hesitate to re-draw structures or to move electron pairs to different locations on the same atom if it helps you to accomplish the purpose at hand. Notice that the structure of one ammonia molecule was re-drawn in the structure on the left so that the curved-arrow formalism could be drawn clearly.

3.6   (a)

$$H_2\ddot{N}-H \qquad :NH_3 \quad \rightleftharpoons \qquad H_2\ddot{N}^- \quad H-\overset{+}{N}H_3$$

3.7   In the pair $H_2O/^-OH$, H_2O is the acid, because it has one more proton than ^-OH; in the other pair, CH_3OH is the acid for the same reason. Thus, the reaction required is one in which the acid H_2O donates a proton to the base $^-OCH_3$.

$$HO-H \qquad :\ddot{O}-CH_3 \quad \rightleftharpoons \quad H\ddot{O}:^- \; + \; H-\ddot{O}-CH_3$$

3.8   (a) The pK_a is the negative logarithm of the dissociation constant. Therefore,

$$pK_a = -\log (5.8 \times 10^{-6}) = 5.24.$$

(c) $-\log (50) = -1.7$. An acid with a dissociation constant of 50 is a strong acid; strong acids have *negative* pK_a values.

 If you need a refresher on determining logarithms with your calculator, here are the keystrokes for the solution to Part (a) on an RPN-style (Hewlett-Packard) calculator:

$$5.8 \;\boxed{E}\; 6 \;\boxed{+/-}\;\boxed{ENTER}\;\boxed{LOG}\;\boxed{+/-}$$

The result is your answer. On other types of calculators, the keystrokes for the same operation are

$$5.8 \;\boxed{EXP}\; 6 \;\boxed{+/-}\;\boxed{LOG}\;\boxed{+/-}$$

3.9   (a) The dissociation constant is $10^{(-pK_a)}$, or $10^{-7.8}$ or 1.58×10^{-8}.
(c) The dissociation constant is $10^{-(-2)} = 1 \times 10^2 = 100$.

 To take antilogaritms with your calculator, first enter the number, for example, 7.8, and change the sign ($\boxed{+/-}$) to give -7.8. Then press the 10^x key to get the answer. If the result is a very small number, you will have to display it in exponential notation to avoid an answer of zero on your calculator.

3.10   (a) The strongest acid in Problem 3.8 is the one with the largest dissociation constant, that is, acid (c).

3.11   (a) The strongest base in the second column of Table 3.1 is the conjugate base of the weakest acid. Because the weakest acid in the table is ammonia, the strongest base is the amide ion $^-NH_2$.

3.12   (a) From Table 3.1, the pK_a of HCN is 9.40, and the pK_a of the ammonium ion is 9.25. In the reaction, ammonia reacts with HCN to give ammonium ion, $^+NH_4$, and cyanide ion, ^-CN. From the procedure in

Study Problem 3.5, the log of equilibrium constant K_{eq} for the reaction of ammonia as a base with the acid HCN is

$$\log K_{eq} = pK_a(^+NH_4) - pK_a(HCN)$$
$$\text{or} \quad \log K_{eq} = 9.25 - 9.40 = -0.15$$

from which K_{eq} is calculated by taking the antilogarithm:

$$K_{eq} = 10^{(-0.15)} = 7.08 \times 10^{-1}.$$

3.13 (a) When ammonia acts as a base toward water, the acidic species on the left is water, and the relevant pK_a from Table 3.1 is 15.7. The other relevant pK_a is that of the ammonium ion, the acid on the right side, which is 9.25.

$$H_2O + NH_3 \rightleftharpoons HO^- + \overset{+}{N}H_4$$

acidic species on
the left side

(b) When ammonia acts as an acid toward the base water, the acidic species on the left side is ammonia, and the relevant pK_a is 35 for ammonia. The other relevant pK_a is that of H_3O^+, which is -1.7.

$$NH_3 + H_2O \rightleftharpoons {^-}NH_2 + H_3O^+$$

acidic species on
the left side

3.14 Use Eq. 3.30 in the text, with $2.303RT = 5.706$ kJ/mol at 298 K:

$$\log K_{eq} = -\Delta G^\circ/(2.303RT) = -(-14.6)/(5.706) = 2.56$$
$$\text{or} \quad K_{eq} = 362$$

In this Solutions Manual, kJ/mol is used as the unit of energy. If you wish to convert kJ/mol to kcal/mol, simply divide by 4.184.

3.16 (a) To complete this calculation, first calculate the equilibrium constant, and then use it to do a mass-action calculation.

$$K_{eq} = 10^{-\Delta G^\circ/2.303RT} = 10^{-(-2.93)/5.706} = 10^{0.513} = 3.26$$

Let x be the concentration of C at equilibrium. The equilibrium constant can then be written as

$$K_{eq} = \frac{[C]}{[A][B]} = \frac{x}{(0.1 - x)(0.2 - x)} = 3.26$$

from which is obtained the quadratic expression

$$x^2 - 0.607x + 0.02 = 0$$

The quadratic formula gives

$$x = \frac{0.607}{2} \pm \frac{\sqrt{0.368 - 0.080}}{2} = 0.303 \pm 0.268$$

Choose the physically meaningful value of x to obtain

$$[C] = 0.035 \ M, [B] = 0.2 - x = 0.165 \ M, \text{ and } [A] = 0.1 - x = 0.065 \ M.$$

3.17   (a) Thiols are more acidic than alcohols (element effect), and a chlorine-substituted alcohol is more acidic than the unsubstituted alcohol (polar effect). Hence the order of decreasing pK_a (increasing acidity) is

$$CH_3CH_2OH \qquad ClCH_2CH_2OH \qquad ClCH_2CH_2SH$$

$$\xrightarrow{\text{decreasing } pK_a \atop \text{increasing acidity}}$$

3.18   (a) Use Eq. 3.38 in the text, with $2.3RT = 5.706$ kJ/mol at 298 K.

$$\Delta G° = 2.3RT(pK_a) = (5.706)(2.66) \text{ kJ/mol} = 15.2 \text{ kJ/mol}$$

3.19   To say that acetic acid is weaker than fluoroacetic acid means that the conjugate base of acetic acid is less stable relative to the un-ionized acid than the conjugate base of fluoroacetic acid is to its un-ionized acid (Fig. 3.2). "Less stable" means that more free energy is required to convert the weaker acid into its conjugate base, and this reduced stability is reflected in the larger standard free energy of ionization for acetic acid.

Solutions to Additional Problems

3.20   Compound (a) is electron-deficient, because the boron has a sextet of electrons. Compound (b) is *not* electron-deficient because each atom has an octet of electrons. Compound (c) is electron-deficient because the carbon has a sextet of electrons.

 It might seem strange that a compound with an unshared electron pair should be electron-deficient. It is the valence-electron count, not the presence or absence of unshared pairs, that determines electron deficiency. Any species is electron-deficient when an atom is surrounded by fewer than four pairs of electrons (or one pair for hydrogen).

3.22   The general approach to all parts is to identify an electron-deficient compound, and use it as a Lewis acid by attacking the electron-deficient atom with an electron pair on the other molecule.

(a)

 Don't forget that you can redraw structures to suit the purposes of the problem at hand.

(c)

The magnesium is electron-deficient even after this reaction occurs, and therefore it can be attacked by a second molecule of the ether:

(e) This is an *intramolecular* Lewis acid-base reaction.

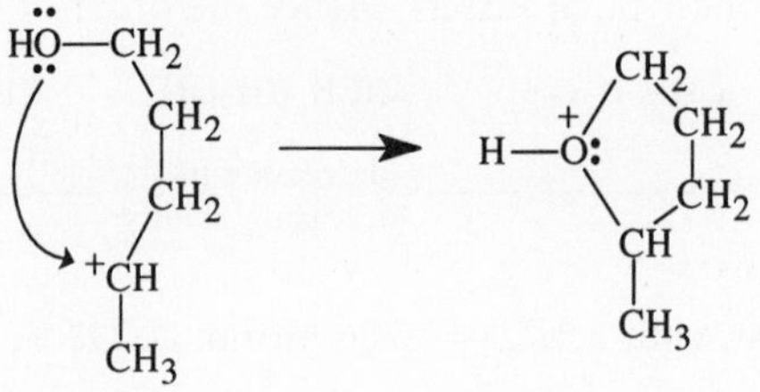

(g) In this case, use the carbon as the attacking atom rather than the oxygen. If the oxygen were to attack, the oxygen in the resulting product would have two positive charges. In contrast, attack of the carbon gives a product in which the corresponding carbon is neutral.

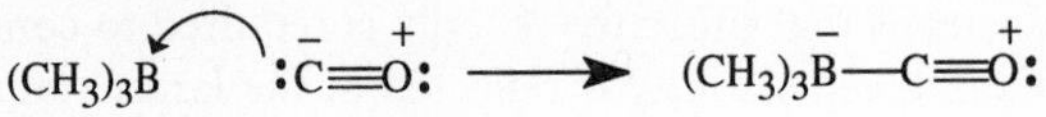

3.23 (a)

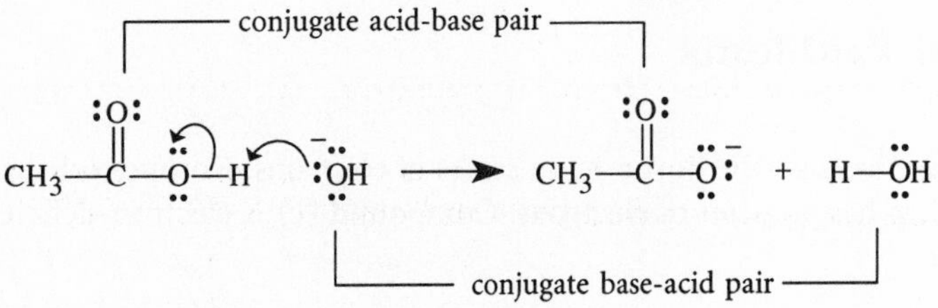

(c) In this example, the acid-base reaction is *intramolecular,* that is, within the same molecule.

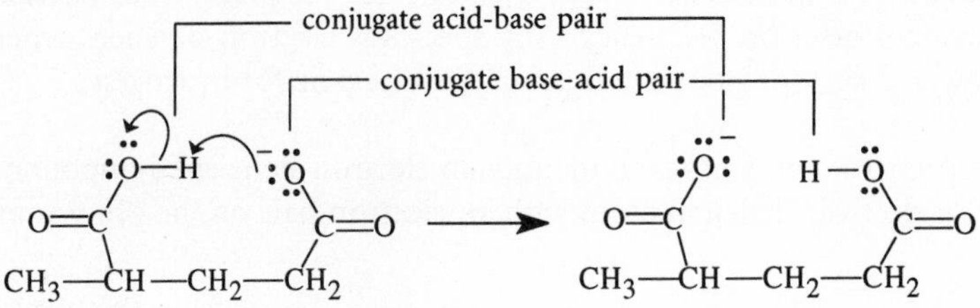

(e)

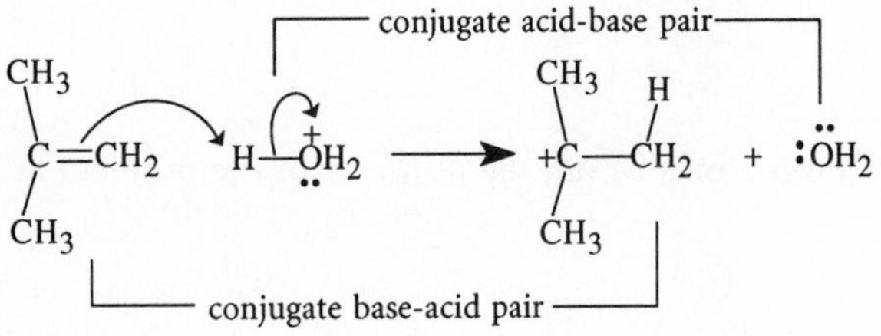

3.24 *Step 1:*

(a) (1), (4), and (5). The reaction is Lewis acid-base reaction; the proton that is transferred is the Lewis acid, and the —OH of the alcohol is the Lewis base. (Note that all Brønsted acid-base reactions are Lewis acid-base reactions.)

(b) and (c)   The conjugate pairs and the curved-arrow formalism are as follows:

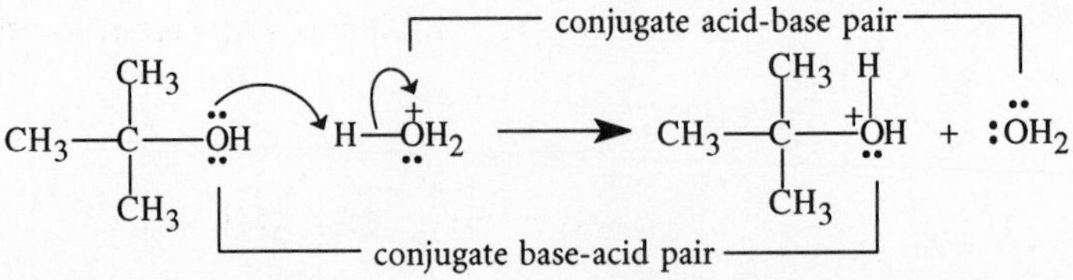

Step 2:

(a) (1), (3) The reaction is a Lewis acid-base dissociation, which is a type of Lewis acid-base reaction.

(c)

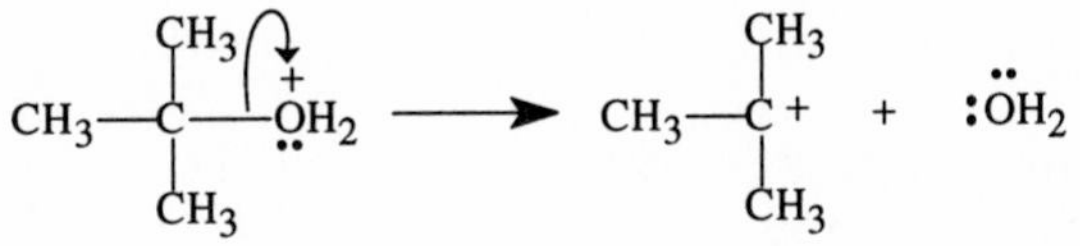

Step 3:

(a) (1), (4), (5). The reaction is a Brønsted acid-base reaction.

(b) and (c)   The conjugate pairs and the curved-arrow formalism are as follows:

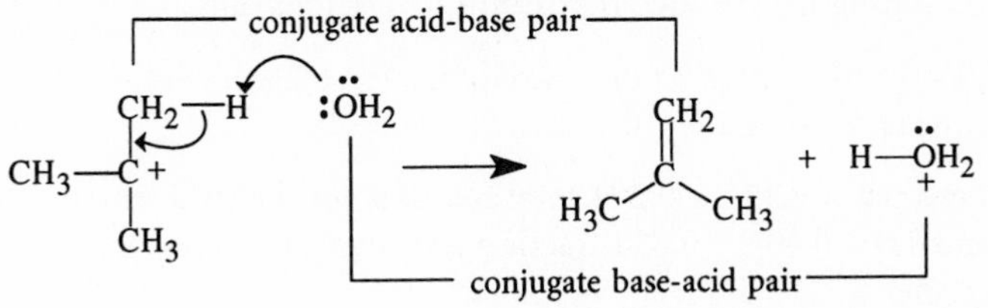

Note that none of the steps as written can be classified as (2), an association reaction of a Lewis base with an electron-deficient Lewis acid. However, the *reverse* of Step 2 is such a reaction.

3.25   (a)

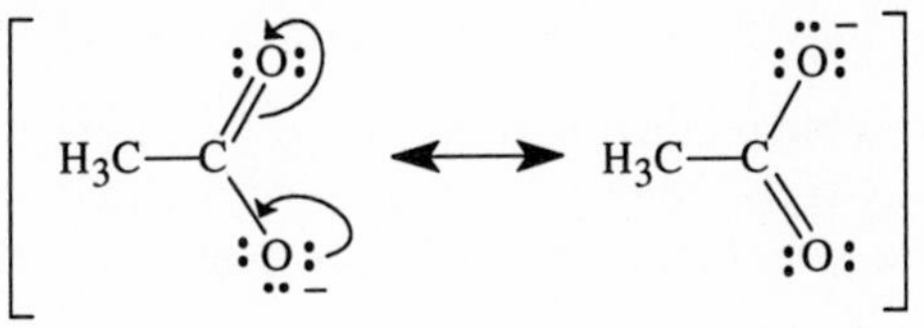

3.26   (a) 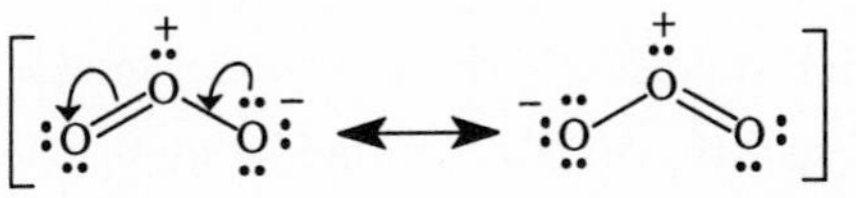   (c)

3.27   (a) 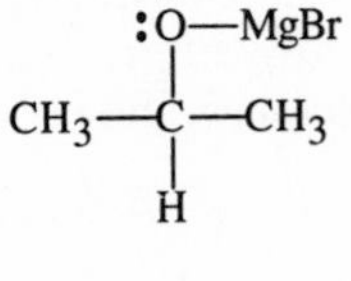   (c)

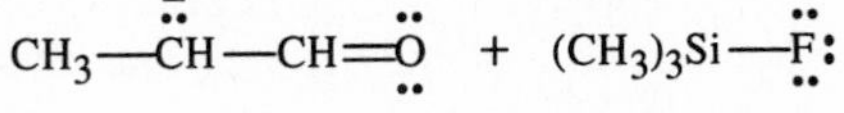

3.28   (a) 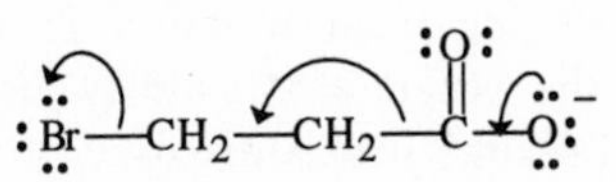   (c) 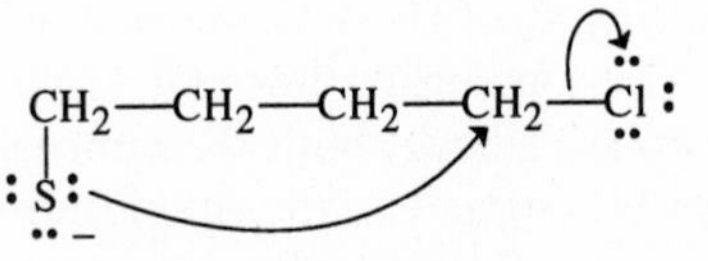   (e)

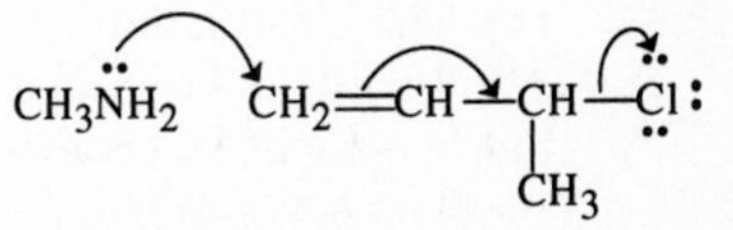

3.29   (a) The curved arrow is going in the wrong direction; it should go from the electrons on the ⁻OH to the acidic hydrogen:

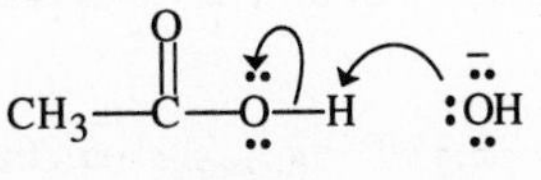

3.30 (a)

3.31 (a) Let K_{eq} be the ratio [*anti*]/[*gauche*]; then $\Delta G°$ for *gauche* $\rightleftarrows$ *anti* is –3.8 kJ/mol. (Because *gauche* has higher energy than *anti*, $\Delta G°$ is negative.) Then use Eq. 3.30 in the text.

$$\log K_{eq} = -\Delta G°/2.3RT = -(-3.8)/5.706 = 0.6667$$
$$\text{or} \quad K_{eq} = 10^{0.6667} = 4.63$$

This is the ratio of the *anti* conformation to any *one gauche* conformation. Taking the total fraction of butane as 1.0, and noting from above that [*anti*] = 4.63[*gauche*],

$$1.0 = [anti] + 2[gauche] = 4.63[gauche] + 2[gauche] = 6.63[gauche]$$
$$\text{or} \quad [gauche] = 1.0/6.63 = 0.151$$

Consequently, there is 2 × 0.151 = 0.302 total fraction *gauche* conformation (that is, about 30% *gauche* conformation) and 1.0 – 0.302 = 0.698 fraction *anti* conformation (that is, about 70% *anti* conformation) in a sample of butane.

3.32 To arrange in order of decreasing pK_a means to arrange in order of *increasing acidity*.

(a) CH_3CH_2OH is less acidic than $ClCH_2CH_2OH$, which is less acidic than Cl_2CHCH_2OH. Chlorine substituents increase acidity because of their electron-withdrawing polar effect; the more chlorines there are, the greater the effect.

(c) The element effect down a column of the periodic table predicts that $HN(CH_3)_2$ should be less acidic than $HP(CH_3)_2$, which should be less acidic than $HAs(CH_3)_2$. Notice that these acidities refer to the reaction in which the *neutral* molecules lose a proton to give the *anionic* conjugate bases (M = N, P, or As):

$$H-\overset{\cdot\cdot}{M}(CH_3)_2 + base\overset{-}{:} \quad \rightleftarrows \quad \overset{-}{:}\overset{\cdot\cdot}{M}(CH_3)_2 + base-H$$

The acidity of $H_2\overset{+}{As}(CH_3)_2$ refers to the reaction in which this positively charged species loses a proton to give the neutral conjugate base $HAs(CH_3)_2$. The cationic acid $H_2\overset{+}{As}(CH_3)_2$ should be more acidic than the neutral acid $HAs(CH_3)_2$ because a positively charged atom is more electronegative than the same neutral atom, and is therefore a better electron acceptor. The better the electron acceptor, the stronger the acid.

(e) CH_3CH_2OH is less acidic than $(CH_3)_2NCH_2CH_2OH$ because of the polar effect of the electronegative nitrogen substituent. In the last compound, not only is the polar effect of the nitrogen present, but also that of the positive charge, which can stabilize the conjugate-base anion. Furthermore, the positively-charged nitrogen is closer to the —OH group. Consequently, $(CH_3)_3\overset{+}{N}$—OH is the most acidic of the three compounds.

3.33 (a) The information needed is the pK_a of H—CN, the acid on the left side of the equation. From Table 3.1, the pK_a of H—CN is 9.40. The equilibrium favors the side with the weaker acid, namely the right side, but by very little, because the two acids have about the same pK_a. Apply the procedure used in Study Problem 3.5 to obtain a quantitative estimate of the equilibrium constant K_a..

$$\log K_{eq} = 9.76 - 9.40 = 0.36$$
$$\text{or} \quad K_{eq} = 10^{0.36} = 2.3$$

3.34 (a) The standard free-energy change is obtained by applying Eq. 3.29 in the text, using the value of K_{eq} calculated in Problem 3.33(a).

$$\Delta G° = -2.3RT \log K_{eq} = -5.706(0.36) = -2.05 \text{ kJ/mol}$$

(c) If the equilibrium concentrations of $(CH_3)_3N$ and HCN are both x, then the equilibrium concentrations of $(CH_3)_3\overset{+}{N}H$ and ^-CN are both $(0.1 - x)$. Substituting in the equilibrium-constant expression,

$$K_{eq} = 2.3 = \frac{[(CH_3)_3\overset{+}{N}H][^-CN]}{[(CH_3)_3N][HCN]} = \frac{(0.1-x)^2}{x^2}$$

This leads to the quadratic expression $x^2 + 0.154x - 0.00769 = 0$, which is solved to give $x = 0.0397\ M$. Thus, at equilibrium, the concentrations of the species on the left side of the equation are both $0.0397\ M$, and the concentrations of the species on the right side are both $0.1000 - 0.0397 = 0.0603\ M$.

3.35   (a) By Eq. 3.27, text p. 103, the compound with the smaller pK_a has the smaller standard free energy of dissociation ΔG_a°. Consequently, phenylacetic acid has the smaller ΔG_a°, and is therefore the stronger acid.
(c) Because the phenyl substituent is acid-strengthening, it stabilizes the conjugate-base anion in the same sense that a chlorine does. Consequently, its polar effect, like that of a chlorine, is electron-withdrawing.

3.36   (a) The ionizations of malonic acid:

First ionization ($pK_a = 2.86$):

Second ionization ($pK_a = 5.70$):

(b) The first pK_a is lower than that of acetic acid because, in the product of the first ionization, the electron-withdrawing polar effect of the un-ionized carboxylic acid group stabilizes the conjugate-base anion on the other. (The two electronegative oxygens are responsible for the polar effect of the carboxylic acid group.) The second pK_a is greater than that of acetic acid because one negative charge in the dianionic product of the second ionization interacts unfavorably (repulsively) with the second negative charge, and the energy of the dianion is thereby raised. (See Fig. 3.2, text page 109.)

(c) As n becomes larger, the polar groups become more remote. Hence, *both* polar effects discussed in part (b) become less significant, and *both* pK_a values approach that of acetic acid; that is, the pK_a values of the two ionizations become more similar. (In terms of the hint, as r in the denominator of the electrostatic law becomes larger, the energy of interaction of one charge with another becomes smaller.)

Even if n is very large, the two pK_a values are not exactly equal. Theoretically, the first pK_a should be 0.6 smaller than the second pK_a (that is, the first K_a should be 4 times larger than the second K_a). The reason is that there are two acidic hydrogens contributing to the first ionization, but only one contributing to the second; and there are two carboxylate ions that can be protonated in the conjugate base of the monoacid, but only one in the conjugate base of the diacid. Thus, the first ionization is four times as likely as the second on purely statistical grounds.

3.37   As Eq. 3.39 in the text shows, the polar effect of the fluorine stabilizes negative charge; however, it *destabilizes* positive charge because the positive ends of the C—F dipoles interact repulsively with the positive charge of the cation, as shown by the following diagram:

Consequently, the product of the second equilibrium is destabilized relative to the neutral starting material. Because it is has a higher energy, the energy required to form it is larger, and the equilibrium constant for its formation is smaller. In other words, Eq. (1) has the equilibrium constant most favorable to the right.

3.39   The equilibrium constant for the reaction shown, K_{eq}, is given by

$$K_{eq} = \frac{[HB][A^-]}{[HA][B^-]}$$

Following the hint, first show that this equals K_{HA}/K_{HB}.

$$K_{HA}/K_{HB} = \frac{[A^-][H_3O^+]}{[HA]} \times \frac{[HB]}{[H_3O^+][B^-]} = \frac{[HB][A^-]}{[HA][B^-]}$$

or $K_{eq} = K_{HA}/K_{HB}$

Then take logarithms of this equation:

$$\log K_{eq} = \log K_{HA} - \log K_{HB} = pK_{HB} - pK_{HA}$$

Taking antilogs of both sides proves the assertion:

$$10^{(\log K_{eq})} = K_{eq} = 10^{(pK_{HB} - pK_{HA})}$$

3.40   (a) Both HI and HCl are much stronger acids than H_3O^+. Consequently, the following equilibrium lies well to the right for both X = Cl and X = I:

$$HX \;+\; H_2O \;\rightleftharpoons\; H_3O^+ \;+\; X^-$$

Since the major acidic species in solution in each case is H_3O^+, and 10^{-3} M of this species is present in each case, both solutions have the same pH value of 3.

Note that the most acidic species that can exist in a solvent is the conjugate acid of the solvent; in water, this is H_3O^+. Likewise, the most basic species that can exist in a solvent is the conjugate base of the solvent; in water this is ^-OH. Because acids or bases stronger than the solvent react to give the conjugate acid or base of the solvent, respectively, their greater acidity or basicity is not reflected in the pH of the resulting solution. This effect is sometimes termed the *leveling effect* of solvent.

3.41   The first step in the reaction is shown in Eq. 3.6 of the text:

The second step involves attack of the Lewis base ^-OH on the electron-deficient Lewis acid BH_3:

3.42   (a) Because bond dissociation energy decreases down a column of the periodic table, the H—At bond should be weaker than the H—I bond. The bond dissociation energy of H—At is smaller.

(b) Because electron affinities decrease down a column of the periodic table, the electron affinity of At should be smaller than that of I.

(c) Because dissociation energies dominate the dissociation constant within a *column* of the periodic table, H—At should be a stronger acid than H—I.

4

Introduction to Alkenes; Reaction Rates

Terms

Concepts

I. Structure, Bonding, and Heat of Formation in Alkenes

A. GENERAL

1. Alkenes (sometimes called olefins) are hydrocarbons that contain carbon-carbon double bonds.
    a. The carbon-carbon double bond consists of a σ bond and a π bond.
    b. The π electrons of a double bond attack Brønsted or Lewis acids.
2. Because alkenes have fewer hydrogens than the corresponding alkanes, they are classified as unsaturated hydrocarbons.

B. HYBRIDIZATION

1. The carbon atoms of a double bond, as well as other trigonal-planar atoms, are sp^2-hybridized.
    a. Two of the three $2p$ orbitals mix with the $2s$ orbital to form three sp^2 orbitals.
    b. One $2p$ orbital remains unhybridized.

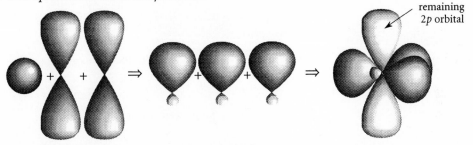

one $2s$ and two of the three $2p$ orbitals   three sp^2 orbitals   trigonal arrangement

2. Hybridization and geometry are related.
    a. Whenever a main-group atom has trigonal-planar geometry, its hybridization is sp^2.
    b. Whenever a main-group atom has tetrahedral geometry, its hybridization is sp^3.
3. Trigonal carbon withdraws (attracts) electrons from alkyl groups.

C. THE π BOND

1. The second bond of a carbon-carbon double bond, called a π bond, arises from the overlap of a $2p$ orbital on one carbon atom with a $2p$ orbital on the other.
    a. Bonding (π) molecular orbitals, resulting from additive overlap of the two carbon $2p$ orbitals, are at lower energy than the isolated p orbitals.
    b. Antibonding molecular orbitals (π^*), resulting from subtractive overlap of the two carbon $2p$ orbitals, are at higher energy than the isolated p orbitals.
    c. The π molecular orbital has a nodal plane; the π^* molecular orbital has two nodal planes.

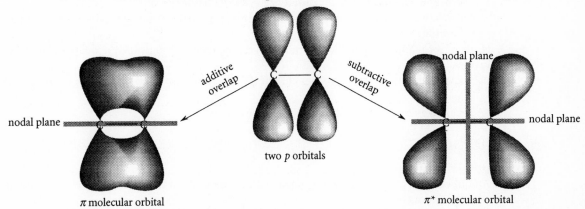

π molecular orbital   two p orbitals π^* molecular orbital

2. The π bond is one bond with two lobes; the π bond thus has electron density both above and below the plane defined by the carbons of the double bond and the directly attached atoms.

D. STEREOISOMERISM

1. Compounds with identical connectivities that differ in the spatial arrangement of their atoms are called stereoisomers.

   a. Rotation about the alkene double bond does not occur under normal conditions. In order for such an internal rotation to occur, the π bond would have to break; thus, some alkenes can exist as *cis* and *trans* isomers.

   b. *Cis* and *trans* isomers are named with the *E,Z* priority system.

E-2-butene (or *trans*-2-butene)   Z-2-butene (or *cis*-2-butene)

2. A stereocenter (stereogenic atom) is an atom at which the interchange of two groups gives a stereoisomer.

3. Isomeric alkenes that differ in the positions of their double bonds are examples of constitutional isomers.

E-2-butene (or *trans*-2-butene)   1-butene

E. RELATIVE STABILITIES OF ALKENE ISOMERS

1. An alkene is stabilized by alkyl substituents on the double bond.

2. The alkene with the greatest number of alkyl substituents on the double bond is usually the most stable one.

3. Heats of formation provide information about how alkene stabilities vary with structure.

F. HEATS OF FORMATION

1. The standard heat of formation ΔH_f° of a compound is the heat of the reaction in which the compound is formed from its elements in their natural states at 1 atm pressure and 25 °C.

   a. A reaction in which heat is liberated is said to be an exothermic reaction.

   b. A reaction in which heat is absorbed is said to be an endothermic reaction.

2. Chemical reactions and their associated energies can be added algebraically (Hess's law of constant heat summation).

   a. The heat of any reaction is the difference in the enthalpies of products and reactants.

   b. The relative enthalpies of many organic compounds are available in standard tables as heats of formation.

3. Heats of formation (enthalpies of formation) can be used to determine the relative stabilities of various bonding arrangements.

   a. Alkenes with more alkyl branches at their double bonds are more stable than isomers with fewer branches

   b. *Trans* alkenes are more stable than their *cis* isomers.

II. *Nomenclature*

A. UNBRANCHED ALKENES

1. An unbranched alkene is named by replacing the *ane* suffix in the name of the corresponding alkane with the ending *ene*.

2. The carbons are numbered from one end to the other so that the double bond receives the lowest number.

$$\overset{6}{H_3C}-\overset{5}{CH_2}-\overset{4}{CH_2}-\overset{3}{CH}=\overset{2}{CH}-\overset{1}{CH_3}\qquad \text{2-hexene}$$

B. BRANCHED ALKENES

1. The names of alkenes with branched chains are derived from their principal chains.
   a. The principal chain is defined as the carbon chain containing the greatest number of double bonds, even if this is not the longest chain.
   b. If more than one candidate for the principal chain have equal numbers of double bonds, the principal chain is the longest of these.

principal chain

2. The principal chain is numbered from the end so as to give the lowest numbers to the double bonds at the first point of difference.
   a. If the name remains ambiguous after determining the correct number for the double bonds, then the principal chain is numbered so that the lowest numbers are given to the branches at the first point of difference.
   b. The position of the double bond is cited in the name after the names of the substituents.
   c. If a compound contains more than one double bond, the *ane* ending of the corresponding alkane is replaced by *adiene* (if there are two double bonds), *atriene* (if there are three double bonds), etc.

a 1,3,7-nonatriene

3. Substituent groups may also contain double bonds.
   a. Some widely occurring groups of this type have special names that must be learned.
   b. The names of other groups are constructed from the name of the parent hydrocarbon by dropping the final *e* and replacing it with *yl*, and are numbered from the point of attachment to the principal chain.

5-(2-propenyl)-1,3,7-nonatriene
or
5-allyl-1,3,7-nonatriene

allyl group or 2-propenyl group

4. Some alkenes have nonsystematic traditional names that are recognized by the IUPAC.

$$Ph-CH=CH_2 \qquad \text{styrene}$$

C. THE CAHN-INGOLD-PRELOG (*E,Z*) SYSTEM

1. The Cahn-Ingold-Prelog system is applied to alkenes by assigning relative priorities to the two groups on each carbon of the double bond.
   a. The designation *Z* (*zusammen*; together) is used when the groups of highest priority are on the same side of the double bond.
   b. The designation *E* (*entgegen*; across) is used when the groups of highest priority are on different sides of the double bond.

groups of highest priority are shown in black

(*Z*)-3-chloro-4-methyl-3-hexene

(*E*)-1-deuteriopropene

2. Examine the atoms directly attached to a given carbon of the double bond.
   a. Assign higher priority to the group containing the atom of higher atomic number (or atomic mass in the case of isotopes).
   b. If the atoms directly attached to the double bond are the same:

   i. Arrange the attached atoms within each set in descending priority order, and make a pairwise comparison of the atoms in the two sets.
   ii. The higher priority is assigned to the atom of higher atomic number (or atomic mass in the case of isotopes) at the first point of difference.
   iii. Double bonds are treated by a special convention, in which the double bond is rewritten as a single bond and the atoms at each end of the double bond are duplicated.
   c. If the sets of attached atoms are identical, move away from the double bond within each group to the next atom following the path of highest priority and repeat part b above.

D. UNSATURATION NUMBER

1. The unsaturation number U of a molecule is equal to the total number of rings and multiple bonds in the molecule.

$$U = \frac{2C + 2 + N - X - H}{2} = \text{number of rings + multiple bonds}$$

2. The number of hydrogens in a fully saturated molecule is equal to $2C + 2 + N - X$.
   a. Every ring or double bond reduces the number of hydrogens from this maximum by two (2).
   b. Each halogen atom in an organic compound reduces the maximum possible number of hydrogens by one (1).
   c. Each nitrogen atom present increases the maximum possible number of hydrogens by one (1).
   d. Each oxygen atom present has no effect on the maximum possible number of hydrogens.

$$U = \frac{2(8) + 2 + (1) - (1) - (8)}{2} = 5 = 1 \text{ ring + 4 double bonds}$$

III. Principles of Reactions

A. REGIOSELECTIVE REACTIONS

1. A reaction that gives only one of several possible constitutional isomers is said to be a regioselective reaction.
2. The most characteristic type of alkene reaction is addition to the carbon-carbon double bond.
3. The regioselectivity observed in the addition reactions of hydrogen halides or water to alkenes is a consequence of several factors:
   a. The rate-limiting transition state of each reaction resembles a carbocation.
   b. The relative stability of carbocations is tertiary > secondary > primary.
   c. The structures and energies of transition states for reactions involving unstable intermediates (such as carbocations) resemble the structures and energies of the unstable intermediates themselves (Hammond's postulate).
4. Reactions involving carbocation intermediates show rearrangements in some cases.

B. REACTION RATES

1. The rate of a chemical reaction can be defined for our purposes as the number of reactant molecules converted into product in a given time.
2. The relative free energies of transition state and reactants determine the reaction rate.
3. Two factors that govern the intrinsic reaction rate are:
   a. the size of the energy barrier (standard free energy of activation $\Delta G^{\circ\ddagger}$)—reactions with larger $\Delta G^{\circ\ddagger}$ are slower.

b. the temperature: reactions are faster at higher temperatures.

4. The equilibrium constant for a reaction tells us absolutely nothing about its rate.

C. CATALYSTS

1. A substance that increases the rate of a reaction without being consumed is called a catalyst.
   a. A catalyst cannot affect the position of an equilibrium but it does affect the rate at which a reaction comes to equilibrium.
   b. A catalyst accelerates both forward and reverse reactions of an equilibrium equally.
2. Catalysts are of two types: heterogeneous and homogeneous.
   a. A catalyst that is soluble in a reaction solution is called a homogeneous catalyst. Acid-catalyzed hydration of alkenes is a reaction involving homogeneous catalysis.
   b. A catalyst that exists in a separate phase from the reactants is called a heterogeneous catalyst. Catalytic hydrogenation of alkenes is a reaction involving heterogeneous catalysis.
3. Enzymes are biological catalysts.

D. MULTISTEP REACTIONS, THE RATE-LIMITING STEP, AND MECHANISMS

1. Each step of a multistep reaction has its own characteristic rate and therefore its own transition state.
2. The slowest step in a multistep chemical reaction is called the rate-limiting step (rate-determining step) of the reaction.
   a. The rate-limiting step is the step with the transition state of highest free energy.
   b. The overall rate of a reaction is equal to the rate of the rate-limiting step.
   c. Anything that increases the rate of this step increases the overall reaction rate.
3. The complete description of a reaction pathway, including any reactive intermediates such as carbocations, is called the mechanism of the reaction.

E. TRANSITION-STATE THEORY

1. As reactants change into products, they pass through an unstable state of maximum free energy called the transition state.
   a. The transition state has a higher energy than either the reactants or products.
   b. The transition state represents an energy barrier to the interconversion of reactants and products.
2. The energy barrier $\Delta G^{\circ\ddagger}$, called the standard free energy of activation, is equal to the difference between the standard free energies of the transition state and reactants.
   a. The size of the energy barrier $\Delta G^{\circ\ddagger}$ determines the rate of a reaction: the higher the barrier, the lower the rate.
   b. This energy barrier is shown graphically in a reaction free-energy diagram in which the pathway of the reaction from reactants to products is called the reaction coordinate.

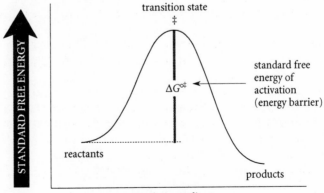

3. In general, molecules obtain the energy required for a reaction from their thermal motions.
4. The energy of a chemical system can be increased by raising the temperature.

F. HAMMOND'S POSTULATE

1. The assumption that the structure of a transition state closely resembles the structure of a corresponding unstable intermediate is known as Hammond's postulate.

a. The structures and energies of the transition states are approximated by the structures and energies of the respective unstable intermediates themselves.

b. It is not the stabilities of the intermediates themselves that determine which of two competing reactions is faster; it is the relative free energies of the transition states for the formation of these intermediates that determine the relative rates of reactions.

G. STRUCTURE AND STABILITY OF CARBOCATIONS

1. The species with a positively charged, electron-deficient carbon is called a carbocation (carbonium ion in older literature).

a. Carbocations are examples of reactive (unstable) intermediates—species that react so rapidly that they never accumulate in more than very low concentrations.

b. Carbocations are powerful electron-deficient Lewis acids and thus are potent electrophiles.

2. The relative stability of carbocations is in the order:

tertiary > secondary > primary

a. A primary carbocation has one alkyl group bound to the electron-deficient carbon.

b. A secondary carbocation has two alkyl groups bound to the electron-deficient carbon.

c. A tertiary carbocation has three alkyl groups bound to the electron-deficient carbon.

primary carbocation   secondary carbocation   tertiary carbocation

3. The explanation for the stabilization of carbocations by alkyl branching is a phenomenon called hyperconjugation, which is the overlap of bonding electrons from the adjacent σ bonds with the unoccupied p orbital of the carbocation.

a. The energetic advantage of hyperconjugation comes from the additional bonding symbolized by the double bond.

H. CARBOCATION REARRANGEMENTS

1. Carbocations can react with nucleophiles or rearrange to more stable carbocations.

2. In a rearrangement, a group or atom from the starting material has moved to a different position in the product.

secondary carbocation   tertiary carbocation

a. Rearrangement of a carbocation is favored by the greater stability of the rearranged ion.

b. Some rearrangements involve a hydride shift—the migration of a hydrogen with its two bonding electrons.

secondary carbocation   tertiary carbocation

Reactions

··

I. Addition Reactions of Alkenes

A. ADDITION OF HYDROGEN HALIDES TO ALKENES

1. Hydrogen halides add to alkenes in a regioselective manner so that the hydrogen adds to the less branched carbon and the halide group to the more branched carbon of the double bond to give products called alkyl halides.

2. The addition of a hydrogen halide to an alkene is a regioselective reaction that occurs in two successive steps.
   a. In the first step, protonation of the alkene double bond occurs at the carbon with the fewer alkyl branches so that the more stable carbocation is formed—the one with the greater number of alkyl branches at the electron-deficient carbon.
   b. In the second step, the halide ion attacks the electron-deficient carbon.

3. Markownikoff's rule: the halogen of a hydrogen halide attaches itself to the carbon of the double bond bearing the least number of hydrogens and the greater number of carbons.
4. Hydrogen halide addition to an alkene is a regioselective reaction because addition involves the transition state that resembles the more stable of the two possible carbocation intermediates.
5. The formation of a carbocation from an alkene is an electron-pair displacement reaction in which the π bond acts as a Brønsted base towards the Brønsted acid H—X.

B. CATALYTIC HYDROGENATION OF ALKENES

1. Catalytic hydrogenation is an addition of hydrogen in the presence of a catalyst and is one of the best ways to convert alkenes into alkanes.
   a. Because hydrogenation catalysts are insoluble in the reaction solution, they are examples of heterogeneous catalysts.
   b. Benzene rings are inert to conditions under which normal double bonds react readily, but under conditions of high temperature and pressure they can be hydrogenated.

C. HYDRATION OF ALKENES

1. The addition of water to the alkene double bond, called alkene hydration, is an acid-catalyzed reaction.
   a. Because the catalyzing acid is soluble in the reaction solution, it is a homogeneous catalyst.
   b. Hydration is a regioselective reaction.

2. Alkene hydration is a multistep reaction.
   a. In the first step of the reaction (the rate-limiting step) the double bond is protonated to give a carbocation.
   b. In the next step of the hydration reaction, the carbocation is attacked by the Lewis base water in a Lewis acid-base association reaction.
   c. Finally, a proton is lost to solvent in another Brønsted acid-base reaction to give an alcohol product and regenerate the catalyzing acid.

3. Because the hydration reaction involves carbocation intermediates, some alkenes give rearranged hydration products.
4. With sulfuric acid in solution, alkene hydration occurs in two distinct stages.
   a. In the first stage sulfuric acid undergoes an addition to an alkene to give an alkyl sulfate as the addition product.
   b. In the second step the alkyl sulfate reacts with water to give an alcohol and regenerate the sulfuric acid.

an alkyl sulfate

Study Guide Links

✓4.1 Different Ways to Draw the Same Structure

The discussion in the text requires you to realize that there are several different ways to draw the same alkene stereoisomer. For example, two equivalent ways to represent *cis*-2-butene are the following:

cis-2-butene, or (Z)-2-butene

If this is not clear, think of the structure as an object. Rotate the structure 180° about an axis through the C=C bond, as follows:

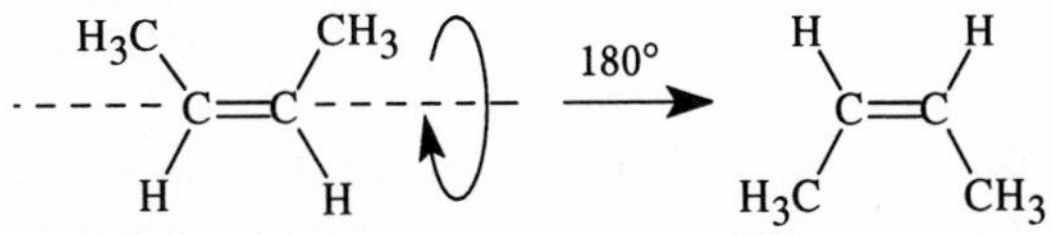

(Of course, when we perform such a rotation, the atom labels, for example, CH₃, are maintained "right side up"; that is, they are not inverted.)

Similarly, two equivalent ways of drawing *trans*-2-butene are as follows:

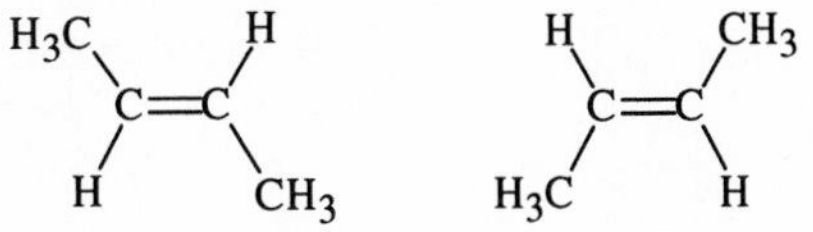

trans-2-butene, or (E)-2-butene

The two structures in text Eq. 4.3 are examples of identical structures drawn differently. Some students think that structures such as these are stereoisomers, but in fact they represent the same molecule. As these structures demonstrate, the ultimate test of identity of two structures is their congruence—the superimposability of each atom of one structure and an identical atom of the other.

✓4.2 Drawing Structures from Names

When asked to draw a complicated structure from a name, as in Problem 4.5a, you should review again the procedure used in Study Problem 2.3. This procedure requires you to be *systematic*. The natural tendency is to try to write the finished structure immediately. Instead, you should take it one step at a time:

1. Write the carbon skeleton of the principal chain; do not be concerned with stereochemistry.
2. Add the substituents to the principal chain.
3. Add the hydrogens.
4. Decide on relative group priorities at the double bonds to which an *E* or *Z* configuration must be assigned. Jot them down on your structure.

(list continues)

5. Redraw your structure with proper stereochemistry. It helps to maintain proper 120° bond angles at the double bonds.

Let's illustrate with an example. Suppose you are asked to draw the structure of (2*E*,4*Z*)-3-isobutyl-2,4-hexadiene. Follow the above steps in order.

1. The principal chain contains six carbons with double bonds at carbons 2 and 4:

$$C-C=C-C=C-C$$

2. There is an isobutyl group at carbon 3. Add the hydrogens within this group, since you won't have to manipulate this group further.

$$
\begin{array}{c}
C-C=C-C=C-C \\
| \\
CH_2 \\
| \\
CH(CH_3)_2
\end{array}
\Big\} \text{ isobutyl group}
$$

3. Add the missing hydrogens.

$$
\begin{array}{c}
CH_3-CH=C-CH=CH-CH_3 \\
| \\
CH_2 \\
| \\
CH(CH_3)_2
\end{array}
$$

4. Indicate the relative priorities of the groups at each double bond. For example, focusing on the leftmost double bond (carbons in boldface below), the relative priorities of the attached groups are indicated by numbers:

(You should complete the priorities for the rightmost double bond.)

5. Redraw the structure with proper stereochemistry.

(2*E*,4*Z*)-3-isobutyl-2,4-hexadiene

You should work Problem 4.5a using this step-by-step procedure.

In many cases, it is easier to work a complicated problem if it can be broken into smaller "chunks." When you learn to bypass the anxiety created by a complex problem and adopt this approach, you have made significant progress toward becoming a good problem solver.

4.3   **Sources of Heats of Formation**

Heats of formation are not obtained by direct measurement. Rather, they are calculated from more readily available combustion data by applying Hess's law. To illustrate, suppose that we want to calculate the heat of formation of *trans*-2-butene from the following known heats of combustion: hydrogen, −241.8 kJ/mol; carbon, −393.5 kJ/mol; and *trans*-2-butene, −2530.0

kJ/mol. The combustion reaction (Sec. 2.7) is the reaction of each species with oxygen to give water (in the combustion of hydrogen), carbon dioxide (in the combustion of carbon), and both water and carbon dioxide (in the combustion of hydrocarbons).

Hess's law allows us to express the formation of *trans*-2-butene from its elements as the sum of three combustion reactions. Identical species on opposite sides of the equations cancel, and the resulting equations and their enthalpies are added algebraically:

Equations: $\Delta H°$ (kJ/mol)

$$4H_2 \; + \; 2O_2 \longrightarrow 4H_2O \qquad\qquad 4(-241.8) = \; -967.2$$

$$4C \; + \; 4O_2 \longrightarrow 4CO_2 \qquad\qquad 4(-393.5) = -1574.0$$

$$4CO_2 + 4H_2O \longrightarrow 6O_2 \; + \; \textit{trans-2-butene} \qquad\qquad +2530.0$$

Sum: $4C \; + \; 4H_2 \longrightarrow$ *trans*-2-butene $\Delta H° = -11.2$ kJ/mol

Note that because four moles of both H_2 and C are required, their respective enthalpies of combustion must be multiplied by 4. Note also that the combustion of *trans*-2-butene must be written in reverse so that the formation equation comes out with *trans*-2-butene on the right. Consequently, the sign of the enthalpy of this combustion is also reversed.

Fortunately, a calculation like this is not necessary every time a heat of formation is needed, because such calculations have already been done to provide the data available in standard tables of heats of formation.

4.4   Free Energy and Enthalpy

To review: If you want to calculate an equilibrium constant, the *standard free energies* ($\Delta G°$ values) of the reactants and products are the relevant quantities. If you want to ask which of two molecules has a less energetic (or more stable) arrangement of bonds, then you need the *standard enthalpies of formation* ($\Delta H°$ values) of the two compounds.

This section discusses why two different types of energy are needed for different purposes. The discrepancy between the enthalpy change for any process (such as a reaction) and the free-energy change lies in something called the *entropy change*. The relationship between the free-energy change, the enthalpy change, and the entropy change is deceptively simple:

$$\Delta G° = \Delta H° - T\Delta S° \qquad\qquad\qquad (SG4.1)$$

In this equation, $\Delta S°$ is the entropy change and T is the absolute temperature in kelvins. What is the meaning of the entropy change?

Entropy measures molecular randomness, or *freedom of motion*. For example, consider the following reaction:

$$\square \longrightarrow H_2C{=}CHCH_2CH_3$$

cyclobutane   1-butene

The experimentally determined $\Delta H°$ for the reaction at 25 °C is −26.53 kJ/mol (−6.34 kcal/mol), and the $\Delta G°$ for the reaction at 25 °C is −40.13 kJ/mol (−9.59 kcal/mol). Solving Eq. SG4.1, $T\Delta S°$ at $T = 298$ K is 13.60 kJ/mol (3.25 kcal/mol); that is, $\Delta S°$ is positive. A positive $\Delta S°$ means that there is more *randomness* or *freedom of motion* in the products than the reactants.

What is it about 1-butene that is more "random" than cyclobutane? Notice that 1-butene has two carbon-carbon single bonds. Internal rotation can occur readily about these bonds, just as in butane. However, in cyclobutane, *internal rotations cannot occur* because the various carbon atoms are constrained into a small ring. (If this is not clear to you, make a model of cyclobutane and convince yourself that internal rotation about its carbon-carbon bonds cannot occur.) Thus, there is *greater freedom of motion* in 1-butene than in cyclobutane. Because of the internal rotations in 1-butene, the hydrogens on the carbons connected by single bonds can move through a greater volume of space than the same hydrogens in cyclobutane. Thus, *their positions are more random.* Hence, the randomness—or *entropy*—in 1-butene is greater than that in cyclobutane.

Because of entropy—the "randomness factor"—the energy that controls the position of a chemical equilibrium ($\Delta G°$) is different from the energy stored in chemical bonds ($\Delta H°$). Eq. SG4.1 shows that an increase of randomness in a reaction—a positive $\Delta S°$—gives the product an additional advantage in a chemical equilibrium *over and above* that which results from the formation of more stable bonds.

✓4.5   Solving Reaction Problems

You'll encounter problems such as Problem 4.18 and 4.21 throughout the text. They are designed to test your understanding of the reactions you'll study. In many cases, students who make a reasonable effort to understand each reaction find that supplying the products of a reaction when the starting materials are given, as in Problem 4.18, is a relatively straightforward exercise. However, problems such as Problem 4.21, in which you are given a product and asked to supply an appropriate reactant (or *two* appropriate reactants!) sometimes presents difficulties. This Study Guide Link discusses a systematic approach to such problems.

Suppose you are asked to give the structure of an alkene that would give the following compound as the major product of H—Br addition:

$$CH_3CH_2CH_2CHCH_3$$
$$|$$
$$Br$$

It is not hard to understand that the bromine of the product must have come from the H—Br. However, there are many hydrogens in the product! Which one came from the H—Br, and which ones were there to start with?

First, recognize that the carbon bearing the bromine must have originally been one carbon of the double bond. It then follows that the other carbon of the double bond must be an *adjacent* carbon (because two carbons involved in the same double bond must be adjacent.). Use this fact to construct *all possible alkenes* that *might* be starting materials. Do this by removing the bromine and a hydrogen from *each adjacent carbon in turn.*

$$\overset{3}{C}H_3\overset{}{C}H_2\overset{2}{C}H_2\overset{1}{C}HCH_3$$
$$|$$
$$Br$$

remove Br from carbon-2 and H from carbon-1 $\Rightarrow$ $CH_3CH_2CH_2CH=CH_2$
                                                              1-pentene

remove Br from carbon-2 and H from carbon-3 $\Rightarrow$ $CH_3CH_2CH=CHCH_3$
                                                              2-pentene

(The symbol $\Rightarrow$ means, "Implies as starting materials.")

Which of these is correct? Or are they both correct? You haven't finished the problem until you've mentally carried out the addition of HBr to *each compound.* Doing this and applying the known regioselectivity of HBr addition leads to the conclusion that the desired alkyl halide could be prepared as the major product from 1-pentene. However, both carbons of the double bond of 2-pentene bear the same number of alkyl groups. Eq. 4.18, text p. 146, indicates that from this starting material we should expect not only the desired product, but also a second product:

$$CH_3CH_2\underset{\underset{\displaystyle Br}{|}}{C}HCH_2CH_3$$

Furthermore, the two products should be formed in nearly equal amounts. Consequently, 1-pentene is the only alkene that will give the desired alkyl halide as the *major* product (that is, the one formed almost exclusively).

Solving this type of problem is something like taking apart an engine or a household appliance to make a repair. Taking it apart is only half the battle. Once it is apart, you then have to put it back together. Suppose you find that the parts seem to fit together in several different ways. You must then decide which of the ways will allow the engine or the appliance to run properly and which will not. Analogously, once you have identified potential starting materials, you must determine whether they really will work, given the known characteristics of the reaction.

✓4.6 Solving Structure Problems

A number of problems ask you to deduce structures that are initially unknown by piecing together chemical data. Problems 4.47 and 4.48 are of this type. This study guide link illustrates a systematic approach to this type of problem by beginning the solution to Problem 4.47.

1. If the formulas of any of the unknowns are given, deduce all the information you can from the formulas. Begin with the unsaturation numbers (Sec. 4.3).

   In Problem 4.47, compound X has an unsaturation number $U = 2$, and compound Y has $U = 1$. This means that X has one ring and one double bond, or two double bonds. (It has to have at least one double bond, because it undergoes addition.) Compound Y results from addition of HBr, because the formula of Y is equal to that of X plus the elements of HBr.

2. Write all the information in the problem in equation form. This process gives you the entire problem at a glance.

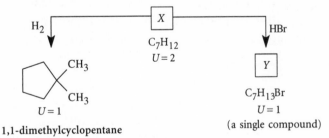

3. If there is a structure given explicitly anywhere in the problem, *even at the end of the problem*, deduce what you can from this structure.

   Notice that 1,1-dimethylcyclopentane is given explicitly in the problem as the catalytic hydrogenation product of X. Barring rearrangements, the structure provides the real key to solving the problem: it provides *the carbon skeleton of compound X*. It follows that one of the degrees of unsaturation of X must be a ring, and the other, as previously deduced, is a double bond. Thus, compound X must be a five-membered ring containing a double bond with two methyl groups on one carbon. The carbon bearing the methyl groups cannot be part of the double bond (why?).

4. Write out *all possibilities* and choose between (or among) them from the evidence presented, if possible.

   This step is one that meets great resistance from many students for some reason. But it is essential. Suppose a detective arrests the first person he or she meets that fits the description of a burglar; chances are, the detective will make a lot of false arrests. Instead, he or she ideally should list *all* the suspects and decide which among them fit all the data. The only "suspects" for *X* are the following:

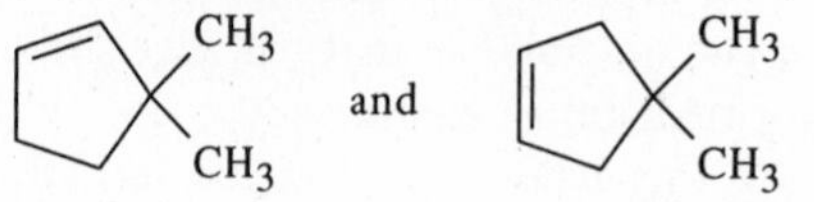

But what about the following structure?

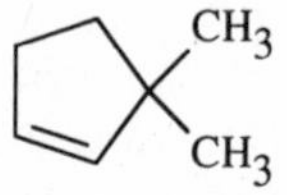

This is the same as *X1*! Don't forget that most structures can be drawn in different ways.

To finish solving the problem, notice that compound *Y* is a *single* compound. Only one of the "suspects" above would give a *single compound* on addition of HBr; the other would give a mixture. What is the solution? Which compound is *X*?

✓4.7   Solving Mechanistic Problems

Chapter 4 has stressed the importance of mechanisms in understanding two alkene reactions: hydrogen halide addition and hydration. The principles learned in these sections, along with the curved-arrow formalism and acid-base reactions covered in Sections 3.1–3.4, can be used to understand reactions you've never seen before. Problem 4.61 asks that you propose a rational series of steps that can account for such a reaction. This reaction has not been presented in this text. It contains several strange elements that might discourage you initially from trying to understand it. First, a ring is formed. Second, the product is an ether; none of the reactions covered so far give ethers as products. Nevertheless, a rational approach can be used to understand this reaction and its mechanism.

Break the problem-solving strategy into steps.

1.   Examine the reactants and starting materials and *label corresponding atoms.*

The result of such an analysis is as follows:

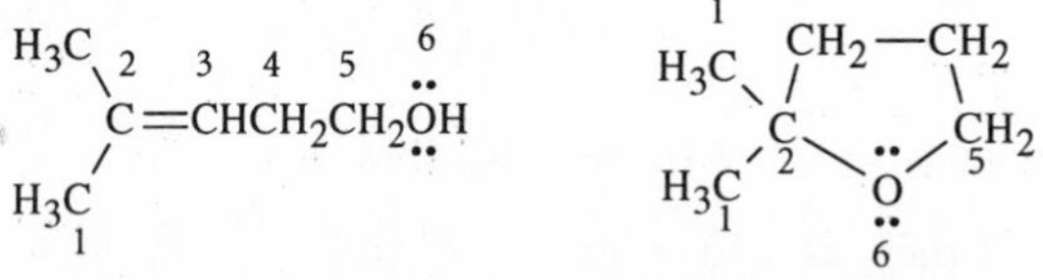

reactant	product

Notice that the atoms at the vertices of the skeletal structure in the product have been explicitly written out. You should always do this so that you are reminded how many hydrogens are located at each carbon. Both CH_3 groups are labeled "1" since they are indistinguishable. Because these CH_3 groups are attached to carbon-2 in the reactant, postulate the same in the product. The identity of the oxygen is clear, since it is the only oxygen in the molecule. Because it is attached to carbon-5 in the starting material, postulate the same for the product. The identities of carbons 3 and 4 follow. Note that the correspondences proposed are hypotheses—but they are *reasonable* hypotheses. You are asked to *propose* a rational mechanism, not prove that it is correct.

2.   Describe what has happened to the various functional groups in the molecule. Does the transformation fit the pattern of any reactions you have seen before?

Observe that the double bond in the starting material is missing in the product. In fact, carbon-3 of the double bond, the less branched carbon, has added a hydrogen, and carbon-2, the more branched carbon of the double bond, has added oxygen-6 from within the same molecule. This is, therefore, an *addition* reaction. It is also a *regioselective* addition in the same sense as hydration or HBr addition: hydrogen has added to the carbon with fewer alkyl branches, and the more electronegative atom (oxygen) has added to the carbon with more alkyl branches. Furthermore, the acid used is not consumed, and is therefore a catalyst. *What reactions have you studied that are acid-catalyzed regioselective additions?* Recall from Sec. 4.10B that hydration is such an addition. The reaction in this problem is indeed very similar to hydration, except that oxygen-6 of the —OH group, rather than water, has taken part in the addition.

3. Write a mechanism using the curved-arrow formalism and principles that you know from studying closely related reactions. Use separate structures for each mechanistic step; that is, do not attempt to represent several mechanistic steps on the same structure.

In hydration, the π electrons attack the acidic hydrogen of the catalyzing acid (see Eq. 4.38a, text p. 163) so that this hydrogen is transferred to the carbon of the double bond that has the smaller number of alkyl branches, and a carbocation is left at the carbon with the greater number of alkyl branches. Such a step would provide the extra hydrogen required at carbon-3. Using the curved-arrow formalism, write this *Brønsted acid-base reaction*:

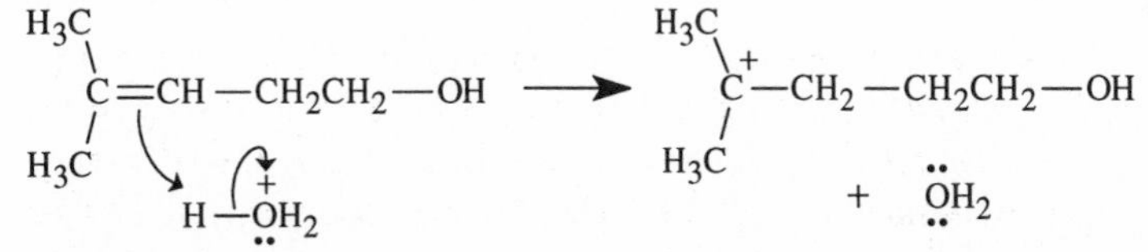

(Because the H_2SO_4 is present in dilute aqueous solution, the catalyzing acid is H_3O^+.) The analysis in Step 1 showed that, in order to form the ring, oxygen-6 must attack carbon-2. Since the electron-deficient carbon is a Lewis acid, and the oxygen is a Lewis base, the situation is ripe for this *Lewis acid-base reaction*.

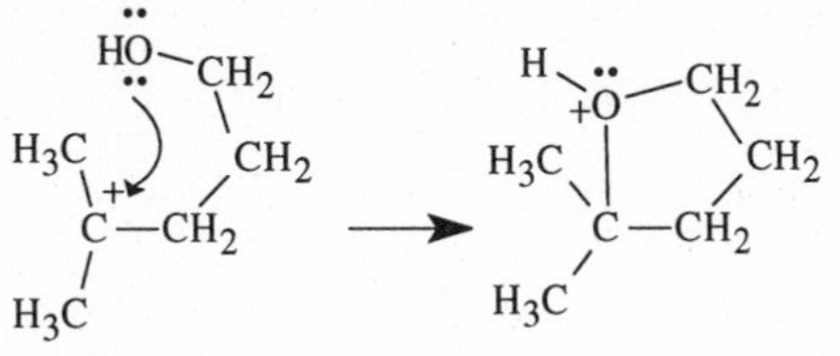

(This step is analogous to the attack of water on the carbocation in hydration; see Eq. 4.38b, text p. 163.) The carbocation is redrawn above so that the oxygen is poised in the correct place to form the ring. (Remember that rotations about carbon-carbon single bonds occur readily; so redrawing the molecule in this way is reasonable.)

Completion of the mechanism requires only loss of the proton from the oxygen. A water molecule, which was generated in the first step as the conjugate base of the acid catalyst H_3O^+, can remove this proton to re-form H_3O^+ in a *Brønsted acid-base reaction*:

product

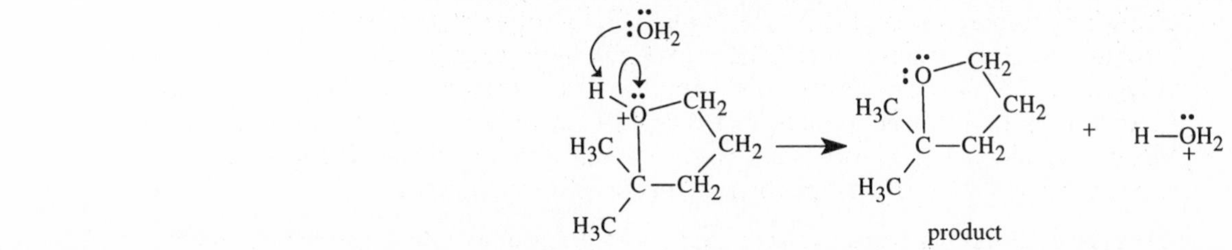

The final product is drawn differently from the one in the problem, but it is the same molecule. (You can prove this to yourself, if necessary, by analyzing the connectivities of the two.) Evidently, this reaction is *virtually identical* to hydration. The only difference is that the nucleophile (Lewis base) is derived from oxygen-6—the oxygen from within the same molecule—rather than water.

Notice that the mechanism consists of three simple acid-base reactions. Acid-base reactions, or closely analogous reactions, play a key role in the mechanisms of many organic reactions.

Writing mechanisms for known reactions, as in this problem, should help you to *understand* reactions rather than *memorize* them. This is the first step toward the development of a skill that is powerful indeed: the ability to *predict* the course of unknown reactions before they are ever run in the laboratory!

One final point: You should practice Step 1 above *every time* you see a new reaction in this text. The ability to postulate a correspondence between atoms of the reactant and product will help you to read and understand the discussions of organic reactions much more efficiently.

Problems 4.61 and 4.62 in the text offer you additional practice in solving mechanistic puzzles. Try your hand at them. But remember: be systematic, and use the curved-arrow formalism correctly.

....

Solutions

Solutions to In-Text Problems

4.1 (a) 1-Pentene cannot exist as *cis, trans* isomers.

(c) 1,3-Pentadiene can exist as *cis, trans* isomers at the bond between carbons 3 and 4; both of these carbons are stereocenters. (*Cis-trans* isomerism is *not* possible at the double bond between carbon-1 and carbon-2.)

(e) In principle, cyclobutene, which is a *cis* alkene, might also exist as a *trans* isomer. However, the model-building exercise should have convinced you that such an isomer is far too strained to exist. The carbons of the double bond are stereocenters.

4.2 (a)

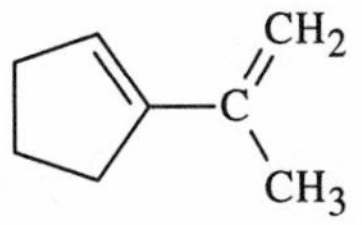

1-isopropenylcyclopentene

(c)

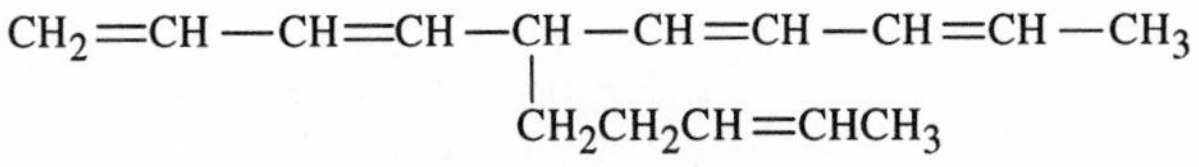

CH_2=CH—CH=CH—CH—CH=CH—CH=CH—CH_3
|
CH_2CH_2CH=$CHCH_3$

5-(3-pentenyl)-1,3,6,8-decatetraene

4.3 (a) 1,3-dimethylcyclopentene (c) 5-allyl-2,6-nonadiene

➠ Remember, the 2-propenyl group is called the *allyl group*; see text page 131.

4.4 (a) First, name the compound without considering stereochemistry: 4-isobutyl-1,4-hexadiene. Next, determine the stereochemistry at the double bond between carbons 4 and 5. (*E,Z* isomerism is not possible at carbon-1). Start by writing the structure with the double bond between carbons 1 and 2 replicated, and decide on the path of highest priority within each group (shown in boldface in the structure below):

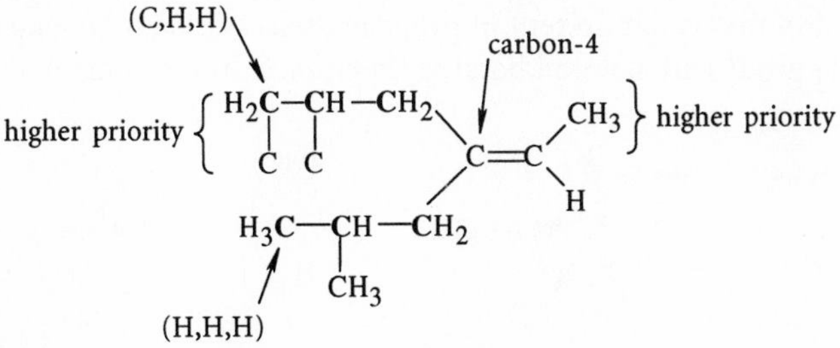

(C,H,H)
higher priority { H_2C—CH—CH_2
C C
carbon-4
C=C CH_3 } higher priority
H
H_3C—CH—CH_2
CH_3
(H,H,H)

As the diagram above shows, the first point of difference occurs at the third carbon out; thus, the allyl group has higher priority than the isobutyl group. Because the groups of higher priority are on the same side of the double bond, the stereochemistry at this double bond is *Z*, and the full name is therefore (*Z*)-4-isobutyl-1,4-hexadiene. No number before the *Z* is necessary when only one double bond has a stereochemical designator.

4.5   (a)

$$H_3C \quad H$$
$$\underset{H}{\overset{}{C}}=\underset{CH_2-CH}{\overset{H}{C}}$$

(2E,7Z)-5-[(E)-1-propenyl]-2,7-nonadiene

4.6   (a) The decision in this case occurs at the oxygen, because it occurs along the path of higher priority; the chlorines cannot affect the decision.

$$CH_3CH_2 \qquad\qquad Cl_2CH$$
$$\quad CH- \qquad vs. \qquad CH-$$
$$CH_3O \qquad\qquad HO$$
$$\quad O(C) \qquad\qquad O(H)$$
(higher priority)

(c)

$$(CH_3)_3C- \qquad\qquad CH_3 \quad C(O,C,H)$$
$$\qquad vs. \qquad CH-$$
$$C(C,C,C) \qquad OCH_3$$

(higher priority)

4.7   (a) Use Eq. 4.8 to obtain $U = 3$.

4.8   (a) The suffix "triene" in the name indicates that the carbon has three double bonds; consequently, the unsaturation number is 3.

4.9   The empirical formula is $C_{(85.60/12.01)}H_{(14.40/1.008)}$ or $C_{7.127}H_{14.29}$ or CH_2. Any compound with this empirical formula can have only one degree of unsaturation. Thus, $U = 1$; it can have *either* one ring *or* one double bond.

4.11   (a) 2-Methylpropene should have the greater dipole moment because the additional CH_3—C bond makes a dipole contribution that is not present in propene itself. (In the following analysis, the H—C bond dipoles of the CH_2 groups are ignored because they cancel in the comparison of the two molecules.)

$$H_3C \qquad\qquad\qquad\qquad\qquad\qquad H_3C$$
$$\quad C=CH_2 \qquad\qquad\qquad\qquad\quad C=CH_2$$
$$H \qquad\qquad\qquad\qquad\qquad\qquad H_3C$$

H—C bond dipole

CH_3—C bond dipole   resultant

CH_3—C bond dipole

CH_3—C bond dipole   resultant (larger)

   Here's a brief refresher on graphical vector addition. (1) The relative lengths of vectors are proportional to their magnitudes. (2) Any vector may be moved at will so long as its angular orientation is not changed. (3) Two parallel vectors can be added by adding their lengths, and the resultant points in the same direction as the original pair. (4) Two antiparallel vectors (that is, two vectors that are parallel but pointing in opposite directions) can be added by subtracting their lengths, and the resultant points in the direction of the longer vector of the constructing a parallelogram in which two sides consist of the vectors themselves and the remaining two sides are parallel to these vectors (dashed lines in the diagram above). The vector addition product, or resultant, is the diagonal of the resulting parallelogram. (6) The resultant of many vectors can be determined as the resultants of successive pairs.

4.12   The enthalpy of reaction is obtained by subtracting the ΔH_f° of the reactant, 1-butene, from that of the product, 2-methylpropene. Therefore, $\Delta H^\circ = -16.90 - (-0.13) = -16.77$ kJ/mol.
(b) The product 2-methylpropene is more stable because its ΔH_f° is more negative.

4.14   The combustion reaction for 1-hexene is

$$H_2C{=}CHCH_2CH_2CH_2CH_3 + 9O_2 \longrightarrow 6CO_2 + 6H_2O$$

The given ΔH° for this reaction is -3770.4 kJ/mol. This is equal to the heats of formation of products minus those of the reactants. The heats of formation of CO_2 and H_2O are the same as the heats of combustion of C and H_2, respectively, because these combustion reactions form CO_2 and H_2O from their elements.

$$-3770.4 \text{ kJ/mol} = 6\Delta H_f^\circ(CO_2) + 6\Delta H_f^\circ(H_2O) - \Delta H_f^\circ(\text{1-hexene}) - 9\Delta H_f^\circ(O_2)$$
$$= 6(-241.8 \text{ kJ/mol}) + 6(-393.5 \text{ kJ/mol}) - \Delta H_f^\circ(\text{1-hexene}) - 9(0)$$

Solving,

$$\Delta H_f^\circ(\text{1-hexene}) = -1450.8 - 2361.0 + 3770.4 = -41.4 \text{ kJ/mol}$$

4.15   (a) Compound (1) is more stable than compound (2) because the double bond in (1) has one more alkyl substituent.

4.16   2,3-Dimethyl-2-butene is not so stable as might be expected because it contains two pairs of *cis* methyl groups. The destabilizing van der Waals repulsions between these methyl groups counteract the stabilizing effect of additional alkyl substitution at the double bond.

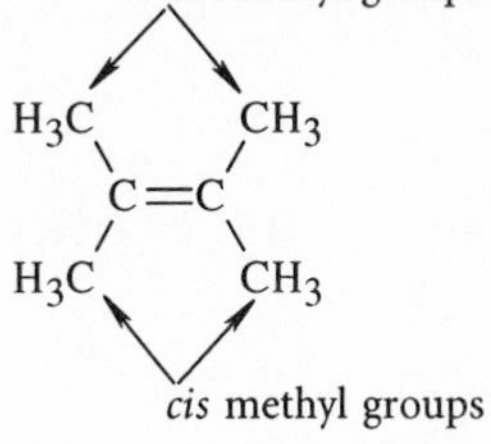

4.17   (a) 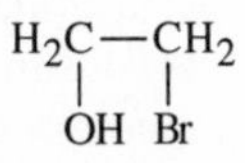   (c)

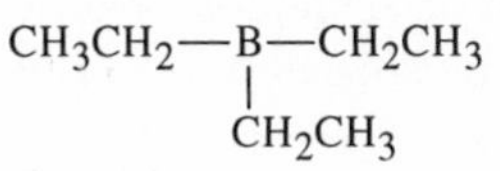

H$_2$C—CH$_2$ CH$_3$CH$_2$—B—CH$_2$CH$_3$
   | | |
  OH Br   CH$_2$CH$_3$

4.18   (a) The bromine goes to the carbon with the methyl branch.

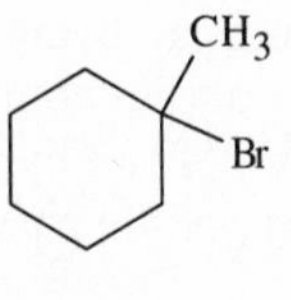

4.19   Assume that the addition of I—N$_3$ follows the pattern of HBr addition. The group attacked by the alkene π electrons, that is, the iodine, goes to the carbon with fewer alkyl branches; consequently, the azide group (N$_3$) goes to the carbon with more alkyl branches.

(solution continues)

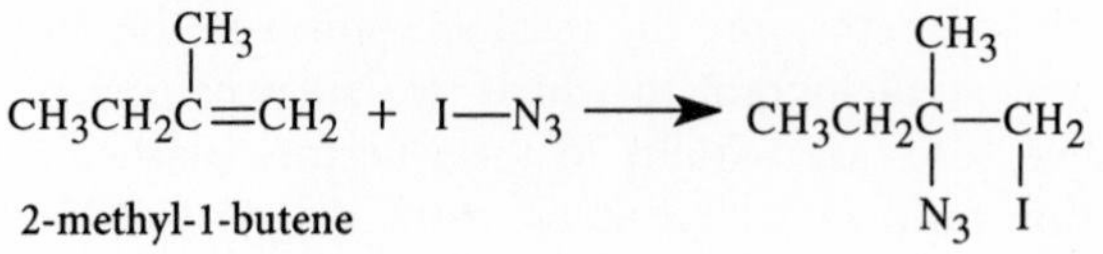

$$CH_3CH_2\overset{\overset{\displaystyle CH_3}{|}}{C}=CH_2 \ + \ I-N_3 \ \longrightarrow \ CH_3CH_2\overset{\overset{\displaystyle CH_3}{|}}{\underset{\underset{\displaystyle N_3}{|}}{C}}-\overset{}{\underset{\underset{\displaystyle I}{|}}{C}}H_2$$

2-methyl-1-butene

4.21 (a)

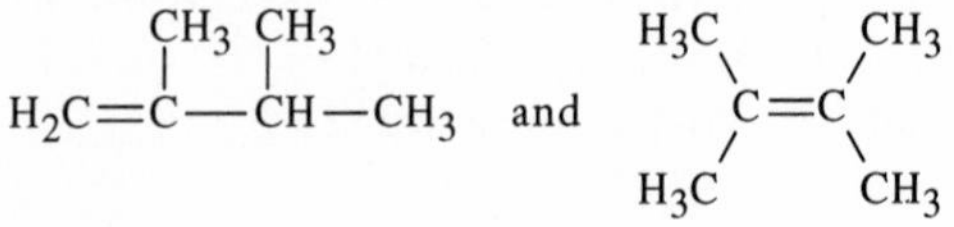

$$H_2C=\overset{\overset{\displaystyle CH_3}{|}}{C}-\overset{\overset{\displaystyle CH_3}{|}}{C}H-CH_3 \quad \text{and} \quad \underset{H_3C}{\overset{H_3C}{>}}C=C\underset{CH_3}{\overset{CH_3}{<}}$$

4.22 Protonation of the double bond gives a carbocation intermediate *A* which can be attacked by bromide ion to give the minor product *B*.

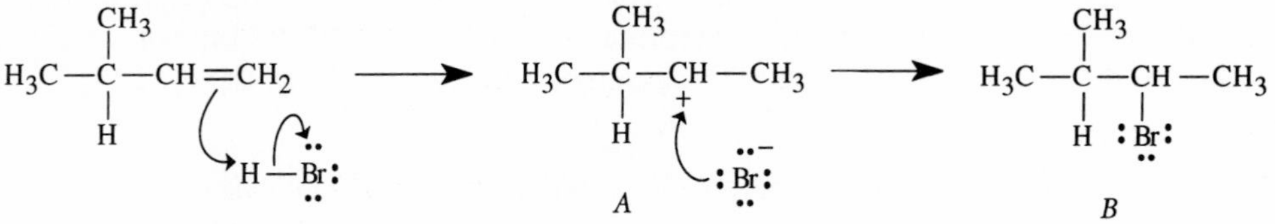

(see structures *A*, *B*)

Carbocation *A* can also rearrange by a hydride shift to give carbocation *C*, which is attacked by bromide ion to give the major product *D*.

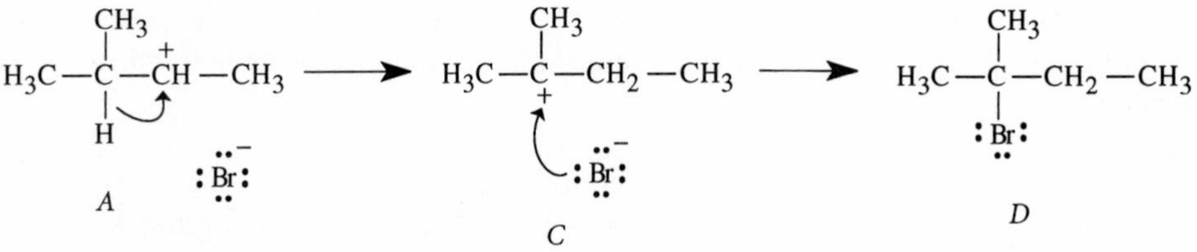

(see structures *A*, *C*, *D*)

4.23 If these alkyl halides could be prepared from alkenes, the alkene starting materials would have to be *A1*, *B1*, and *C1* for *A*, *B*, and *C*, respectively:

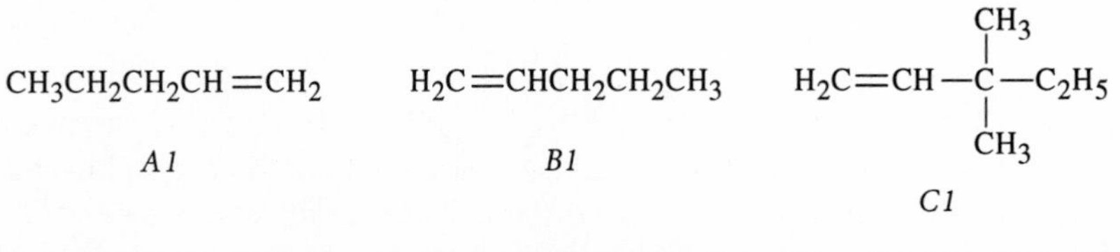

$$CH_3CH_2CH_2CH=CH_2 \qquad H_2C=CHCH_2CH_2CH_3 \qquad H_2C=CH-\overset{\overset{\displaystyle CH_3}{|}}{\underset{\underset{\displaystyle CH_3}{|}}{C}}-C_2H_5$$

$$A1 \qquad\qquad\qquad B1 \qquad\qquad\qquad C1$$

(Note that *B1* is the same as *A1*; it is simply drawn differently.) Reaction of *A1* (or *B1*) with HBr would give *B* but not *A* because the bromine goes to the carbon of the double bond with the alkyl branch. Hence, compound *B*, but not compound *A*, could be prepared by HBr addition to alkene *B1*. Addition of HBr to alkene *C1* would not give solely compound *C*, but would also give one or more rearrangement products.

4.24 The reaction-free energy diagram shown in Fig. SG4.1, p. 71, meets the criteria in the problem.

4.26 (a) The curved arrows show where bonds are being formed and broken. Thus, there is a partial bond between the HO⁻ and the carbon, and a partial bond between the CH₃ and the Br. (Notice that the locations of the different groups are not defined by the curved-arrow formalism; the arrangement shown below is arbitrary. The important points are the partial bonds and the partial charges.)

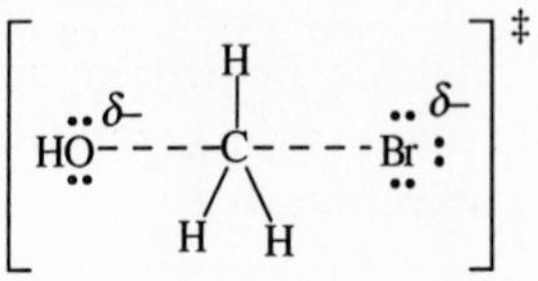

$$\left[\ \ddot{H}\ddot{O}\overset{\delta-}{----}\overset{\overset{\displaystyle H}{|}}{\underset{\underset{\displaystyle H\ \ H}{}}{C}}----\overset{\delta-}{\ddot{\ddot{Br}}} \ \right]^{\ddagger}$$

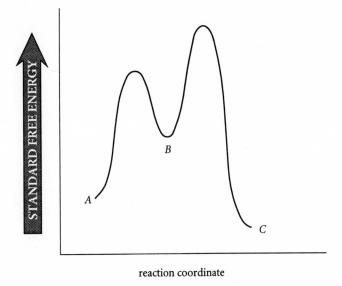

reaction coordinate

Figure SG4.1   *Reaction-free energy diagram for the solution to Problem 4.24.*

4.27 Because the transition states of the two reactions should resemble the respective carbocation intermediates (Hammond's postulate), examine the structures of the two carbocations and determine which should be more stable.

$$H_3C—\overset{+}{\underset{\underset{CH_3}{|}}{C}}—CH_3 \quad Br^-$$

carbocation intermediate
from HBr addition to
2-methylpropene

$$H_3C—\overset{+}{C}H—CH_2—CH_3 \quad Br^-$$

carbocation intermediate
from HBr addition to
trans-2-butene

The carbocation intermediate in the addition of HBr to 2-methylpropene is tertiary, and is thus more stable than the carbocation intermediate in the addition of HBr to *trans*-2-butene, which is secondary. If the transition states for the two addition reactions resemble the carbocation intermediates, the transition state for the reaction of 2-methylpropene is also more stable. The relative rates of the two reactions are governed by the *differences* between the standard free energies of the transition states and their respective starting materials. If the two alkenes do not differ appreciably in energy, then the relative rates of the two reactions are governed by the relative free energies of the two transition states. Under this assumption, addition of HBr to 2-methyl propene is faster, because the transition state for this reaction has lower free energy.

4.28   (a) The product is hexane, $CH_3CH_2CH_2CH_2CH_2CH_3$. (Hydrogen adds to both double bonds.)

4.29   Five alkenes with the formula C_6H_{12} that give hexane as the product of catalytic hydrogenation:

$$H_2C{=}CHCH_2CH_2CH_2CH_3 \qquad CH_3CH{=}CHCH_2CH_2CH_3 \qquad CH_3CH_2CH{=}CHCH_2CH_3$$

1-hexene   (*E*)- and (*Z*)-2-hexene   (*E*)- and (*Z*)-3-hexene

4.30   The mechanism of the reaction in Eq. 4.39:

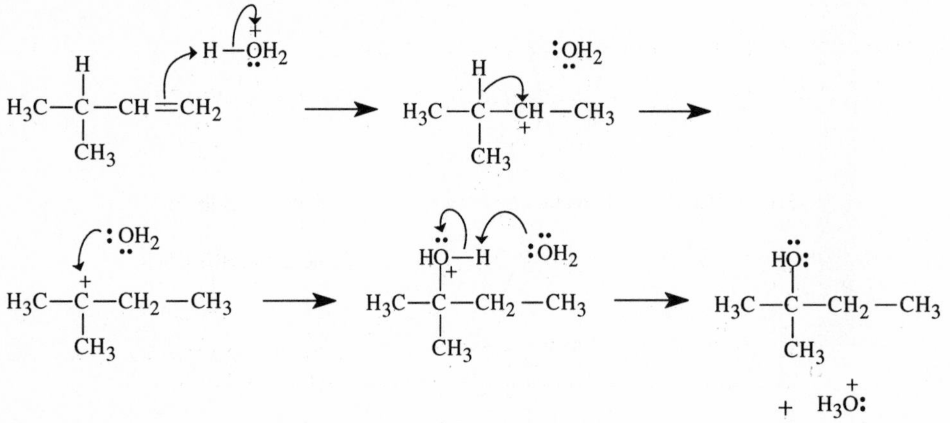

The rearrangement takes place because a more stable (tertiary) carbocation is formed from a less stable (secondary) one.

4.32   In the hydration of propene, the —OSO_3H group of sulfuric acid goes to the central carbon. This group is then replaced by the —OH group to give isopropyl alcohol.

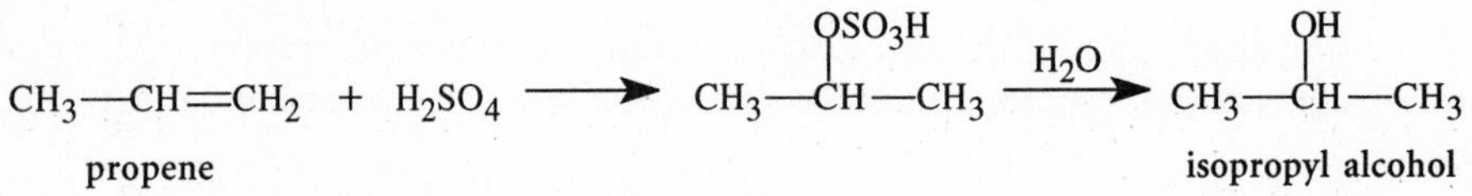

A secondary carbocation is a reactive intermediate in the reaction; the bisulfate (—OSO_3H) group goes to the central carbon because it attacks this carbocation:

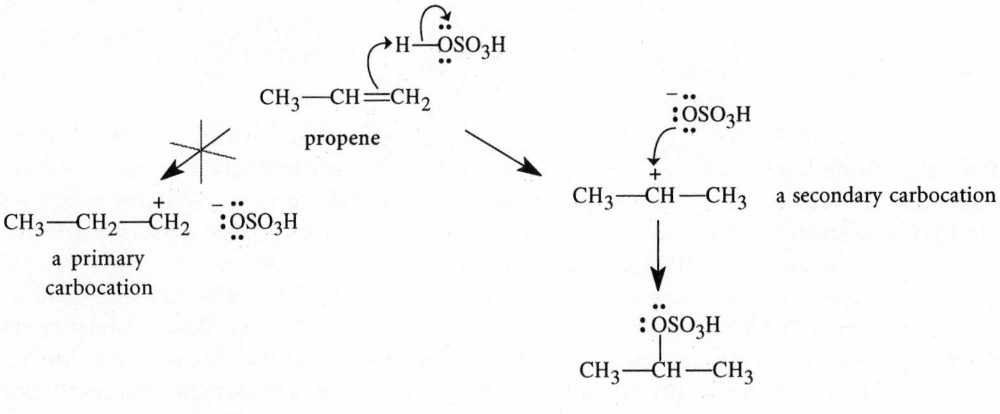

4.34   The alcohol formed in the hydration of methylenecyclobutane:

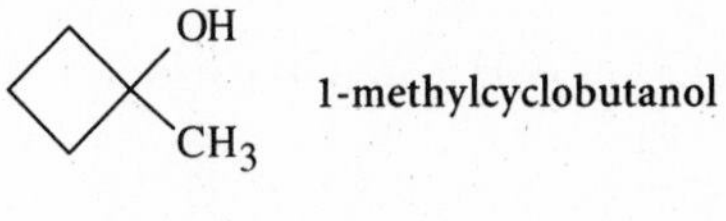

1-methylcyclobutanol

Solutions to Additional Problems

4.35 The structures and substitutive names of the methylpentene isomers:

$$H_2C=CCH_2CH_2CH_3$$
$$|$$
$$CH_3$$

2-methyl-1-pentene

$$H_2C=CHCHCH_2CH_3$$
$$|$$
$$CH_3$$

3-methyl-1-pentene

$$H_2C=CHCH_2CHCH_3$$
$$|$$
$$CH_3$$

4-methyl-1-pentene

$$CH_3C=CHCH_2CH_3$$
$$|$$
$$CH_3$$

2-methyl-2-pentene

$$CH_3CH=CCH_2CH_3$$
$$|$$
$$CH_3$$

3-methyl-2-pentene
(*E*) and (*Z*) isomers

$$CH_3CH=CHCHCH_3$$
$$|$$
$$CH_3$$

4-methyl-2-pentene
(*E*) and (*Z*) isomers

4.37 (a) Each of the following methylpentene isomers should give mostly one product when it reacts with HBr: 2-methyl-1-pentene; 4-methyl-1-pentene; 2-methyl-2-pentene; and both stereoisomers of 3-methyl-2-pentene. In contrast, 3-methyl-1-pentene should give a rearrangement product as well as the product of normal addition; and each of the stereoisomers of 4-methyl-2-pentene can give two products of normal addition as well as a rearrangement product.

4.38 (a) Arranging the alkenes in order of increasing heats of formation is the same as arranging them in order of decreasing stability. The major factors that govern alkene stability are (1) the number of alkyl branches on the double bond; and whether the alkene is *cis* or *trans*. The order of increasing heats of formation:

(*E*)-3-methyl-2-pentene $\approx$ 2-methyl-2-pentene $\approx$ (*Z*)-3-methyl-2-pentene $<$ (*E*)-4-methyl-2-pentene $\approx$

2-methyl-1-pentene $<$ (*Z*)-4-methyl-2-pentene $<$ 3-methyl-1-pentene $\approx$ 4-methyl-1-pentene

The reason that 2-methyl-2-pentene is somewhat less stable than (*E*)-3-methyl-2-pentene is that the latter compound has two methyl groups in a *cis* relationship, whereas the former compound has a methyl and a somewhat larger ethyl group in a *cis* relationship. The larger the groups involved in a *cis* relationship, the greater the van der Waals repulsions and the more the molecule is destabilized.

4.39 (a)   (c)   (e)   (g)

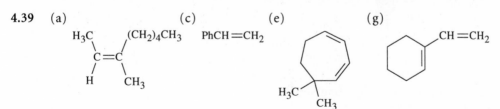

(c) PhCH=CH$_2$

4.40 (a) 6-methyl-1-heptene   (c) 3-ethylcyclopentene   (e) 2-isopropyl-1-methyl-1,3-cyclohexadiene
(g) 1-(1-cyclopentenyl)cyclopentene   (h) (*E*)-5-[(*Z*)-1-propenyl]-1,6-nonadiene

4.41 First draw the structure corresponding to the name, and then rename it if necessary.

(a)

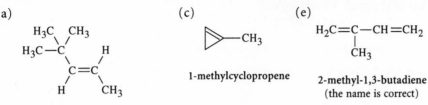

trans-4,4-dimethyl-2-pentene
or (*E*)-4,4-dimethyl-2-pentene

(c) 1-methylcyclopropene

(e) 2-methyl-1,3-butadiene
(the name is correct)

4.42 (a) *Z*   (c) *E*; note in this case that the two branches of the ring are treated as separate substituents; the fact that the two branches are tied into a ring has no effect on their relative priorities.

4.43 (a) Either of the following two alkenes would react with HBr to give the alkyl halide product shown in the problem:

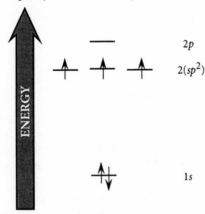

1-methylcyclopentene   methylenecyclopentane

4.44 (a) Cyclohexane and 1-hexene are constitutional isomers (C), because they have the same molecular formulas but different connectivities.
(c) (*E*)- and (*Z*)-2-hexene are stereoisomers (S)
(e) Cyclopentane (C_5H_{12}) and cyclopentene (C_5H_{10}) are not isomers (N) because they have different molecular formulas.

4.45 Boron trifluoride (BF_3) has trigonal-planar geometry. Like the carbons of ethylene, which also have the same geometry, the boron of BF_3 is sp^2-hybridized. The hybridization diagram for boron is as follows:

ENERGY

— 2p

↑ ↑ ↑ 2(sp^2)

↑↓ 1s

4.46 (a) A hydrogen-carbon $1s$-sp^3 σ bond.
(c) One bond is a carbon-carbon sp^2-sp^2 σ bond; the other is a $2p$-$2p$ π bond.
(e) A carbon-hydrogen sp^2-$1s$ σ bond.

4.47 The process to be used in working this problem is thoroughly explained in Study Guide Link 4.6. The structures of *X* and *Y* are as follows:

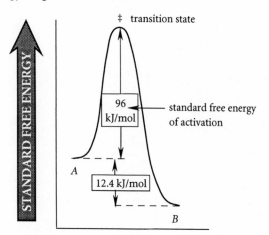

X Y

4.49 A bad idea! You would save the company millions by pointing out that hydration of propene gives 2-propanol (isopropyl alcohol), not 1-propanol. (Why?)

$$CH_3CH\!=\!CH_2 \ + \ H_2O \ \xrightarrow{\ \text{acid}\ } \ \underset{\underset{OH}{|}}{CH_3CHCH_3}$$

propene   2-propanol

4.50 Before drawing the reaction-free energy diagram, convert K_{eq} into a free energy. (Assume 25 °C or 298 K.)

$$\Delta G° = -2.3RT \log K_{eq} = -5.706 \log(150) = -12.4 \text{ kJ/mol}$$

(a)   The reaction-free energy diagram is as follows:

‡ transition state

STANDARD FREE ENERGY

96 kJ/mol ⟶ standard free energy of activation

A

12.4 kJ/mol

B

reaction coordinate

(b) The standard free energy of activation for the reaction in the B to A direction is 12.4 kJ/mol + 96 kJ/mol = 108.4 kJ/mol. This follows geometrically from the diagram in part (a); it is the difference between the standard free energies of B and the transition state.

4.52 The first step of hydration, protonation of the double bond, is rate-limiting (text page 163), and the reactive intermediate that results from this process is a carbocation. According to Hammond's postulate, the transition state closely resembles this intermediate.

CH₃

carbocation intermediate from
protonation of 1-methylcyclohexane

4.54 Following the hint, an energy diagram for the relationships involved is shown in Fig. SG4.2 on the following page. As this diagram emphasizes, the enthalpy "distance" from $5C + 5H_2$ to $5CO_2 + 5H_2O$ is the same in both cases. (The enthalpy of O_2 required for combustion is ignored, and in any case is zero by convention.) Because this total $\Delta H°$ is known from the 3-methyl-1-butene case, the required enthalpy x can be calculated. According to the diagram, then,

$$-28.95 - 3147.75 = -36.32 + x$$

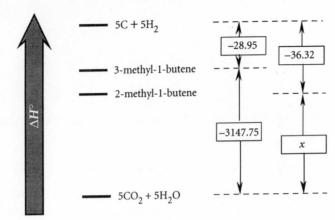

Figure SG4.2   *Energy diagram to accompany the solution for Problem 4.54. All enthalpies are in kJ/mol.*

Solving, $x = -3140.38$ kJ/mol. Thus, 3140.38 kJ/mol of heat is liberated when 2-methyl-1-butene is burned. Less energy is liberated than when 3-methyl-1-butene is burned because 2-methyl-1-butene has lower energy than 3-methyl-1-butene. (Why?)

4.56   When the two hydrogenation equations are written as shown below and Hess's law is applied, the H_2 molecules in the two equations cancel and the butanes in the two equations also cancel; as a result, the sum of the enthalpies of the two reactions is equal to the enthalpy difference between the two 2-butenes.

$$cis\text{-2-butene} + \cancel{H_2} \longrightarrow \cancel{butane} \qquad \Delta H° = -119.66 \text{ kJ/mol}$$
$$\cancel{butane} \longrightarrow trans\text{-2-butene} + \cancel{H_2} \quad \Delta H° = x \text{ kJ/mol}$$

$$cis\text{-2-butene} \longrightarrow trans\text{-2-butene} \qquad \Delta H° = -4.18 \text{ kJ/mol}$$

Therefore,

$$-119.66 + x = -4.18$$

or $\qquad x = +115.48$ kJ/mol.

Because the hydrogenation equation for *trans*-2-butene is written in the reverse direction, x is the negative of the enthalpy of hydrogenation of *trans*-2-butene. Consequently, $\Delta H°$ for the hydrogenation of *trans*-2-butene is -115.48 kJ/mol.

4.58   If you made the models as requested you found that it is very difficult to close the ring containing the double bond with the E (or *trans*) configuration, whereas the ring containing the Z (or *cis*) double bond can be closed without difficulty. (If this conclusion is not clear, be sure you built a model of the *cyclic* alkenes and not a pair of the acyclic heptenes.) Thus, (E)-cycloheptene is less stable than (Z)-cycloheptene. Because the relative heats of formation of two isomers are the same as their relative stabilities, it follows that (Z)-cycloheptene has the smaller (less positive or more negative) heat of formation.

4.60   (a) The data show that methyl groups are more electron-donating than hydrogens. That is, the CH_3—C= bond dipole is larger than the H—C= bond dipole. This follows because each CH_3—C bond dipole in the first structure, and each H—C bond dipole in the second structure, is parallel to a C—Cl bond dipole within the same structure. As shown in the following vector diagram, in which R= CH_3 or R = H, the parallel dipoles reinforce to give a larger resultant A; and two such resultants add to give a resultant B for the entire molecule.

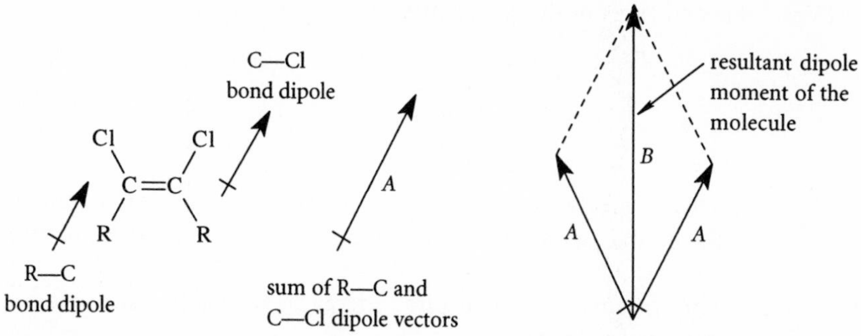

(See the note following the solution to Problem 4.11 on graphical vector addition, if necessary.) It follows from this diagram that the larger is the R—C bond dipole, the larger is the molecular dipole moment. Because the compound containing the methyl groups has the larger dipole moment, it follows that the CH_3—C bond dipole is larger than the H—C bond dipole. In other words, a methyl group is more electron donating than a hydrogen towards a double bond.

(b) Assume that the major contributors to the molecular dipole moment are the CH_3—C and the C—Cl bond dipoles as indicated in Part (a). These bond dipoles reinforce in the E isomer, and are nearly opposed in the Z isomer. Consequently, the E isomer has the greater dipole moment.

4.61 Study Guide Link 4.7 discusses the solution to this problem in detail.

4.63 Protonation occurs at carbon-5 (see numbered structure below) to form a tertiary carbocation. (Protonation at carbon-1 would have given a secondary carbocation.) The tertiary carbocation resulting from protonation then adds to the other double bond (the only choice) to give a secondary carbocation, which is attacked by water.

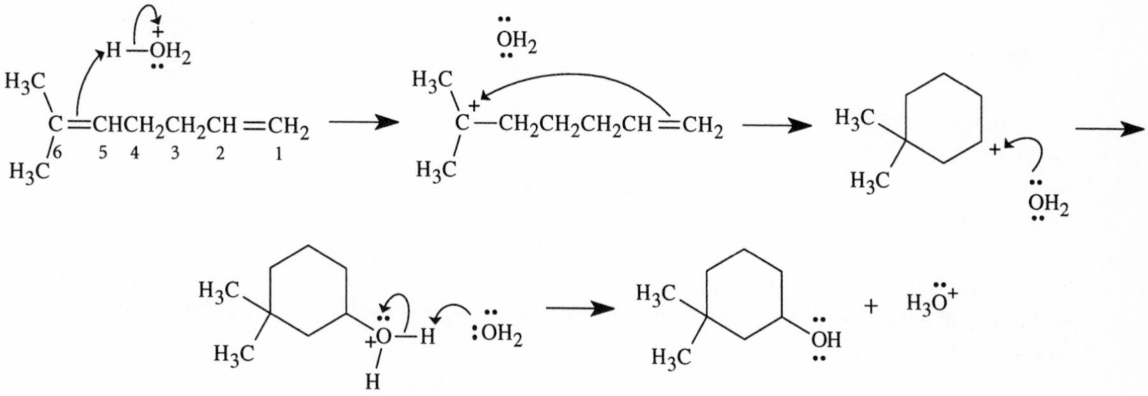

4.65 (a) We refer to the alkene on the left, 2,3-dimethyl-1-butene, as A, and the alkene on the right, 2,3-dimethyl-2-butene, as B. Compound B is favored at equilibrium because it has the smaller (less positive) ΔG_f°. (Notice that this alkene has more alkyl substituents on its double bond.) The equilibrium constant is obtained from Eq. 3.30. Define K_{eq} as $[B]/[A]$:

$$\log K_{eq} = -\Delta G_f^\circ/2.3RT = -\Delta G_f^\circ/5.706$$
$$= -(75.86 - 79.04)/5.706 = 3.18/5.706 = 0.557$$
$$\text{or} \qquad K_{eq} = 10^{0.557} = 3.61$$

(b) The equilibrium constant tells us *nothing* about how rapidly this interconversion takes place. For example, in the absence of a catalyst, the rate of this interconversion is essentially zero. (However, this reaction does occur at a convenient rate in the presence of an acid catalyst, as suggested by Problem 4.64.)

4.66   (a) First draw the structures of the two compounds:

$$H_2C=CHCH_2CH_3 \qquad H_2C=\overset{\displaystyle |}{\underset{\displaystyle CH_3}{C}}CH_3$$

1-butene   2-methylpropene

2-Methylpropene has two alkyl substituents on its double bond, whereas 1-butene has only one. Because alkyl branches on the double bond stabilize an alkene, 2-methylpropene is the more stable of the two alkenes.

(b) The reaction with the smaller standard free energy of activation ($\Delta G^{\circ\ddagger}$) is faster. Consequently, hydration of 2-methylpropene is faster.

(c) Let 1-butene = A, and let 2-methylpropene = B. The reaction-free energy diagrams for hydration are as follows. (Only the first, rate-limiting, step of the mechanism is shown in each diagram.)

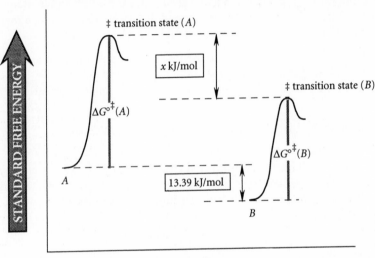

reaction coordinates

These diagrams require the following equality:

$$13.39 + \Delta G^{\circ\ddagger}(A) = x + \Delta G^{\circ\ddagger}(B)$$

Rearranging,

$$\Delta G^{\circ\ddagger}(A) - \Delta G^{\circ\ddagger}(B) = x - 13.39$$

The difference between the standard free energies of activation is given in the problem as 22.84 kJ/mol. Hence,

$$x = 22.84 + 13.39 = 36.23 \text{ kJ/mol}$$

Thus, the transition state for hydration of 1-butene is 36.23 kJ/mol *less stable* than the transition state for hydration of 2-methylpropene. From Sec. 4.7B, the mechanism of hydration involves a carbocation intermediate, and the first step, protonation of the double bond to form this intermediate, is rate-limiting. In the case of 2-methylpropene, the carbocation intermediate is tertiary; in the hydration of 1-butene, it is secondary. Tertiary carbocations are more stable than secondary carbocations. Because the transition states resemble the respective carbocations in structure and energy (Hammond's postulate), the transition state for formation of the more stable carbocation is the more stable transition state, as observed.

5

Addition Reactions of Alkenes

Terms

Concepts

I. Reactions Involving Alkenes

A. ADDITION REACTIONS

1. The characteristic reaction of alkenes is addition to the double bond which occurs by:
   a. mechanisms involving carbocation intermediates (addition of hydrogen halides, hydration)
   b. mechanisms involving cyclic ion intermediates (oxymercuration, halogenation)
   c. mechanisms involving free-radical intermediates (free-radical addition of HBr, polymerization)
   d. concerted mechanisms (hydroboration, glycol formation, ozonolysis)
2. Many addition reactions are regioselective because of their mechanisms.

B. OTHER REACTIONS

1. Some useful transformations of alkenes involve additions followed by other transformations.
   a. Oxymercuration followed by reduction affords alcohols.
   b. Hydroboration followed by oxidation affords alcohols.
   c. Ozonolysis followed by:
      i. treatment with $(CH_3)_2S$ affords aldehydes and ketones by cleavage of the double bond.
      ii. treatment with H_2O_2 affords ketones and carboxylic acids by cleavage of the double bond.
   d. Addition of OsO_4 or $KMnO_4$ followed by hydrolysis affords glycols.
2. Alkenes such as ethylene and propene are produced by cracking alkanes at high temperature.

II. Writing Organic Reactions

A. REACTION EQUATIONS

1. The most thorough way to write a reaction is to use a complete, balanced equation.
2. In many cases it is convenient to abbreviate reactions by showing only the organic starting materials and the major organic products; reaction conditions are written over and/or under the arrow.

B. PRODUCT YIELDS

1. The theoretical yield is the maximum amount of product that can be formed in a reaction if there are no losses or side-reactions.
2. The actual yield is the amount of product isolated from a reaction.
3. The percentage yield is the actual yield divided by the theoretical yield times 100%.

III. Bond Dissociations and Free Radicals

A. BOND DISSOCIATION ENERGIES

1. The bond dissociation energy of a covalent bond measures the energy required to break the bond homolytically to form two free radicals.
   a. For a bond between two atoms A—B, the bond dissociation energy is defined as the enthalpy $\Delta H°$ of the homolytic reaction.
   b. The lower a bond dissociation energy is, the lower is the temperature required to rupture the bond in question and form free radicals at a reasonable rate.
2. The $\Delta H°$ of reactions can be calculated or estimated by subtracting the bond dissociation energies of the bonds formed from those of the bonds broken.

B. TYPES OF BOND DISSOCIATIONS

1. An electron-pair bond may break so that one of the bonding atoms receives both electrons and the other becomes electron-deficient; this type of bond breaking is called heterolytic cleavage or heterolysis.

$$-\overset{|}{\underset{|}{C}}\!\!-\!\!\ddot{\underset{\cdot\cdot}{I}}\text{:} \xrightarrow{\text{heterolytic bond dissociation}} -\overset{|}{\underset{|}{C}}{}^+ \quad + \quad \text{:}\ddot{\underset{\cdot\cdot}{I}}\text{:}^-$$

a carbocation   an anion

2. An electron-pair bond may break so that each bonding atom retains one electron of the chemical bond; this type of bond breaking is called homolytic cleavage or homolysis.
   a. Any species with at least one unpaired electron is called a free radical.
   b. A different curved-arrow formalism is used for homolysis, called the fishhook formalism, in which electrons are moved individually rather than in pairs.

$$\text{homolytic bond dissociation}$$

a free radical   a free radical

C. Free Radicals and Reactions

1. Free radicals behave as reactive intermediates—they react before they can accumulate in significant amounts.
2. Free-radical reactions typically involve three types of steps: initiation steps, propagation steps, and termination steps.
3. Free radicals undergo four types of reaction:
   a. addition to a double bond

   $$R\cdot + \;\;C{=}C \longrightarrow R{-}C{-}C\cdot$$

   b. atom abstraction—a free radical removes an atom from another molecule, and a new free radical is formed

   $$R\cdot + \;H{-}C{-}C{-} \longrightarrow R{-}H + \cdot C{-}C{-}$$

   c. β-scission (the reverse of addition to a double bond)

   $$-C{-}C{-}C\cdot \longrightarrow -C\cdot + \;\;C{=}C$$

   d. recombination (the reverse of bond rupture) to form a covalent bond

   $$R\cdot + R\cdot \longrightarrow R{-}R$$

4. If a reaction occurs in the presence of heat, light, or a known free-radical initiator (peroxides, AIBN, etc.), but does not occur in its absence, it is fairly certain that the reaction involves free-radical intermediates.
5. Free-radical reactions that involve propagation steps and their associated initiation and termination steps are said to be free-radical chain reactions.
   a. For each free radical consumed in the propagation steps, one is produced.
   b. The free radicals involved in the propagation steps of a chain reaction are said to propagate the chain.

D. Free-Radical Chain Reaction Mechanisms

1. In initiation steps, the free radicals that take part in subsequent steps of the reaction are formed from a molecule that readily undergoes homolysis, called a free-radical initiator.
   a. The initiator is in effect the source of free radicals.
   b. Sometimes heat or light initiates a free-radical reaction in which the additional energy promotes homolysis of the free-radical initiator.
   c. A second initiation step is required in the free-radical addition of HBr to alkenes (atom abstraction): the removal of a hydrogen atom from HBr by the free radical that was formed in the first initiation step.

2. In propagation steps, even though products are formed there is no net consumption or destruction of free radicals.
    a. The free-radical by-product of one propagation step serves as the starting material for another.
    b. Some propagation steps are atom-abstraction reactions.
3. In termination steps, free radicals are destroyed by recombination reactions.
    a. In a recombination reaction, the reverse of a homolytic reaction, two free radicals come together to form a covalent bond.
    b. The recombination reactions of free radicals are in general highly exothermic; they have very favorable (negative) $\Delta H°$ values.

E. EXPLANATION OF THE PEROXIDE EFFECT

1. In the addition of HBr to alkenes in the presence of peroxides, the rather large bromine atom reacts at the least branched carbon of the double bond because otherwise it would experience van der Waals repulsions with the hydrogens in the branches.
    a. Any effect on a reaction that can be attributed to van der Waals repulsions is termed steric effect.
    b. Because the reaction with the transition state of lower energy is the faster reaction, reaction of the Br atom at the less branched carbon of the alkene double bond (giving the more branched free radical) is faster than reaction at the more branched carbon.
2. The stability of free radicals is in the order tertiary > secondary > primary.
    a. The effect of branching on free-radical stability is considerably smaller than the effect of branching on carbocation stability.
    b. The stability order implies that free radicals, like alkenes and carbocations, are stabilized by alkyl group substitution at sp^2-hybridized carbons.
3. The "peroxide effect" is not limited to peroxides; any good free-radical initiator will bring about the same effect.

Reactions

I. Reactions of Alkenes with Halogens

A. ADDITION OF CHLORINE AND BROMINE

1. Halogens undergo addition to alkenes to give compounds with halogens on adjacent carbons called vicinal dihalides.

$$\text{C=C} \xrightarrow{\text{Cl}_2} \text{Cl—C—C—Cl} \qquad \text{a vicinal dihalide}$$

2. The rapid disappearance of the red bromine color during bromine addition is a useful test for alkenes.
3. Bromine addition to alkenes can involve a reactive intermediate called a bromonium ion.
    a. In a bromonium ion, a positively charged bromine with an octet of electrons is bonded to two other atoms.

a bromonium ion

    b. Attack of the bromide ion at either of the carbons bound to bromine in the bromonium ion completes the addition of bromine.

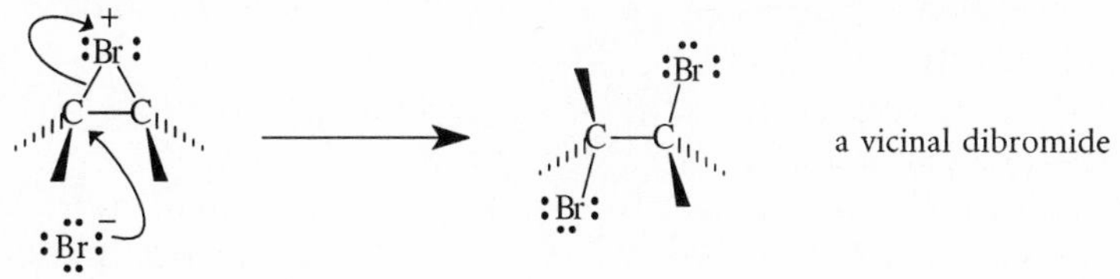

a vicinal dibromide

B. Formation of Halohydrins

1. When an alkene is brominated in a solvent containing a large amount of water, water attacks the bromonium ion.

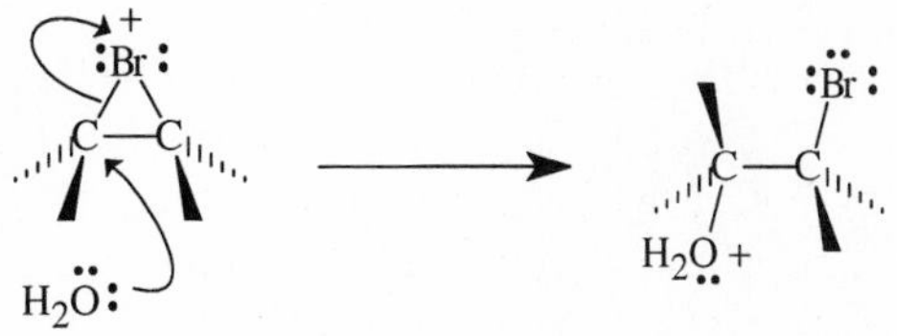

2. Loss of a proton from the oxygen gives a compound containing both an —OH and a —Br group called a bromohydrin.

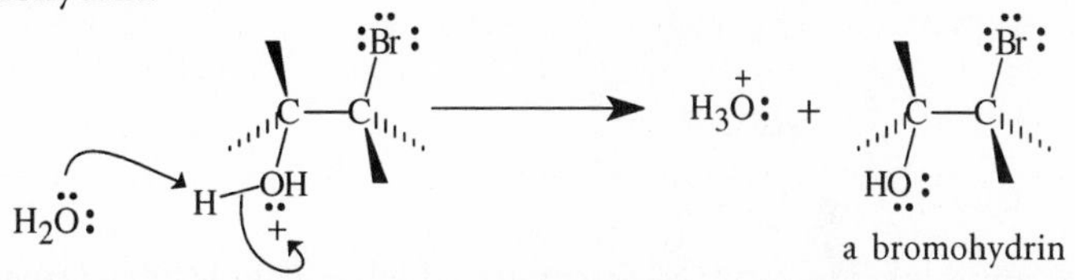

a bromohydrin

    a. Bromohydrins are members of the general class of compounds called halohydrins, which are compounds containing both a halogen and an —OH group.

    b. The reaction is completely regioselective when one carbon of the alkene is bonded to two alkyl substituents and the other is bonded to two hydrogens.

C. Free-Radical Addition of Hydrogen Bromide to Alkenes

1. Alkenes react with hydrogen bromide in the presence of peroxides so that the bromine adds to the less branched carbon of the double bond.

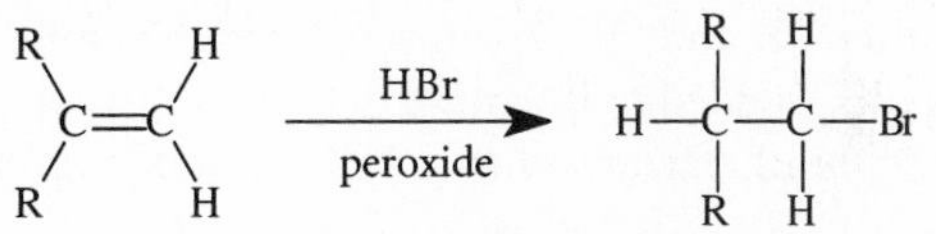

    a. This reversal of regioselectivity in HBr addition is termed the peroxide effect.

    b. This mechanism involves reactive intermediates called free radicals.

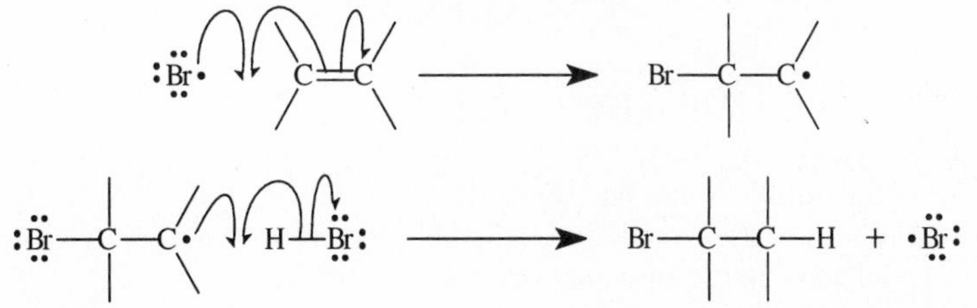

2. The regioselectivity of HI or HCl addition to alkenes is not affected by the presence of peroxides.

II. Conversion of Alkenes into Alcohols

A. Oxymercuration-Reduction of Alkenes

1. Alkenes react with mercuric acetate in aqueous solution to give addition products in which an —HgOAc (acetoxymercuri) group and an —OH (hydroxy) group derived from water have added to the double bond; this reaction is called oxymercuration.

    a. The first step of the reaction mechanism involves the formation of a cyclic ion called a mercurinium ion.

a mercurinium ion

b. The mercurinium ion is attacked by the solvent water occurs exclusively at the more branched carbon, even if that carbon has only one alkyl substituent.

c. The addition is completed by transfer of a proton to the acetate ion that is formed.

2. Oxymercuration adducts are easily converted into alcohols by treatment with the reducing agent sodium borohydride ($NaBH_4$) in base. In this reaction, the carbon-mercury bond is replaced by a carbon-hydrogen bond.

3. The overall reaction sequence, called an oxymercuration-reduction, results in the net addition of the elements of water (H and OH) to an alkene double bond in a regioselective manner.
   a. The —OH group is added to the more branched carbon of the double bond.
   b. The product is free of rearrangements and other side reactions that are encountered in hydration.

B. HYDROBORATION-OXIDATION OF ALKENES

1. Borane BH_3 adds regioselectively to alkenes so that the boron becomes bonded to the less branched carbon of the double bond, and the hydrogen becomes bonded to the more branched carbon.
   a. The addition of BH_3 to alkenes, called hydroboration, occurs in a single step without intermediates, that is, by a concerted mechanism.
   b. One borane molecule can add to three alkene molecules; the product is a trialkylborane.

2. Trialkylboranes are oxidized to alcohols using basic hydrogen peroxide, H_2O_2. The net result of this transformation is replacement of the boron by an —OH in each alkyl group.

$$\left(H-\overset{|}{\underset{|}{C}}-\overset{|}{\underset{|}{C}} \right)_3 B \xrightarrow[\text{HO}^-]{\text{H}_2\text{O}_2} \quad 3\ H-\overset{|}{\underset{|}{C}}-\overset{|}{\underset{|}{C}}-\text{OH} \quad + \quad \text{borate salts}$$

3. The overall reaction sequence, called a hydroboration-oxidation, results in the net addition of the elements of water (H and OH) to an alkene in a regioselective manner.
   a. The —OH grouxp is added to the least branched carbon atom of the double bond.
   b. The product is free of rearrangements and other side reactions that are encountered in hydration.

$$\text{H}_3\text{C}-\overset{\overset{\displaystyle\text{CH}_3}{|}}{\underset{\underset{\displaystyle\text{CH}_3}{|}}{C}}-\text{CH}=\text{CH}_2 \xrightarrow[\text{2) H}_2\text{O}_2/\text{HO}^-]{\text{1) BH}_3} \text{H}_3\text{C}-\overset{\overset{\displaystyle\text{CH}_3}{|}}{\underset{\underset{\displaystyle\text{CH}_3}{|}}{C}}-\text{CH}_2-\overset{\overset{\displaystyle\text{OH}}{|}}{\text{CH}_2}$$

C. Conversion of Alkenes into Glycols

1. Alkenes are readily converted into vicinal glycols, compounds that have an —OH group on each of two adjacent carbons, using either osmium tetroxide (OsO_4) or alkaline potassium permanganate ($KMnO_4$).

$$\overset{}{\underset{}{C}}=\overset{}{\underset{}{C} \xrightarrow{\text{OsO}_4 \text{ or KMnO}_4}} \quad -\overset{\overset{\displaystyle\text{HO}}{|}}{\underset{|}{C}}-\overset{\overset{\displaystyle\text{OH}}{|}}{\underset{|}{C}}- \quad \text{a vicinal glycol}$$

   a. Reaction with OsO_4 involves a cyclic intermediate called an osmate ester, which is formed in a cycloaddition reaction.

$$\overset{}{\underset{}{C}}=\overset{}{\underset{}{C}} \xrightarrow{\text{OsO}_4} \quad \text{an osmate ester}$$

   A glycol is formed when the cyclic osmate ester is treated with water and a mild reducing agent such as sodium bisulfite ($NaHSO_3$).

$$\xrightarrow{\text{NaHSO}_3} \quad + \quad \text{reduced osmium}$$

   b. Reaction with $KMnO_4$ forms the glycol product spontaneously, and the MnO_2 by-product precipitates; no other reagents are necessary.

$$\xrightarrow[\text{HO}^-]{\text{KMnO}_4} \quad + \quad MnO_2 \quad \text{brown precipitate}$$

   The reaction with $KMnO_4$ is the basis for a well-known qualitative test for alkenes, called the Baeyer test for unsaturation, in which the brilliant purple color of the permanganate ion is replaced by a murky brown precipitate of manganese dioxide (MnO_2).

D. Comparison of Methods for the Synthesis of Alcohols from Alkenes

1. Hydration of alkenes is a useful industrial method for preparing a few alcohols, but it is not a good laboratory method.
2. Hydroboration-oxidation is a good laboratory method which gives an alcohol in which the —OH group has been added to the least branched carbon of the double bond.

3. Oxymercuration-reduction is a good laboratory method which gives an alcohol in which the —OH group has been added to the more branched carbon of the double bond.

III. *Other Reactions*

A. OZONOLYSIS OF ALKENES

1. The reaction of an alkene with ozone to yield products of double-bond cleavage is called ozonolysis.
2. Ozone, O_3, adds to alkenes at low temperature in a concerted cycloaddition reaction to yield unstable cyclic compounds called molozonides.

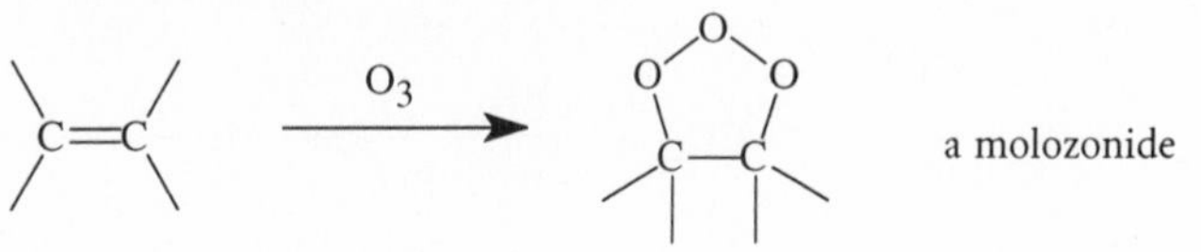

a molozonide

3. The molozonide cycloaddition product is unstable and spontaneously forms the ozonide.

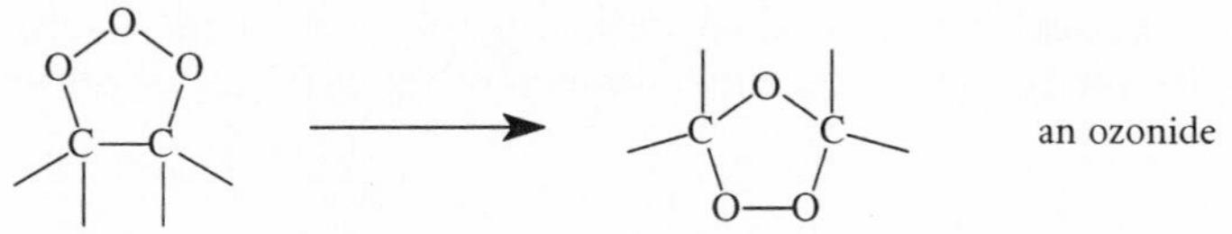

an ozonide

4. If the ozonide is treated with dimethyl sulfide, $(CH_3)_2S$, the ozonide is split into products that contain a C=O group.
   a. If a carbon of the double bond in the starting material bears a hydrogen, an aldehyde is formed.
   b. If a carbon of the double bond bears no hydrogens, a ketone is formed.

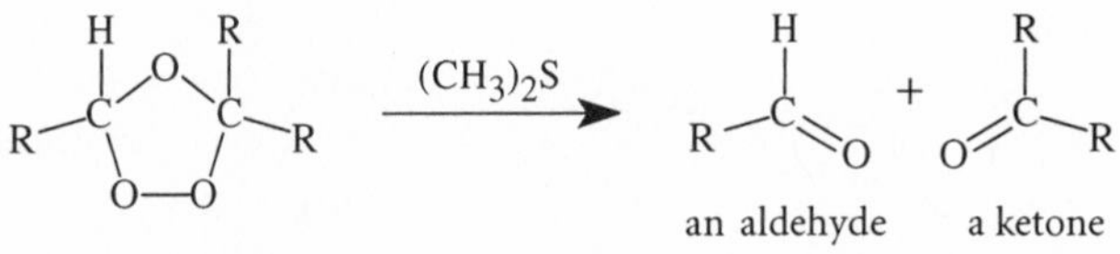

an aldehyde   a ketone

5. If the ozonide is simply treated with water, hydrogen peroxide (H_2O_2) is formed as a by-product and carboxylic acids are formed instead of aldehydes.

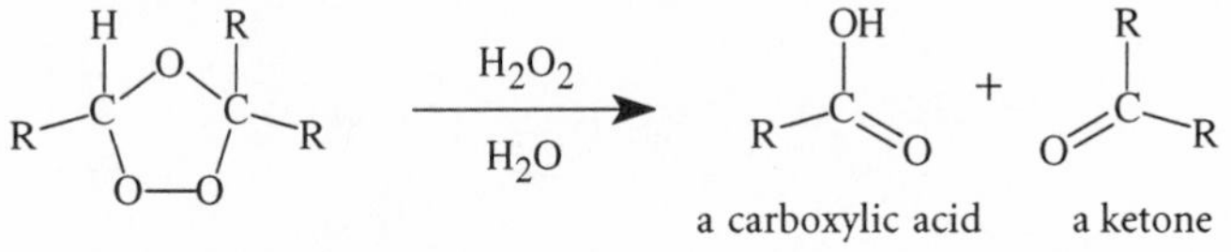

a carboxylic acid   a ketone

B. FREE-RADICAL POLYMERIZATION OF ALKENES

1. In the presence of free-radical initiators such as peroxides, many alkenes (monomers) react to form polymers in a reaction called polymerization.

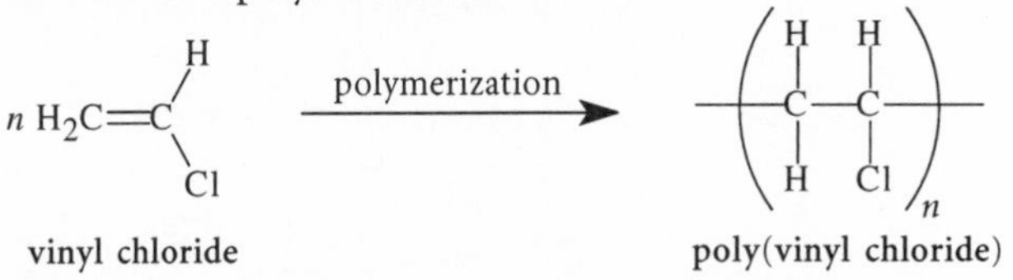

vinyl chloride   poly(vinyl chloride)

   a. Polymers are very large molecules composed of repeating units.
   b. Polyethylene is an example of an addition polymer—a polymer in which no atoms of the monomer unit have been lost as a result of the polymerization reaction.
   c. Polyethylene formation occurs by a free-radical mechanism and is thus a free-radical polymerization.

C. THERMAL CRACKING OF ALKENES

1. Simple alkenes are produced industrially from alkanes in a process called cracking, which breaks larger alkanes into a mixture of smaller hydrocarbons, some of which are alkenes.

a. Ethylene, the alkene of greatest commercial importance, is produced by a process called thermal cracking.
b. Cracking involves free-radical reactions.
2. In cracking, ethyl radicals form ethylene by loss of a hydrogen atom in a process called β-scission, in which the cleavage occurs one carbon away from the radical site.

$$H-CH_2-CH_2 \longrightarrow H\cdot + CH_2-CH_2$$

Study Guide Links

√5.1   How to Study Organic Reactions

You have now studied several organic reactions, and perhaps it is apparent that you are going to study many more. How can you keep all these reactions straight? How can you recall just the right reaction to use in solving a particular problem?

There are four keys to learning reactions; each of these points is discussed further below.

1. Be active when you read the text.
2. Be organized.
3. Review frequently.
4. Study in small chunks.

Be active when you read the text. To see what this means, let's go through one section of the text right now. Open your text to Section 5.3A, "Oxymercuration-Reduction of Alkenes." This section is fairly typical of those in the text that describe important reactions. First comes a statement of the reaction. When you read this, don't just stare at the page. Instead, *be active!* Try to make a hypothesis about where the different atoms come from. (The use of color in the equations is designed to help you with this.) For example, where does the —OH group in the product of text Eq. 5.11 originate? Where do the two —OAc groups end up? If you think you have the right idea, write the same reaction using a different alkene starting material.

Next, the text provides a few practical facts about the reaction, for example, some information about the solvent or the reaction conditions. These details can be very important, even crucial (in which case the text will say so), or they might be added for completeness or for your later use as reference material. In most cases, you should read them but not bother to memorize them, because the object of our first pass through the chapter is to understand the reaction itself.

Next follows a discussion of the *mechanism* of the reaction. What should you do with the mechanism? What you *should not* do is memorize it. What you *should do* is to follow each step of the mechanism to see whether your hypothesis about where the various atoms come from in the product is correct, and how they get there. Look for Lewis acid-base reactions. Look for points in common with other mechanisms you have studied. One purpose of mechanisms is to help you see certain unifying ideas that seemingly different reactions have in common. (You may have noticed, for example, that carbocation theory keeps creeping into the discussion of many of the addition reactions you have studied.) As you go further along in the text, you might even try to write a mechanism on your own before looking at the one in the text. (Use Study Guide Link 4.7 to help you.) Comparing your mechanism to the one in the text will help you to refine your mechanism-writing skills. Finally, after studying the mechanism, *be active!* Start with a different alkene starting material and write the mechanism on your own.

Sometimes there are further applications of the reactions or additional facts—sort of a "wrap-up." For example, Eq. 5.15 exemplifies the point that no rearrangements are observed in oxymercuration.

Finally come the problems. The answers to the problems marked with an asterisk are provided in this manual; the answers to the remaining problems are provided to your instructor in an instructor's supplement. Your instructor may post these for you, or may want you to try your hand at these problems without any assistance. You will find that each nonasterisked problem is of the same type as the asterisked problem that precedes it.

Work problems immediately when you come to them, and work them *with the answer book closed!* These provide a valuable test over your understanding of the chemistry involved.

Put a mark by the ones you find difficult; re-work these problems before an exam. Another strategy is to work the problems that are asterisked initially; then use the others as a "pre-test" for exam study when you think you understand the material.

As you read the section, you may come across certain unfamiliar terms. For example, in the very first sentence of Section 5.3A is the term "addition." Do you know what this means? *The great temptation is to skip over unfamiliar terms, hoping that they will become clear in context.* This strategy often works when reading a novel, but you *can't* do this in scientific reading! You *must* understand each term before proceeding to the next. Ask yourself, "What is an addition reaction?" If you don't know, don't feel that you're intellectually deficient— just find out the answer. *Everyone* requires a number of repetitions in order to learn new ideas. (The author can't keep repeating the definition, because it would make the book even larger than it is!) How do you find out? *Use the index!* Or, in some cases, a cross-reference to an earlier section will be given. Look it up! If you get in the habit of taking this approach, you will find yourself retaining more, and you ultimately will have to do less work. If an unfamiliar term is one you have had to look up before, make yourself a glossary: Write the term and its definition, and review it periodically. You have probably noticed that each chapter in this Study Guide opens with a list of terms. A good use of this list is to see how many of the terms you can define without looking them up. It particularly helps to *write* the definition if you are having a problem putting a particular definition into words.

If you do what has been suggested, you'll notice that you have been very active. You have a piece of scratch paper on your desk, and it should be filled with notes and structures.

Two other suggestions: First, *do not* underline (or highlight) large sections of the text! This represents an effort to be active, when in fact it represents in many cases *physical* activity coupled with *mental* passivity. If there is something you really want to emphasize, then by all means, highlight away! Otherwise, don't bother. It's a waste of time and effort. Second, if you outline the chapter, don't go into an "autopilot" mode; think about what you write, and don't passively copy sections of the text. Pretend you are a teacher and that you're writing the outline for someone who can't understand the text at all. This Study Guide has also provided chapter outlines for you. In most cases, topics have been reorganized and regrouped in these outlines so that you think about topics in a different way. For example, part of the outline might be called "Reactions;" another part, "Rules;" etc. This should be useful to you if you want to overview the chapter quickly, but is not meant to substitute for summaries that you write for yourself. Some students use these outlines for a "first pass" over a chapter before reading the chapter in detail.

Be organized. After you have studied a reaction, write it down in a general form for later review. One good way is to fold a sheet of paper in half. Write the reactants on the left and the products on the right. Write two arrows, one on each side. Use "R" groups or "dangling bonds" to make the reaction as general as possible. Let's illustrate. A review sheet for oxymercuration–reduction might look as shown in Fig. SG5.1 on the following page. (Some people prefer to use "flash cards," with one side of a reaction written on the front of the card, and the other on the back.) Each time you learn a new reaction, enter it on your sheet. Notice above that not only the individual reactions of oxymercuration and reduction have been entered, but also a summary reaction. The choice of three "R" groups is arbitrary; you might instead use an alkene starting material R—CH=CH$_2$. The point here is to use enough "R" groups that the regioselectivity of the reaction is apparent. Notice also that not all reactions are balanced; the goal is to focus on what happens to the organic compound. Each time you finish a study session, fold that page in half. With the left side in view, complete the right on a piece of scratch paper. Then, with the right side in view, complete the starting materials. (Notice that the arrow bearing the reaction conditions is repeated on both sides of the fold for this purpose.) You should find that this process helps you learn reactions both ways! This knowledge will become particularly valuable when you start combining reactions in multistep syntheses. Next, ask yourself why the reaction is reasonable. For the oxymercuration step, you should recognize that the reaction is an addition. Go through the mechanism. What

fold

Figure SG5.1 *Part of a typical reaction review sheet for Chapter 5.*

is the electrophile? Why does it react with the alkene π electrons? Why is the reaction regioselective? (Note that the text does not cover the mechanism of the reduction step.) Then write a specific example of the reaction using R groups of your choice.

If you do this *every time you read,* and *if you understand what you read,* then what you are doing here is *not* memorizing the reactions, at least not in a rote fashion, because you will have already learned them by careful study. Rather, you are cataloging them for use as a future study tool. If you can't complete one side or the other of a reaction, look at the missing side. If it looks unfamiliar, return to the text and focus exclusively on the part that is unfamiliar.

Review frequently. Review your reaction sheet *every time you study!* Go back about three assignments for each review, dropping older material from your review. Then review the older material again once every week or two, or right before an exam or quiz. (Notice that this gives you something *active* to do in preparing for an examination.) Do the same thing with the list of terms. You will find that relatively few reactions and terms are covered in each assignment, and that learning each day's assignment takes relatively little time. If you get behind, however, the number of reactions and terms that must be learned will grow rapidly!

Many students and professors suffer from adherence to a theory that one of the author's colleagues calls the "immunization theory of learning." The student's version of this theory is: "I've studied this once, and therefore I should never have to see it again . . . *ever.*" The professor's version of the theory is, "Students studied this once, and therefore they should automatically know it. There is no need for me to review, and I should be annoyed if a student does not have total recall of everything he/she has ever been taught!" Those who adhere to these ideas deny a fundamental caveat about learning: *Continued reinforcement is one of the best ways to learn.* The author is frequently amazed how often students are disappointed in themselves because they didn't learn something on the first pass. You should *expect* to forget material, and you should expect to re-learn it, probably more than once. Each relearning, however, takes less time and brings with it deeper understanding.

Study in small chunks. Set realistic goals for each study session and pursue them regularly. If you can only afford six hours per week, it is far better to study six times per week for one hour than one time for six hours. If you can allow twelve hours per week, then break it into six two-hour sessions. Notice how this strategy is built into the discussion of the reaction sheet above: you should study your reaction sheet over and over again, but confine your concentration to relatively small parts of it.

If you have ever played an organized sport, or studied a musical instrument seriously, you should understand the "small chunk" strategy well. Does the coach of a winning team hold one ten-hour practice per week, or five two-hour practices? Why? Because the mental and physical learning capacities of most individuals erode significantly as fatigue sets in. Does a skilled performer on a musical instrument practice once per week for twelve hours, or six times per week for two hours?

What is remarkable is that when you *stay organized* and *study in small chunks*, your capacity for further study increases! Just as an athlete's endurance increases with regular workouts, your ability to concentrate for longer periods of time will increase. Just as an athlete's speed and skill increase with each workout, you will also find that you will become more efficient: you will accomplish much more in less time!

Obviously, the suggestions given here for learning reactions are not the only ones that will work. But if you are at a loss about how to study organic chemistry on your own, try these suggestions, and then refine them to suit your own needs. If you have some other great ideas, send them to the author! Perhaps they will appear in the *next* edition of this Study Guide!

√5.2 Mechanism of Ozonolysis

You may have noticed that several mechanisms, such as bromonium ion formation and mercurinium ion formation, have been presented as stepwise processes to show how these mechanisms relate to mechanisms involving carbocations. Let us take this approach with ozonolysis.

The electrophilic oxygen—the electron-deficient oxygen—is a powerful Lewis acid, because there is a strong drive for an electron-deficient electronegative atom to complete its octet. If this oxygen accepts the π electrons of the alkene, a carbocation is generated:

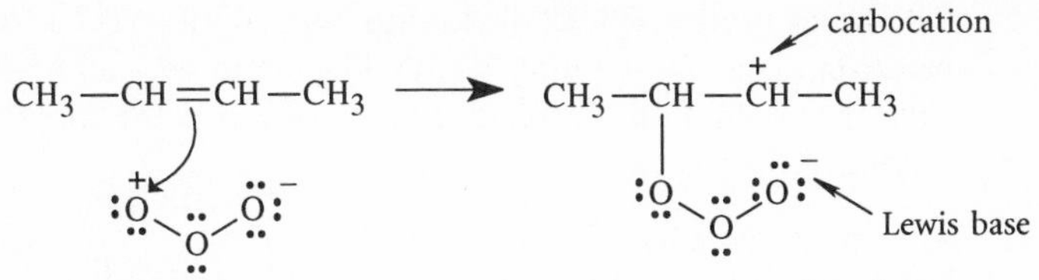

(SG5.3a)

The carbocation, a Lewis acid, is attacked by the negatively charged oxygen, which is a Lewis base. This completes the addition.

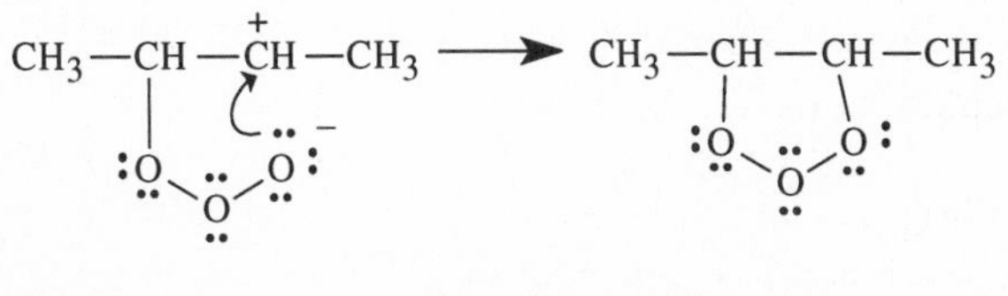

(SG5.3b)

Even though the mechanism is concerted, thinking of it as a two-step process, as shown above, helps to understand why the reaction takes place.

The following discussion gives more detail on the conversion of the molozonide to the ozonide. The molozonide decomposes in a "reverse cycloaddition."

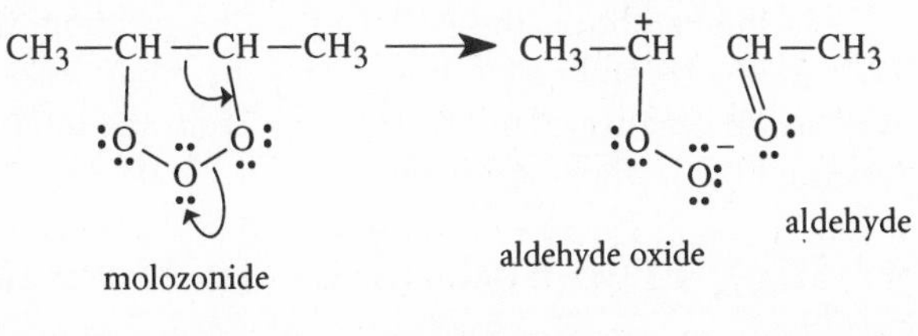

(SG5.4a)

This occurs because the O—O bond is a very weak bond, and oxygen, an electronegative atom, readily accepts an electron pair. Notice that the "aldehyde oxide" is a molecule a lot like ozone: it has an electron-deficient (Lewis acid) end, that is, the carbocation, and an electron-rich (Lewis base) end, that is, the —O⁻. Thus, it should not be surprising that it *reacts* a lot like ozone. It simply adds to the double bond of the aldehyde. Now, there are two ways that this addition can occur. One way is simply the reverse of Eq. SG5.4a. In the other way, the aldehyde adds in the opposite sense:

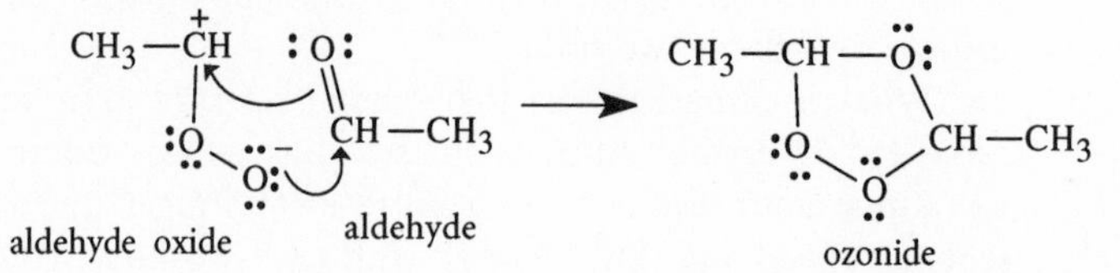

aldehyde oxide   aldehyde   ozonide

(SG5.4b)

In summary, the molozonide-to-ozonide conversion involves a fragmentation-cycloaddition sequence of reactions.

✓5.3   Bond Dissociation Energies and Heats of Reaction

Study Problem 5.3 in the text shows how bond dissociation energies can be used to calculate heats of reactions: subtract the bond dissociation energies of the bonds formed from those of the bonds broken. This study guide link shows why this process works.

Essentially, this is a Hess's law calculation. This can be illustrated by using bond dissociation energies to calculate the $\Delta H°$ of the following gas-phase reaction:

$$\cdot CH_3 + Cl_2 \longrightarrow Cl-CH_3 + Cl\cdot \qquad (SG5.5)$$

The term "bond dissociation energy" will be abbreviated with the acronym "BDE" for convenience. The bond formed is a H_3C-Cl bond, with BDE = 356 kJ/mol (85 kcal/mol). The bond broken is a $Cl-Cl$ bond, with BDE = 247 kJ/mol (59 kcal/mol). The $\Delta H°$ for the reaction, then, is 247 – 356 = –109 kJ/mol, or 59 – 85 = –26 kcal/mol.

The reaction in Eq. SG5.5 can be viewed as the sum of the following two reactions:

$$Cl_2 \longrightarrow 2Cl\cdot \qquad (SG5.6a)$$

$$\cdot CH_3 + Cl\cdot \longrightarrow Cl-CH_3 \qquad (SG5.6b)$$

Sum: $\cdot CH_3 + Cl_2 \longrightarrow Cl-CH_3 + Cl\cdot \qquad (SG5.5)$

The $\Delta H°$ of Eq. SG5.5 is then the sum of the $\Delta H°$ values of the two reactions, SG5.6a and SG5.6b:

	$\Delta H°$
$Cl_2 \longrightarrow 2Cl\cdot$	BDE of Cl_2
$\cdot CH_3 + Cl\cdot \longrightarrow Cl-CH_3$	–BDE of CH_3-Cl
Sum: $\cdot CH_3 + Cl_2 \longrightarrow Cl-CH_3 + Cl\cdot$	(BDE of Cl_2) – (BDE of CH_3-Cl)

As shown above, the $\Delta H°$ of the first reaction, by definition, is equal to the BDE of Cl_2. Because the second reaction is simply the bond dissociation of CH_3-Cl written in reverse, its $\Delta H°$ is the negative BDE of CH_3-Cl. Thus, the total $\Delta H°$ is then the BDE of the bond broken less the BDE of the bond formed.

✓5.4   Writing Free-Radical Chain Mechanisms

A free-radical chain mechanism should be written when a reaction is initiated by peroxides, AIBN, light, or other initiators—or, as in Problem 5.44, when you are asked to write one! Suppose you are asked to write a free-radical chain mechanism for the following free-radical addition reaction.

$$BrCCl_3 \; + \; H_2C{=}CHCH_2CH_2CH_3 \; \xrightarrow[\text{peroxides}]{\text{light}} \; Cl_3C{-}CH_2{-}\underset{|}{\overset{}{C}}HCH_2CH_2CH_3 \qquad (SG5.7)$$
$$Br$$

This is a reaction that has not been presented in the text. What is a rational process for writing a free-radical chain mechanism?

The first thing to notice is *which fragments have added to the alkene.* In the present case, —CCl$_3$ has added to one carbon of the double bond, and —Br to the other. The identities of these fragments indicate which bond has been ruptured in the initiation step(s). Thus, the Br—CCl$_3$ bond is broken.

The next thing to notice is *which group is bound to the carbon with fewer alkyl substituents.* In this case, the —CCl$_3$ group is attached at this carbon, and the —Br at the other carbon. This indicates which radical fragment adds first. In this case ·CCl$_3$ is the radical that adds first to the alkene. Why is this so? Remember that the first radical that adds to the double bond must go to the carbon of the double bond with *fewer alkyl substituents* so that the unpaired electron ends up on the carbon with *more alkyl substituents.* In summary, then, the group at the carbon of the product with fewer alkyl substituents is the radical fragment that first adds to the alkene, *and is therefore the radical that must be generated in the initiation step(s).* In order for ·CCl$_3$ to be generated in the initiation steps, an alkoxy radical (derived from the peroxide) must abstract a bromine atom from BrCCl$_3$:

$$RO{-}OR \longrightarrow 2\,RO\cdot \qquad\qquad (SG5.8a)$$
$$\text{alkoxy radical}$$

$$RO\cdot \quad Br{-}CCl_3 \longrightarrow RO{-}Br \; + \; \cdot CCl_3 \qquad (SG5.8b)$$

The ·CCl$_3$ radical then adds to the double bond at the carbon with fewer (in this case, zero) alkyl substituents.

$$Cl_3C\cdot \quad H_2C{=}CHCH_2CH_2CH_3 \longrightarrow Cl_3C{-}CH_2{-}\overset{\cdot}{C}HCH_2CH_2CH_3 \qquad (SG5.8c)$$

The resulting radical propagates the chain by abstracting a bromine atom from Br—CCl$_3$ and forming another ·CCl$_3$ radical:

$$Br{-}CCl_3$$
$$Cl_3C{-}CH_2{-}\overset{\cdot}{C}HCH_2CH_2CH_3 \longrightarrow Cl_3C{-}CH_2{-}\underset{|}{\overset{}{C}}HCH_2CH_2CH_3 \; + \; \cdot CCl_3 \qquad (SG5.8d)$$
$$Br$$

Reactions SG5.8c and SG5.8d are the propagation steps of the chain reaction. Notice that *the sum of the propagation steps is the overall reaction,* that is, Eq. SG5.7.

Notice also that *each propagation step uses up one radical and produces another.* It is worth making this point because some students write the mechanism of a reaction such as this in the following *incorrect* way:

$$CH_2{=}CHCH_2CH_2CH_3 \longrightarrow Cl_3C{-}CH_2{-}\underset{|}{\overset{}{C}}HCH_2CH_2CH_3 \qquad (SG5.9)$$
$$Cl_3C\cdot \quad \cdot Br \qquad\qquad\qquad\qquad Br$$

Why can't this be right? Because this mechanism destroys radicals rather than propagating them. If this mechanism were correct, there would have to be a separate initiation step for every molecule of alkene that reacts. Furthermore, such a mechanism requires a very improbable collision between three species, two of which (the radicals) are present at very low concentration! The whole idea of a chain mechanism is that the initiation step has to provide only a very small number of free radicals because each propagation step replaces the radical consumed with a new one. Consequently, in a chain mechanism, free radicals (or the initiators from which they are generated) do not have to be present in high concentration.

√5.5   Solving Structure Problems

Before working Problem 5.51 in the text, be sure you have read Study Guide Link 4.6 (also called "Solving Structure Problems").

Problem 5.51 contains a lot of information. The temptation is to solve this problem from the top down. It bears repeating that this is not usually the best way to deal with this type of problem. What is known about *A*? Its empirical formula can be determined from the elemental analysis. (Do this now.) The reaction with Br_2, the fact that it undergoes hydrogenation, and the fact that it reacts with ozone all suggest that *A* is an alkene. Immediately you should look for a *structure*. There are two structures in the problem: one given in words, the other explicitly. The hydrogenation product of *A* is 1-isopropyl-4-methylcyclohexane:

1-isopropyl-4-methylhexane

This tells us *immediately* the carbon skeleton of *A*. Only the double bonds are missing. This structure further implies that *A* contains a ring. Since the empirical formula of *A* is known, the number of carbons (10) in this hydrogenation product also reveals the molecular formula of *A*. (What is the formula?) Since the molecular formula of *A* is known, its unsaturation number can be calculated. Remember that the ring accounts for one degree of unsaturation. How many double bonds does compound *A* have?

The second structure given in the problem is the ozonolysis product. Remember that ozonolysis produces *two* C=O groups for every double bond in the molecule. However, in this ozonolysis product, there are *three* C=O groups. But there are also *nine carbons*. Evidently, one carbon and one C=O group are missing. This means that the missing carbon is part of the missing C=O group, and that a one-carbon fragment was liberated by ozonolysis as a second product which was evidently not recovered. This conclusion is consistent with the unsaturation number, which indicates that there are two double bonds. Two double bonds, after ozonolysis, gives four C=O groups.

This information establishes a partial structure for *A:* compound *A* must contain a $=CH_2$ group. Such a group can come from the known carbon skeleton in only two ways:

1   or   *2*

Since there are two double bonds in *A*, one additional double bond must be provided to convert either *1* or *2* into possible structures for *A*. From the ozonolysis product, this double bond must be *within* the ring, because ozonolysis opens the ring. Now it is time to *write all the possibilities*. (Recall that this point was stressed in Study Guide Link 4.6.) From compound *1* the only possibilities for *A* are the following:

1-1   or   *1-2*

From compound *2* the only possibilities are the following:

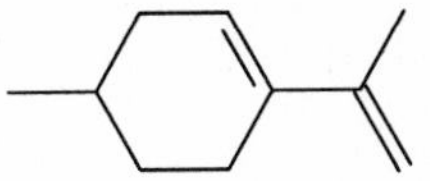

| 2-1 | or | 2-2 | or | 2-3 |

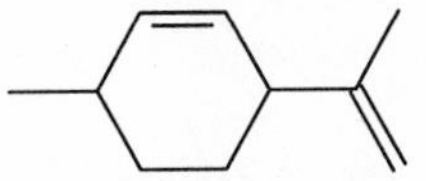

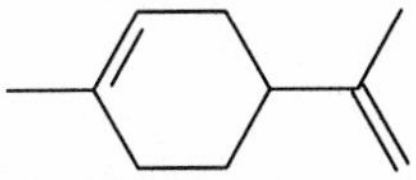

There are two ways to finish the problem. One is to apply ozonolysis to each of the five structures and see which one gives the observed product. Notice that the observed product contains two groups of the following type:

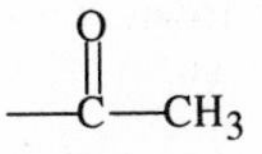

Which structures would give *two* of these groups after ozonolysis? You should finish the problem by deciding between these.

The second way to finish the problem is to reverse ozonolysis mentally in all possible ways. The carbons involved in double bonds have to be the ones "labeled" with oxygens.   These are carbons 1, 2, and 3 in the following structure of the ozonolysis product:

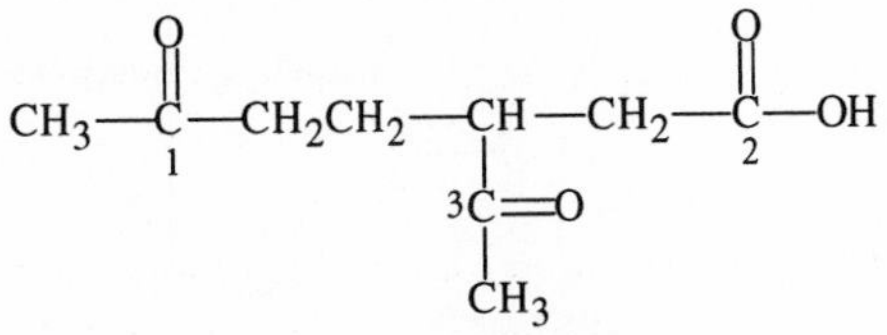

(And remember, one carbon is missing, which will be referred to as carbon-4.) Once again, *write out all possibilities.* All possible connections are:

1.   C1 with C2, C3 with C4
2.   C1 with C3, C2 with C4
3.   C1 with C4, C2 with C3

And all but one can be ruled out by recalling that compound *A* has to contain a six-membered ring. The only structure for *A* that has a six-membered ring is obtained by connecting C1 to C2. Hence, C3 and C4 are also connected. You should finish reconstructing compound *A*. (The final solution is in the Solutions part of this chapter.)

    You can see that working a problems such as this is like solving a big puzzle. You need to understand the reactions involved, but you need something else: the ability to reason analytically. If you keep working problems of this sort, you can learn to develop this reasoning ability. This ability can help you not only in chemistry; it will help you in your career (whatever it is)—and it can even help you solve ordinary day-to-day problems.

Solutions

Solutions to In-Text Problems

5.1   (a) The reaction occurs by attack of the azide ion on an iodonium ion, a cyclic cation analogous to the bromonium ion. Attack occurs at the carbon with the alkyl substituents, because this opens the weakest bond.

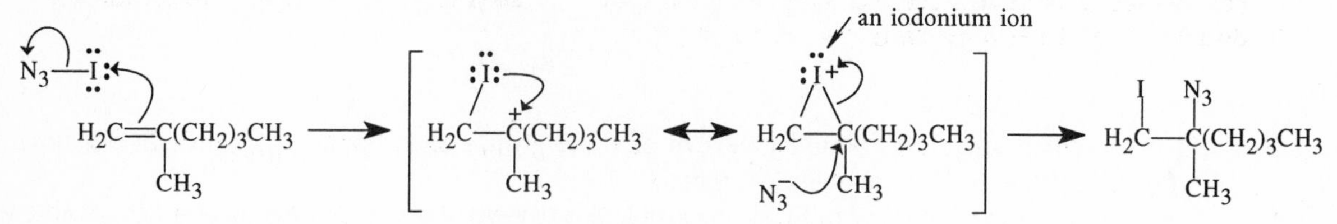

5.2   The alkene that would react with Br_2 in H_2O to give the chlorohydrin *B* in Study Problem 5.1:

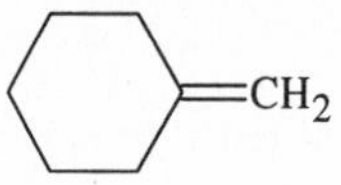

methylenecyclohexane

5.3   (a)

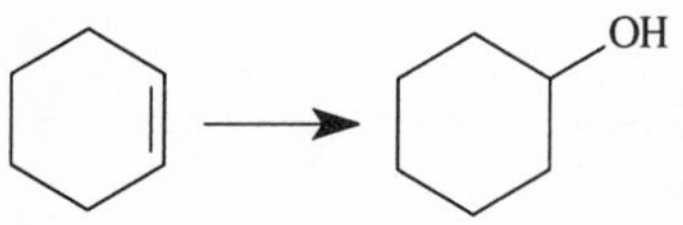

(c)

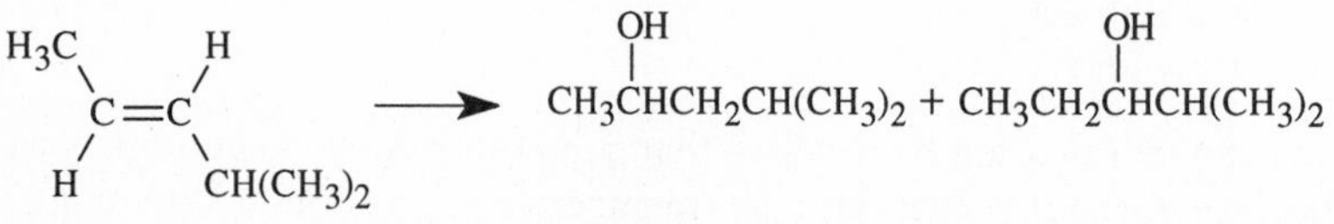

In this case two products are formed because the carbons of the alkene double bond have the same number of alkyl substituents (one); consequently, there is no basis for regioselectivity.

5.4   (a) Either of the following alkenes would give the alcohol in part (a) as the major product:

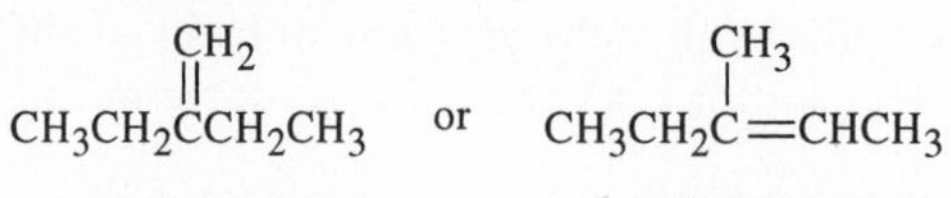

$$CH_3CH_2\overset{\overset{\displaystyle CH_2}{\|}}{C}CH_2CH_3 \quad \text{or} \quad CH_3CH_2\overset{\overset{\displaystyle CH_3}{|}}{C}=CHCH_3$$

2-ethyl-1-butene   3-methyl-2-pentene

(either (*E*) or (*Z*) stereoisomer)

5.5   The products of hydroboration-oxidation are the same as the ones in parts (a) and (c) of Problem 5.3. In (a), the alkene is symmetrical; consequently, regioselectivity has no meaning for this alkene. In part (c), the carbons of the double bond have the same number of alkyl substituents; consequently, there is no reason to expect significant regioselectivity.

5.6   The products are the same for parts (a) and (b) of both problems because both alkenes are symmetrical; consequently, regioselectivity has no meaning. For part (c) of both problems, the carbons of the double bond have the same number of alkyl substituents; consequently, there is no basis for expecting substantial regioselectivity in either reaction, although there might be small differences in the relative amounts of corresponding products. In part (d), the carbons of the double bond have different numbers of alkyl substituents. In Problem 5.3(d), the —OH group goes to the carbon bearing the greater number of alkyl substituents; in Problem 5.5(d), the —OH group goes to the carbon bearing the smaller number of alkyl substituents.

5.7   (a) The alkene required is methylenecyclohexane. (The structure is in the solution to Problem 5.2 above.)

5.8   (a) Because *cis*-2-butene is a symmetrical alkene, the two reactions give the same product.
      (c) Because 3-methyl-1-pentene has a different number of alkyl substituents at the two carbons of its double bond, the two reactions give different products:

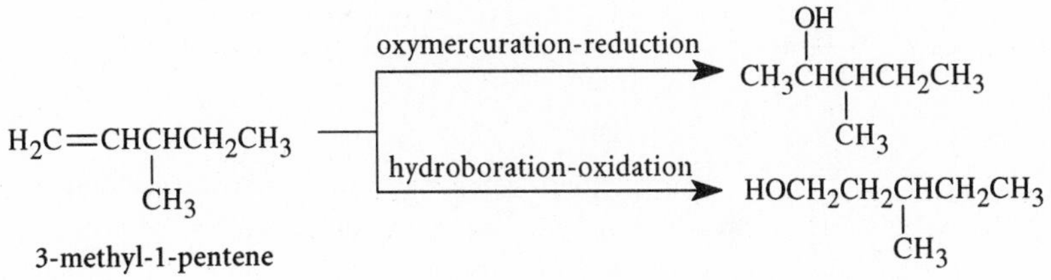

5.9   (a)

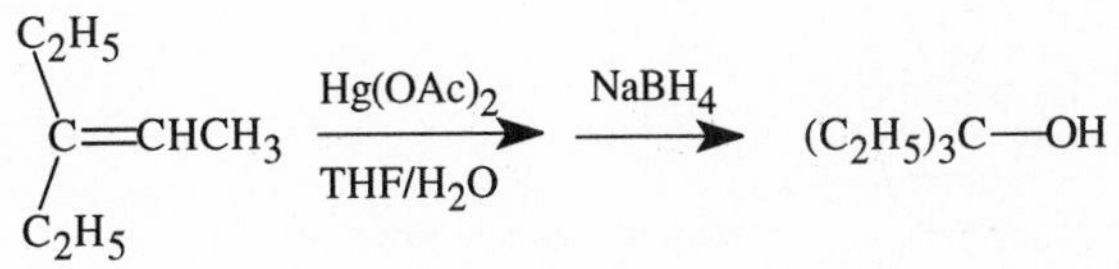

      (c)

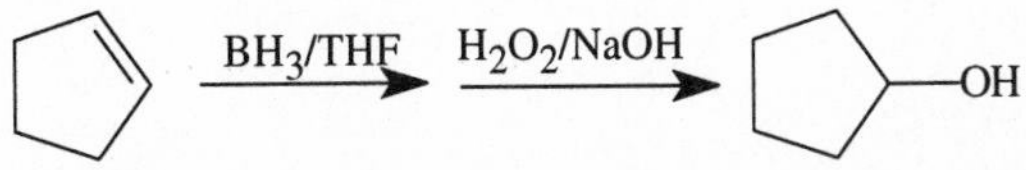

Although hydroboration-oxidation is used in this solution, oxymercuration-reduction would in principle work equally well because the alkene is symmetrical, and regioselectivity is therefore not an issue.

5.10   (a)   (c)

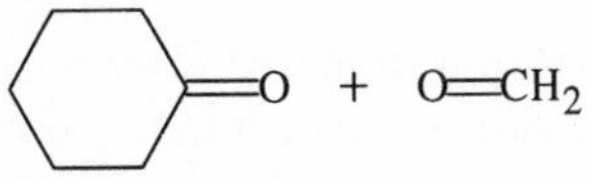

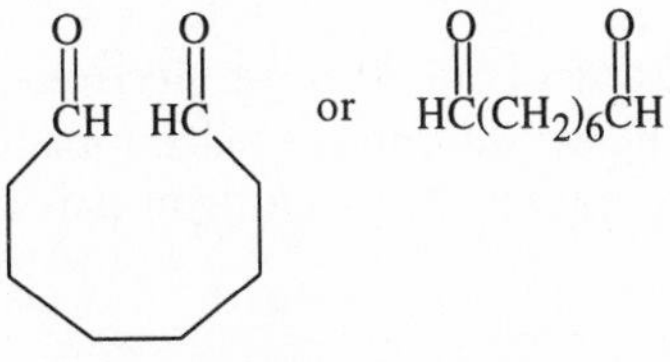

      (d) There is no reaction of ozone with 2-methylpentane because it is an alkane and has no double bonds.

5.11   (a)   (c)

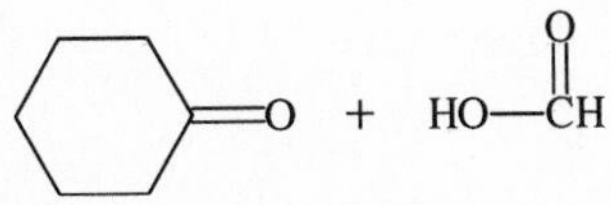

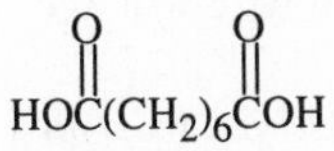

      (d) For the reason given in the previous problem, there is no reaction with ozone.

5.12 (a) (c)

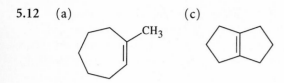

5.13 As you might have discovered when you worked part (b) of the previous problem, whether an alkene is *E* or *Z* cannot be determined from the products of ozonolysis.

5.14 (a) (c)

$$\underset{\overset{|}{CH_3}}{\overset{OH}{\underset{|}{H_3C-C}}}\overset{OH}{\underset{|}{-CH_2}}$$

5.15 (a)

$$HOCH_2CH{=}CHCH_2OH \quad or \quad \underset{\overset{|}{OH}}{HOCH_2CHCH{=}CH_2} \quad or \quad H_2C{=}CH-CH{=}CH_2$$

5.16 Reason by analogy to Eq. 5.35a:

cyclic manganate ester

5.17 (a) (c)

$$\underset{(CH_3)_2\overset{\bullet}{C}-CH_2}{\overset{:\overset{\bullet\bullet}{Br}:}{}}$$

$$\overset{\bullet}{C}H_2-CH{=}CH-CH{=}CH_2$$

> Notice from part (c) that resonance structures can be derived with the fishhook formalism just as with the curved-arrow formalism.

5.18 Any reaction that involves splitting of one or more electron-pair bonds into unpaired electrons, or any reaction that involves the joining of two unpaired electrons to give an electron-pair bond, is homolytic. Any reaction involving the movement of only electron-pair bonds and/or unshared pairs is heterolytic.

(a) (c)

$$H-\overset{\bullet\bullet}{O}H \longrightarrow H\bullet + \bullet\overset{\bullet\bullet}{O}H \qquad CH_2{=}CH_2 \quad H-\overset{\bullet\bullet}{Br}: \longrightarrow \overset{+}{C}H_2-CH_3 \quad :\overset{\bullet\bullet}{Br}:^-$$

(homolytic) (heterolytic)

5.19 Two free-radical intermediates involved in the reaction are

$$\overset{\bullet}{CH_3(CH_2)_3CHCH_2Br} \qquad \bullet Br$$

A *B*

Recombination of *A* with *A*, *B* with *B*, and *A* with *B*, gives rise to three recombination products:

CH$_3$(CH$_2$)$_3$CHCH$_2$Br

CH$_3$(CH$_2$)$_3$CHCH$_2$Br

Br—Br

CH$_3$(CH$_2$)$_3$CHCH$_2$Br (with Br substituent)

$A + A$ $B + B$ $A + B$

5.21   (a)   (c)

CH$_3$CH$_2$CHCH$_3$ (with Br substituent)

cyclopentane with CH$_3$ and Br substituents

5.22   (c) The initiation steps are shown in Eq. 5.43, text p. 197, and Eq. 5.45, text p. 198, for a typical peroxide. The propagation steps are as follows:

[reaction scheme showing cyclopentane radical with CH$_3$ and Br, reacting with HBr and H—Br to give product + •Br]

The termination steps are analogous to those shown in the solution to Problem 5.19. One is the recombination of two bromine atoms to form Br$_2$. The other two are as follows:

[radical recombination reaction schemes]

Notice that when dealing with radical species it is necessary only to write the unpaired electron; unshared pairs (for example, those on bromine in the previous solution) need not be written explicitly when they are not involved in the reaction.

5.23   (a) Two Cl—CH$_2$ bonds are formed (335 kJ/mol each); a Cl—Cl bond (247 kJ/mol) and a carbon-carbon π bond ($\approx$ 270 kJ/mol) are broken. Subtract the bond dissociation energies of the bonds formed from those of the bonds broken to obtain the desired estimate for $\Delta H°$.

$$\Delta H° = 247 + 270 - 2(335) = -153 \text{ kJ/mol}$$

The value used in this solution for the bond dissociation energy of the π bond (270 kJ/mol) is in the middle of the range of values given in Table 5.3. If you used a different value, your answer will, of course, be correspondingly different.

5.24   Abstraction of bromine would form a C—Br bond (bond dissociation energy $\approx$ 280 kJ/mol); abstraction of a hydrogen forms a C—H bond (bond dissociation energy $\approx$ 420 kJ/mol). The hydrogen is abstracted because a stronger bond is formed. (The bond broken, an H—Br bond, is the same in either case.)

5.25   (a) The heat of formation of any species is the heat absorbed when one mole of the species is formed from the elements in their natural states. Molecular hydrogen (H$_2$) is the "element in its natural state" from which the hydrogen atom is formed; consequently, the heat of formation of the hydrogen atom is half the bond dissociation energy of H_2, that is, $435 \div 2$, or 218 kJ/mol.

(b) By the same reasoning, the heat of formation of the chlorine atom is half the bond dissociation energy of Cl_2, that is, $247 \div 2$, or 124 kJ/mol.

5.27 To obtain the polymer structure, add one end of the double bond in one molecule to the other end of the double bond in another molecule and continue this process indefinitely.

(a)

$$\left[\text{CH} - \text{CH}_2 \right]_n \quad \text{polypropylene}$$
$$\text{CH}_3$$

5.28 (a) In the initiation step, a radical derived from homolysis of a peroxide or other initiator adds to the double bond of the alkene so as to form the secondary free radical:

$$RO \cdot \quad CH_2 = CH \longrightarrow RO - CH_2 - \overset{\cdot}{C}H$$
$$CH_3 \qquad\qquad\qquad CH_3$$

This radical adds to another molecule of propene, and the process is repeated indefinitely:

$$RO - CH_2 - \overset{\cdot}{C}H \quad CH_2 = CH \xrightarrow{CH_2=CHCH_3} RO - CH_2 - CH - CH_2 - \overset{\cdot}{C}H$$
$$CH_3 \qquad\quad CH_3 \qquad\qquad\qquad\qquad\qquad CH_3 \qquad\quad CH_3$$

Any radical recombination reaction might serve as a termination reaction, for example the following:

$$2\,RO\left[CH_2 - CH\right]CH_2 - \overset{\cdot}{C}H \longrightarrow RO\left[CH_2 - CH\right]CH_2 - CH - CH - CH_2\left[CH - CH_2\right]OR$$
$$\qquad\quad CH_3 \Big]_n \quad CH_3 \qquad\qquad\qquad CH_3 \Big]_n \quad CH_3 \;\; CH_3 \qquad CH_3 \Big]_n$$

5.29 One of several possibilities is for octane to rupture at the bond between carbon-2 and carbon-3:

$$CH_3 - \overset{\cdot}{C}H_2 \quad CH_2(CH_2)_4CH_3 \longrightarrow CH_3 - \overset{\cdot}{C}H_2 \;+\; \overset{\cdot}{C}H_2(CH_2)_4CH_3$$

The ethyl radical then undergoes β-scission to give ethylene and a hydrogen atom (reaction 1 below). Two hydrogen atoms can recombine to give H_2, but H_2 can also be formed when a hydrogen atom abstracts another hydrogen atom from any one of the CH_2 groups of octane to generate a new radical (reaction 2 below).

(1) $$H - CH_2 - \overset{\cdot}{C}H_2 \longrightarrow H \cdot \; + \; CH_2 = CH_2$$

(2) $$H \cdot \quad H - \overset{\cdot}{C}H - CH_3 \longrightarrow H_2 + \; \cdot CH - CH_3 \longrightarrow \text{further reactions}$$
$$\qquad (CH_2)_5 - CH_3 \qquad\qquad\qquad (CH_2)_5 - CH_3$$

Solutions to Additional Problems

5.31 (a)

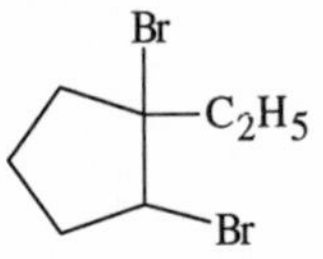

(b)

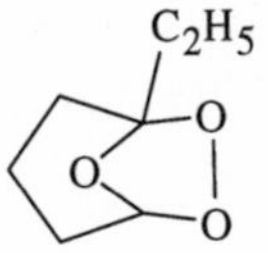

(c)

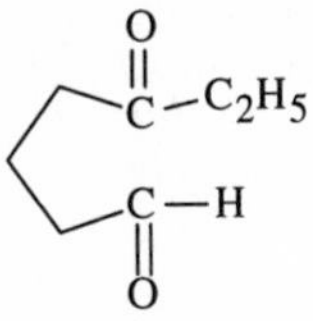

(d)

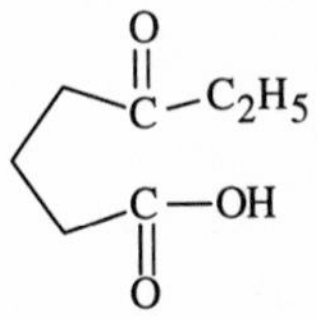

(e) (f)

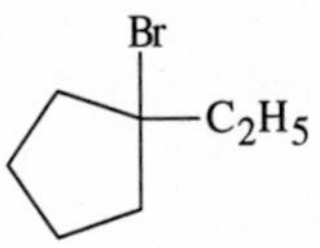

(g) 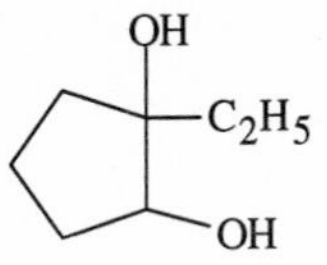 (h) 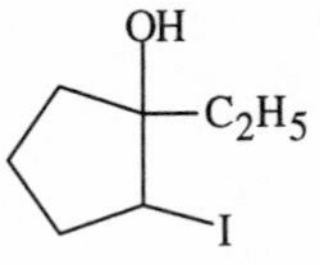 (i) 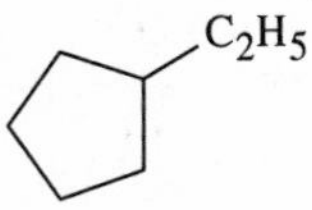 (j) 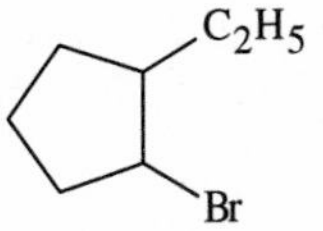 (k)

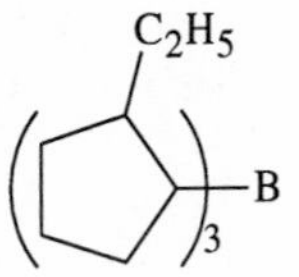

(l) 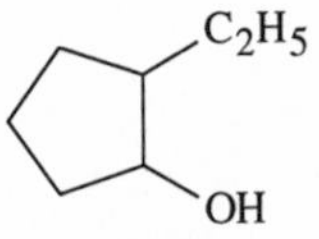 (m) 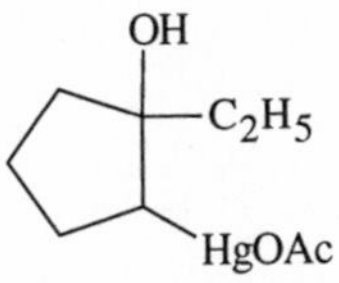 (n) 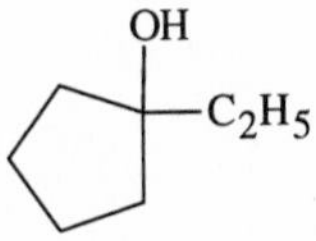 (o) 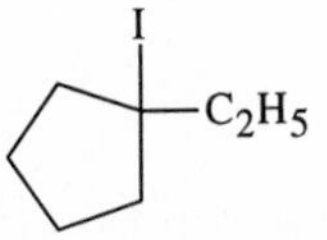 (p)

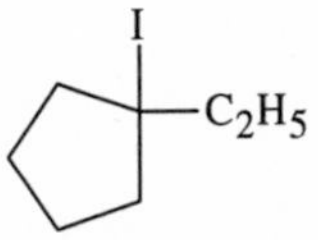

Notice in part (p) that there is no "peroxide effect" on HI addition; that is, free radical initiators such as peroxides or AIBN have no effect on the regioselectivity of HI addition.

5.33 (a) Any symmetrical alkene with six carbons is an acceptable answer. Three examples are the following:

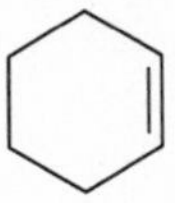

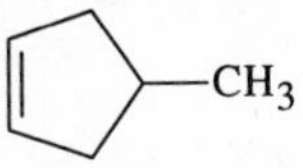

$CH_3CH_2CH = CHCH_2CH_3$

(b) Any cyclic alkane with five carbons is an acceptable answer. Two examples are the following:

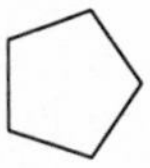

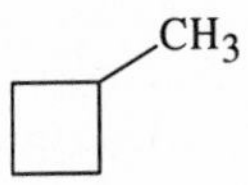

(c)

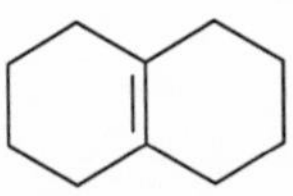

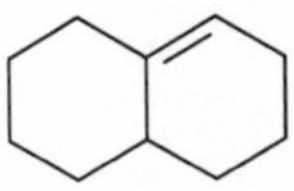

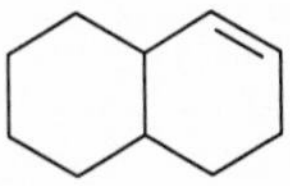

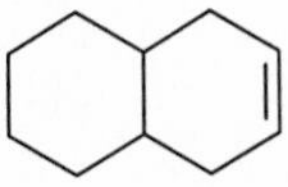

(d)

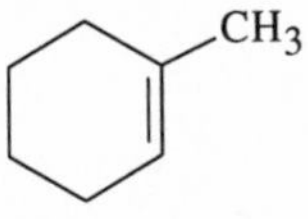

and

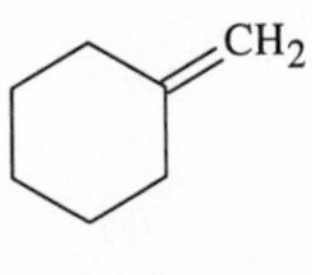

(e) 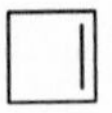 (f)

$C_2H_5 - CH = CH - C_2H_5$

(E) and (Z) isomers

(g) Any symmetrical cyclic alkene with five carbons is an acceptable answer. Three examples are the following:

(solution continues)

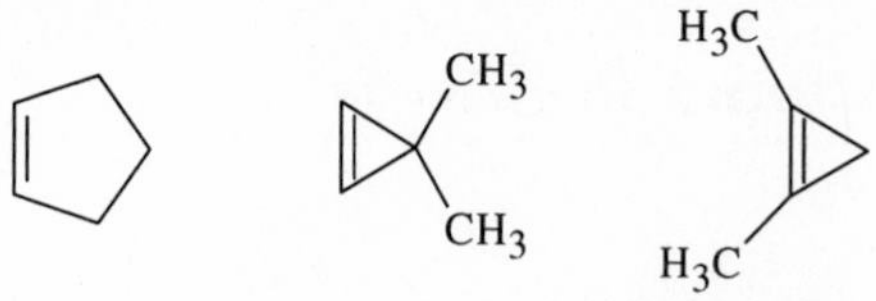

5.35 (a)

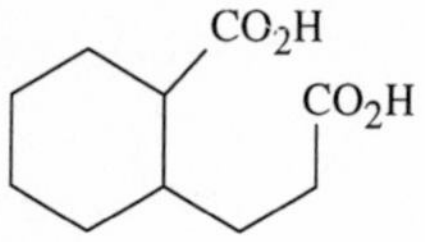

(c)

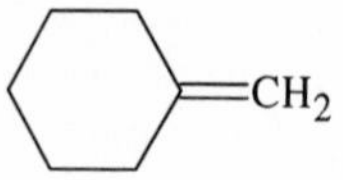

5.36 (a)

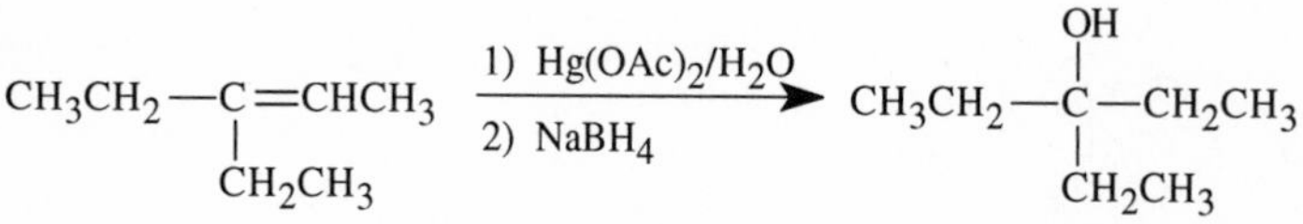

$$CH_3CH_2-\underset{\underset{CH_2CH_3}{|}}{C}=CHCH_3 \xrightarrow[\text{2) NaBH}_4]{\text{1) Hg(OAc)}_2/\text{H}_2\text{O}} CH_3CH_2-\underset{\underset{CH_2CH_3}{|}}{\overset{\overset{OH}{|}}{C}}-CH_2CH_3$$

(c)

$$H_2C=CHCH_2CH_2CH_2CH_3 \xrightarrow[\text{2) H}_2\text{O}_2/\text{NaOH}]{\text{1) BH}_3/\text{THF}} HOCH_2CH_2CH_2CH_2CH_2CH_3$$

(e)

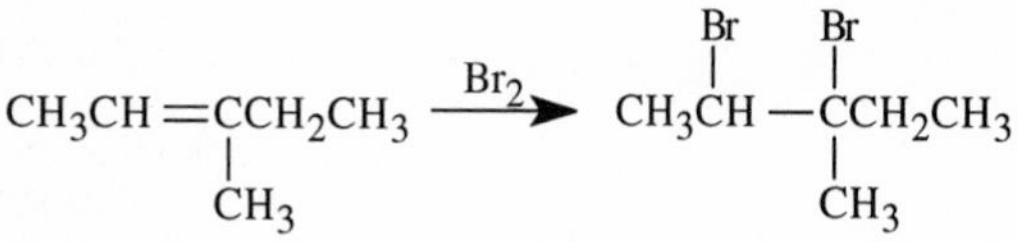

$$CH_3CH=\underset{\underset{CH_3}{|}}{C}CH_2CH_3 \xrightarrow{\text{Br}_2} CH_3\overset{\overset{Br}{|}}{C}H-\underset{\underset{CH_3}{|}}{\overset{\overset{Br}{|}}{C}}CH_2CH_3$$

(f)

$$CH_3CH_2\underset{\underset{CH_2}{||}}{C}CH_2CH_3 \xrightarrow[\text{2) NaHSO}_3/\text{H}_2\text{O}]{\text{1) OsO}_4} CH_3CH_2\underset{\underset{CH_2OH}{|}}{\overset{\overset{OH}{|}}{C}}CH_2CH_3$$

(h)

$\xrightarrow[\text{2) (CH}_3)_2\text{S}]{\text{1) O}_3}$ $H-\overset{\overset{O}{||}}{C}(CH_2)_4\overset{\overset{O}{||}}{C}-CH_3$

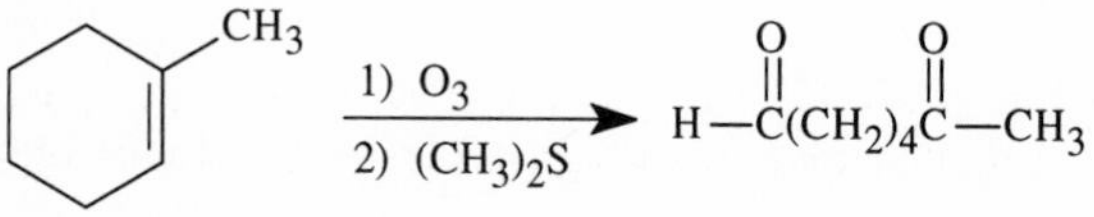

(j)

$$H_2C=CHCH_2CH_2CH_3 \xrightarrow{\text{HBr, peroxides}} BrCH_2CH_2CH_2CH_2CH_3$$

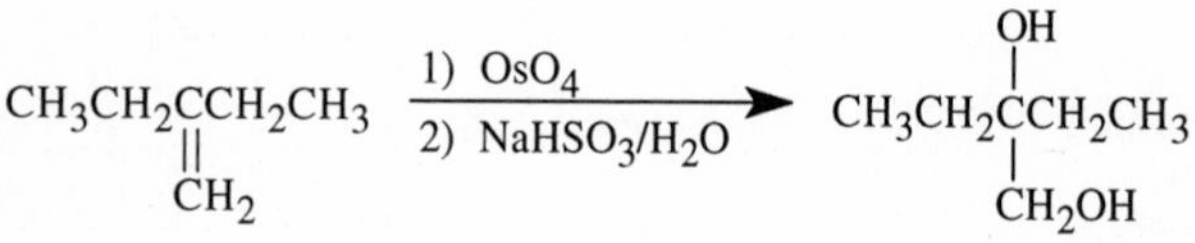

(l)

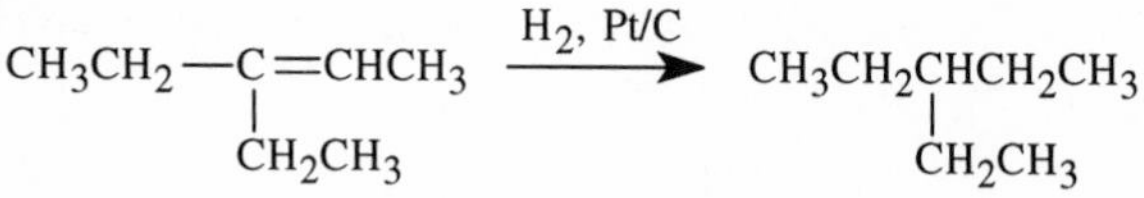

$$CH_3CH_2-\underset{\underset{CH_2CH_3}{|}}{C}=CHCH_3 \xrightarrow{\text{H}_2\text{, Pt/C}} CH_3CH_2\underset{\underset{CH_2CH_3}{|}}{CH}CH_2CH_3$$

5.37 (a)

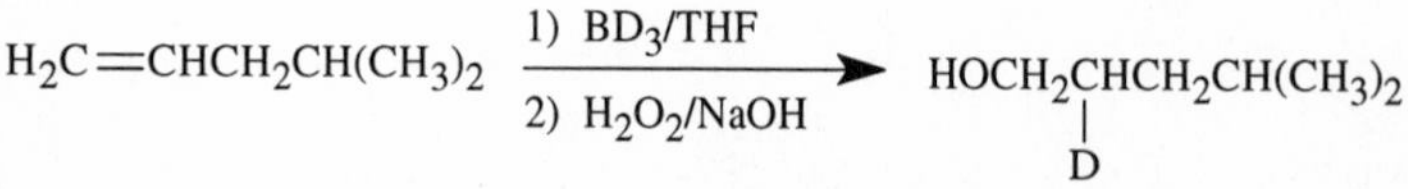

$$H_2C=CHCH_2CH(CH_3)_2 \xrightarrow[\text{2) H}_2\text{O}_2/\text{NaOH}]{\text{1) BD}_3/\text{THF}} HOCH_2\underset{\underset{D}{|}}{CH}CH_2CH(CH_3)_2$$

(b)

$$H_2C=CHCH_2CH(CH_3)_2 \xrightarrow[\text{2) NaBD}_4]{\text{1) Hg(OAc)}_2/\text{H}_2\text{O}} D-CH_2\underset{\underset{}{}}{\overset{\overset{OH}{|}}{CH}}CH_2CH(CH_3)_2$$

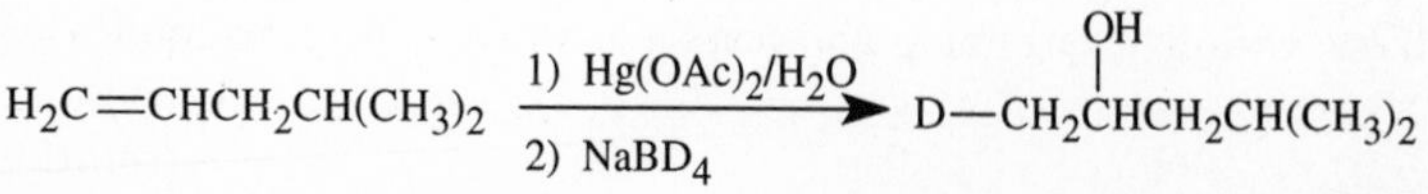

5.39 (a) In an inert solvent, the only nucleophile available to attack the bromonium ion intermediate is bromide ion. Hence, the product is the dibromoalkane:

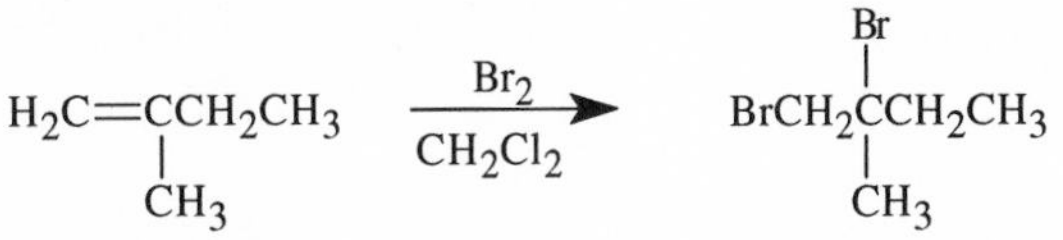

2-methyl-1-butene   1,2-dibromo-2-methylbutane

(b) When water is present, H_2O is available as a nucleophile to attack the bromonium ion as shown in Eq. 5.5a, text page 177. The product is the following bromohydrin:

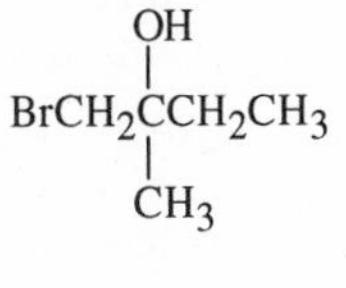

1-bromo-2-methyl-2-butanol

(c) When the bromination is carried out in methanol, then methanol, like water in part (b), is the nucleophile present in greatest concentration; consequently, it attacks the bromonium ion to give an ether:

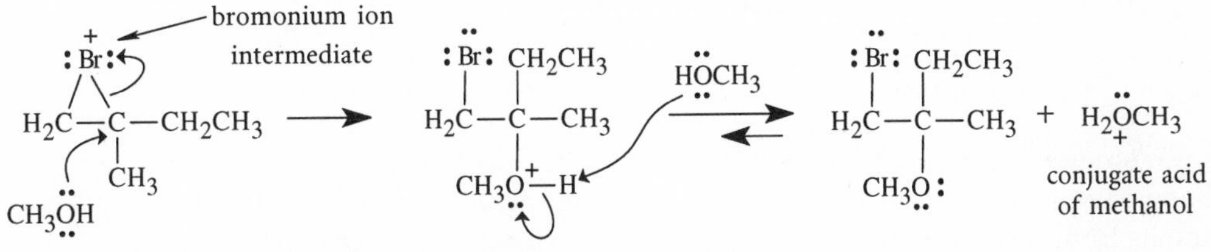

(d) When the solution contains both methanol and bromide ion, both nucleophiles compete for the bromide ion. Which one "wins" the competition depends on their intrinsic effectiveness as nucleophiles and on their relative concentrations. If the bromide concentration is high enough, the product will be largely the dibromoalkane shown in the solution to part (a). At lower bromide concentrations the product will be a mixture of this dibromoalkane and the ether shown in the solution to part (c).

5.41 (a) Only the first two products are formed in the absence of peroxides, and only the third is formed in the presence of peroxides. Different products are formed because different mechanisms and reactive intermediates are involved.

(b) The mechanism for the formation of the first two products is identical to that shown for reaction of the same alkene with HCl in Eqs. 4.26 and 4.27a–b, text page 152, except, of course, that HBr is used instead of HCl. The first product results from a carbocation rearrangement, and the second from normal regioselective ("Markownikoff") addition. The third product is the consequence of a free-radical addition mechanism, the propagation steps of which are as follows:

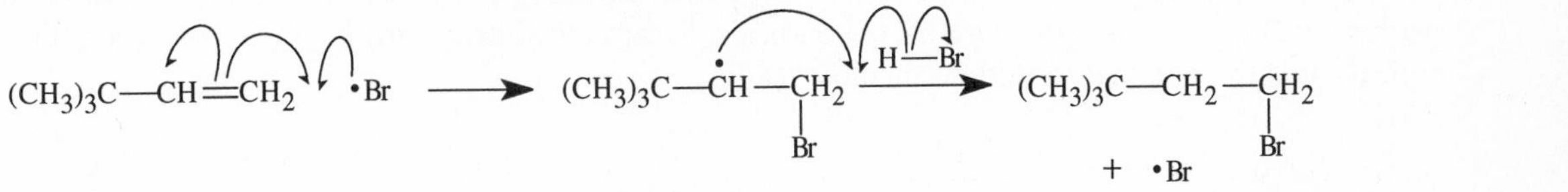

(The initiation steps are shown in Eqs. 5.43 and 5.45 on pp. 197–198 of the text.)

(c) Peroxide-promoted addition is in competition with normal addition and rearrangement, processes that should occur at the same rate at which they occur in the absence of peroxides. The fact that *only* the product of peroxide-promoted addition is observed, then, means that this process is much faster than the other, competing, processes.

5.42   (a)

(b) Because each carbon of the double bond has one alkyl group, there is no preference for one product over the other on the basis of free radical stability. However, steric effects (van der Waals repulsions) in the transition state are also important in free-radical HBr addition (see text pages 201–202). Addition of the bromine atom to the double bond at the carbon atom that is closer to the large *tert*-butyl group, which results in product *A*, generates significant van der Waals repulsions; however, addition of the bromine atom to the double bond at the carbon that is more remote from the *tert*-butyl group, which results in product *B*, generates less severe van der Waals repulsions. Consequently, product *B* is formed in greatest amount.

5.43   (a)   Reactive intermediate: Product: (c)   Reactive intermediate: Product:

a bromonium ion   a mercurinium ion

5.44   (a)   (Be sure to read Study Guide Link 5.4.) Because the CF₃ group has added to the carbon of the double bond with no alkyl branches, the radical •CF₃ must be the first one that adds to the alkene. The radical that results from this addition abstracts an iodine atom from CF₃I to give the product.

(b) Addition of •CF₃ to 2-methyl-1-pentene occurs at an *unbranched* carbon to give a *tertiary* free radical; addition of •CF₃ to (*E*)-4-methyl-2-pentene must occur at a *branched* carbon to give a *secondary* free radical. Because addition to 2-methyl-1-pentene gives the more stable free radical, and because there should be less severe van der Waals repulsions in the transition state, this addition should be the faster of the two.

5.45   (a)   Because carbon-carbon bonds are weaker than carbon-hydrogen bonds (see Table 5.3), a carbon-carbon bond should break more easily. Breaking of a bond to any of the methyl groups (pathway (1) below) gives a methyl radical and a tertiary free radical. Breaking of the central carbon-carbon bond (pathway (2) below) gives two tertiary free radicals. Because the latter process gives the more stable radicals, it is the one that should occur most readily.

more stable combination of radicals

(b) The faster a reaction, the lower the temperature at which it occurs at a given rate. The cracking of 2,2,3,3-tetramethylbutane is the faster reaction, and therefore occurs at lower temperature. The cracking of 2,2,3,3-tetramethylbutane (as shown in part (a)) is faster because it gives two tertiary radicals; the cracking of ethane, in contrast, gives two methyl radicals. The bond that breaks to give the radical with the greater number of alkyl branches is the weaker bond, and is therefore the bond that is more easily broken.

(c) The $\Delta H°$ for the reaction in part (a) is the difference between the heats of formation of products and reactants. Thus,

$$\Delta H° = 2\Delta H°_f(\cdot C(CH_3)_3) - \Delta H°_f(2,2,3,3\text{-tetramethylbutane})$$
$$= 2(48.5) - (-225.9) = 322.9 \text{ kJ/mol}$$

The $\Delta H°$ for the cracking of ethane is

$$\Delta H° = 2\Delta H°_f(\cdot CH_3) - \Delta H°_f(\text{ethane})$$
$$= 2(145.6) - (-84.7) = 375.9 \text{ kJ/mol}$$

This calculation shows that the cracking of ethane is 53 kJ/mol less favorable than the cracking of 2,2,3,3-tetramethylbutane. This demonstrates quantitatively the assertion in part (b), namely, that 2,2,3,3-tetramethylbutane undergoes cracking more readily. The underlying assumption is that the free energies of activation and enthalpies of the cracking reaction are similar.

5.46   (a) Ozonolysis breaks the polymer at its double bonds:

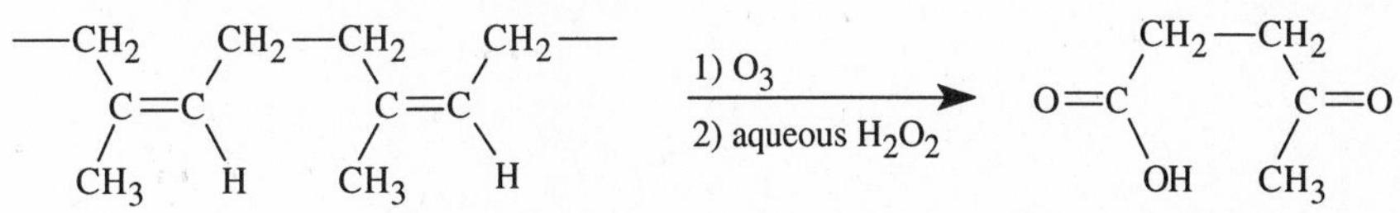

(b) Because E and Z isomers of alkenes give the same ozonolysis products, gutta-percha is evidently either the all-E stereoisomer of natural rubber or a polymer containing both E and Z alkene units. (In fact, it is the all-E stereoisomer.)

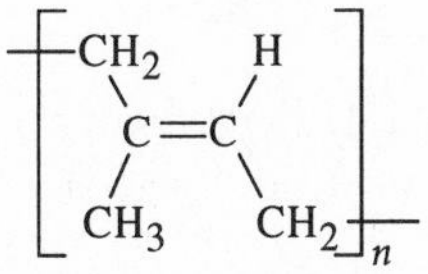

gutta-percha

5.47   (a) Because the H—CN bond is stronger than the O—H bond, the abstraction of a hydrogen atom from HCN by the *tert*-butoxy radical is endothermic by 519 − 435 = 84 kJ/mol. (These numbers are from Table 5.3 with the bond energy of CH_3O—H as an approximation for that of $(CH_3)_3CO$—H.) Because both initiation reactions are highly endothermic, the reaction is not likely to generate a high enough concentration of radicals to initiate a chain reaction.

(b) In this reaction, formation of a C—H bond releases 418 kJ/mol of energy; breaking the H—CN bond requires 519 kJ/mol of energy. Consequently, this step is endothermic by 101 kJ/mol. Free-radical chain reactions with highly endothermic propagation steps generally do not occur. (See text page 200.)

5.49   In the general reaction, the C—H bond of methane and the X—X bond of the halogen are broken; the C—X bond of the alkyl halide and the H—X bond of the hydrogen halide are formed. The overall energetics of the reactions are as follows, in kJ/mol:

	Breaking of C—H bond	Breaking of X—X bond	Formation of CH_3—X bond	Formation of H—X bond	Overall energy change ($\Delta H°$)
X = Cl	439	247	−356	−431	−101
X = Br	439	192	−293	−368	−30
X = I	439	151	−238	−297	+55

These calculations show that chlorination and bromination are exothermic, whereas iodination is endothermic. Consequently, iodination does not occur because it is energetically unfavorable.

(b) Since the iodination reaction is energetically unfavorable, its reverse is energetically favorable. Consequently, H—X (from the trace of acid) reacts with CH_3—I as follows:

$$H—X \;+\; CH_3—I \;\longrightarrow\; I—X \;+\; CH_4$$

If X = I, then I—X = I_2. If X is some other group, then the following exchange can occur:

$$2\,I—X \;\rightleftharpoons\; I_2 \;+\; X_2$$

In either case, iodine is formed.

5.50 (a) The structure of polystyrene:

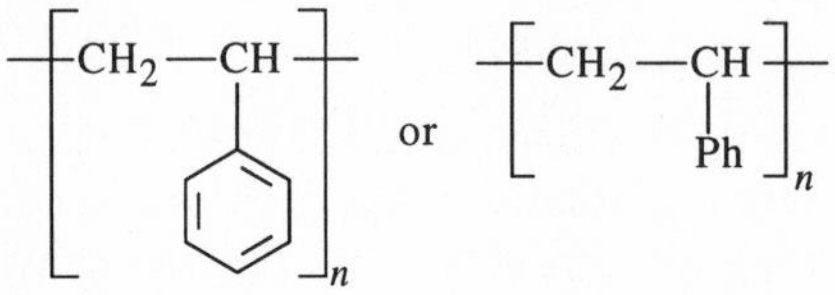

(b) Because both "ends" of 1,4-divinylbenzene can be involved in polymer formation, addition of 1,4-divinylbenzene serves to connect, or *crosslink,* polymer chains. Such a crosslink is shown with bolded bonds in the following structure:

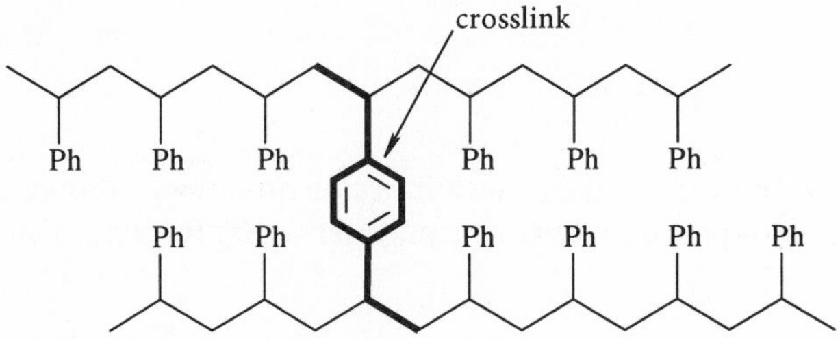

Notice that because only a small amount of 1,4-divinylbenzene is used, divinylbenzene does not polymerize with itself.

Crosslinks are introduced into polymers to increase their strength and rigidity.

5.51 A detailed discussion of the approach to solving this problem is found in Study Guide Link 5.5. Compound *A* is limonene, a natural product obtained from the oils of lemons and oranges.

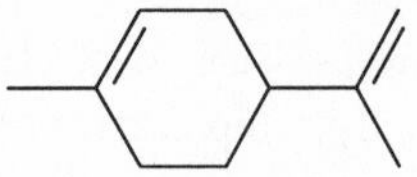

limonene
(compound *A*)

5.53 (a) Protonation of the double bond gives a tertiary carbocation which is attacked by one of the oxygens within the same molecule.

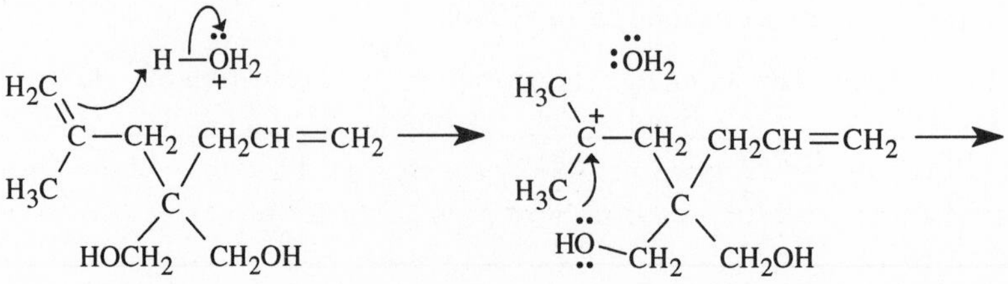

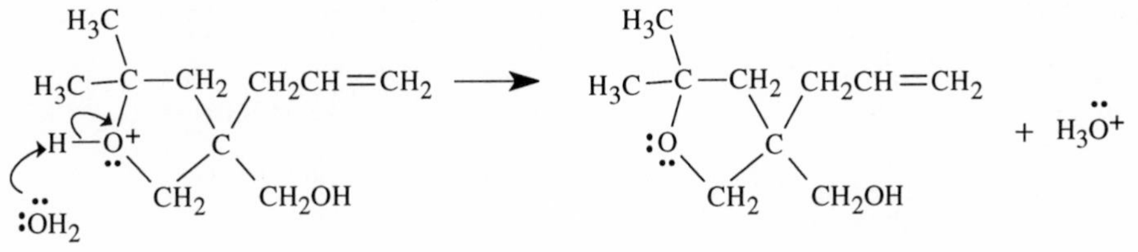

(c) The oxygen of a hydroxy group introduced in one oxymercuration reaction attacks a mercurinium ion within the same molecule. Two products are formed because the mercurinium ion can be attacked at either of two carbons.

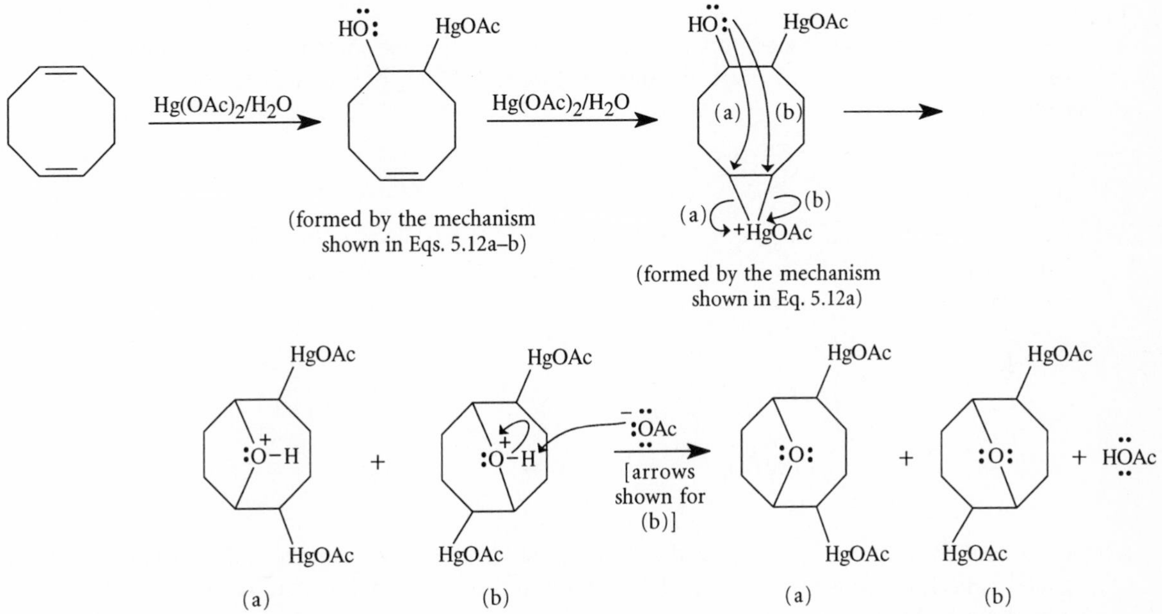

(formed by the mechanism shown in Eqs. 5.12a–b)

(formed by the mechanism shown in Eq. 5.12a)

[arrows shown for (b)]

(a)   (b)   (a)   (b)

Reaction with NaBH$_4$ results in the formation of the products shown in the problem by replacement of the —HgOAc groups with —H.

(d) In the initiation step, an ethylthio radical is formed by reaction of C$_2$H$_5$SH with a peroxy radical:

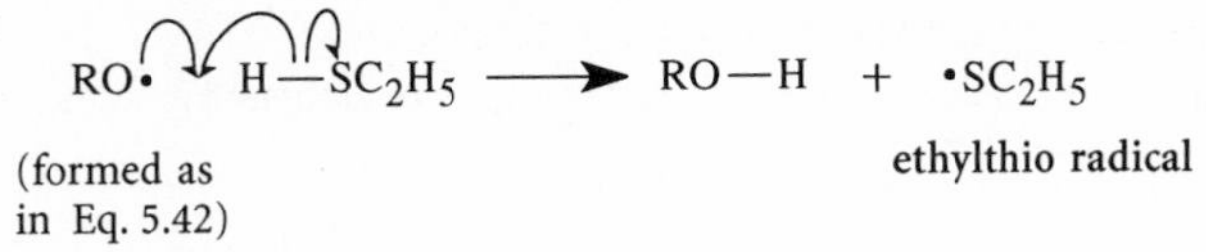

$$RO\cdot \quad H—SC_2H_5 \longrightarrow RO—H \; + \; \cdot SC_2H_5$$

(formed as in Eq. 5.42)   ethylthio radical

The propagation steps of the chain reaction are as follows:

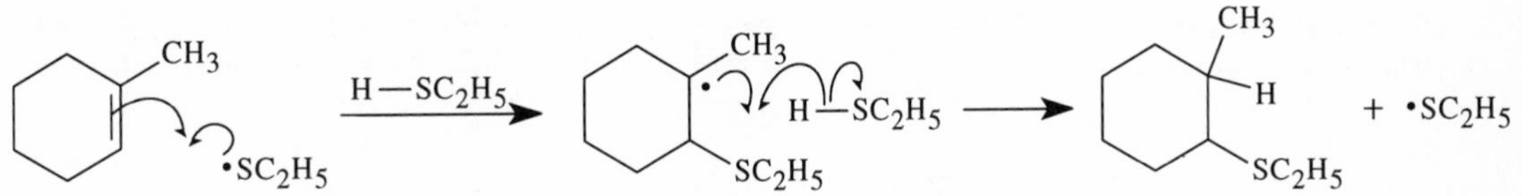

H—SC$_2$H$_5$

$\cdot$SC$_2$H$_5$

5.54   (a) Homolysis of the O—Cl bond followed by a β-scission reaction produces the C=O double bond and opens the ring. Abstraction of a chlorine atom from starting material by the resulting radical propagates the chain.

(solution continues)

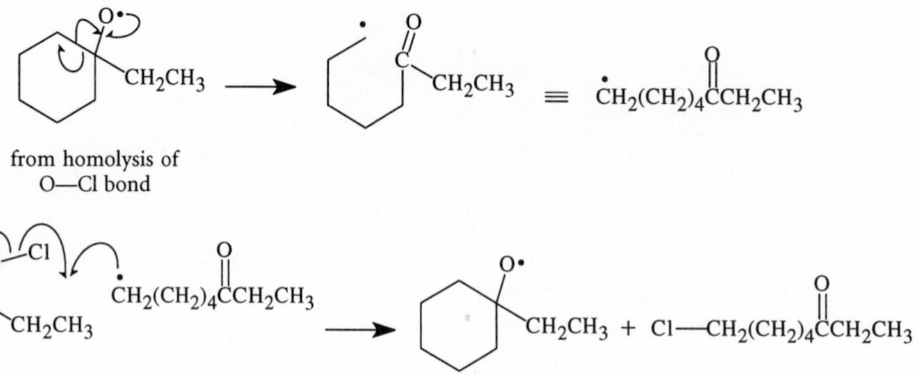

from homolysis of
O—Cl bond

$\equiv$ $\cdot CH_2(CH_2)_4\overset{O}{\overset{\|}{C}}CH_2CH_3$

$CH_2(CH_2)_4\overset{O}{\overset{\|}{C}}CH_2CH_3$

$CH_2CH_3 + Cl—CH_2(CH_2)_4\overset{O}{\overset{\|}{C}}CH_2CH_3$

(b) The minor product comes from a process exactly like the one shown in (a). The major product comes from a β-scission reaction in which the isopropyl radical rather than a ring-opened primary radical is produced:

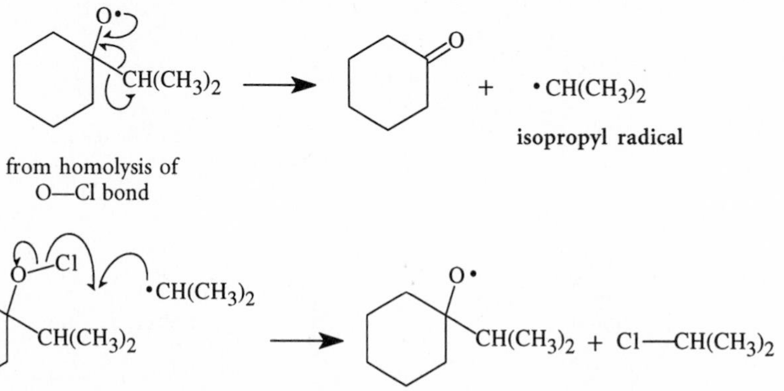

from homolysis of
O—Cl bond

$+$ $\cdot CH(CH_3)_2$

isopropyl radical

$CH(CH_3)_2 + Cl—CH(CH_3)_2$

In this case, formation of the isopropyl radical, which is *secondary,* competes with formation of the ring-opened radical, which is *primary.* The data in the problem show that formation of the secondary radical is favored, undoubtedly because it is more stable.

5.55   The starting alkene contains six carbons; acetone, the ozonolysis product of B, contains three carbons. Because acetone is the only product formed, it is reasonable to suppose that *two equivalents* of it are produced. The elemental analysis of B shows that it has the empirical formula CH_2; thus, B is evidently an alkene that contains one double bond. From the ozonolysis data, the structure of B is

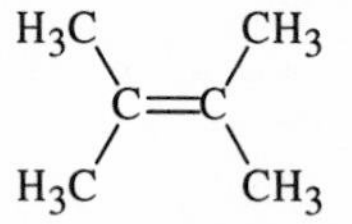

$$\underset{H_3C}{\overset{H_3C}{>}}C=C\underset{CH_3}{\overset{CH_3}{<}}$$ compound B

Comparison of the carbon connectivities of compound B and the alkene starting material shows that a rearrangement has occurred during the transformation of the alkene starting material to B. A source of the rearrangement is a reaction that involves carbocations; the HBr addition is such a reaction. Indeed, the reaction shown in Eq. 4.25, text page 151, with HBr instead of HCl, shows that the structure of compound A is likely to be

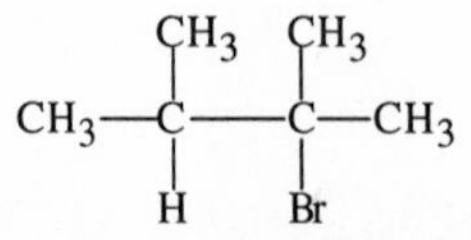

$$CH_3—\underset{H}{\overset{CH_3}{\underset{|}{\overset{|}{C}}}}—\underset{Br}{\overset{CH_3}{\underset{|}{\overset{|}{C}}}}—CH_3$$ compound A

This structure has the same carbon connectivity as compound B. In the reaction that converts A to B, the strong base in the problem removes a proton and a bromide ion is expelled as KBr. Given the structure of compound B shown above, the following analysis is reasonable:

$$CH_3-\underset{\underset{\displaystyle H}{|}}{\overset{\overset{\displaystyle CH_3}{|}}{C}}-\underset{\underset{\displaystyle Br}{|}}{\overset{\overset{\displaystyle CH_3}{|}}{C}}-CH_3$$

removed by the base ⟋   ⟍ lost as Br⁻

A curved-arrow mechanism consistent with this analysis is the following:

$$CH_3-\underset{\underset{\displaystyle H}{|}}{\overset{\overset{\displaystyle CH_3}{|}}{C}}-\underset{\underset{\displaystyle :\ddot{B}r:}{|}}{\overset{\overset{\displaystyle CH_3}{|}}{C}}-CH_3 \longrightarrow \underset{\displaystyle H_3C}{\overset{\displaystyle H_3C}{}}C=C\underset{\displaystyle CH_3}{\overset{\displaystyle CH_3}{}} + (CH_3)_3C-\ddot{O}-H + :\ddot{B}r:^-$$

$$(CH_3)_3C-\ddot{O}:^-$$

You may wonder why only the tertiary hydrogen is removed by the base. In fact, a competing reaction occurs in which one of the primary hydrogens is removed to give a second product:

$$(CH_3)_3C-\ddot{O}:^-$$

$$CH_3-\underset{\underset{\displaystyle H}{|}}{\overset{\overset{\displaystyle CH_3}{|}}{C}}-\underset{\underset{\displaystyle :\ddot{B}r:}{|}}{\overset{\overset{\displaystyle CH_3}{|}}{C}}-\overset{\overset{\displaystyle H}{|}}{\underset{\underset{\displaystyle }{|}}{C}}H_2 \longrightarrow H_3C-\underset{\underset{\displaystyle H}{|}}{\overset{\overset{\displaystyle H_3C}{|}}{C}}-\overset{\overset{\displaystyle CH_3}{|}}{C}=CH_2 + (CH_3)_3C-\ddot{O}-H + :\ddot{B}r:^-$$

The problem, however, focused on only one of the products for simplicity.

Introduction to Stereochemistry

Terms

Concepts

I. *Chirality and Symmetry*

A. SYMMETRY AND MIRROR IMAGES

1. The symmetry of any object can be described by certain symmetry elements, which are lines, points, or planes that relate equivalent parts of an object.

   a. A plane of symmetry is sometimes called an internal mirror plane.

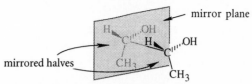

b. If a molecule has a plane of symmetry, it is not chiral.
2. Some molecules are congruent to their mirror images.
   a. The congruence of a molecule and its mirror image shows that they are identical.
   b. If the two mirror images are congruent, the molecule is achiral; if not, the molecule is chiral.

B. ASYMMETRIC CARBON

1. An asymmetric carbon is a carbon bonded to four different groups.
   a. An asymmetric atom is another type of stereocenter, or stereogenic atom.

$$\text{asymmetric atom} \searrow \underset{\underset{H}{\overset{Br}{\big|}}}{C} \swarrow \text{stereocenter}$$

Cl······C······CH$_3$

   b. All asymmetric carbons are stereocenters, but not all stereocenters are asymmetric carbons.
      i. A stereocenter is an atom at which the interchange of two groups gives a stereoisomer.
      ii. The carbon atoms involved in double bonds of *E,Z* isomers are also stereocenters.
2. A molecule that contains only one asymmetric carbon is chiral.
   a. Chiral molecules lack certain symmetry elements such as a plane of symmetry.
   b. Molecules that are not chiral are said to be achiral.
3. An asymmetric carbon is not a necessary condition for chirality; some chiral molecules contain no asymmetric atoms.

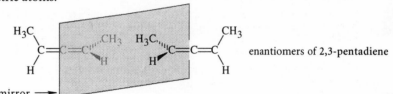

enantiomers of 2,3-pentadiene

4. A *meso* compound is a single achiral compound that contains asymmetric atoms.

$$\text{asymmetric atom} \longrightarrow \qquad \longleftarrow \text{asymmetric atom}$$

***meso*-2,4-dibromopentane**

   a. If a molecule contains *n* asymmetric carbons, then it has 2^n stereoisomers unless there are *meso* compounds.
   b. If there are *meso* compounds, then there are fewer than 2^n stereoisomers.

II. *Isomeric Relationships and Nomenclature*

A. ISOMERS

1. Isomers have the same molecular formula.
   a. Constitutional isomers have different atomic connectivities.
   b. Stereoisomers are molecules that have the same atomic connectivity but differ in the arrangement of their atoms in space.
      i. The study of stereoisomers and the chemical effects of stereoisomerism is called stereochemistry.
      ii. The *cis-trans* (or *E,Z*) isomers of alkenes are examples of stereoisomers.

2-ethyl-1-butene   ←— constitutional isomers —→   (*Z*)-3-methyl-2-pentene   ←— stereoisomers —→   (*E*)-3-methyl-2-pentene

2. Two types of stereoisomers are
   a. enantiomers, which are molecules that are related as object and noncongruent mirror image.
   b. diastereomers, which are stereoisomers that are not enantiomers and differ in all of their physical properties.

Two nonidentical molecules
↓
Same molecular formula? —— No —→ The molecules are *not* isomers.
↓ Yes
Same connectivity? —— No —→ The molecules are *constitutional isomers.*
↓ Yes
Mirror images? —— No —→ The molecules are *diastereomers.*
↓ Yes
The molecules are *enantiomers.*

B. ENANTIOMERS

1. Molecules that are noncongruent mirror images are called enantiomers.
   a. Enantiomers must not only be mirror images; they must also be noncongruent mirror images.
   b. Molecules that can exist as enantiomers are said to be chiral.

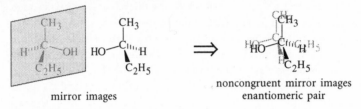

mirror images   noncongruent mirror images / enantiomeric pair

2. In order for a pair of chiral molecules with more than one asymmetric carbon to be enantiomers, they must have different configurations at every asymmetric carbon.

C. RACEMATES

1. A mixture containing equimolar amounts of two enantiomers is called a racemate or racemic mixture.
   a. Racemates typically have physical properties that are different from those of the pure enantiomers.
   b. The optical rotation of the racemate is zero; a racemate is optically inactive.
2. The process of forming a racemate from a pure enantiomer is called racemization.
3. The separation of a pair of enantiomers is called an enantiomeric resolution.
   a. In this process, a pair of enantiomers are allowed to react with a chiral resolving agent, a process that forms diastereomers, which, unlike enantiomers, have different physical properties.
   b. After separating the resulting diastereomers, the pure enantiomers are separated from the resolving agent.
4. Salt formation is such a convenient reaction that it is often used for the enantiomeric resolution of amines and carboxylic acids.

D. CONFORMATIONAL STEREOISOMERS

1. Enantiomers that are interconverted by a conformational change are called conformational enantiomers.
   a. The two *gauche* conformations of butane are examples of conformational enantiomers.

← conformational enantiomers →

2. Diastereomers that are interconverted by a conformational change are called conformational diastereomers.
   a. *Anti-* and either *gauche*-butane are conformational diastereomers.

← conformational diastereomers →

3. A compound which, under ordinary conditions, consists of rapidly interconverting enantiomers is considered to be achiral.
   a. Some amines undergo rapid interconversion of stereoisomers in a process called amine inversion.

← amine inversion →

   b. Amine inversion is another example of racemization.

E. NOMENCLATURE OF ENANTIOMERS: THE *R,S* SYSTEM

1. Cahn-Ingold-Prelog priority rules used to assign *E* and *Z* conformations to alkene stereoisomers are used to assign the stereochemical configuration, or arrangement of atoms, at each asymmetric carbon in a molecule in the following steps:
   a. Identify an asymmetric carbon and the four different groups bound to it.

four different groups
$$\begin{cases} D \\ H \\ CH_2CH_3 \\ CH_2CH_2CH_3 \end{cases}$$

   b. Assign priorities to the four different groups according to the rules given below. The convention used in the text is that the highest priority = 1.
      i. Assign higher priority to the atom of higher atomic number (or atomic mass in the case of isotopes) at the first point of difference.
      ii. If two or more of the attached atoms are identical, move away from the chiral atom within each group to the next atom following the path of highest priority and repeat part *i* above.
      iii. Multiple bonds are treated by a special convention, in which the multiple bond is rewritten as a single bond and the atoms at each end of the multiple bond are replicated.

   c. Arrange the attached atoms in descending priority order.
   d. View the molecule along the bond from the asymmetric carbon to the group of lowest priority, that is, with the asymmetric carbon nearer and the lowest priority group farther away.
   e. Consider the clockwise or counterclockwise order of the remaining group priorities.
      i. If the priorities of these groups decrease in the clockwise direction, the asymmetric carbon is said to have the *R* configuration.
      ii. If the priorities of these groups decrease in the counterclockwise direction, the asymmetric carbon is said to have the *S* configuration.

f. A stereoisomer is named by indicating the configuration of each asymmetric carbon before the systematic name of the compound.

(1*R*,2*S*,4*R*)-1-chloro-2,4-dimethylcyclohexane

2. Assigning an *R* or *S* configuration to every asymmetric carbon in a molecule specifies its absolute stereochemical configuration, or absolute stereochemistry.

III. *Physical Properties and Representations of Stereoisomers*

A. POLARIZED LIGHT AND OPTICAL ACTIVITY

1. A pair of enantiomers have identical physical properties except for their optical activities. The optical rotations of a pair of enantiomers have equal magnitudes but opposite signs.
   a. Plane-polarized light, or simply polarized light, is light with an electric field that oscillates in only one plane; polarized light is obtained by passing ordinary light through a polarizer.
   b. If plane-polarized light is passed through one enantiomer of a chiral substance, the plane of polarization of the emergent light is rotated.
2. A substance that rotates the plane of polarized light is said to be optically active.
   a. Individual enantiomers of chiral substances are optically active.
   b. Optical activity is measured in a device called a polarimeter.
      i. If the sample rotates the plane of polarized light in the clockwise direction, the optical rotation is given a plus sign (+) and the sample is said to be dextrorotatory.
      ii. If the sample rotates the plane of polarized light in the counterclockwise direction, the optical rotation is given a minus sign (−) and the sample is said to be levorotatory.
   c. The number of degrees α that the analyzer must be turned to measure the angle of rotation is the optical rotation of the sample.
      i. The observed optical rotation α is proportional to both the concentration *c* (in grams per milliliter) of the sample and the length *l* (in decimeters) of the sample container; the dependence of optical activity on concentration is sometimes called Biot's law.

$$\alpha = [\alpha]\, c\, l \qquad \text{Biot's law}$$

      ii. The specific rotation $[\alpha]$ is used as the standard measure of optical activity and is a constant for a particular compound; it is also a function of temperature and wavelength.

specific rotation $\longrightarrow [\alpha]_D^{20}$

3. There is no simple relationship between the sign of optical rotation and absolute configuration.
4. A sample of a pure chiral compound uncontaminated by its enantiomer is said to be enantiomerically pure.
5. Optical activity and chirality formed the logical foundation for the postulate of tetrahedral bonding geometry at carbon.

B. STEREOCHEMICAL CORRELATION

1. The absolute configurations of most organic compounds are determined experimentally by using chemical reactions to correlate them with other compounds of known absolute configurations in a process called stereochemical correlation.
2. The most secure way of relating absolute configurations is to use reactions that do not break the bonds at the asymmetric carbon.

C. FISCHER PROJECTIONS

1. The structures of chiral compounds can be drawn in planar representations called Fischer projections.
   a. All vertical bonds are assumed to be oriented away from the observer.
   b. All horizontal bonds are assumed to be oriented toward the observer.

c. The asymmetric carbons themselves are not drawn, but are assumed to be located at the intersections of vertical and horizontal bonds.

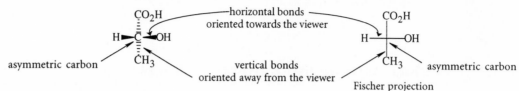

2. Several valid Fischer projections can be drawn for most chiral molecules which are derived by the folowing rules:

a. A Fischer projection may be turned 180° in the plane of the paper

equivalent Fischer projections

b. A Fischer projection may not be turned 90°

not equivalent Fischer projections

c. A Fischer projection may not be lifted from the plane of the paper and turned over

not equivalent Fischer projections

d. The three groups at either end of a Fischer projection may be interchanged in a cyclic permutation; that is, all three groups can be moved at the same time in a closed loop so that each occupies an adjacent position

all are equivalent Fischer projections

3. It is particularly easy to recognize enantiomers and *meso* compounds from the appropriate Fischer projections, because planes of symmetry in the actual molecules reduce to lines of symmetry in their projections.

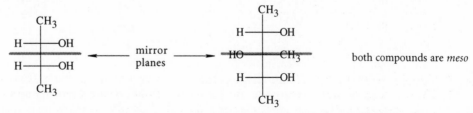

both compounds are *meso*

Study Guide Links

✓6.1   Finding Asymmetric Carbons in Rings

Finding asymmetric carbons in rings requires that each arm of a ring be treated as a separate group. To illustrate, consider the carbon of the structure below that is indicated with an asterisk.

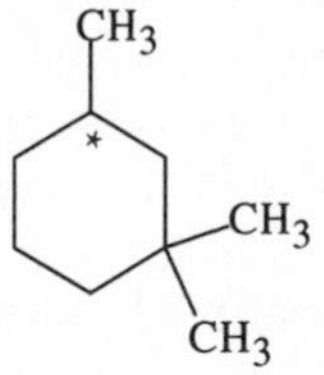

This carbon bears a CH_3 group and a hydrogen. (The hydrogen is not shown explicitly, because this is a skeletal structure). The other two groups are the two "arms" of the ring. *Even though the two groups are tied together into a ring, they can be considered as separate groups.* Proceeding counterclockwise from the asterisked carbon, the connectivity sequence is —CH_2—CH_2—; proceeding clockwise, the connectivity sequence is —CH_2—$C(CH_3)_2$—; a difference occurs at the second carbon out. Hence, all the groups on the asterisked carbon are different, and thus this carbon is an asymmetric carbon. Notice that we continue as far as necessary along each branch to find a difference.

The following structure, in contrast, has no asymmetric carbon:

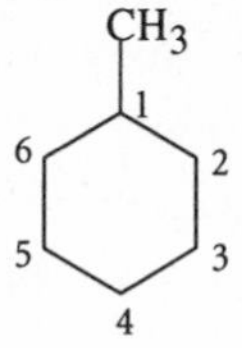

For example, carbon-1 bears a methyl, a hydrogen, and the two "arms" of the ring. Proceeding around the ring from carbon-1, at carbon-2 and carbon-6 the arms are the same (—CH_2— groups); at carbon-3 and carbon-5 the arms are the same (again, —CH_2— groups); and the two arms join at carbon-4. If no difference is found when proceeding from the carbon in question to the point at which the arms of the ring join, the two arms of the ring are identical.

✓6.2   Stereocenters and Asymmetric Carbons

Stereochemical terminology changes as chemists acquire a more rigorous understanding of the geometrical aspects of stereochemistry. The purpose of this section is to alert you to a problem that still exists with the term *stereocenter*. The definition used in this text—"an atom at which interchange of two groups gives a stereoisomer"—is the one accepted by all experts in the field of stereochemistry. A number of texts, however, have defined *stereocenter* to be an atom with four different groups attached. This is the same definition that is used in this text for *asymmetric carbon*. Thus, some texts do not distinguish between the terms *asymmetric carbon* and *stereocenter*. As shown in the text, however, all symmetric carbons are stereocenters, but the converse is not true: not all stereocenters are asymmetric carbons. The cases discussed in Sec. 4.1B of the text demonstrate this point: the carbons of the double bond in *E* and *Z* isomers are stereocenters but are not asymmetric carbons. Because so many texts misdefine this term, you should be alert to the fact that you may see it misused.

A number of chemists have objected to the use of the term *asymmetric carbon* because a carbon atom, of course, is not "asymmetric." In fact, many chiral *molecules* themselves are not "asymmetric;" they do have certain types of symmetry. Perhaps a better term for asymmetric carbon would be *asymmetrically substituted carbon*. Other terms you might hear are "chiral center" and "chiral carbon," but objections have been raised to these also. The term *asymmetric carbon* has been so widely used that everyone knows what it means, and this text has continued to use it. The important point is that, whatever we call it, it is *not* the same as a stereocenter.

√6.3 Using Perspective Structures

The most foolproof way to assign absolute configurations from a perspective structure (that is, a structure in which the bonds are represented as lines, wedges, and dashed wedges) is to build a model and use it to assign configurations. However, once you gain experience in doing this, you will probably be able to use perspective structures themselves to assign configurations.

Before trying to use perspective structures, be sure you understand how to interpret them. This was discussed in Chapter 1, p. 15, of the text. To review:

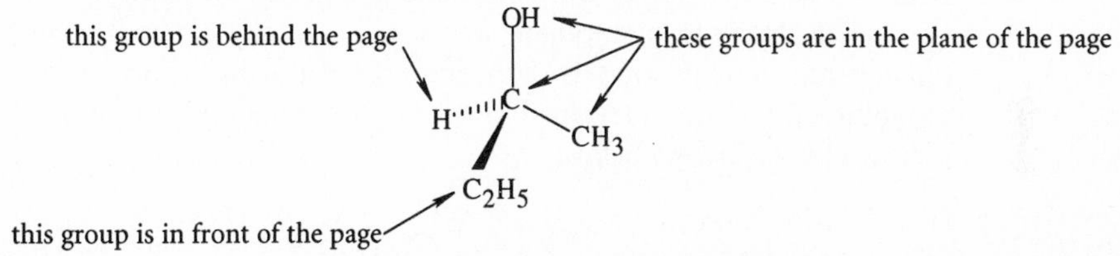

The groups with ordinary bonds, along with the central carbon, are in the plane of the page. The group connected to the central carbon by the solid wedge is in front of the page. The group connected to the central carbon by the dashed wedge is behind the page.

If you use journals, monographs, or other texts, you will find that there are variations on the conventions used for representing perspective. In some cases, solid bonds rather than solid wedges are used; dashed bonds rather than dashed wedges are used. Thus, all of the following structures mean the same thing:

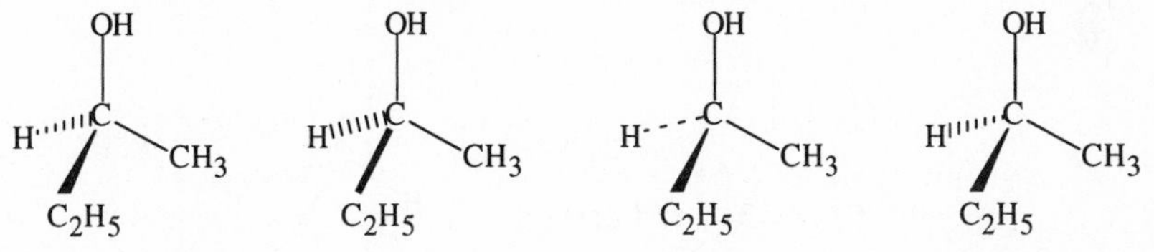

Notice that the last structure is the same as the first except for the orientation of the dashed wedge. Although this last convention is widely used, this text uses a convention in which the thick end of the dashed wedge is nearest the observer, because it more accurately conveys perspective.

If you have a choice in the matter, it is probably easiest to assign the configuration of an asymmetric carbon using perspective formulas if you place the group of lowest priority behind the page—that is, on a dashed wedge. The structure above has been drawn in this manner. When you do this, you are automatically viewing the molecule in the proper manner for assigning configuration:

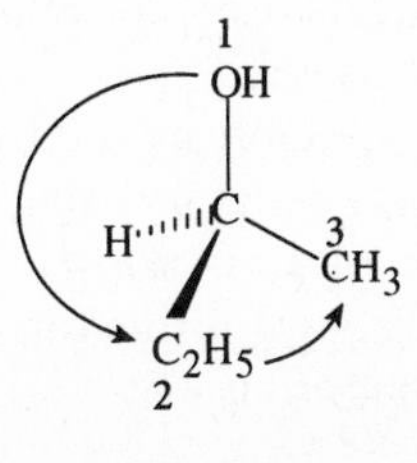

The H is farthest from the observer, and we are almost looking down the C—H bond. Of course, it doesn't matter whether the dashed bond is slanted to the right (as in the text, Sample Problem 6.2) or to the left, as above.

If you are confronted with a structure that is *not* drawn in this standard manner, and if you need to assign a configuration, then either you have to build a model, or you have to learn to manipulate the perspective structure mentally. Let's consider these two situations in turn.

If you build a model, keep your model as simple as possible. *Never build more of a model than necessary to solve the problem at hand.* (For example, if a molecule contains a methyl group and you don't have to deal explicitly with the hydrogens, then simply use a carbon.) When you are asked to assign configuration, a handy device is to dedicate four atoms of your model set for this purpose. Take an atom and paint a "1" on it with "white-out" (the stuff used for covering typing errors). Paint a "2" on another atom; a "3" on another; and a "4" on another. Then, when you see a perspective formula you can't interpret directly, first assign relative priorities to the groups in the structure. Then build a model using a carbon and only your four labeled atoms. Just make sure they go in the same positions of relative priority as the groups of the structure you are trying to interpret. Then manipulate your model so that group "4" is farthest away from you, and assign the configuration.

The ability to manipulate structures mentally generally comes by working problems in stereochemistry with models. However, the following short exercise might help you develop this ability. Start with a tetrahedral model, and imagine turning it 120° in the plane of the page as shown by the arrows below:

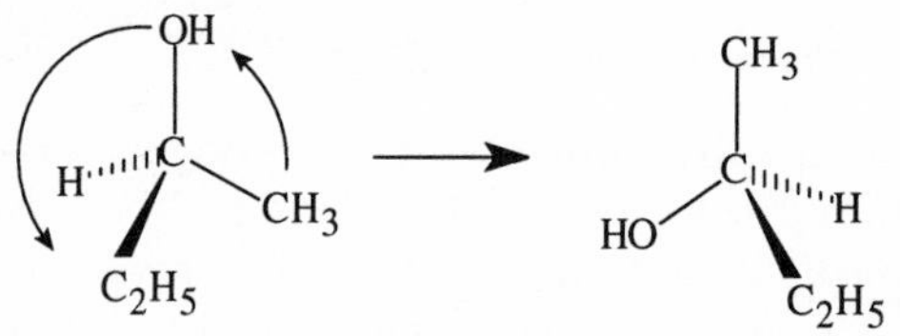

Now imagine spinning it about each bond in turn so that each group moves into the adjacent position, as follows:

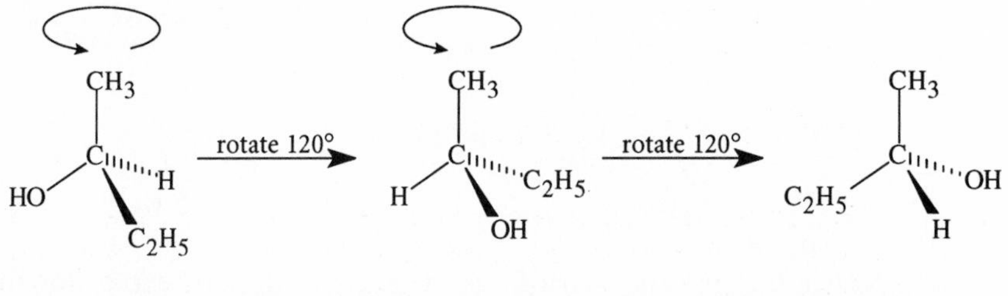

Then practice turning the structure about the other bonds and draw the result. Use a model to check yourself. If you spend a few minutes carrying out these or similar exercises, you should quickly become more proficient in visualizing perspective structures in three dimensions.

6.4 Terminology of Racemates

In earlier literature, the term *racemate* referred only to the *solid state* of a racemic chiral compound in which a single crystal consists of equal amounts of the two enantiomers. The term has been broadened by conventional usage (or some would say mis-usage) to include liquids and solutions as well.

Two other commonly used terms also originated from solid-state chemistry. In a *racemic compound,* a racemate exists in a single crystal form that contains equal numbers of molecules of each enantiomer. In a *racemic mixture,* the solid is a mixture of two crystal types, one for each enantiomer. The two forms can be distinguished by an analysis of their melting behavior as a function of composition.

Nowadays the terms *racemic mixture* and *racemate* are often used interchangeably to indicate *any* equimolar mixture of enantiomers.

6.5   Center of Symmetry

The manipulation shown at the bottom of page 246 in the text demonstrates that *meso*-2,3-butanediol is congruent to its mirror image, and is therefore achiral. You may have noticed that the conformation of 2,3-butanediol used in this equation has *no internal mirror plane.*

How can a molecule be achiral if it has no internal mirror plane? This molecule has one of the other chirality-excluding symmetry elements that are less frequently encountered. The symmetry element in this molecule is called a **center of symmetry.** A center of symmetry is a point in the exact center of an object. More precisely, a center of symmetry is a point such that *all* straight lines drawn through the point touch equivalent and indistinguishable parts of the objects at identical distances. Thus, the two oxygens lie on a line (heavy line) through the center of symmetry (heavy dot) and are equidistant from the center.

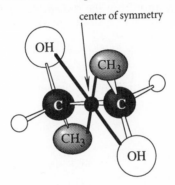

The two methyl groups are related in the same way, as are the hydrogens, the two asymmetric carbons, or any other equivalent elements of the molecule. The *anti* conformation of butane is another molecule that owes its absence of chirality to a center of symmetry.

Except for planes of symmetry, centers of symmetry are the most frequently occurring symmetry elements in achiral molecules.

6.6   Conformational Enantiomers

Although butane undergoes internal rotation too rapidly for its enantiomeric *gauche* forms to be isolated, chemists have succeeded in preparing other compounds in which internal rotation about a single bond is so slow that conformational enantiomers can be isolated and purified. The following compound is an example:

conformational enantiomers
can be isolated at room temperature.

The two benzene rings are forced to occupy perpendicular planes; if they were to occupy the same plane, substituents on one ring would "bump into" those on the other, and very severe van der Waals repulsions would result. The conformation of the compound shown above is chiral. (You can verify its chirality by drawing or constructing its mirror image and showing that it is noncongruent.) Interconversion of the two enantiomeric conformations requires a rotation about the central bond, indicated by the circular arrow, which takes the rings into the same plane. Because of the van der Waals repulsions just noted, this process requires such high energy that it cannot occur at room temperature.

In contrast, the compound in which the —CO_2H groups are replaced by hydrogens and the chlorine substituents are replaced by fluorines *cannot* be separated into its enantiomers at room temperature. In this case, racemization requires a planar conformation in which two small atoms (F and H) are close. Because van der Waals repulsions are much smaller here, this process requires less energy, and racemization occurs readily at room temperature.

✓6.7 Additional Manipulations of Fischer Projections

Two other "tricks" can prove useful in handling Fischer projections. The first has to do with assignment of configuration. If the lowest-priority group is in one of the vertical positions, the configuration can be assigned directly, as shown in the text. But what if you are dealing with a Fischer projection in which the lowest-priority group is in a horizontal position? Of course, you could re-draw the molecule in a new Fischer projection, or you could simply build a model and assign configuration from the model. But these procedures are too time-consuming; there is an easier way:

If the lowest-priority group is in a horizontal position, simply assign the configuration using the other three groups—but, since the assignment will be wrong, reverse it!

Let's illustrate with the following Fischer projection by assigning the configuration at the carbon marked with an asterisk:

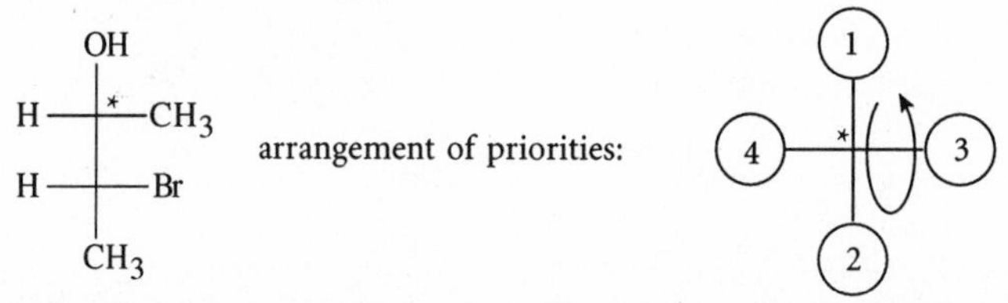

Groups 1, 2, and 3 lie in a counterclockwise pattern. Normally, this would indicate an *S* configuration. But since the viewing convention has been violated, this is not correct. Hence, the configuration is *R*.

Since there are only two possible outcomes, knowing the wrong answer is as good as knowing the correct one. (The author tells his students that this is one of the few situations in which two wrongs make a right.)

The second "trick" stems from the definition of a stereocenter. By definition, the structure of a stereoisomer can be generated by interchanging two groups at a stereocenter. This works as well in Fischer projections as it does in three-dimensional models. To illustrate, let's generate the enantiomer of the compound shown above. In an enantiomer, the configuration of *every* asymmetric carbon must be reversed. To do this, simply interchange any two groups at *each* asymmetric carbon:

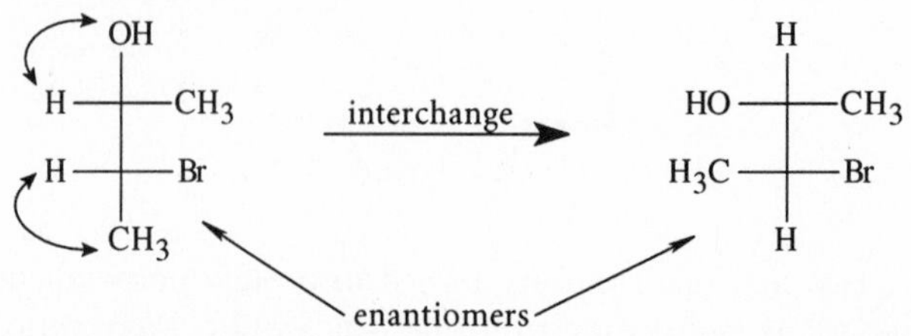

These two Fischer projections certainly do not appear to be mirror images. Nevertheless, they are Fischer projections of enantiomers. (Try to perform allowed manipulations of the second Fischer projection so that it is a mirror image of the first.)

Solutions

Solutions to In-Text Problems

6.1 (a) Chiral   (c) Chiral

6.2 (a) Chiral.
(b) Achiral (neglecting the writing).
(c) Achiral, assuming that it is sharpened with a symmetrical point and that there is no writing on the side.
(d) Achiral (neglecting the internal organs and any superficial characteristics).
(e) Achiral.
(f) Chiral. Actually, the author, a right-hander, didn't know this until a reviewer, a southpaw, commented on how badly the scissor industry discriminates against left-handers.

6.3 (a) A plane of symmetry in methane is a plane containing any two C—H bonds and bisecting the other two. (See illustration below.)
(b) A plane of symmetry in a sphere is any plane containing the center of the sphere.
(c) A plane of symmetry is any plane containing the tip of the cone and a diameter of the base. (See illustration below)
(d) A plane of symmetry is any plane containing the tip and the center of the base.
(e) A plane of symmetry is the plane containing the two C—Cl bonds and bisecting the angle between the two C—H bonds (see illustration below); or any plane containing the two C—H bonds and bisecting the angle between the two C—Cl bonds

(a)   (c)   (e)

6.4 The asymmetric carbons are identified with an asterisk (*).

(a)   (c)

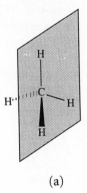

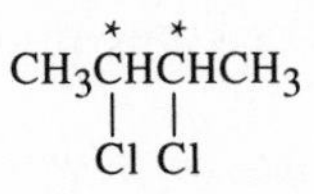

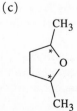

6.5 (a)   (c)

6.6 (a) The asymmetric carbon in the given stereoisomer of malic acid has the *S* configuration.
(c) The asymmetric atom is a nitrogen, and it has the *S* configuration.

6.7 Halve the concentration. If the rotation at the first concentration is +10°, the rotation at the lower concentration should read +5°; if the rotation at the first concentration is –350°, the rotation at the lower concentration should be –175°, which is the same as +185°; if the rotation at the higher concentration is +370°, the rotation at the lower concentration should be +185°. To decide between the last two possibilities, halve the concentration again.

6.8 (a) A concentration of 0.1 *M* is the same as a concentration of (0.1 mol/L)(150 g/mol)(0.001 L/mL) = 0.015 g/mL. Using this number for *c* in Eq. 6.1,

$$[\alpha] = \frac{\alpha}{cl} = \frac{+0.20 \text{ deg}}{(0.015 \text{ g mL}^{-1})(1 \text{ dm})} = +13.3 \text{ deg mL g}^{-1} \text{ dm}^{-1}$$

or $[\alpha] = +13.3°$

(b) Because the two enantiomers are present in equal amounts, their rotations cancel, and the observed rotation is zero.
(c) Because the concentration is halved, the observed rotation is also halved, that is, is 0.10°.
(d) The specific rotation is the same, because by definition this parameter is independent of concentration.
(e) The specific rotation of *L* is –13.3°; enantiomers have specific rotations of equal magnitudes and opposite signs.
(f) In this solution, the 0.005 mole of *L* cancels the rotation of 0.005 mole of *D*. The resulting rotation is as if only 0.005 mole, or a concentration of 0.05 mol/L, of *D* were present. This amount of *D* corresponds to a concentration of (0.05 mol/L)(150 g/mol)(0.001 L/mL) = 0.0075 g/mL. Since this concentration is half of the original concentration, the observed rotation is also half of the original, that is, 0.10°.

6.10 Begin with the perspective structure of the reactant; in the product, corresponding groups are in corresponding positions, because the reaction breaks none of the bonds to the asymmetric carbon.

(S)-(+)-enantiomer

(–)-enantiomer;
has the *R* configuration

This analysis shows that the (–)-enantiomer of the product has the *R* configuration. Notice that the "change of configuration" caused by this reaction is merely an artifact of the relative group priorities.

6.12 (a) In order for a molecule to have a *meso* stereoisomer, it must have more than one asymmetric atom, and it must be divisible into constitutionally identical halves (that is, halves that have the same connectivity relative to the dividing line). By these criteria, compound (a) does possess a *meso* stereoisomer.

$$CH_3CH-\overset{\overset{\displaystyle H}{|}}{\underset{\underset{\displaystyle H}{|}}{C}}-CHCH_3$$

(with Cl above each CH₃CH and H above the central carbon, and a vertical dashed dividing line through the center)

Notice that the dividing line may pass through one or more atoms.

(c) *Trans*-2-hexene cannot exist as a *meso* stereoisomer because it has no asymmetric atoms.

6.13 As suggested by the hint, the internal mirror plane passes through the central carbon atom, the hydrogen, and the OH group. This carbon can have either of two configurations. The only requirement for a *meso* compound is that the asymmetric carbons at the end of the structure must have opposite configurations.

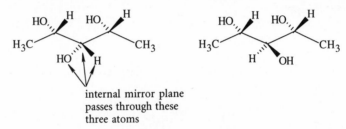

internal mirror plane
passes through these
three atoms

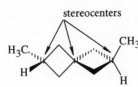 Stereocenters in the center of *meso* compounds, such as the central carbons of the structures above, are sometimes termed *pseudoasymmetric centers* because the only difference between two of the groups attached to such a carbon is their stereochemical configurations. Although there is a configurational notation for such carbons, it is not discussed in the text.

6.14 A resolving agent must be *chiral* and it must be *enantiomerically pure*. Only *A* meets both of these criteria.

6.16 The compounds in (a) and (c) are both chiral. The molecule in (a) has only one stereocenter, which is the same as its asymmetric carbon. The molecule in (c) has three stereocenters, only one of which is an asymmetric carbon.

(a)

(structure with label: asymmetric carbon and a stereocenter)

(c)

(structure with labels: stereocenter and asymmetric carbon; stereocenters)

(e) This compound is chiral and has three stereocenters, none of which is an asymmetric carbon.

stereocenters

(ring structure with H₃C and CH₃ substituents, H atoms)

This example shows that a tetrahedral stereocenter is *not* the same thing as an asymmetric carbon.

6.17   (a) In *anti* butane, the two —CH₂— carbons are stereocenters, because interchange of a CH₃ group and an H gives a stereoisomer (the *gauche* conformation).

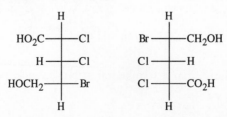

anti butane   *gauche* butane

   As the term *stereocenter* is normally used, it is not applied to individual, rapidly interconverting conformations. Consequently, butane has no stereocenters. However, under conditions of very low temperature at which the conformations of butane are stable, or (more practically) for molecules that consist of conformations that do not interconvert rapidly, the term can be applied.

6.18   The three conformations of *meso*-2,3-butanediol:

Conformation *A* is achiral; it is congruent to its mirror image. (This was demonstrated on text page 246.) Conformations *B* and *C* are enantiomers. This is not obvious from the way they are drawn above, but if you will make a model of each, you should be able to see that they are noncongruent mirror images.

(b) Because its conformations interconvert rapidly, *meso*-2,3-butanediol cannot be optically active. As the text indicates, molecules that consist of rapidly interconverting enantiomers are said to be achiral. However, at very low temperatures, conformations *B* and *C* could in principle be isolated; each would be optically active, and the two conformations would have rotations of equal magnitudes and opposite signs.

6.19   (a) The carbon stereocenter is fixed in the *S* configuration, but the nitrogen stereocenter undergoes inversion between *R* and *S* configurations. Consequently, the molecule is a mixture of C(*S*),N(*S*) and C(*S*),N(*R*) conformations, that is, a rapidly equilibrating mixture of diastereomers.

(b) Each of the rapidly interconverting species in part (a) has an enantiomer: C(*S*),N(*S*) has an enantiomer C(*R*),N(*R*), and C(*S*),N(*R*) has an enantiomer C(*R*),N(*S*). Because the inversion of the amine does not affect the carbon stereocenter, it would be possible to resolve the racemate of this compound into enantiomeric sets of rapidly interconverting diastereomers. The compound in part (a) itself *is* one such set.

6.20   (a) Many Fischer projections are possible for a compound with three asymmetric carbons. The two below are related by a 180° rotation in the plane of the page.

If yours don't look like these, try to use the allowed manipulations of Fischer projections to see whether yours are equivalent. (There are eighteen possible correct Fischer projections!)

6.21 (a) The way to solve a problem of this sort is to manipulate one of the Fischer projections so that two groups in one projection are in the same positions as the corresponding groups in the other; then the relationship between the two molecules should be clear. Thus, manipulate the projection on the right so that the OH groups are in the vertical positions. It then becomes apparent that this structure and the one on the left in the problem are enantiomers.

$$
\begin{array}{c}
CH_3 \\
HO_2C-\!\!\!-OH \\
HO-\!\!\!-CO_2H \\
CH_3
\end{array}
\quad\xrightarrow[\text{permutations}]{\text{two cyclic}}\quad
\begin{array}{c}
OH \\
H_3C-\!\!\!-CO_2H \\
HO_2C-\!\!\!-CH_3 \\
OH
\end{array}
\qquad
\begin{array}{c}
OH \\
HO_2C-\!\!\!-CH_3 \\
H_3C-\!\!\!-CO_2H \\
OH
\end{array}
\quad\leftarrow\;\;
\begin{array}{l}\text{left structure}\\\text{in problem}\end{array}
$$

$$\underbrace{\qquad\qquad\qquad\qquad\qquad\qquad}_{\text{enantiomers}}$$

6.22 (a) *R*
(c) The upper asymmetric carbon has the *R* configuration; the lower one has the *S* configuration. (Notice that this is a *meso* compound; hence, once you determine the configuration of one asymmetric carbon, the other must have the opposite configuration.)

6.23 Pasteur isolated the chiral stereoisomers, that is, the (2R,3R) and the (2S,3S) stereoisomers. (Because they are optically active, they must be chiral.) Yet to be isolated was the *meso* stereoisomer. Like all *meso* compounds, *meso*-tartaric acid is achiral and therefore is optically inactive.

$$
\begin{array}{c}
CO_2H \\
H-\!\!\!-OH \\
HO-\!\!\!-H \\
CO_2H
\end{array}
\qquad
\begin{array}{c}
CO_2H \\
HO-\!\!\!-H \\
H-\!\!\!-OH \\
CO_2H
\end{array}
\qquad
\begin{array}{c}
CO_2H \\
H-\!\!\!-OH \\
H-\!\!\!-OH \\
CO_2H
\end{array}
$$

$$\underbrace{\qquad\qquad\qquad\qquad}_{\text{discovered by Pasteur}}\qquad\qquad \text{\textit{meso}-tartaric acid}$$

Solutions to Additional Problems

6.24 (a)   (c)

In part (c), the two carbons of the double bond are technically stereocenters, but, because a *trans*-cyclohexene is too unstable to exist, the fact that they are stereocenters has no practical significance.

6.25 (a) First, draw the structure of 3,4,5,6-tetramethyl-4-octene.

$$
\begin{array}{c}
\quad\;\; CH_3\; CH_3 \\
\quad\;\;\; |\quad\;\; | \\
CH_3CH_2\overset{3}{C}H\overset{4}{C}=\overset{5}{C}\overset{6}{C}HCH_2CH_3 \\
\quad\;\; |\qquad\;\; | \\
\quad\;\; CH_3\quad\; CH_3
\end{array}
\qquad \text{3,4,5,6-tetramethyl-4-octene}
$$

There are six stereoisomers of this compound: (3S,4Z,6S), (3R,4Z,6R), (4Z,*meso*), (3S,4E,6S), (3R,4E,6R), and (4E,*meso*).
(b) Carbons 3, 4, 5, and 6 are stereocenters.
(c) Carbons 3 and 6 are asymmetric carbons.

6.27   The asymmetric carbons are marked with asterisks.

(a)   (c)   (e)

(no asymmetric carbons)

6.28   (a) The asymmetric carbon has the *R* configuration.
(c) The asymmetric atom is a phosphorus, and it has the *R* configuration. (Notice that the —N(CH$_3$)$_2$ group is the group of lowest priority.)

6.29   (a) and (b)   The enantiomers of ibuprofen:

(S)-ibuprofen   (R)-ibuprofen

 Don't forget that there are many correct ways to draw a given perspective structure. The structures above were drawn with the group of lowest priority in a "rear" (dashed wedge) position, because it is easiest to assign configuration in such a situation. If you wonder whether one of your structures is the same as one of those above, build a model of each and test them for congruence. If they are congruent, they are the same.

6.30   Start with an arbitrary Fischer projection of the proper configuration and rotate it 180° in the plane of the paper:

A   *B*

Perform cyclic permutations of *A* at the bottom carbon to generate two new projections:

A

Next, perform a cyclic permutation of *A* at the top carbon to generate a new projection *A′* and perform two cyclic permutations on the lower carbon of *A′* to generate a total of three new projections:

$$
\begin{array}{ccc}
\text{C}_2\text{H}_5 & \text{C}_2\text{H}_5 & \text{C}_2\text{H}_5 \\
\text{HO}\!-\!\!|\!-\!\text{H} & \text{HO}\!-\!\!|\!-\!\text{H} & \text{HO}\!-\!\!|\!-\!\text{H} \\
\text{CH}_3\!-\!\!|\!-\!\text{OH} & \text{H}\!-\!\!|\!-\!\text{CH}_3 & \text{HO}\!-\!\!|\!-\!\text{H} \\
\text{H} & \text{OH} & \text{CH}_3 \\
A'
\end{array}
$$

Then do a cyclic permutation on the top carbon of *A′* to give *A″*, and perform two cyclic permutations on the lower carbon of *A″* to get three more projections:

$$
\begin{array}{ccc}
\text{OH} & \text{OH} & \text{OH} \\
\text{H}\!-\!\!|\!-\!\text{C}_2\text{H}_5 & \text{H}\!-\!\!|\!-\!\text{C}_2\text{H}_5 & \text{H}\!-\!\!|\!-\!\text{C}_2\text{H}_5 \\
\text{CH}_3\!-\!\!|\!-\!\text{OH} & \text{H}\!-\!\!|\!-\!\text{CH}_3 & \text{HO}\!-\!\!|\!-\!\text{H} \\
\text{H} & \text{OH} & \text{CH}_3 \\
A''
\end{array}
$$

Now repeat the same routine on structure *B*:

$$
\begin{array}{ccc}
\text{H} & \text{H} & \text{H} \\
\text{HO}\!-\!\!|\!-\!\text{CH}_3 & \text{HO}\!-\!\!|\!-\!\text{CH}_3 & \text{HO}\!-\!\!|\!-\!\text{CH}_3 \\
\text{HO}\!-\!\!|\!-\!\text{C}_2\text{H}_5 & \text{H}\!-\!\!|\!-\!\text{OH} & \text{C}_2\text{H}_5\!-\!\!|\!-\!\text{H} \\
\text{H} & \text{C}_2\text{H}_5 & \text{OH} \\
B
\end{array}
$$

$$
\begin{array}{ccc}
\text{OH} & \text{OH} & \text{OH} \\
\text{CH}_3\!-\!\!|\!-\!\text{H} & \text{CH}_3\!-\!\!|\!-\!\text{H} & \text{CH}_3\!-\!\!|\!-\!\text{H} \\
\text{HO}\!-\!\!|\!-\!\text{C}_2\text{H}_5 & \text{H}\!-\!\!|\!-\!\text{OH} & \text{C}_2\text{H}_5\!-\!\!|\!-\!\text{H} \\
\text{H} & \text{C}_2\text{H}_5 & \text{OH} \\
B'
\end{array}
$$

$$
\begin{array}{ccc}
\text{CH}_3 & \text{CH}_3 & \text{CH}_3 \\
\text{H}\!-\!\!|\!-\!\text{OH} & \text{H}\!-\!\!|\!-\!\text{OH} & \text{H}\!-\!\!|\!-\!\text{OH} \\
\text{HO}\!-\!\!|\!-\!\text{C}_2\text{H}_5 & \text{H}\!-\!\!|\!-\!\text{OH} & \text{C}_2\text{H}_5\!-\!\!|\!-\!\text{H} \\
\text{H} & \text{C}_2\text{H}_5 & \text{OH} \\
B''
\end{array}
$$

This process shows that there are a total of eighteen valid Fischer projections.

6.32 Either of the following isomeric heptanes is a correct answer:

$$
\begin{array}{cc}
\text{CH}_3 & \text{CH}_3 \\
| & | \\
\text{C}_2\text{H}_5\!-\!\text{CH}\!-\!\text{CH}_2\text{CH}_2\text{CH}_3 & \text{C}_2\text{H}_5\!-\!\text{CH}\!-\!\text{CH(CH}_3)_2 \\
\text{3-methylhexane} & \text{2,3-dimethylpentane}
\end{array}
$$

6.34 (a) True. 2-Chlorohexane and 3-chlorohexane are examples of chiral constitution isomers.
(c) False. An achiral molecule is congruent to its mirror image. (See Fig. 6.1 for an example.)
(e) False. For example, 2-chlorohexane has no diastereomer. In order for a chiral molecule to have a diastereomer it must have two or more stereocenters.

(g) False. *Meso* compounds have asymmetric carbons and are not chiral.

(i) True. By definition, interchanging two groups at a stereocenter generates a stereoisomer.

(k) False. Every chiral compound is optically active, although it is possible in some cases for the optical activity to be too small to be detected experimentally.

(m) False, because certain achiral molecules have symmetry elements other than planes of symmetry. (See text page 230.)

(o) True, because the presence of a plane of symmetry rules out chirality.

6.35   The isomeric alkyl chlorides $C_6H_{13}Cl$ with the carbon skeleton of 3-methylpentane:

$$
\begin{array}{cccc}
\underset{|}{CH_2Cl} & \underset{|}{CH_3} & Cl\ CH_3 & \underset{|}{CH_3} \\
CH_3CH_2CHCH_2CH_3 & ClCH_2CH_2CHCH_2CH_3 & CH_3CHCHCH_2CH_3 & CH_3CH_2CCH_2CH_3 \\
A & B & C & \underset{|}{Cl} \\
& & & D
\end{array}
$$

Compounds *B* and *C* are chiral. The four compounds *A–D* are constitutional isomers. Compound *B* can exist as a pair of enantiomers; compound *C* can exist as two diastereomeric pairs of enantiomers (four stereoisomers total).

6.37   The conformations of isopentane (generated by clockwise rotations of the rear carbon):

A   *B*   *C*

(a) Conformations *A* and *B* are chiral; they are conformational enantiomers.

(b) These two conformations interconvert so rapidly that they cannot be isolated. As the text states on page 254, a molecule that consists of rapidly interconverting enantiomers is considered to be achiral.

(c) The heats of formation of *A* and *B* are identical, and are less than the heat of formation of *C*, because *C* contains two *gauche*-methyl interactions, whereas *A* and *B* each have only one. The heats of formation of *A* and *B* are equal because enantiomers have identical physical properties.

6.38   Compound *B* can undergo amine inversion, which rapidly interconverts its enantiomeric forms. Compound *A* cannot undergo inversion because it does not have an unshared electron pair; hence, its two enantiomeric forms cannot interconvert and can therefore be isolated.

6.39   (a) A 0.75 *M* solution of this alkene contains (0.75 mol/L)(146.2 g/mol)(0.001 L/mL) = 0.1096 g/mL. Using Eq. 6.1, with $l = 1$ dm, the observed rotation is

$$(-76 \text{ deg mL g}^{-1} \text{ dm}^{-1})(0.1096 \text{ g mL}^{-1})(1 \text{ dm}) = -8.33 \text{ degrees}$$

Note that the specific rotation of the *S* alkene is negative because it is the enantiomer of the *R* alkene, which has a positive specific rotation.

6.40   Salts that are either identical or enantiomers should have identical solubilities. The solution to the problem then hinges on determining the relationship between each pair of salts. If they have the same configurations at corresponding asymmetric carbons, they are identical and have identical solubilities; if they have different configurations at both of the corresponding asymmetric carbons, they are enantiomers and have identical solubilities, because enantiomers have identical properties. If they have different configurations at one asymmetric carbon and the same configuration at the other, they are diastereomers and have different solubilities, because diastereomers have different properties. For *A* and *B*:

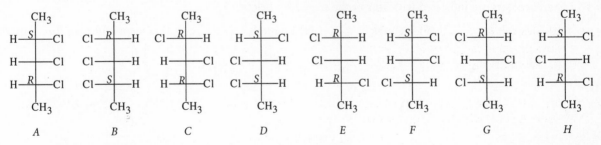

The two salts are therefore diastereomers and should have different solubilities in methanol. Similar analyses give the following conclusions:

A and *C* are diastereomers and should have different solubilites.
A and *D* are enantiomers and have identical solubilities.
B and *C* are enantiomers and have identical solubilities.
B and *D* are diastereomers and should have different solubilities.
C and *D* are diastereomers and should have different solubilities.

6.41 Because hydrogenation causes two different groups to become identical, the carbon that was asymmetric before the reaction is not asymmetric after the reaction. Because the product is achiral, it cannot be optically active.

$$CH_2{=}CH{-}CH{-}CH_2CH_3 \xrightarrow{\ H_2,\ catalyst\ } CH_3CH_2{-}CH{-}CH_2CH_3$$
$$\underset{CH_3}{|} \qquad\qquad\qquad\qquad \underset{CH_3}{|}$$

chiral   achiral

6.42

A and *B:*	diastereomers	*A* and *C:*	identical
A and *D:*	diastereomers	*A* and *E:*	diastereomers
B and *C:*	diastereomers	*B* and *D:*	enantiomers
B and *E:*	diastereomers	*C* and *D:*	diastereomers
C and *E:*	diastereomers	*D* and *E:*	diastereomers

6.43 (a) and (b)   To be sure that every possibility is covered, draw all $2^3 = 8$ possibilities:

CH₃	CH₃	CH₃	CH₃	CH₃	CH₃	CH₃	CH₃
H—*S*—Cl	Cl—*R*—H	Cl—*R*—H	H—*S*—Cl	Cl—*R*—H	H—*S*—Cl	Cl—*R*—H	H—*S*—Cl
H——Cl	Cl——H	H——Cl	Cl——H	Cl——H	H——Cl	H——Cl	Cl——H
H—*R*—Cl	Cl—*S*—H	H—*R*—Cl	Cl—*S*—H	H—*R*—Cl	Cl—*S*—H	Cl—*S*—H	H—*R*—Cl
CH₃	CH₃	CH₃	CH₃	CH₃	CH₃	CH₃	CH₃
A	*B*	*C*	*D*	*E*	*F*	*G*	*H*

By turning projections 180° in the plane of the page you can show that $A = B$, $C = E$, $D = F$, and $G = H$. Consequently, there are four stereoisomers of this compound.

(c)   The stereoisomers that do *not* have internal mirror planes are chiral, that is, structures *C* (=*E*) and *D* (=*F*). Their mirror images are noncongruent. Structures *A* (=*B*) and *G* (=*H*) are *meso*.

(d) Carbon-3 is a stereocenter in *A* (=*B*) and *G* (=*H*) because interchange of the H and the Cl in each case gives a diastereomer. Carbon-3 is *not* a stereocenter in structures *C* (=*E*) and *D* (=*F*). (See the note following the solution to Problem 6.13 on p. 123 of this manual.)

6.44 (a) If the geometry is tetrahedral, there are two enantiomeric stereoisomers (drawn on the top of the next page in Fischer projection):

$$Cl\!-\!\!\begin{array}{c} F \\ | \\ \hline | \\ Br \end{array}\!\!-\!I \qquad I\!-\!\!\begin{array}{c} F \\ | \\ \hline | \\ Br \end{array}\!\!-\!Cl$$

(b) For square-planar geometry, there are also two stereoisomers. These are found by making all possible pairwise switches between adjacent groups and ruling out identities. Note that switching nonadjacent groups gives the same structure, because these are *truly planar* structures (not Fischer projections of nonplanar structures), and can therefore lifted out of the plane of the page and turned over.

$$Cl\!-\!\!\begin{array}{c} F \\ | \\ C \\ | \\ Br \end{array}\!\!-\!I \qquad Cl\!-\!\!\begin{array}{c} I \\ | \\ C \\ | \\ Br \end{array}\!\!-\!F$$

(c) For pyramidal geometry, there are three diastereomeric sets of enantiomers (six stereoisomers total):

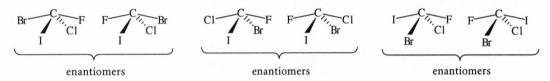

enantiomers   enantiomers   enantiomers

6.46   The infallible test for chirality is to make the mirror image and test it for congruence:

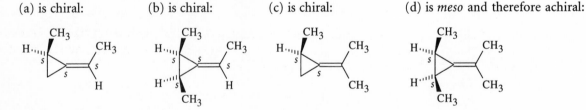

Because the structure is not congruent to its mirror image, the compound is chiral.

6.47   Stereocenters are indicated with an *s* in the structures below.

(a) is chiral: (b) is chiral: (c) is chiral: (d) is *meso* and therefore achiral:

In part (b), identification of the stereocenter at the junction of the ring and the double bond is tricky. Breaking the two bonds of the ring and interchanging them gives a stereoisomer:

6.48   The optical rotations and melting points show that Fischer obtained enantiomers as a result of this transformation. Consider what each of the two possible geometries predicts for an interchange of the two groups. For tetrahedral carbon, the interchange of any two groups at an asymmetric carbon gives enantiomers. Consequently, the experimental result is consistent with tetrahedral geometry.

enantiomers

If the molecule were pyramidal, and if Fischer *happened* to be dealing with a stereoisomer in which the —CO_2H and —$CONH_2$ groups were at opposite (rather than adjacent) corners of the pyramid, he would also have obtained enantiomers from his experiment:

enantiomers

(That these two structures are enantiomeric can be seen by comparing the two structures after turning either one 180° about a vertical axis through the asymmetric carbon.)

   But if Fischer *happened* to be dealing with a stereoisomer in which the —CO_2H and —$CONH_2$ groups were at adjacent corners, the exchange should have given diastereomers. Thus, the conversion of one stereoisomer into its enantiomer could have been the result of two things occurring simultaneously: (1) pyramidal geometry, and (2) the fortuitous choice of a particular stereoisomer. Consequently, Fischer's result was consistent with either tetrahedral or pyramidal geometry.

6.49   Neither result taken alone rules out pyramidal geometry. However, for an atom with two identical groups and pyramidal geometry to be achiral, the two identical groups must be at the opposite corners of the pyramid.

achiral

If the two groups had been at opposite corners of the pyramid in the *first* experiment, then in the *second experiment with the same starting material*, the two groups must have been at adjacent corners, and this transformation would have yielded a chiral compound:

a chiral compound

(Be sure to convince yourself that the molecule on the right is chiral. Do this by drawing (or making a model of) its mirror image and show that the two are noncongruent.) On the other hand, if the starting compound had tetrahedral geometry, then making *any* two groups identical would give an achiral compound. Since this was the experimental result, pyramidal geometry was ruled out, and tetrahedral geometry thus remained the only (and, as it turned out, correct) possibility.

7

Cyclic Compounds and Stereochemistry of Reactions

Terms

Concepts

I. Cycloalkanes

A. MONOCYCLIC COMPOUNDS

1. A compound that contains a single ring is called a monocyclic compound.
2. The relative stabilities of the monocyclic alkanes can be determined from their heats of formation.
3. Of the cycloalkanes with ten or fewer carbons, cyclohexane has the smallest heat of formation per CH_2; thus, cyclohexane is the most stable of these cycloalkanes.

4. Cyclohexane has the same stability as a typical unbranched alkane (−20.7 kJ/mol or −4.95 kcal/mol per CH_2 group).
5. Cyclopentane has somewhat higher energy than cyclohexane.
6. Cyclobutane and cyclopropane are the least stable of the monocyclic alkanes, cyclopropane being the least stable cycloalkane.
7. Rings larger than cyclopropane are puckered.

B. BICYCLIC AND POLYCYCLIC COMPOUNDS

1. Bicyclic compounds share two or more common atoms (called bridgeheads) between two rings.
   a. A bicyclic compound is classified as a fused bicyclic compound when the bridgehead carbons are adjacent.

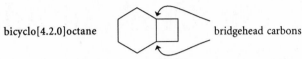

bicyclo[4.2.0]octane   bridgehead carbons

   b. A bicyclic compound is classified as a bridged bicyclic compound when the bridgehead carbons are not adjacent.

bicyclo[2.2.2]octane   bridgehead carbons

2. Spirocyclic compounds share a single common atom between two rings.

spiro[4.3]octane   common atom

3. The name of a polycyclic structure is comprised of the following components in the given order:
   a. The substituents and their positions are indicated as in monocyclic compounds; the bridges are numbered beginning at a bridgehead and proceeding around the rings (in decreasing size).
   b. The appropriate prefix *bicyclo*, *tricyclo*, *tetracyclo*, etc., indicates the number of rings present.
   c. Brackets contain, in decreasing size, the number of carbon atoms in each bridge.
   d. The name of the open-chain hydrocarbon containing the same total number of carbon atoms.

H_3C— CH_3 3,8-dimethylbicyclo[4.3.0]nonane

4. Polycyclic compounds contain many rings joined at common atoms.

C. CIS AND TRANS RING FUSION

1. Two rings in a fused bicyclic compound can be joined in a *cis* or *trans* arrangement.
2. There are two stereoisomers of bicyclo[4.4.0]decane, commonly called decalin.
   a. In *cis*-decalin, two —CH_2— groups of ring *B* (shaded circles) are *cis* substituents on ring *A*; likewise, two —CH_2— groups of ring *A* (shaded squares) are *cis* substituents on ring *B*.

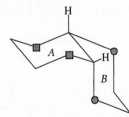

   b. In *trans*-decalin, the —CH_2— groups to the ring fusion are *trans*-diequatorial and the bridgehead hydrogens are *trans*-diaxial.

3. Each cyclohexane ring in *cis*-decalin can undergo the chair flip.

4. In *trans*-decalin, the cyclohexane rings can assume twist-boat conformations but cannot flip into their alternate chair conformations.
5. *Trans*-decalin is more stable than *cis*-decalin because it has fewer 1,3-diaxial interactions.
6. In rings of six atoms or greater, both *cis*- and *trans*-fused isomers are well known, but the *trans*-fused ones are more stable because 1,3-diaxial interactions are minimized.
7. *Trans* fusion introduces ring strain in rings of fewer than six atoms.

D. TRANS-CYCLOALKENES
1. A double bond at a bridgehead in small rings introduces a great amount of strain, and requires twisting the molecule about the double bond, thus weakening the overlap of *p* orbitals involved in the π bond.
2. In a bicyclic compound, a bridgehead atom contained solely within small rings (seven or fewer atoms) cannot be part of a double bond (Bredt's rule).
3. Bicyclic compounds that have bridgehead double bonds in rings of more than seven atoms are more stable and can be isolated.

E. STEROIDS
1. Steroids comprise a particularly important class of the many naturally occurring compounds having fused rings.
2. In many cases all ring fusions are *trans*; all-*trans* ring fusion causes a steroid to be conformationally rigid and relatively flat.
3. Many steroids have methyl groups, called angular methyls, at carbons 10 and 13.
4. The face of the steroid over which the angular methyl groups occupy axial positions is called the β-face, whereas the other face is called the α-face.

F. PLANAR REPRESENTATION OF CYCLIC COMPOUNDS
1. Planar structures of cyclic compounds can be used for situations in which conformational details are not important.
   a. The bond of an up substituent (towards the observer) is represented by a solid wedge.
   b. The bond of a down substituent (away from the observer) is represented by a dashed wedge.

2. A cyclohexane ring is usually drawn in a slightly tilted and rotated perspective so that all of its bonds are visible.
   a. Three alternating carbons (referred to as up carbons) define a plane that is above the plane defined by the other three carbons (referred to as down carbons).
   b. Two perspectives are commonly used for cyclohexane rings; a rotation of either perspective by an odd multiple of 60° about an axis through the center of the ring gives the other perspective.
      i. Draw two parallel bonds slanted to the left for one perspective, or slanted to the right for the other perspective.

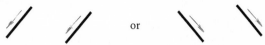

        or

      ii. Connect the tops of the slanted bonds with two more bonds in a uneven "V" arrangement.

        or

      iii. Connect the bottoms of the slanted bonds with the remaining two bonds in an uneven "inverted V" arrangement.

        or

3. Planar structures are particularly useful for assessing the chirality of cyclic compounds.

II. *Conformations of Cyclohexane*

A. CHAIR CONFORMATIONS

1. The most stable conformation of cyclohexane is called the chair conformation because of its resemblance to a lawn chair.
   a. Bonds on opposite sides of the ring are parallel.
   b. There are two types of hydrogens.
      i. If a model of a chair cyclohexane rests on a table, six C—H bonds are perpendicular to the plane of the table; these hydrogens are called axial hydrogens.
      ii. Six C—H bonds point outward along the periphery of the ring; these hydrogens are called equatorial hydrogens.

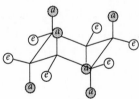

      iii. If an axial hydrogen is up on one carbon, the two neighboring axial hydrogens are down, and vice versa. The same is true of the equatorial hydrogens.
   c. In the chair conformation, all bonds are staggered.
2. The stability of cyclohexane is due to the fact that all of its bonds can be staggered without compromising the tetrahedral carbon geometry.
3. When a cyclohexane molecule undergoes internal rotations, a change in the conformation of the ring occurs.
4. Movement of each down-carbon up and up-carbon down changes one chair conformation into another completely equivalent chair conformation.
   a. The equatorial hydrogens become axial, and the axial hydrogens become equatorial.
   b. The up carbons become down carbons, and the down carbons become up carbons.
5. The interconversion of two chair forms of cyclohexane is called the chair interconversion or the chair flip.

6. The chair flip causes axial and equatorial hydrogens to change positions rapidly so that, averaged over time, these hydrogens are equivalent and indistinguishable.

B. Boat Conformations

1. The boat conformation is the result of simultaneous internal rotations about all carbon-carbon bonds except those to one carbon.
2. The boat conformation is not a stable conformation of cyclohexane; it contains two sources of instability:
   a. Certain hydrogens are eclipsed.
   b. The "flagpole" hydrogens experience van der Waals repulsion.

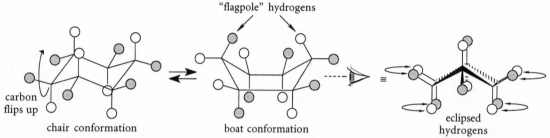

C. Twist-boat Conformation

1. The boat conformation undergoes very slight internal rotation that results in another stable conformation of cyclohexane called a twist-boat conformation.
2. The twist-boat conformation is less stable than the chair conformation.
3. A boat conformation itself can be thought of as the transition state for interconversion of two twist-boat conformations.

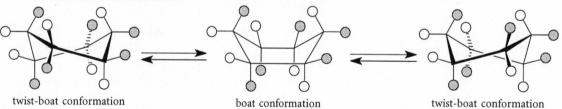

| twist-boat conformation | boat conformation | twist-boat conformation |

III. *Conformational Analysis*

A. Monosubstituted Cyclohexanes

1. The investigation of molecular conformations and their relative energies is called conformational analysis.
2. Methylcyclohexane is a mixture of two conformational diastereomers.
   a. Diastereomers have different energies, one form is more stable than the other, and the more stable form is present in greater amount.
   b. The equatorial conformation of a substituted cyclohexane is more stable than the axial conformation.
   c. When the methyl group is in an axial position, van der Waals repulsions occur between one of the methyl hydrogens and the two axial hydrogens on the same face of the ring.
      i. Such unfavorable interactions between axial groups are called 1,3-diaxial interactions.
      ii. A methyl-hydrogen 1,3-diaxial interaction "costs" the same amount of energy (3.7 kJ/mol) as a *gauche*-butane interaction.

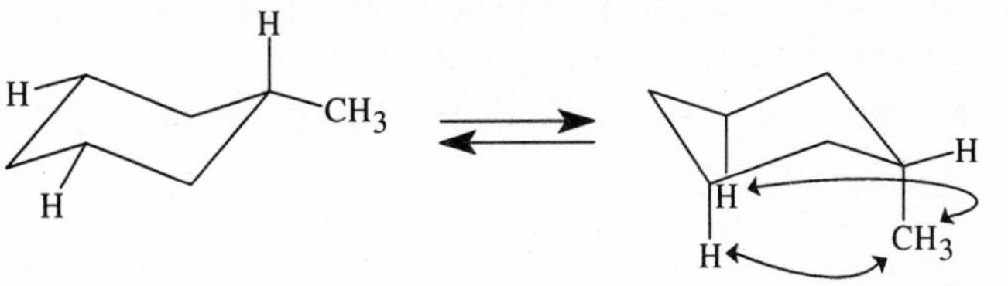

1,3-diaxial interactions

3. The larger the substituent group on a cyclohexane ring, the more the equatorial conformation is favored.

B. DISUBSTITUTED CYCLOHEXANES; *CIS-TRANS* ISOMERISM IN CYCLIC COMPOUNDS

1. The *trans* designation is used with cyclic compounds when two substituents have an up-down relationship.

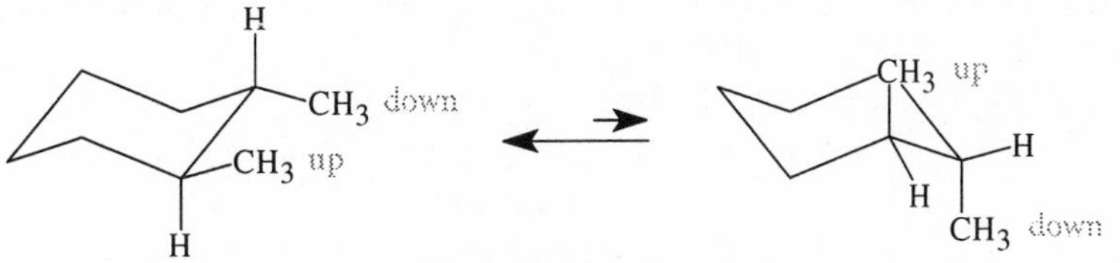

2. The *cis* designation is used with cyclic compounds when two substituents have an up-up or down-down relationship.

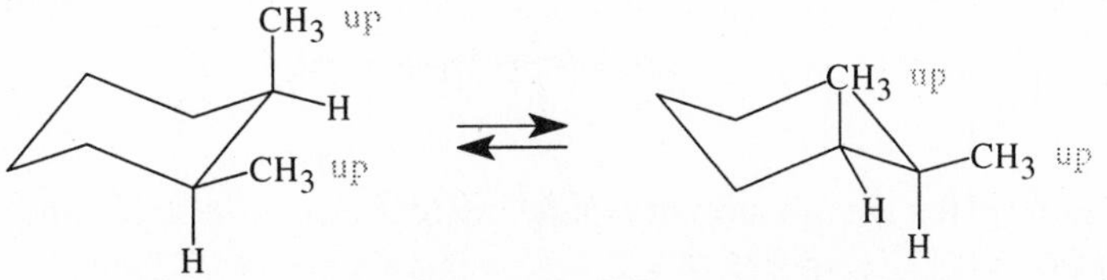

3. The *cis* and *trans* designations specify the relative stereochemical configurations of two asymmetric carbons in the ring which are not affected by the conformation equilibrium.

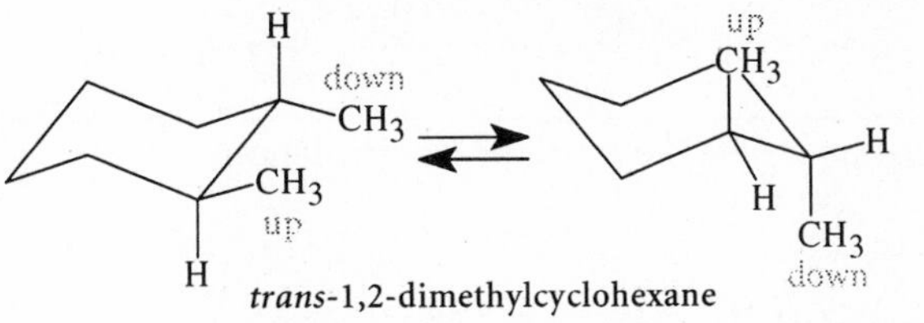

trans-1,2-dimethylcyclohexane 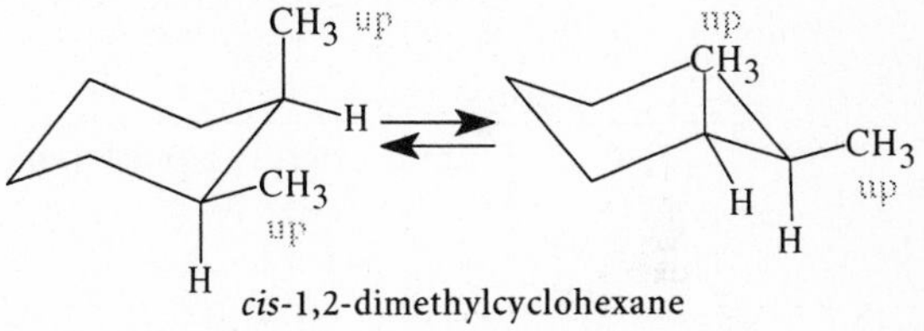   *cis*-1,2-dimethylcyclohexane

4. When a ring contains more than two substituents, the *R,S* system must be used to indicate configuration.

5. When two groups on a substituted cyclohexane conflict in their preference for the equatorial position, the preferred conformation can usually be predicted from the relative conformational preferences of the two groups.

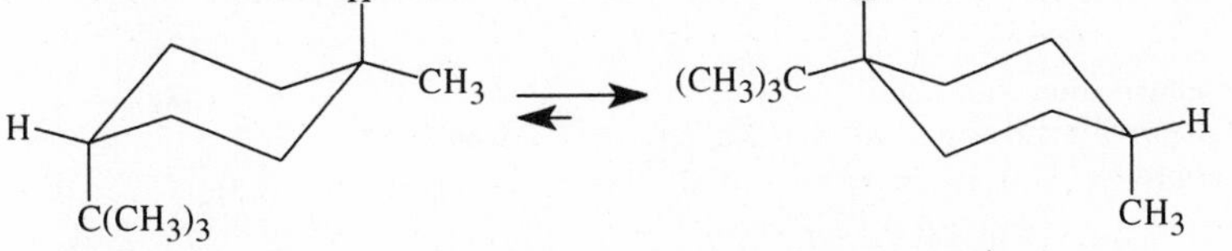

favored because the large *tert*-butyl group is in an equatorial position

C. STEREOCHEMICAL CONSEQUENCES OF THE CHAIR FLIP

1. Some molecules are rapidly interconverting mixtures of diastereomers.

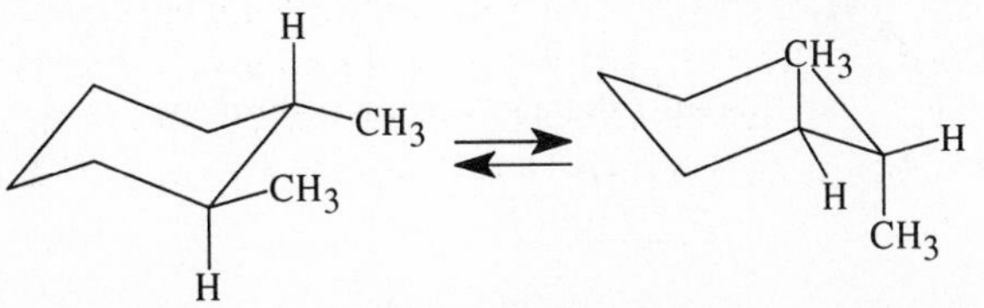

conformational diastereomers

a. Such molecules can be resolved into enantiomers; that is, they can be isolated in optically active form.

b. The chirality of such molecules is evident from their planar structures.

noncongruent mirror images

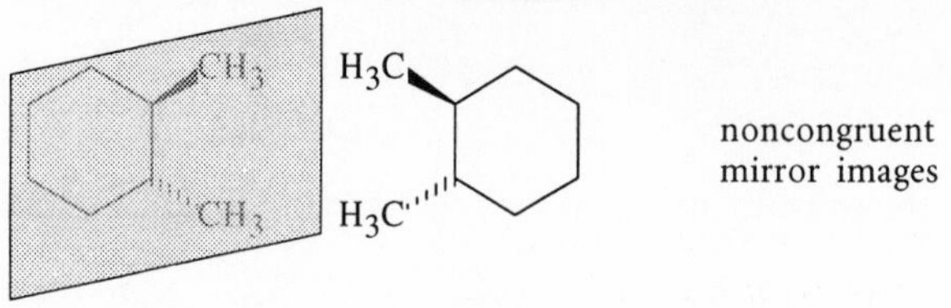

2. Some molecules are rapidly interconverting mixtures of enantiomers.

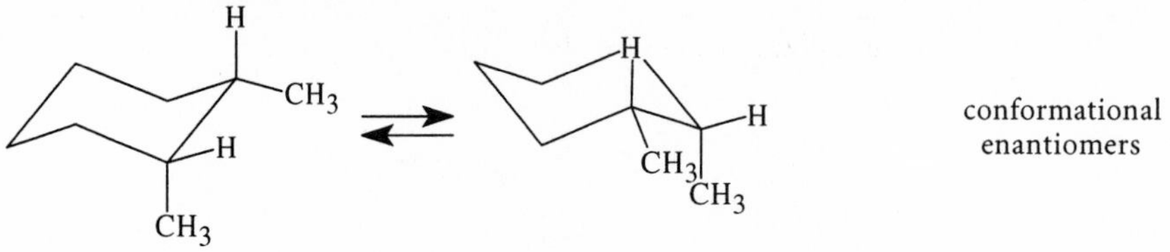

conformational enantiomers

a. Such molecules are considered to be achiral.

b. That such molecules are *meso* compounds can be seen from their planar structures.

mirror plane

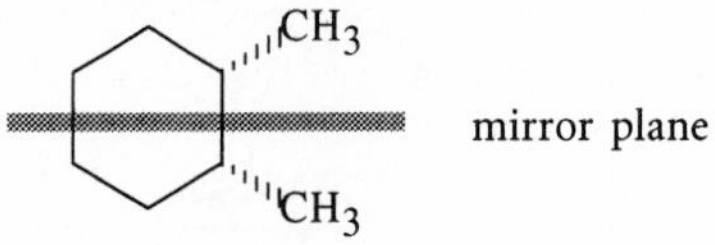

3. Some molecules are mixtures of rapidly interconverting *meso* conformations.

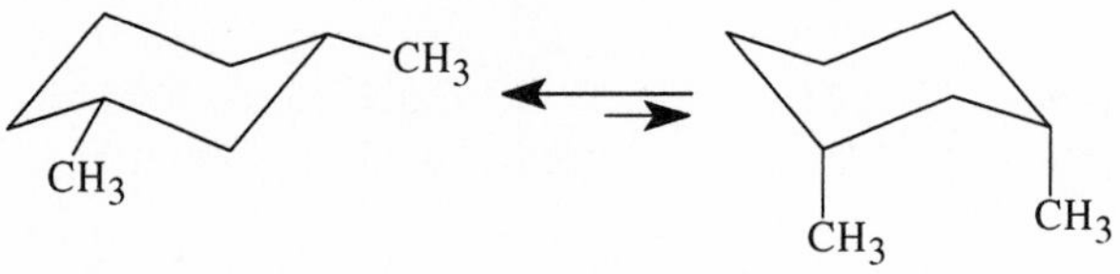

D. CYCLOPENTANE

1. Cyclopentane exists in a puckered conformation, called the envelope conformation.

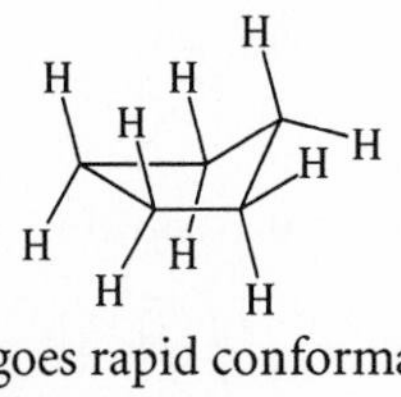

2. The envelope conformation undergoes rapid conformational changes in which each carbon alternates as the "point" of the envelope.

3. Substituted cyclopentanes also exist in envelope conformations, and the substituents adopt positions that minimize van der Waals repulsions with neighboring groups.

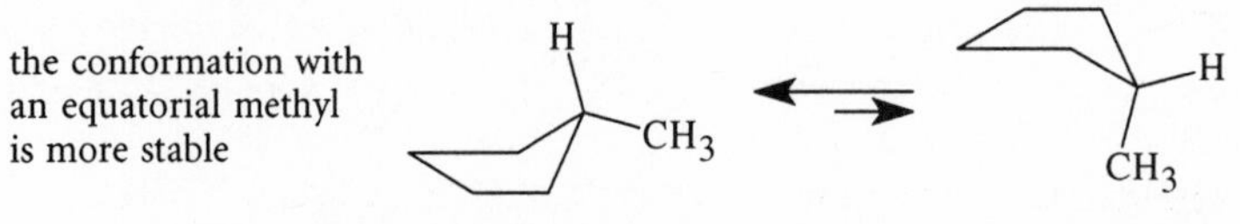

the conformation with an equatorial methyl is more stable

4. When a cyclopentane ring has two or more substituent groups, *cis* and *trans* relationships between the groups are possible, just as in cyclohexane.

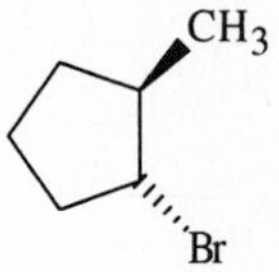

trans-1-bromo-2-methylcyclopentane

E. CYCLOBUTANE

1. Cyclobutane consists of two puckered conformations in rapid equilibrium.

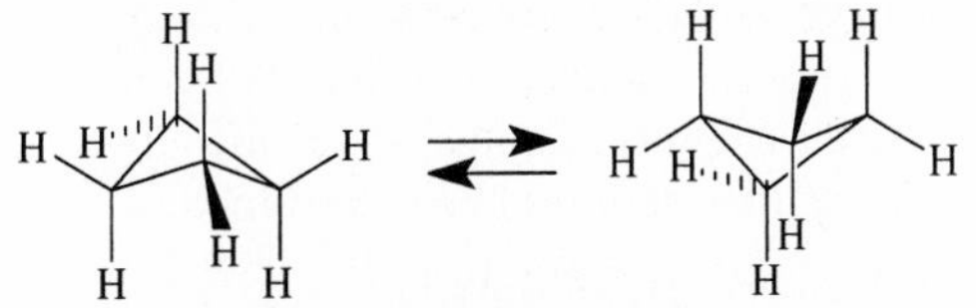

2. Angle strain, or ring strain, is the increase in energy that results when the angles between carbon-carbon bonds are constrained by the size of the ring to be much smaller than the optimum tetrahedral angle.

F. CYCLOPROPANE

1. Cyclopropane has a planar carbon skeleton.

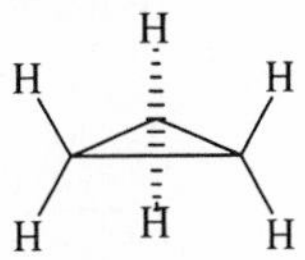

2. Neither angle strain nor eclipsing interactions between hydrogens in cyclopropane can be relieved by puckering.
3. Each carbon-carbon bond of cyclopropane is bent in a "banana" shape between the two carbons.
4. Bent bonds reduce angle strain but do so at a cost of less effective overlap between the carbon orbitals.

IV. Stereochemistry and Chemical Reactions

A. GENERAL

1. When chiral products are formed from achiral starting materials, both enantiomers of a pair are always formed at identical rates; the product is always a racemate.

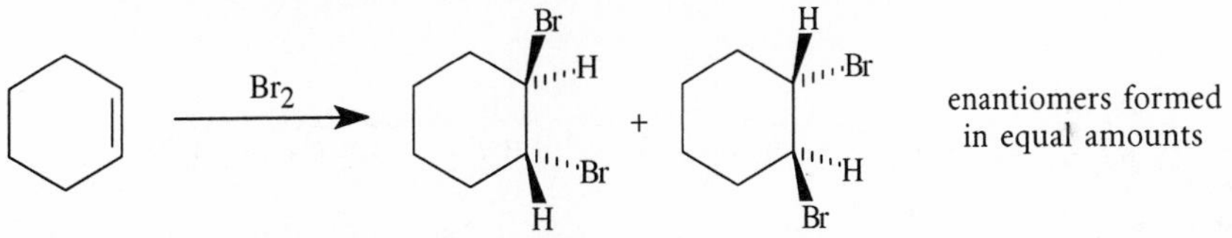

enantiomers formed
in equal amounts

a. The enantiomeric transition states have identical free energies, as do the enantiomeric starting materials.
b. Optical activity never arises spontaneously in the reactions of achiral compounds.
2. Enantiomers have identical reactivities with achiral reagents because enantiomers have the same physical properties.
3. A pair of enantiomers differ in their chemical or physical behavior only when they interact with other chiral objects or forces because they are involved in diastereomeric interactions.
a. Enantiomers have different reactivities with a chiral reagent because diastereomers formed have different properties.
b. The transition state for the reaction of one enantiomer is the diastereomer of the transition state for the reaction of the other.
c. Most enzymes accept only one enantiomer of a chiral substrate.
4. When diastereomeric products can be formed in a reaction, they are in principle formed at different rates and different amounts.

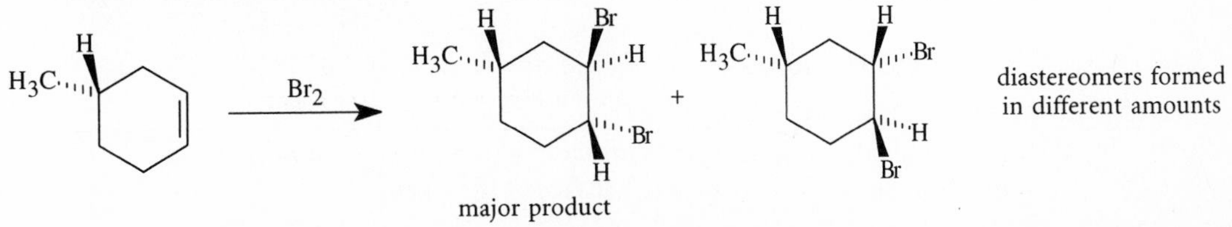

diastereomers formed
in different amounts

major product

    a. Diastereomers are formed in different amounts because they are formed through diastereomeric transition states; one transition state has lower standard free energy than its diastereomer.

    b. Diastereomers in general have different reactivities towards any reagent, whether the reagent is chiral or achiral because diastereomers have different free energies.

5. Each carbon at which a chemical change occurs must be a stereocenter in the product in order for the stereochemistry of the reaction to be determined.

B. SYN AND ANTI ADDITION TO ALKENES

1. The stereochemistry of an addition can be determined only when the stereochemically different modes of addition give rise to stereochemically different products, that is, when both carbons of the double bond become carbon stereocenters in the product.

2. The *syn* and *anti* modes of addition can be distinguished by analyzing the stereochemistry of the products.

    a. In a *syn* addition, two groups add to a double bond from the same side or face.

    b. In an *anti* addition, two groups add to a double bond from opposite sides or faces.

C. REACTIONS AT ASYMMETRIC CARBONS

1. A substitution reaction, a reaction in which one group is replaced by another, can occur in two stereochemically different ways.

    a. Retention of configuration occurs when substitution results in the substituting group and the leaving group having the same relative stereochemical positions.

    b. Inversion of configuration occurs when substitution results in the substituting group and the leaving group having different relative stereochemical positions.

    c. Loss of configuration occurs when there is a mixture of both retention and inversion of configuration.

2. A reaction in which particular stereoisomer(s) of the product are formed to the exclusion of other(s) is said to be a stereoselective reaction.

V. Alkene Addition Reactions

A. BROMINE ADDITION TO ALKENES

1. Bromine addition to alkenes is in many cases a stereoselective *anti* addition.

    a. A bromonium ion can form at either face of the alkene.

b. The formation of a bromonium ion followed by backside attack of bromide is a mechanism that accounts for the observed *anti* addition of Br_2 to alkenes.

B. Hydroboration–Oxidation of Alkenes

1. Hydroboration is a stereoselective *syn* addition and is a direct consequence of the concerted mechanism of the reaction.
2. The oxidation of organoboranes is a substitution reaction that occurs with retention of configuration.
3. The hydroboration-oxidation of an alkene brings about the net *syn* addition of the elements of H—OH to the double bond.

C. Catalytic Hydrogenation of Alkenes

1. Catalytic hydrogenation of most alkenes is a stereoselective *syn* addition.

D. Oxymercuration–Reduction of Alkenes

1. Oxymercuration-reduction is not a stereoselective reaction.
2. Oxymercuration of alkenes is typically a stereoselective *anti* addition.
3. The mercury is replaced by hydrogen with loss of stereochemical configuration.
4. The reaction is regioselective and is very useful in situations in which stereoselectivity is not an issue.

E. Formation of Glycols from Alkenes

1. The formation of glycols from alkenes with either alkaline $KMnO_4$ or OsO_4 is a stereoselective *syn* addition.

VI. Stereochemistry and Mechanism

1. Stereochemistry provides important details about the mechanism of a reaction.
2. Mechanisms are postulated that are consistent with all known evidence.
3. Mechanisms are revised when necessary to accommodate new experimental evidence.
4. No mechanism is ever proved.

Study Guide Links

··

✓7.1   Manipulating Cyclohexane Rings

··

Although you can often resort to models to visualize the stereochemical relationships between chair conformations of substituted cyclohexanes, it is useful to be able to carry out on paper manipulations like the ones used in the text. To gain greater facility with such manipulations, try the following exercises, using *cis*-1,2-dimethylcyclohexane as an example. Draw the result of each manipulation, and then check your drawing against the results shown. *Be sure to follow each manipulation with a model if there is any question about it!*

First, turn the structure 180° about a horizontal ("*x*") axis:

Now turn the original structure 180° about a "*y*" axis:

Practice turning the structure in 60° increments about the same axis:

Finally, turn the original structure 180° about an axis perpendicular to the page ("*z*" axis):

If you can readily perform manipulations like these without recourse to models, you can solve virtually any problem involving cyclohexane stereochemistry.

Let's use these skills to solve the following problem. What is the stereochemical relationship between the two chair conformations of *trans*-1,3-dimethylcyclohexane?

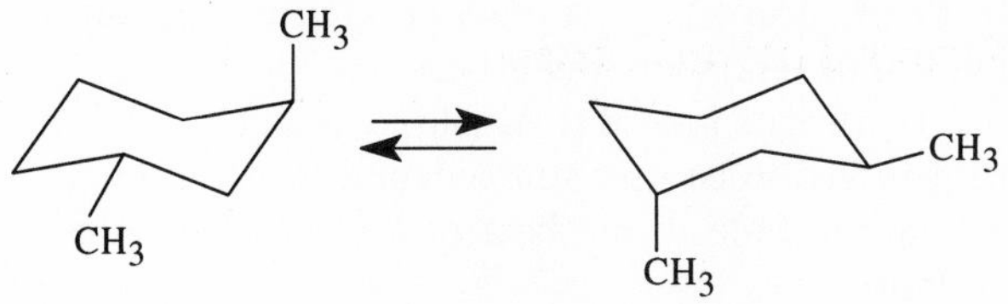

(SG7.1)

There are three possible relationships between these conformations: they could be conformational diastereomers; they could be conformational enantiomers; or they could be identical. The strategy here is to orient one of the methyl groups in one structure in the same way as a methyl group in the other structure. Then compare the relative positions of the remaining methyl groups. To do this, rotate the second structure about an axis perpendicular to the page, as follows:

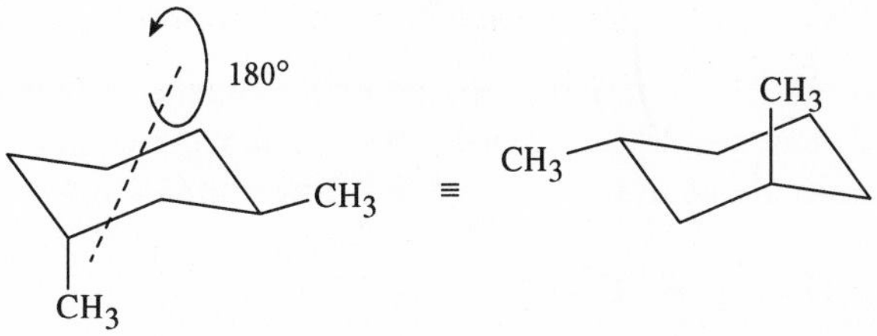

Finally, rotate the resulting structure 60° about an axis in the page, as follows:

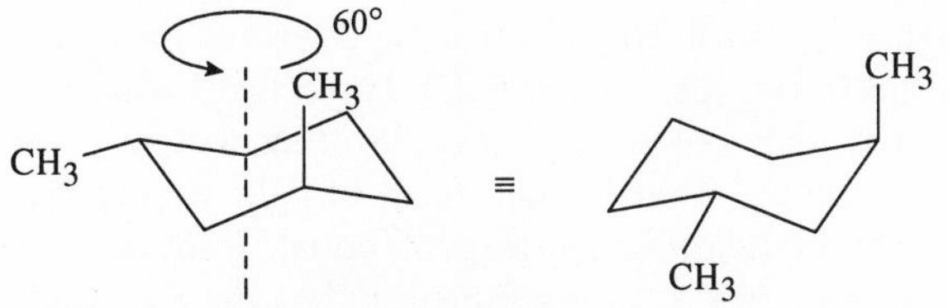

This structure is identical to the first conformation in Eq. SG7.1. Consequently, the chair flip in this case interconverts *identical species.*

It is worth noting that some people, even after practice, have difficulty with manipulations like these because they have inherent difficulty with spatial perception. (Don't assume you are such a person until you have practiced!) This has nothing to do with your intelligence! If you are such a person, don't hesitate to use models. The answer you get may take a little more time, but it is just as correct.

7.2   Alkenelike Behavior of Cyclopropanes

The large amount of strain in cyclopropanes causes them to be more reactive than ordinary cycloalkanes. For example, cyclopropanes can be hydrogenated, although not so readily as alkenes.

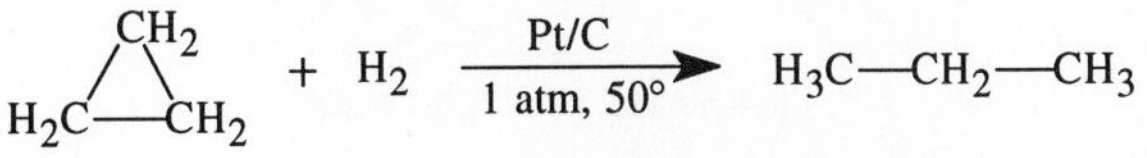

Although reactions like this suggest that cyclopropanes react like alkenes, an alternative view is that alkenes are cycloalkanes with "two-membered rings."

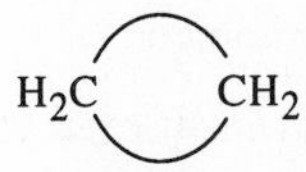

ethylene viewed as a
cyclic alkane

In fact, chemists have shown with quantum calculations that a view of the alkene double bond as two "bent bonds" (similar to those in cyclopropane) is just as valid as the picture adopted by this text in which the two bonds are different, one a σ bond and one a π bond. (Many model sets represent double bonds as "bent bonds.")

If ethylene indeed is visualized as a "two-membered ring"—that is, *cycloethane*—it would be expected to be very strained, and that this strain would be reflected in a very large heat of formation. Indeed, the heat of formation of ethylene per CH_2 is even greater than that of cyclopropane.

	cyclobutane	cyclopropane	"cycloethane" (ethylene)
ΔH_f° per CH_2 (kJ/mol):	6.90	17.8	26.2
(kcal/mol):	1.65	4.25	6.25

In the conventional π model of the alkene double bond, the high energy of the bonds in ethylene resides completely in the relatively weak π bond. In the bent-bond model, the two bonds are equivalent. They are stronger than a π bond, but weaker than a σ bond.

7.3   Optical Activity

Optical activity is an important example of the principles discussed in the text. Plane-polarized light is actually the *vector sum of right and left circularly polarized light,* two forms of light in which the electric field vector propagates as a right- and left-handed helix, respectively. (See Fig. SG7.1.) These two forms of light are noncongruent mirror images— that is, they are enantiomers. In other words, plane-polarized light is "enantiomeric light." Let's call the right-handed form of light *R,* and the left-handed form *S.* Suppose that plane polarized light—consisting of equal amounts of *R* and *S* forms—is passed through an enantiomerically pure sample that has the *S* configuration. When the light interacts with the sample, two types of interactions occur: *S* sample with *S* light, and *S* sample with *R* light. *These two interactions are diastereomeric,* because an *S,S* combination is the diastereomer of an *S,R* combination. Because these two interactions have different energies, one is stronger than the other. The form of light with the stronger interaction is retarded in its passage through the sample. This in turn causes the plane of the polarized light to rotate—that is, it causes optical activity. *The enantiomeric forms of light differ when they interact with a chiral object.*

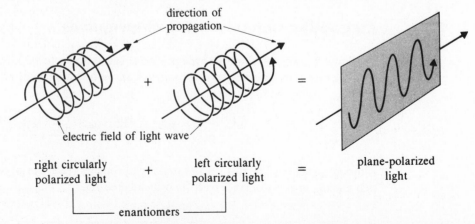

Figure SG7.1   *The electric fields of right and left circularly polarized light propagate through space as helices, and the two add vectorially to give plane-polarized light. Notice that the two helices are chiral; consequently, plane-polarized light, their vector-addition product, is a racemate.*

√7.4   Reactions of Chiral Molecules

Notice carefully that Sec. 7.8A of the text deals only with reactions of *achiral molecules* that give enantiomeric pairs of products. What happens when *chiral* molecules react to give stereoisomeric products? No general principles cover this situation. In some cases, stereoisomers of the product may be formed in equal amounts; in other cases, one stereoisomer may predominate. The result can be determined by the reaction mechanism, that is, the stereochemical nature of chemical changes that occur at stereocenters. (A situation of this sort is discussed in Sec. 7.9B.) In other cases reactants or intermediates are racemized in the reaction, in which case the principles of Sec. 7.7A apply to each enantiomer of the racemate.

It is even possible for chiral molecules to react in some cases to give achiral products even when a stereocenter does not undergo a chemical change. Thus, the following hydrogenation reaction involving a chiral alkene gives an achiral alkane because the asymmetric carbon in the alkene is no longer an asymmetric carbon in the alkane product:

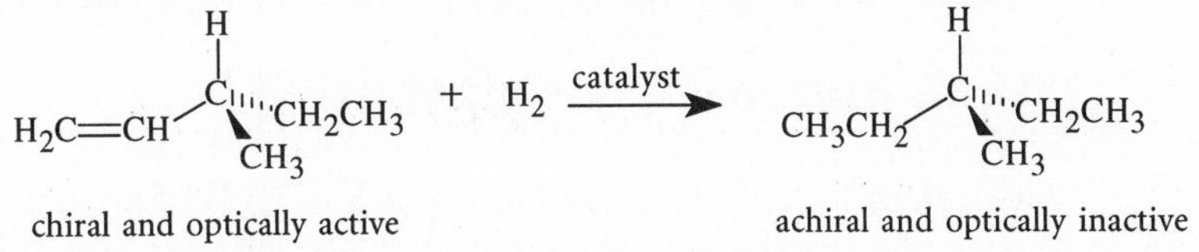

$$H_2C=CH-\overset{\overset{\displaystyle H}{|}}{\underset{\underset{\displaystyle CH_3}{|}}{C}}^{\text{''}}CH_2CH_3 \ + \ H_2 \ \xrightarrow{\text{catalyst}} \ CH_3CH_2-\overset{\overset{\displaystyle H}{|}}{\underset{\underset{\displaystyle CH_3}{|}}{C}}^{\text{''}}CH_2CH_3$$

chiral and optically active   achiral and optically inactive

Optical activity (and chirality) is lost in this reaction because the product is not a chiral molecule. In *this particular case,* the reaction causes a symmetry to occur within the structure of the product that does not exist in the reactant. Optical activity would *not* be lost in the same reaction of a different alkene in which the product is chiral:

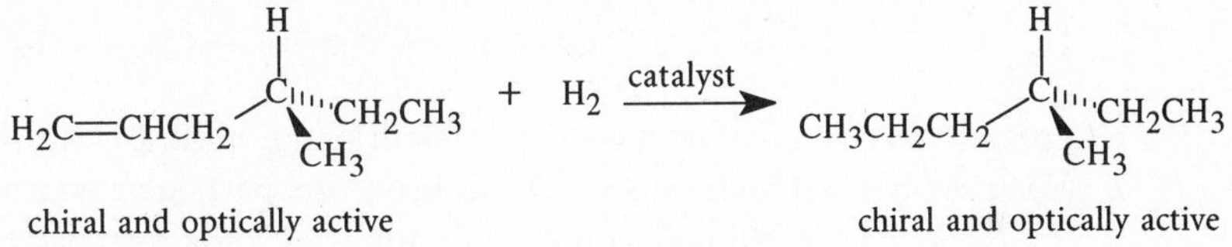

$$H_2C=CHCH_2-\overset{\overset{\displaystyle H}{|}}{\underset{\underset{\displaystyle CH_3}{|}}{C}}^{\text{''}}CH_2CH_3 \ + \ H_2 \ \xrightarrow{\text{catalyst}} \ CH_3CH_2CH_2-\overset{\overset{\displaystyle H}{|}}{\underset{\underset{\displaystyle CH_3}{|}}{C}}^{\text{''}}CH_2CH_3$$

chiral and optically active   chiral and optically active

(See also Problem 6.41, text p. 268.)

√7.5   Analysis of Reaction Stereochemistry

Notice carefully that a given type of reaction might yield diastereomers with one starting material, enantiomers with another, and no stereoisomers at all with a third. For example, the bromine addition to cyclohexene in Study Problem 7.4 gives a pair of diastereomers, and one of these is formed as a pair of enantiomers. In bromine addition to propene, only a pair of enantiomers is formed. In bromine addition to ethylene, no stereoisomers at all are formed. For a given reaction, no single stereochemical result applies to all compounds. Each reaction must be analyzed individually.

7.6   Stereoselective and Stereospecific Reactions

The purpose of this section is to introduce you to additional terminology that you may hear in discussions of reaction stereochemistry. First, be sure that you understand the meaning of *stereoselective.* A *stereoselective reaction* is one that gives certain stereoisomer(s) of the product to the exclusion of others. Note that some texts define a stereoselective reaction to be a reaction that gives *only one* of the possible stereoisomers. This is not a generally accepted definition. Bromine addition to *cis*-2-butene, for example, gives equal amounts of two enantiomers; however, the reaction is stereoselective because almost none of the other possible stereoisomer (the *meso* diastereomer) is formed.

(continues)

Some texts define **stereospecific reactions**. A stereospecific reaction is one in which the stereochemistry of the product depends on the stereochemistry of the starting material. Thus, bromine addition to 2-butene is stereospecific because *cis*-2-butene gives the racemic product, whereas *trans*-2-butene gives the *meso* product. Before the experimental result was known, it would have been conceivable that both *cis*- and *trans*-2-butenes might have given the same diastereomer of the product—say, the racemate. (Of course, this result would have demanded a different mechanism.) If this had been the result, bromine addition would be a stereoselective reaction, but it would not be stereospecific. *All stereospecific reactions are stereoselective, but not all stereoselective reactions are stereospecific.* There actually are examples of stereoselective reactions that are not stereospecific, but none have been presented to this point in the text.

One problem with the term *stereospecific* is that it has been defined by some texts to mean something quite different. Some texts define *stereospecific* to mean "100% stereoselective within limits of detection." The same sense can be conveyed by saying simply that a reaction is "highly stereoselective." Chemists seem to be reaching a consensus that this definition of *stereospecific* is at best not very useful, and, at worst, confusing.

✓7.7   When Stereoselectivity Matters

The stereoselectivity of a reaction (or lack of it) is of no concern if stereochemically different modes of a reaction give the same products. Let's see how this statement would apply in the case of bromine addition to alkenes.

The addition of bromine to ethylene does not give stereoisomers; hence, it makes absolutely no difference in the outcome of the reaction whether the reaction is a *syn* or *anti* addition or a mixture of the two.

$$CH_2{=}CH_2 \xrightarrow{Br_2} Br{-}CH_2{-}CH_2{-}Br$$

Furthermore, we cannot even tell whether the addition is *syn* or *anti*. It is often assumed (reasonably) that additions which occur in one particular way on certain compounds occur in the same way on similar compounds for which the stereoselectivity cannot be determined. Thus, because addition of bromine is an *anti* addition with other alkenes, we assume that it is an *anti* addition for ethylene as well. (*Anti* addition could be verified by examining bromine addition to isotopically substituted ethylenes:

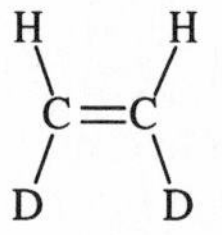

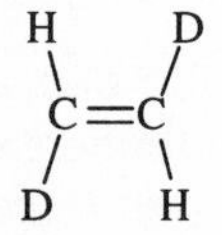

Most people would accept that *anti* addition to the deuterium-substituted ethylenes is proof that ethylene itself undergoes *anti* addition.)

In the addition of bromine to propene, stereoisomers are formed:

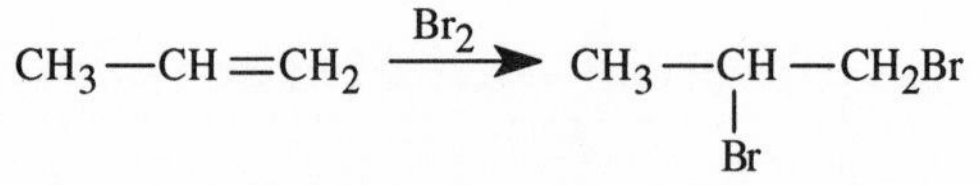

$$CH_3{-}CH{=}CH_2 \xrightarrow{Br_2} CH_3{-}\underset{\underset{Br}{|}}{CH}{-}CH_2Br$$

Two *enantiomers* are possible, but they *must* be formed in the same amounts, because the reagents are achiral. Because carbon-1 of the product is not a stereocenter, the question of a *syn* or *anti* addition is irrelevant to the composition of the product mixture in the addition of bromine to propene.

Remember the point made in Sec. 7.9A of the text: the stereochemistry of an addition is relevant only when *both* carbons of the double bond become stereocenters as a result of the addition. Thus, one can study the mode of bromine addition to *cis*- and *trans*-2-butene, as in Sec. 7.9C, because the two carbons of the double bond in the addition product are stereocenters. In a practical sense, bromine addition would not be as useful when applied to the 2-butenes or to cyclohexene if it were not stereoselective, because a mixture of diastereomers would be formed.

The text showed that oxymercuration-reduction is not a stereoselective reaction. The lack of stereoselectivity matters only when it is applied to an alkene that would give a mixture of diastereomeric products. Many reactions that are not stereoselective are useful as long as their use is confined to situations in which stereoselectivity is not an issue.

Solutions

Solutions to In-Text Problems

7.1   Use the equation $\Delta G° = -2.3RT \log K_{eq}$ by supplying the appropriate energy difference in Fig. 7.5 as the $\Delta G°$ value and solving for K_{eq}. Note that $2.3RT$ at 298 K is 5.706 kJ/mol. The standard free energy difference of the twist-boat and chair forms is given in the figure as 15.9 kJ/mol.

$$15.9 \text{ kJ/mol} = (-5.706 \text{ kJ/mol}) \log K_{eq}$$
$$\log K_{eq} = \frac{15.9 \text{ kJ/mol}}{-5.706 \text{ kJ/mol}} = -2.79$$
$$K_{eq} = 1.62 \times 10^{-3}$$

Thus, the ratio of twist-boat concentration to chair concentration is 0.00162. In other words, there is about 616 times as much chair form as there is twist-boat form in any sample of cyclohexane at 25 °C.

7.3   (a) Because there are *two* fluorine-hydrogen 1,3-diaxial interactions in the axial conformation of fluoro-cyclohexane (just as there are two methyl-hydrogen 1,3-diaxial interactions in the axial conformation of methylcyclohexane), the energetic cost of a single 1,3-diaxial fluorine-hydrogen interaction is half of the energy difference between the axial and equatorial forms, or 0.5 kJ/mol (0.13 kcal/mol).

(b) The *gauche* conformation of 1-fluoropropane has a single *gauche* fluorine-methyl interaction:

This is approximately the same as a *single* hydrogen-fluorine 1,3-diaxial interaction in the axial conformation of fluorocyclohexane. From part (a), this is 0.5 kJ/mol, or 0.13 kcal/mol.

7.4   Because there is rapid rotation about the bond between the ring and the CH_2 of the ethyl group, the ethyl group can rotate about this bond so that the van der Waals radii of the methyl group and the axial hydrogens do not overlap.

In the conformation on the right, the interactions of the axial hydrogens with the two hydrogens of the ethyl substituent are the same as they would be with the two hydrogens of a methyl substituent.

7.5   (a) The two chair conformations of *cis*-1,3-dimethylcyclohexane:

(solution continues)

Don't forget that there are many valid ways to draw a *cis*-1,3-disubstituted cyclohexane. If the structures you drew look different from those above, try to manipulate your structures mentally to see whether they can be made to look identical to the ones above. (Study Guide Link 7.1 provides some exercises that should be helpful.)

7.6 (a) Two boat conformations of *cis*-1,3-dimethylcyclohexane:

7.7 The given conformation of *cis*-1,3-dimethylcyclohexane has no 1,3-diaxial methyl-hydrogen interactions. The given conformation of *trans*-1,3-dimethylcyclohexane has two 1,3-diaxial methyl-hydrogen interactions, each of which contributes 3.7 kJ/mol to the energy of the *trans* isomer. Consequently, the given conformation of *trans*-1,3-dimethylcyclohexane is 7.4 kJ/mol more energetic (less stable) than that of *cis*-1,3-dimethylcyclohexane.

7.9 (a)   (c)

cis-1,3-dimethylcyclohexane

(1*R*,2*S*,3*R*)-2-chloro-1-ethyl-
3-methylcyclohexane

7.10 (a) The large *tert*-butyl group assumes the equatorial position:

(c) The larger isopropyl group assumes the equatorial position:

7.11 As stated at the bottom of text page 290, a molecule is chiral (that is, can be isolated in optically active form) when its planar structure is noncongruent to its mirror image.

(a) *Trans*-1,3-dimethylcyclohexane can be isolated in optically active form. [See structures below part (e)].

(c) *Cis*-1,4-dimethylcyclohexane is achiral, and therefore cannot be optically active.

(e) *Cis*-1-ethyl-3-methylcyclohexane can be isolated in optically active form.

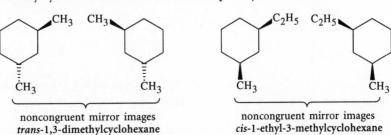

noncongruent mirror images
trans-1,3-dimethylcyclohexane

noncongruent mirror images
cis-1-ethyl-3-methylcyclohexane

7.12   (a) *Trans*-1,4-dimethylcyclohexane does *not* have asymmetric carbons.
(b) There are two stereocenters in *trans*-1,4-dimethylcyclohexane, because interchange of the methyl and hydrogen groups at either one gives *cis*-1,4-dimethylcyclohexane, a stereoisomer.

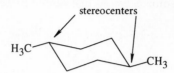

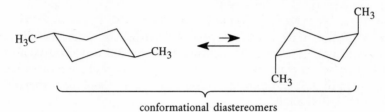

(c) The two chair conformations of *trans*-1,4-dimethylcyclohexane are conformational diastereomers:

conformational diastereomers

(d) *Trans*-1,4-dimethylcyclohexane is not chiral because it has an internal mirror plane.

trans-1,4-dimethylcyclohexane

7.13   (a) Enantiomers   (c) Identical molecules

7.14   (a) In planar *trans*-1,3-dibromocyclobutane, the carbon-bromine bonds are directed in opposite directions; hence, their bond dipoles would cancel, and the net dipole moment of the molecule would be zero. In puckered *trans*-1,3-dibromocyclobutane, the bond dipoles do not cancel.

<table>
<tr><td>planar trans-1,3-dibromocyclobutane:
bond dipoles cancel</td><td>puckered trans-1,3-dibromocyclobutane:
bond dipoles do not cancel</td></tr>
</table>

7.15   (a) *Cis*-1,2-dimethylcyclopropane is achiral; it is a *meso* compound.

cis-1,2-dimethylcyclopropane

7.16   (a) Bicyclo[2.2.2]octane

7.17   A fused bicyclic compound, by definition, has a "zero-carbon bridge." Consequently, compound (b) is a fused bicyclic compound; compound (a) is a bridged bicyclic compound.

7.18   The two labeled carbons in the diagram of *cis*-decalin on the following page are axial substituents in the other ring; consequently, the hydrogens on these carbons are involved in 1,3-diaxial interactions (dashed lines in the structure.). There are, as in other axially-substituted alkylcyclohexanes, two 1,3-diaxial

lines in the structure.). There are, as in other axially-substituted alkylcyclohexanes, two 1,3-diaxial interactions per axial carbon, except that one is common to both carbons; hence, there are a total of three 1,3-diaxial interactions. Because there are no 1,3-diaxial interactions in *trans*-decalin, it is more stable than *cis*-decalin by $3 \times 3.7 = 11.1$ kJ/mol.

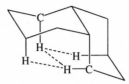

7.19   (a) The model of *cis*-bicyclo[3.1.0]hexane is much more easily built. The smaller is a ring, the closer is the dihedral angle of the *trans* bonds to 180°. Forcing this dihedral angle to much smaller values introduces strain within the ring. Such a large dihedral angle causes the ends of *trans* bonds to be too far apart to be readily bridged by only one carbon.

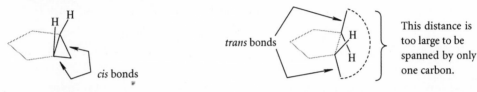

   Be sure that you build the models called for in this and the following problem. Models are much more revealing, and provide you with much greater insight, than simply reading verbal descriptions.

7.20   (a) Although both molecules have bridgehead double bonds, the double bond in compound *B* is more twisted, and a model of this molecule is more difficult to build. Consequently, compound *B* is less stable and therefore would have the greater (more positive or less negative) heat of formation.

7.21   (a) Because a nail and a hammer are achiral objects, all interactions of these objects with any chiral object (such as a hand) and its enantiomer are identical. Hence, assuming equal strength, a right and a left hand should be equally adept at driving a nail. (This assumes, of course, that the person driving the nail doesn't accidentally bend it into some chiral shape.) In contrast, when it drives a screw, a chiral hand interacts with the helical (and chiral) threads of the screw. For this reason, the right and left hands of the same person should differ in their ability to drive a screw.

7.22   The following stereoisomers of 2,3-dibromobutane could form in the bromination of *cis*-2-butene:

	CH₃		CH₃		CH₃

Because the (2R,3R) and the (2S,3S) stereoisomers are enantiomers, they must be formed in identical amounts. The *meso* compound is a diastereomer of these, and should therefore be formed in a different amount. (Sec. 7.9C of the text shows that very little of the *meso* compound is formed.)

7.24   Write all possible diastereomers resulting from addition to one enantiomer of the starting material. These are in principle formed in different amounts. (Problem 7.25b considers this case further.)

Each of these structures has an enantiomer; drawing these gives four additional diastereomers, which are also formed in different amounts. (Diastereomers are *always* formed in different amounts in principle.)

The process that gives *A* is enantiomeric to the process that gives *A´*, and therefore the two processes must occur with identical rates. Because the same amounts of the enantiomeric alkenes are involved in each process (because the starting alkene is racemic), the enantiomers of each pair are formed in identical amounts. Thus, *A* and *A´* are formed in identical amounts; *B* and *B´* are formed in identical amounts; and so on.

7.25   (a) Because the bromonium-ion mechanism involves *anti* addition, the bromination of cyclohexene must give *trans*-1,2-dibromocyclohexane as a product.

The equation above shows the formation of one enantiomer; the other enantiomer is formed in the same amount from the same bromonium-ion intermediate. (Show this process with the curved-arrow formalism.)

7.26   (a) Because hydroboration-oxidation is a net *syn* addition, the H and OH are added to the same face of the alkene. This addition can occur in two different ways to give enantiomeric products, which are formed in identical amounts.

7.27 The presence of deuteriums in Problem 7.26 gives rise to diastereomeric products. If the starting material is not isotopically substituted, then the two products are enantiomeric, and the same pair of enantiomers is formed whether the *cis*- or *trans*-alkene is used as the starting material.

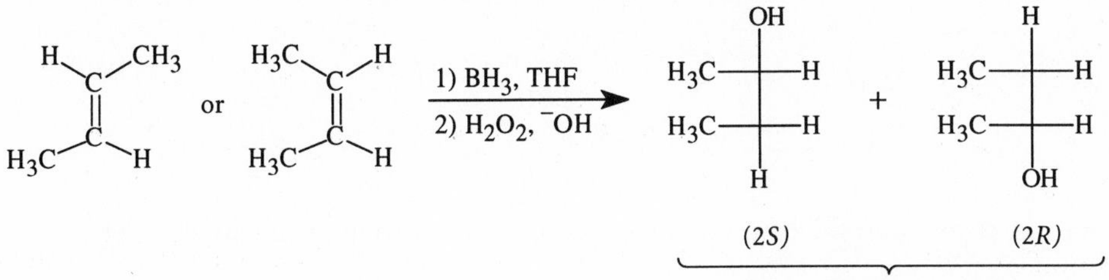

7.28 (a)

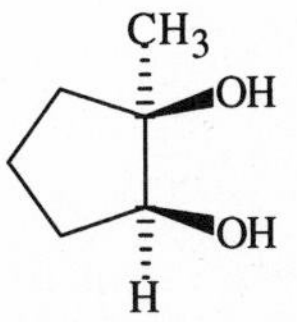

CH₃

(and its enantiomer)

(c)

OH

H——CH₃

H——CH₃

OH

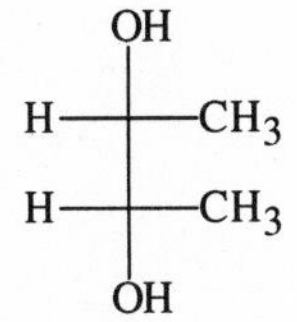

meso-2,3-butanediol

Solutions to Additional Problems

7.29 (a) Two of four that could be drawn are the following:

bicyclo[2.2.0]hexane

bicyclo[2.1.1]hexane

(b)

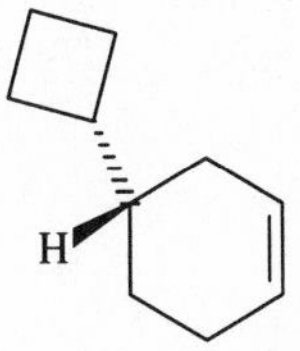

(S)-4-cyclobutylcyclohexene

7.31 Methylcyclohexane and (*E*)-4-methyl-2-hexene are *constitutional isomers*. Hence, they have the same molecular mass but different properties.

(a) A molecular mass determination would not distinguish between the two compounds because their molecular masses are identical.

(b) Uptake of H_2 over a catalyst would distinguish between the two compounds because the alkene would take up hydrogen in the presence of a catalyst, whereas the cycloalkane would not.

(c) Reaction with alkaline $KMnO_4$ would distinguish between the two compounds because only the alkene would react rapidly to give a brown MnO_2 by-product (Baeyer test).

(d) Determination of the empirical formula would not distinguish between the two compounds because both compounds are isomers; they have the same formula, and hence, the same empirical formula.

(e) Because constitutional isomers have different physical properties, determination of the heat of formation would distinguish between the two compounds. (Which would have the greater heat of formation?)

(f) Enantiomeric resolution would distinguish between the two because (*E*)-4-methyl-2-hexene is chiral, and can be resolved into enantiomers, whereas methylcyclohexane, an achiral compound, cannot be.

(g) Bromine addition would distinguish between the two, because the alkene would react with bromine, but the cycloalkane would not.

7.32 (a)   (c)

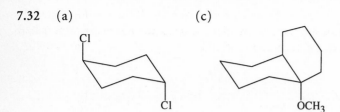

7.33 (a) In this compound the large *tert*-butyl group must assume the equatorial position.

7.34 (a) The equilibrium constant is given as 2.07 for the ratio [equatorial]/[axial]. The standard free-energy difference is calculated by applying Eq. 3.29, text p. 103:

$$\Delta G° = -2.3RT \log (2.07) = -(5.706 \text{ kJ/mol})(0.3160) = -1.80 \text{ kJ/mol}$$

In other words, for the equilibrium *axial ⇄ equatorial,* the equatorial form has the lower standard free energy, and is thus more stable.

7.35 (a) Compound *A* can be prepared by hydroboration-oxidation from either *cis-* or *trans-*3-hexene because the alkene is symmetrical and regioselectivity is not an issue.

$$CH_3CH_2CH=CHCH_2CH_3 \xrightarrow[\text{2) } H_2O_2, \text{ }^-OH]{\text{1) } BH_3/THF} \overset{OH}{CH_3CH_2\underset{|}{C}HCH_2CH_2CH_3}$$

3-hexene   3-hexanol
                                 (compound *A*)

Hydroboration-oxidation of 1-methylcyclopentene would give compound *D*, not compound *B*, because this reaction sequence results in a *syn* addition.

1-methylcyclopentene   compound *D*

There is no other alkene that would give either compound *B* or compound *C* as the *only* constitutional isomer and stereoisomer.

7.36 (1a–c) The reaction gives a racemic mixture, that is, an equimolar mixture of enantiomers.

$$\overset{R}{\underset{|}{CH_3CH_2CH_2\overset{Br}{\underset{CH_2CH_3}{C}}CH_3}} + \overset{S}{\underset{|}{CH_3CH_2CH_2\overset{Br}{\underset{CH_2CH_3}{C}}CH_3}}$$

(solution continues)

If you have an early printing of the text, note that there is an extra methyl group in the alkene that doesn't belong because of a compositor error following proof. The alkene starting material should be the following:

$$CH_3CH_2CH_2C{=}CH_2$$
$$\underset{\displaystyle CH_2CH_3}{|}$$

(1d)   The two enantiomers have identical melting and boiling points.

(3a–b) The products are the following two diastereomers:

$$\overset{R}{CH_3CH_2CH}{-}\overset{R}{CHCH_2OH} \quad + \quad \overset{R}{CH_3CH_2CH}{-}\overset{S}{CHCH_2OH}$$
$$\underset{\displaystyle CH_3}{|} \quad \underset{\displaystyle CH_3}{|} \qquad\qquad\qquad \underset{\displaystyle CH_3}{|} \quad \underset{\displaystyle CH_3}{|}$$

(3c)   The diastereomeric products are formed in different amounts.
(3d)   The two diastereomers should have different melting and boiling points.

(5a)   The starting alkene is racemic, and is therefore an equimolar mixture of two enantiomers. To predict the products, make the prediction for each enantiomer separately and then combine the results. Starting arbitrarily with the *S* enantiomer, the products are the following two diastereomers:

$$\overset{S\ \ S}{CH_3CHCHCH_2Br} \quad + \quad \overset{S\ \ R}{CH_3CHCHCH_2Br}$$
$$\underset{\displaystyle Ph\ \ Br}{|\quad|} \qquad\qquad\qquad \underset{\displaystyle Ph\ Br}{|\quad|}$$

$$A \qquad\qquad\qquad\qquad B$$

The *R* enantiomer of the alkene starting material gives a corresponding pair of diastereomers:

$$\overset{R\ \ R}{CH_3CHCHCH_2Br} \quad + \quad \overset{R\ \ S}{CH_3CHCHCH_2Br}$$
$$\underset{\displaystyle Ph\ Br}{|\quad|} \qquad\qquad\qquad \underset{\displaystyle Ph\ Br}{|\quad|}$$

$$C \qquad\qquad\qquad\qquad D$$

(5b)   Compounds *A* and *B* are diastereomers, as are compounds *C* and *D*, compounds *A* and *D*, and compounds *B* and *C*; compounds *A* and *C* are enantiomers, as are compounds *B* and *D*.
(5c)   The diastereomers are formed in different amounts; the enantiomers are formed in identical amounts.
(5d)   Any pair of diastereomers have different melting points and boiling points; any pair of enantiomers have identical melting points and boiling points.

(7a–d) Because catalytic hydrogenation is a stereoselective *syn* addition, only one product, *cis*-decalin, is formed.

cis-decalin

7.37   (a)   (c)

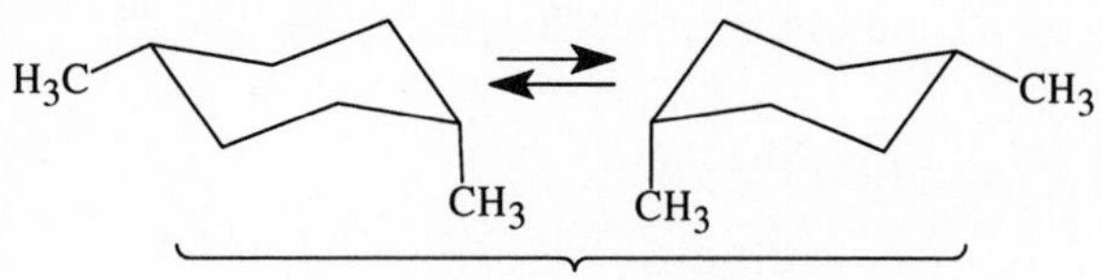

identical structures   diastereomers

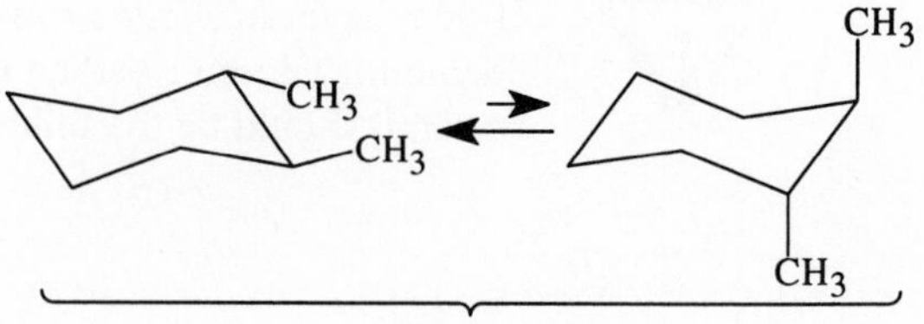

7.38   (a) Note that all ring junctions are *trans.*

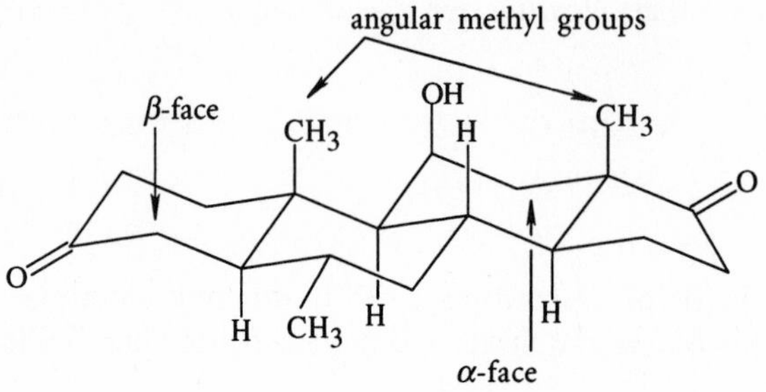

7.39   (a)

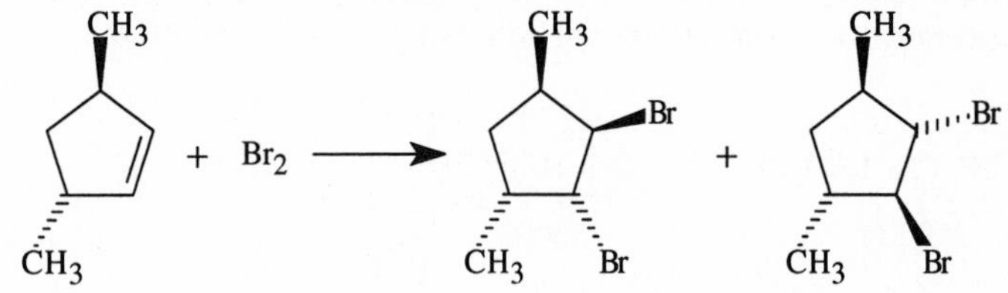

7.40   The two dibromides are the two diastereomeric *trans*-disubstituted *trans*-decalin derivatives:

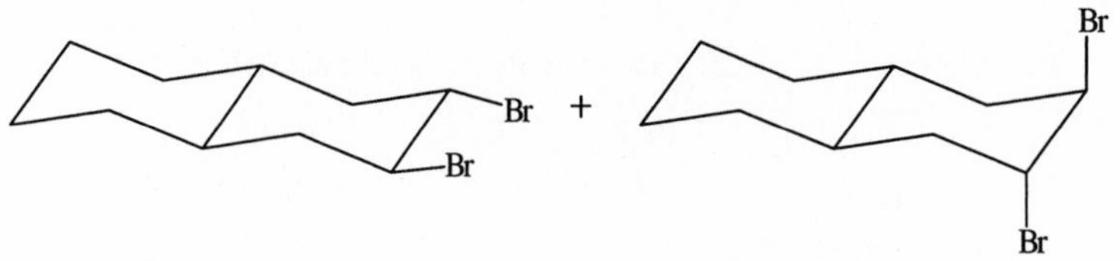

7.42   Of the five alkenes, only (b), (c), and (d) are chiral and could be optically active. Of these, (b) and (c) would give 1,4-dimethylcyclohexane as a hydrogenation product, which, regardless of whether it is *cis* or *trans,* is achiral and therefore optically inactive. [See the solution to Problem 7.12(d).] Hydrogenation of (d) would give optically active *trans*-1,3-dimethylcyclohexane. The answers are therefore (b) and (c).

7.43   Only properties that involve interaction with another chiral object or force will be different. Consequently, properties (a), (b), (d), (e), (g) are identical for the two enantiomers of 2-pentanol. The two enantiomers should differ in properties (c), (f), and (h).

7.44   Because this is an amine, both the chair flip and amine inversion can occur. Note that the amine inversion interchanges the axial and equatorial positions of the electron pair and the hydrogen on the nitrogen *without* flipping the ring. Thus, four chair conformations are in rapid equilibrium:

(solution continues)

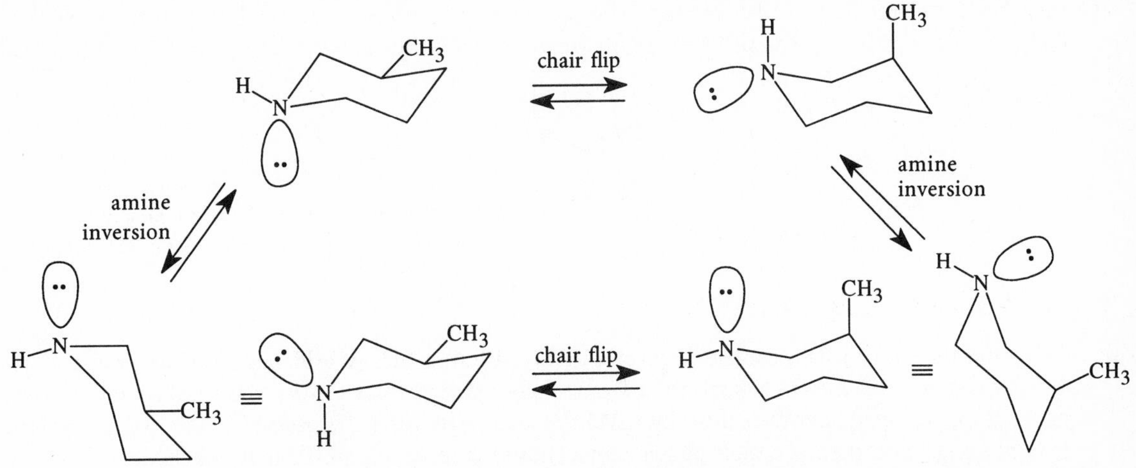

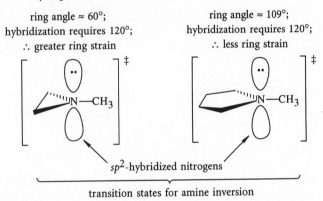

7.46   In the transition state for inversion of both amines, the nitrogen is *sp²*-hybridized. The C—N—C angle required by this hybridization is 120°. This angle is much more difficult to achieve within a three-membered ring, which has an optimal C—N—C angle of 60°, than it is in within a five-membered ring, which has an optimal C—N—C angle of 109°. Hence, there is much more angle strain in the transition state for inversion of 1-methylaziridine than there is in the transition state for inversion of 1-methyl-pyrrolidine. For this reason, the transition state for inversion of the aziridine derivative has a very high energy, and nitrogen inversion in this derivative is very slow. (This is a very unusual situation, since inversion of most amines is very rapid; see Sec. 6.10B.)

ring angle ≈ 60°;
hybridization requires 120°;
∴ greater ring strain

ring angle ≈ 109°;
hybridization requires 120°;
∴ less ring strain

sp²-hybridized nitrogens

transition states for amine inversion

7.47   The *cis* isomer of the starting alkene was used. This is shown by starting with the *cis* isomer and imagining the addition of the two —OH groups to the same face of the double bond. (Recall from text p. 320 that glycol formation with $KMnO_4$ is a *syn* addition.)

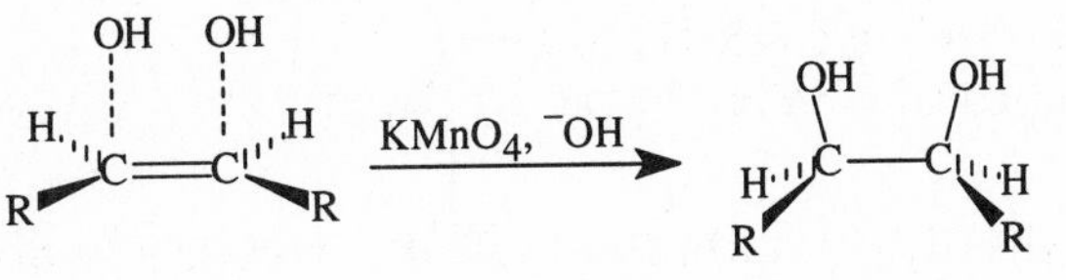

OH   OH

H,,,   C=C   ,,,H
R   R

$KMnO_4$, ⁻OH →

OH   OH

H,,,C—C,,,H
R   R

meso

As this example and Sec. 7.9C show, there is a relationship between the *stereochemistry of the alkene*, the *stereochemistry of the reaction*, and the *stereochemistry of the product*. If we specify any two of these, the other is determined.

7.49   First, convert the Fischer projection of the product into a three-dimensional perspective. (Build a model if necessary.) Then internally rotate about the central bond until the two —CO_2^- groups have a dihedral angle of 180°, as they do in the starting material. Finally, examine the relationship of the —D and —OD groups. This reveals that the product results from *anti* addition of the —D and the —OD to the alkene:

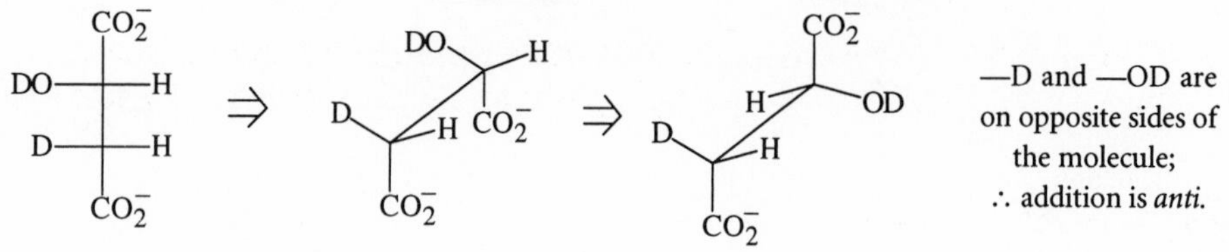

7.50 Following the hint in the problem, interpret each chair structure as a planar "line-and-wedge" structure. If these two structures are congruent, then the two chair structures are *either* identical *or* are conformational stereoisomers. Decide between these by using the chair structures themselves. If the two chair structures are not congruent, then use their planar representations to decide on their relationship.

(a)

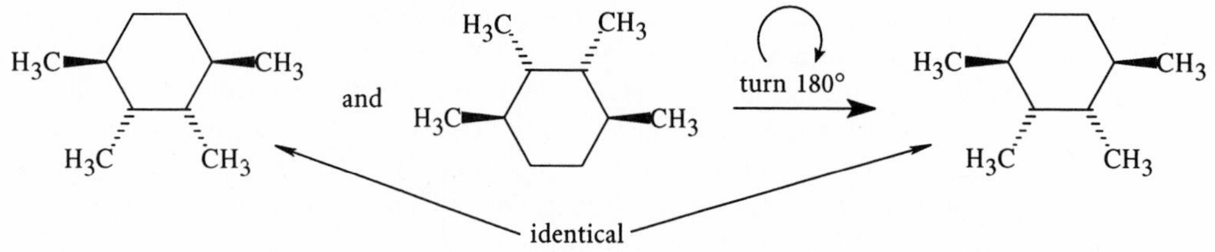

Because the planar representations are identical, the two chair structures are either identical or they are conformational stereoisomers. Examination of the chair structures shows that they noncongruent mirror images. (Use models if necessary!) Consequently, the two structures are conformational enantiomers.

(b)

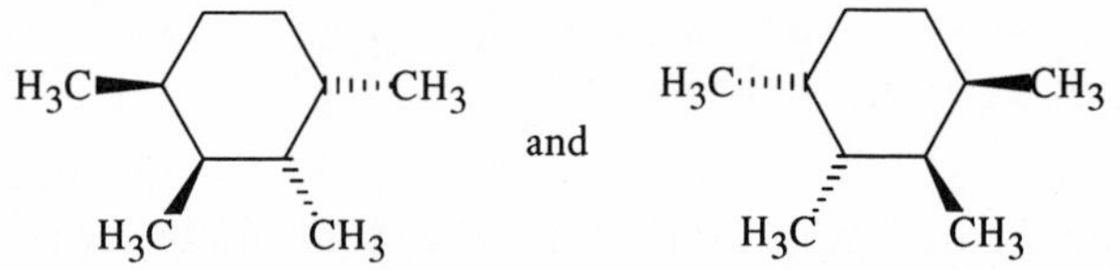

These two planar structures are noncongruent mirror images and are therefore enantiomers. (Notice that their enantiomeric relationship cannot be seen from the chair structures until either of them is subjected to the chair flip!) The chair structures shown are diastereomeric conformations of enantiomeric molecules.

(c)

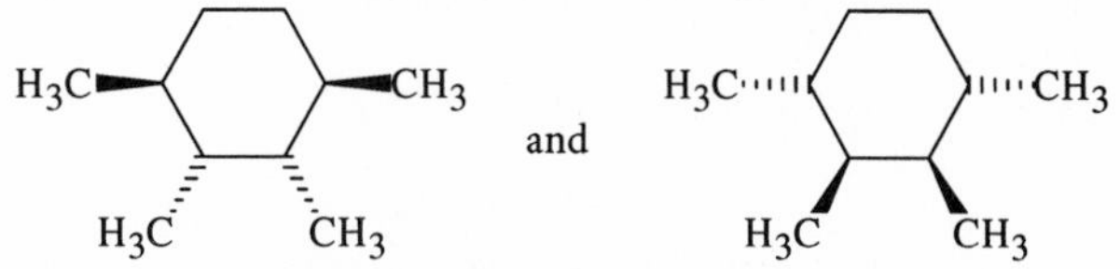

That the two planar structures are congruent can be determined by lifting one out of the page and turning it over onto the other (so that "up" bonds become "down" bonds and vice-versa). Consequently, the two chair conformations are either identical or are conformational stereoisomers. In fact, they are identical conformations drawn in different ways. Again, use models if necessary to demonstrate this to yourself.

Remember that the use of models to solve problems in which two structures are compared requires *two models,* one of the original structure and another of the structure with which it is compared.

Problem 7.50 is a particularly good test of your ability to manipulate cyclohexane rings. If you were able to solve it without a great deal of difficulty, you have in fact mastered the art of handling chair conformations!

7.51   (a) Hydration involves a carbocation intermediate. Because a carbocation can be attacked by nucleophiles at either face of its vacant *p* orbital, a mixture of stereoisomeric hydration products is expected.

Notice that protonation of the double bond by D_3O^+ has been assumed to occur from the upper face of the double bond; protonation at the opposite face followed by attack of D_2O as shown above would yield the enantiomers of *A* and *B*.

(b) D_2O (and thus D_3O^+) rather than H_2O must be used to distinguish the hydrogen added to the double bond from the one that is there to begin with.

(c) The two hydrogens referred to in part (b) could also be distinguished if the one in the starting alkene is replaced by a deuterium, and H_3O^+ is used as the acid, that is, if an alkene of the following structure is used:

7.52   (a) The two chiral stereoisomers are the following enantiomeric pair:

(b) The only candidate structures are those in which the number of axial chlorines equals the number of equatorial chlorines. In addition, the two chlorines within each of the three pairs that have a 1,4-relationship must differ in whether their positions are axial or equatorial. Thus, each of the following stereoisomers has two identical chair conformations.

identical

identical

7.54   The true statements about the relationship of *cis*- and *trans*-decalin (see text p. 297) are (c) and (d).

7.55   In order for a compound to exist in a relatively high-energy twist-boat conformation, there must be a good reason for it to avoid the chair. Indeed, compounds *B* and *D* have very severe 1,3-diaxial interactions between a *tert*-butyl group and a methyl group in the chair conformations shown. Compound *D*, a *cis*-decalin derivative, can avoid its 1,3-diaxial interaction by undergoing a chair flip. However, compound *B*, a *trans*-decalin derivative, cannot undergo the chair flip (see text p. 297). Hence, the only way it can avoid the 1,3-diaxial interaction is for the ring containing the *tert*-butyl group to assume a twist-boat conformation. The correct answer, then, is *B*.

7.56   In conformation *A*, there is a severe van der Waals repulsion between the methyl group and an axial hydrogen on the other ring. (Note that this is different from, and much worse than, a 1,3-diaxial interaction.) In conformation *B*, the offending interaction is absent. The methyl-hydrogen interaction is so unfavorable that the twist-boat conformation *B* is actually the more stable one.

severe van der Waals
repulsion——→ H₃C
                      H

7.58   Within each structure, count the 1,3-diaxial interactions between the methyl group and axial ring hydrogens four carbons away (counting the methyl as one carbon). These are shown with double-headed arrows in the structures of *A* and *B* below. (There are no methyl-hydrogen 1,3-diaxial interactions in compound *C*.)

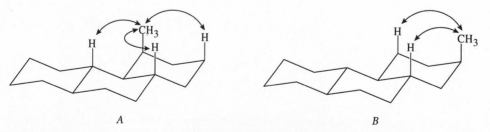

A   *B*

Each interaction shown adds 3.7 kJ/mol to the heat of formation relative to that of *C*. Consequently, *C* has the lowest heat of formation; the heat of formation of *B* is (2 × 3.7) = 7.4 kJ/mol greater; and the heat of formation of *A* is (3 × 3.7) = 11.1 kJ/mol greater than that of *C*.

7.59 There are two distinguishable ways in which the organoborane can approach the π bond of a *cis*-2-butene molecule. Approach from one face gives a transition state in which there is severe van der Waals repulsion between a methyl group of the alkene and a methyl group on the organoborane (structure *A* below). Approach from the opposite face gives a transition state in which the van der Waals repulsion is between a *hydrogen* of the alkene and a methyl group of the borane (structure *B* below).

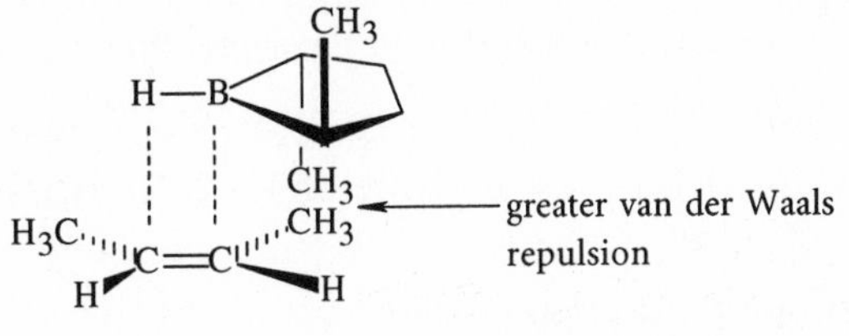

Because the repulsion in *B* involves the smaller group (hydrogen), *B* is the transition state of lower energy. As a result, it is this transition state that leads to the product which, on oxidation with alkaline H_2O_2, gives (*S*)-2-butanol:

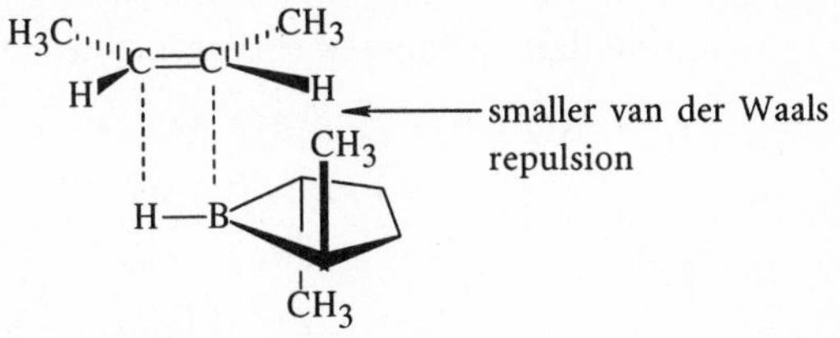

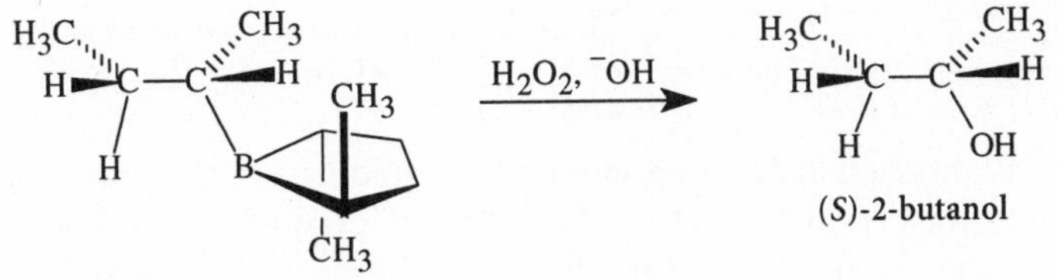

organoborane product from
transition state *B*

(*S*)-2-butanol

7.61 (a)

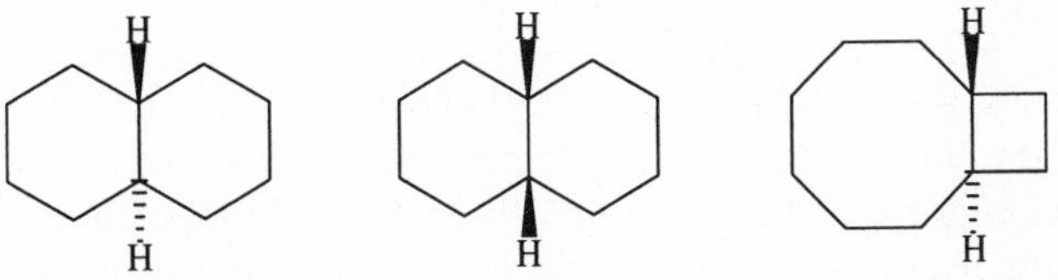

C   *A*   *B*

———— increasing heat of formation ————→

Trans-decalin, *C*, is more stable than *cis*-decalin, *A* (see Problem 7.18, text p. 298). However, there is no angle strain in either compound. Compound *B* is least stable because of the strain in its four-membered ring. You can see from Table 7.1, text p. 272, that the strain in a cyclobutane ring ($4 \times 6.9 = 27.6$ kJ/mol) is far more destabilizing that the three 1,3-diaxial interactions ($3 \times 3.7 = 11.1$ kJ/mol) in *cis*-decalin.

(b)

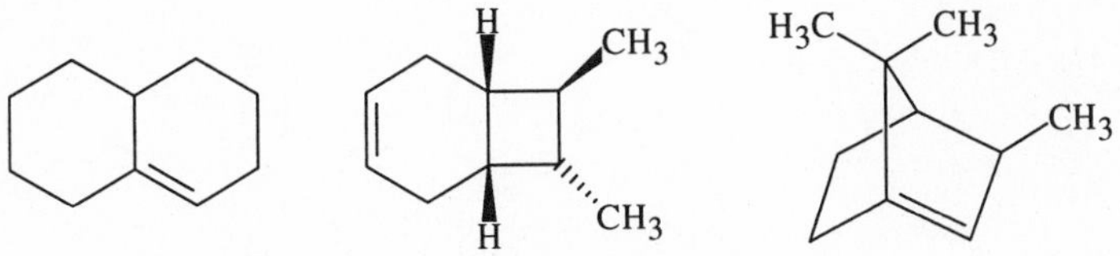

C   *B*   *A*

———— increasing heat of formation ————→

The ring strain in *B* makes it less stable than *C*; and the twisted double bond in *A* (violation of Bredt's rule) makes it so unstable that it cannot be isolated.

7.62   (a) Conformation *A* contains four 1,3-methyl-hydrogen diaxial interactions; the total energy cost of these is $(4 \times 3.7) = 14.8$ kJ/mol. Conformation *B* contains two methyl-hydrogen 1,3-diaxial interactions, at an energy cost of $(2 \times 3.7) = 7.4$ kJ/mol, and the methyl-methyl 1,3-diaxial interaction, at an energy cost to be determined. Let the energy cost of a methyl-hydrogen 1,3-diaxial interaction be $\Delta G°(\text{Me-H}) = 3.7$ kJ/mol, and let the energy cost of a methyl-methyl 1,3-diaxial interaction be $\Delta G°(\text{Me-Me})$. It is given that the $\Delta G°$ of the overall equilibrium, $\Delta G°(\text{eq})$, equals 8.4 kJ/mol and that conformation *A* has lower energy. Assuming that all energies except for the 1,3-diaxial interactions are about the same, then

$$\Delta G°(\text{eq})= \Delta G°(B) - \Delta G°(A) = 8.4 \text{ kJ/mol} = 2\Delta G°(\text{Me-H}) + \Delta G°(\text{Me-Me}) - 4\Delta G°(\text{Me-H})$$

or $$8.4 \text{ kJ/mol} = \Delta G°(\text{Me-Me}) - 2\Delta G°(\text{Me-H}) = \Delta G°(\text{Me-Me}) - 2(3.7 \text{ kJ/mol})$$

Solving,

$$\Delta G°(\text{Me-Me}) = 8.4 \text{ kJ/mol} + 7.4 \text{ kJ/mol} = 15.8 \text{ kJ/mol}$$

This calculation shows that a methyl-methyl 1,3-diaxial interaction is more than four times as costly in energy terms as a methyl-hydrogen 1,3-diaxial interaction. This should make sense, because a methyl group is much larger than a hydrogen.

(b) Conformation *C* has no 1,3-diaxial interactions, whereas conformation *D* has one methyl-methyl 1,3-diaxial interaction, which, from part (a), costs 15.8 kJ/mol, and two methyl-hydrogen 1,3-diaxial interactions, which costs $(2 \times 3.7) = 7.4$ kJ/mol. Thus, conformation *D* is $(15.8 + 7.4) = 23.2$ kJ/mol less stable than conformation *C*. The standard free energy change for the equilibrium is therefore 23.2 kJ/mol

7.64   (a) Two rings in a bridged bicyclic compound can be joined in three stereochemically different ways which might be termed (*in,in*), (*in,out*), and (*out,out*). Assuming hydrogens at both bridgeheads, these ways can be represented schematically as follows.

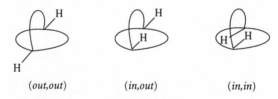

(*out,out*)   (*in,out*)   (*in,in*)

(b) As these simple diagrams suggest, the (*in,out*) and (*in,in*) patterns require that one and two hydrogens, respectively, occupy the region of space *within* a ring. This is sterically impossible for small rings. For small rings, this would actually require such a distortion of the tetrahedral structure that all four bonds to the bridgehead carbon with the *in* configuration would lie on one side of a plane through this carbon. (If the impossibility of achieving this sort of structure is not clear, try to build a model of (*in,out*)-bicyclo[2.2.2]octane.) The larger the rings, the larger the space within the ring, and for very large rings, the "inner" hydrogens can be accommodated easily within the ring and the tetrahedral bonding geometry of the bridgehead carbons can be achieved without significant strain. Hence, the (*in,out*) and (*in,in*) stereoisomers of the 77-carbon bicyclic alkane are stable.

. . . .

Introduction to Alkyl Halides, Alcohols, Ethers, Thiols, and Sulfides

Terms

Concepts

I. *Nomenclature of Alkyl Halides, Alcohols, Ethers, Thiols, and Sulfides*

A. GENERAL

1. Organic compounds are named by both common and substitutive nomenclature.
2. Common nomenclature, or radicofunctional nomenclature, is generally used only for the simplest and most common compounds.
   a. Do not confuse the allyl group with the vinyl group.

   $$H_2C=CH-CH_2- \qquad\qquad H_2C=CH-$$
   allyl group vinyl group

   b. Do not confuse the benzyl group with the phenyl group.

   benzyl group phenyl group

3. In substitutive nomenclature, the name is based on the principal group and principal chain; other groups are cited as substituents.
   a. Identify the principal group: the chemical group on which the name is based and cited as a suffix in the name.
      i. Hydroxy (—OH) and thiol (—SH) groups can be cited as principal groups.
      ii. A compete list of principal groups and their relative priorities are summarized in Appendix I of the text.
      iii. If a compound does not contain a principal group, it is named as a substituted hydrocarbon.
   b. Identify the principal carbon chain on which the name is based by applying the following criteria in the order given below until a decision can be made (an extensive list is given in Appendix I of the text):
      i. greatest number of principal groups;
      ii. greatest number of double and triple bonds;
      iii. greatest length;
      iv. greatest number of other substituents.
   c. Number the principal chain consecutively from one end by applying the following criteria in the order given below until there is no ambiguity:
      i. lowest number for the principal group;
      ii. lowest numbers for multiple bonds, with double bonds having priority over triple bonds;
      iii. lowest numbers for other substituents;
      iv. lowest number for the substituent cited first in the name.
   d. Begin construction of the name with the name of the hydrocarbon corresponding to the principal chain.
      i. Cite the principal group by its suffix and number; its number is the last one cited in the name.
      ii. Cite the names and numbers of the other substituents in alphabetical order at the beginning of the name.
4. Common and substitutive nomenclature should not be mixed.
5. Cyclic compounds with rings that contain more than one type of atom are called heterocyclic compounds.

B. ALKYL HALIDES

1. In an alkyl halide, a halogen atom is bonded to the carbon of an alkyl group.
2. Alkyl halides are classified as methyl, primary, secondary, or tertiary, depending of the number of alkyl groups attached to the carbon bearing the halogen. (Grey arrows show bonds to alkyl groups.)

H_3C—F H_3C \ CH—CH_2—Br / H_3C (cyclohexyl)—Cl H_3C—C—I with CH_3 up and CH_3 down

type of alkyl halide:	methyl	primary	secondary	tertiary
number of alkyl groups:	none	1	2	3

3. The common name of an alkyl halide is constructed from the name of the alkyl group followed by the halide as a separate word; the methyl trihalides are called haloforms.

H_3C—F H_3C \ CH—CH_2—Br / H_3C (cyclohexyl)—Cl H_3C—C—I (CH_3, CH_3) $CHCl_3$

methyl fluoride	isobutyl bromide	cyclohexyl chloride	*tert*-butyl iodide	chloroform

4. The IUPAC substitutive name of an alkyl halide is constructed by applying the rules of alkane and alkene nomenclature.
   a. Halogens are always treated as substituents.
   b. The halogens substituents are named fluoro, chloro, bromo, or iodo, respectively.

H_3C—F H_3C \ CH—CH_2—Br / H_3C (cyclohexyl)—Cl H_3C—C—I (CH_3, CH_3) $CHCl_3$

fluoromethane	1-bromo-2-methylpropane	chlorocyclohexane	2-iodo-2-methylpropane	trichloromethane

C. ALCOHOLS AND THIOLS

1. In an alcohol, a hydroxy group, —OH, is bonded to the carbon of an alkyl group.
2. In a thiol, a sulfhydryl group, —SH, also called a mercapto group, is bonded to an alkyl group; thiols, or mercaptans, are the sulfur analogs of alcohols.
3. Alcohols are classified as methyl, primary, secondary, or tertiary, depending of the number of alkyl groups attached to the carbon bearing the hydroxy group. (Grey arrows show bonds to alkyl groups.)

H_3C—OH H_3C \ CH—CH_2—OH / H_3C (cyclohexyl)—OH H_3C—C—OH (CH_3, CH_3)

type of alcohol:	methyl	primary	secondary	tertiary
number of alkyl groups:	none	1	2	3

4. Compounds that contain two alcohol groups are called glycols.

HO—CH_2—CH_2—OH

ethylene glycol

5. Compounds that contain an alcohol group on two adjacent carbons are called vicinal glycols.
6. In common nomenclature:
   a. The common name of an alcohol is derived by specifying the alkyl group to which the —OH group is attached, followed by the separate word alcohol.
   b. Thiols are named in the common system as mercaptans.

H_3C—OH H_3C \ CH—CH_2—SH / H_3C (cyclohexyl)—OH H_3C—C—SH (CH_3, CH_3)

methyl alcohol	isobutyl mercaptan	cyclohexyl alcohol	*tert*-butyl mercaptan

7. In substitutive nomenclature:
   a. The name of an alcohol is constructed by dropping the final *e* from the name of the parent alkane and adding an *ol* suffix.
   b. To name an alcohol containing more than one —OH group, the suffixes *diol*, *triol*, etc., are added to the name of the appropriate alkane without dropping the final *e*.
   c. For simple thiols, the —SH group is the principal group; the name is constructed by adding a *thiol* suffix to the name of the parent alkane (the final *e* is retained).

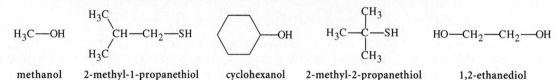

| methanol | 2-methyl-1-propanethiol | cyclohexanol | 2-methyl-2-propanethiol | 1,2-ethanediol |

D. ETHERS AND SULFIDES

1. In an ether, an oxygen is bonded to two alkyl or aryl groups, which may or may not be identical.
2. A thioether, or sulfide, is the sulfur analog of an ether.
3. In common nomenclature:
   a. The common name of an ether is constructed by citing as separate words the two groups attached to the ether oxygen in alphabetical order, followed by the word *ether*.
   b. The common name of a sulfide is constructed by citing as separate words the two groups attached to the sulfur in alphabetical order, followed by the word *sulfide* (or *thioether* as in older literature).

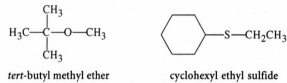

tert-butyl methyl ether   cyclohexyl ethyl sulfide

   c. The names of the following common heterocyclic ethers and sulfides should be learned:

| furan | tetrahydrofuran (THF) | thiophene | 1,4-dioxane | oxirane |

4. In substitutive nomenclature:
   a. Ethers and sulfides are never cited as principal groups; alkoxy groups (—OR) and alkylthio groups (—SR) are always cited as substituents.
   b. An —SR group is named by adding the suffix *thio* to the name of the R group; the final *yl* is not dropped.

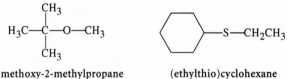

2-methoxy-2-methylpropane   (ethylthio)cyclohexane

5. Oxirane is the parent compound of a special class of heterocyclic ethers, called epoxides, that contain three-membered rings.
   a. Most epoxides are named substitutively as derivatives of oxirane.
   b. The atoms of the epoxide ring are numbered consecutively with the oxygen receiving the number 1 regardless of the substituents present.

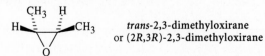

trans-2,3-dimethyloxirane
or (2R,3R)-2,3-dimethyloxirane

II. *Physical Properties of Alkyl Halides, Alcohols, Ethers, Thiols, and Sulfides*

A. STRUCTURE

1. Alkyl halides, alcohols, thiols, ethers, and sulfides have bond angles at carbon which are very close to the tetrahedral value of 109.5°.
2. Oxygen and sulfur have four "groups": two electron pairs and two alkyl groups or hydrogens.
   a. The unshared electron pairs on sulfur occupy orbitals derived from quantum level 3 that take up more space than those on oxygen, which are derived from quantum level 2.
   b. The angle at sulfur is generally found to be closer to 90° than the angle at oxygen.
3. Within a column of the periodic table, bonds to atoms of higher atomic number are longer.
4. Within a row of the periodic table, bond lengths decrease towards higher atomic number.
5. For a given molecular mass, alkyl halide molecules have smaller volumes than alkane molecules.

B. HYDROGEN BONDING

1. The hydrogen bond results from the combination of two factors: a weak covalent interaction between a hydrogen on the donor atom and unshared electron pairs on the acceptor atom, and an electrostatic attraction between oppositely charged ends of two dipoles.
   a. In the liquid state, hydrogen bonding is a force of attraction between molecules.
   b. In the gas phase, hydrogen bonding is much less important and, at low pressures, does not exist.
   c. In the case of alcohols, hydrogen bonding is a weak association of the O—H proton of one molecule with the oxygen of another.
2. Formation of a hydrogen bond requires two partners:
   a. the hydrogen-bond donor (the atom to which the hydrogen is fully bonded).
      i. The best hydrogen-bond donor atoms in neutral molecules are oxygens, nitrogens, and halogens.
      ii. A hydrogen-bond donor is analogous to a Brønsted acid; thus, all strong proton acids are also good hydrogen-bond donors.
   b. the hydrogen-bond acceptor (the atom to which the hydrogen is partially bonded).
      i. The best hydrogen-bond acceptors in neutral molecules are the electronegative first-row atoms oxygen, nitrogen, and fluorine.
      ii. The hydrogen-bond acceptor is analogous to a Brønsted base; thus, all strong Brønsted bases are also good hydrogen-bond acceptors.
   c. Some atoms are donors but not acceptors.

3. The hydrogen bond between two molecules resembles the same two molecules poised to undergo a Brønsted acid-base reaction.

C. BOILING POINTS

1. Noncovalent association of molecules in the liquid state increases their boiling points; the most important forces involved in intermolecular associations are
   a. hydrogen bonding ➠ hydrogen-bonded molecules have greater boiling points;
   b. attractive van der Waals forces (greatest for larger, more extended molecules), which are influenced by :
      i. molecular size ➠ larger molecules have greater boiling points;
      ii. molecular shape ➠ more extended, less spherical molecules have greater boiling points;
   c. attractive interactions between permanent dipoles ➠ molecules with permanent dipole moments have higher boiling points.
2. The polarity of a compound affects its boiling point.
   a. Most alkyl halides, alcohols, and ethers are polar molecules; that is, they have permanent dipole moments.

     b. When molecules with the same shape and molecular mass are compared, in many cases the more polar molecule has the higher boiling point.

  3. A higher boiling point results from:

     a. greater attraction between molecules in the liquid state.

     b. larger intermolecular attractions.

  4. In the case of alkyl halides, the effects of molecular volumes and polarity oppose each other.

     a. They nearly cancel in the case of alkyl chlorides.

     b. Alkyl bromides and iodides have lower boiling points than the alkanes of about the same molecular mass.

  5. The boiling points of alcohols, especially alcohols of lower molecular mass, are unusually high in comparison to those of other organic compounds.

     a. The unusual trends in the boiling points of alcohols are the results of hydrogen bonding.

     b. In order to vaporize a hydrogen-bonded liquid, energy is required to break the hydrogen bonds between molecules.

D. SAFETY HAZARDS OF ETHERS

  1. Samples of ethers can accumulate dangerous quantities of explosive peroxides and hydroperoxides by autoxidation, a spontaneous oxidation by oxygen in air.

  2. Ethers have low flash points. A flash point is the minimum temperature at which a compound is ignited by a small flame under standard conditions.

III. Solvents in Organic Chemistry

A. CLASSIFICATION OF SOLVENTS

  1. Solvents, liquids used to dissolve one or more compounds, can be classified in three ways, which are not mutually exclusive:

     a. A solvent is classified as protic or aprotic, depending on its ability to donate hydrogen bonds.

       i. A protic solvent consists of molecules that can act as hydrogen-bond donors.

       ii. Solvents that cannot act as hydrogen-bond donors are termed aprotic solvents.

     b. A solvent is classified as polar or apolar, depending on the size of its dielectric constant.

       i. A polar solvent has a high dielectric constant; an apolar solvent has a low dielectric constant ε.

       ii. A polar solvent separates, or shields, ions effectively from each other.

       iii. The electrostatic law shows that when the dielectric constant ε is large, the magnitude of the energy of interaction between the ions in solution is small.

     c. A solvent is classified as donor or nondonor, depending on its ability to act as a Lewis base.

       i. Donor solvents consist of molecules that can donate unshared electron pairs, that is, molecules that can act as Lewis bases.

       ii. Nondonor solvents cannot act as Lewis bases.

  2. Molecular polarity and solvent polarity are different concepts.

     a. Molecular polarity, or dipole moment, is a property of individual molecules.

     b. Solvent polarity, or dielectric constant, is a property of many molecules acting together.

  3. Polar molecules are attracted to each other because they can align in such a way that the negative end of one dipole is attracted to the positive end of the other.

  4. When a polar molecule contains a hydrocarbon portion of even moderate size its polarity has little effect on its physical properties.

B. SOLUBILITY

  1. Three solvent properties contribute to the solubility of ionic compounds:

     a. polarity (high dielectric constant), by which solvent molecules separate ions of opposite charge;

     b. proticity (hydrogen-bond donor capacity), by which solvent molecules solvate anions;

     c. electron-donor ability, by which solvent molecules solvate cations through Lewis-base and ion-dipole interactions; unshared electron pairs in a solvent molecule act as Lewis bases (electron donors) toward an electron-deficient cation.

  2. Solubility is the factor that determines whether a given compound will dissolve in a particular solvent.

     a. For a covalent compound, the best rule of thumb for selecting a solvent is "like dissolves like."

    b. For ionic compounds, the best solvents are polar, protic, donor solvents.

3. Ionic compounds in solution can exist in several forms, two of which are ion pairs and associated ions.

    a. In an ion pair, each ion is closely associated with an ion of opposite charge.

    b. Dissociated ions move more or less independently in solution and are surrounded by several solvent molecules called collectively the solvent shell or solvent cage of the ion.

4. The ability of a solvent to shield ions from one another is measured by its dielectric constant ε.

5. Three solvent properties enhance the solubility of ionic compounds:

    a. highly polar solvents separate ions from each other;

    b. protic solvents stabilize anions by hydrogen bonding;

    c. donor solvents stabilize cations by donor interactions.

6. Solvation is a term used to describe the favorable interaction of a dissolved molecule with solvent; when solvent molecules interact favorably with an ion, they are said to solvate the ion. Hydrogen bonding and donor interactions are two solvation mechanisms for ions.

7. The effects of solvents on chemical reactions are closely tied to the principles of solubility.

C. CROWN ETHERS AND IONOPHORE ANTIBIOTICS

1. Crown ethers are heterocyclic ethers containing a number of regularly spaced oxygen atoms that interact with cations through the same mechanism used by donor solvents.

    a. Because the metal ion must fit within the cavity, the crown ethers have some selectivity for metal ions according to size.

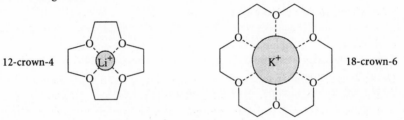

    b. Because crown ethers have significant solubilities in hydrocarbon solvents, they can be used to dissolve salts in hydrocarbon solvents.

2. Ionophores are molecules that form strong complexes with specific ions; crown ethers are one type of ionophore.

3. Crown ethers and other ionophores form complexes with cations by creating artificial solvation shells for them.

4. The ionophore antibiotics form strong complexes with metal ions in much the same way as crown ethers.

    a. An antibiotic is a compound that interferes with the growth or survival of one or more microorganisms.

IV. *Acidity and Basicity*

A. ACIDITY OF ALCOHOLS AND THIOLS

1. Alcohols and thiols are weak acids.

    a. Typical primary alcohols have pK_a values near 15–16 in aqueous solution.

    b. Thiols, with pK_a values near 10–11, are substantially more acidic than alcohols.

2. The conjugate bases of alcohols are called alkoxides, or alcoholates.

    a. The common name of an alkoxide is constructed by deleting the final *yl* from the name of the alkyl group and adding the suffix *oxide*.

    b. In substitutive nomenclature, the suffix *ate* is simply added to the name of the alcohol.

3. The conjugate bases of thiols are called mercaptides in common nomenclature and thiolates in substitutive nomenclature.

4. The relative acidities of alcohols and thiols are a reflection of the element effect.

C. Structural Effects on Alcohol Acidity

1. The acidities of alcohols are in the order:

    methyl > primary > secondary > tertiary

   a. The acidity order in solution is due primarily to the effectiveness with which alcohol molecules solvate their conjugate-base anions.
   b. The relative order of acidity of different types of alcohols is reversed in the gas phase (that is, in the absence of solvent) compared to the relative order of acidity in solution.
2. The acidity of alcohols is reduced by branching near the —OH group and increased by electron-withdrawing substituents.
   a. Substituted alcohols and thiols show the same type of polar effects on acidity as do substituted carboxylic acids.
   b. The polar effects of electronegative groups are more important when these groups are closer to the —OH group.
   c. Branched alcohols are more acidic than unbranched ones in the gas phase because α-alkyl substituents stabilize alkoxide ions.

D. Basicity of Alcohols, Thiols, and Ethers

1. Alcohols, thiols, and ethers are weak Brønsted bases, and react with strong acids to form positively charged conjugate-acid cations that have negative pK_a values.
   a. Alcohols and ethers do not differ greatly from water in their basicities; thiols and sulfides, however, are much less basic.
   b. Alcohols and thiols, like water, are amphoteric substances; that is, they can both gain and lose a proton.
2. Tertiary alkoxides are more basic than primary alkoxides; that is, tertiary alcohols have higher pK_a values than primary alcohols.
3. Water, alcohols, and ethers are good Lewis bases.

· ·

Reactions

· ·

I. Substitution Reactions of Alkyl Halides

A. Formation of Grignard and Organolithium Reagents

1. Compounds that contain carbon-metal bonds are called organometallic compounds.
   a. A Grignard reagent is an organometallic compound of the form R—MgX, where X = Br, Cl, or I.
   b. Organolithium reagents are organometallic compounds of the form R—Li.
2. Both Grignard and organolithium reagents are formed by adding the corresponding alkyl or aryl halides to rapidly stirred suspensions of the appropriate metal.

$$R\text{—}X \ + \ Mg \ \xrightarrow{\text{ether}} \ R\text{—}Mg\text{—}X \qquad (X = Cl, Br, \text{or } I)$$
    Grignard reagent

$$R\text{—}X \ + \ 2\,Li \ \xrightarrow{\text{hydrocarbon}} \ R\text{—}Li \ + \ LiX \qquad (X = Cl, Br, \text{or } I)$$
    organolithium reagent

   a. Reaction of alkyl halides with magnesium metal yields Grignard reagents.
      i. Ether solvents must be used for the formation of Grignard reagents; the ether solvent plays a crucial role in their formation because the ether associates with the metal in a Lewis acid-base interaction.

$$
\begin{array}{c}
\overset{\displaystyle Br}{|}\\
\underset{C_2H_5}{\overset{C_2H_5}{\diagdown}}\ddot{O}: \longrightarrow Mg \longleftarrow :\ddot{O}\underset{C_2H_5}{\overset{C_2H_5}{\diagup}}\\
R
\end{array}
$$

    ii. Grignard reagents are formed on the surface of the magnesium metal.

    iii. The formation of Grignard reagents is believed to involve radical intermediates.

  b. Reaction of alkyl halides with lithium yields organolithium reagents.

    i. Lithium reagents are typically formed in hydrocarbon solvents.

    ii. Organolithium reagents are thought to be formed on the surface of the lithium metal in which a fresh metal surface is continuously exposed to the alkyl halide.

3. Grignard and organolithium reagents behave as strong Brønsted bases and react violently with acids, including water and alcohols, to give alkanes. (See Part II.B of this outline.)

4. All reactions of Grignard and organolithium reagents can be understood in terms of the polarity of the carbon-metal bond.

  a. Because carbon is more electronegative than either magnesium or lithium, the negative end of the carbon-metal bond is the carbon atom.

    i. A carbon bearing three bonds, an unshared electron pair, and a negative formal charge is termed a carbon anion, or carbanion.

    ii. Carbanions are powerful Brønsted bases because their conjugate acids, the corresponding alkanes, are extremely weak acids.

  b. Grignard and organolithium reagents react as if they were carbanions; however, they are not true carbanions because they have covalent carbon-metal bonds.

$$
\underset{\text{polarized bond}}{\overset{\delta-}{-}\overset{|}{\underset{|}{C}}\overset{\delta+}{-}MgBr} \quad \longleftarrow \text{reacts as if it were} \longrightarrow \quad \underset{\text{carbanion}}{-\overset{|}{C}:^-} \; + \; {}^+MgBr
$$

B. FREE-RADICAL HALOGENATION OF ALKANES

1. Alkanes react with bromine and chlorine in the presence of heat or light in free-radical substitution reactions to give alkyl halides.

2. The mechanism of this reaction in fact follows the typical pattern of other free-radical chain reactions; it has initiation, propagation, and termination steps.

  a. The reaction is initiated when a halogen molecule absorbs energy from heat or light and dissociates homolytically into halogen atoms.

$$
X\!-\!X \xrightarrow[\text{or heat}]{\text{light}} X\cdot \; + \; X\cdot \quad (X = Cl \text{ or } Br) \qquad \text{initiation step}
$$

  b. The ensuing chain reaction has the following propagation steps.

$$
R\!-\!H \;\; \cdot X \longrightarrow R\cdot \; + \; H\!-\!X
$$

$$
R\cdot \;\; X\!-\!X \longrightarrow R\!-\!X \; + \; X\cdot
$$

    propagation steps

  c. Termination steps result from the recombination of radical species.

$$
R\cdot \;\; X\cdot \longrightarrow R\!-\!X \qquad \text{one of several possible termination steps}
$$

3. Free-radical halogenations with chlorine and bromine proceed smoothly, halogenation with fluorine is violent, and halogenation with iodine does not occur.

II. Acid-Base Reactions of Alcohol, Thiols, Grignard and Organolithium Reagents

A. FORMATION OF ALKOXIDES AND MERCAPTIDES

1. Alkoxides can be formed from alcohols with stronger bases such as sodium hydride, NaH, which is a source of the hydride ion, H:⁻.

$$R—\overset{..}{\underset{..}{O}}H \;+\; NaH \longrightarrow R—\overset{..}{\underset{..}{O}}:^- Na^+ \;+\; H_2$$

2. Sodium metal reacts with an alcohol to afford a solution of the corresponding sodium alkoxide; the rate of this reaction depends strongly on the alcohol.

3. Because thiols are much more acidic than water or alcohols, they, unlike alcohols, can be converted completely into their conjugate-base mercaptide anions by reaction with one equivalent of hydroxide or alkoxide.

4. Thiols form insoluble mercaptides with many heavy-metal ions, such as Hg^{2+}, Cu^{2+}, and Pb^{2+}.

$$R—\overset{..}{\underset{..}{S}}H \;+\; NaOH \;\xrightarrow{H_2O}\; R—\overset{..}{\underset{..}{S}}:^- Na^+ \;+\; H_2O$$

$$2\,R—\overset{..}{\underset{..}{S}}H \;+\; Hg^{2+} \;\xrightarrow{H_2O}\; R\overset{..}{S}—Hg—\overset{..}{S}R \;+\; 2\,H_3O^+$$

B. PROTONOLYSIS OF ORGANOMETALLIC REAGENTS

1. Any Grignard or organolithium reagent reacts vigorously with even relatively weak acids, such as water and alcohols, to give the conjugate-base hydroxide or alkoxides and the conjugate-acid hydrocarbon of the carbanion; this reaction is an example of protonolysis.

$$R—MgX \;+\; R'—OH \longrightarrow R—H \;+\; Mg^{2+} \;+\; X^- \;+\; R'O^-$$

2. A protonolysis is a reaction with the proton of an acid that breaks chemical bonds.

   a. Protonolysis of Grignard reagents is also useful, because it provides a method for the preparation of hydrocarbons from alkyl halides.

   b. A particularly useful variation of this reaction is the preparation of hydrocarbons labeled with the hydrogen isotopes deuterium (D, or ^2H) or tritium (T, or ^3H) by reaction of a Grignard reagent with the corresponding isotopically labeled water.

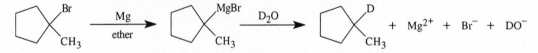

Study Guide Links

√8.1   Common Nomenclature

Sometimes you may see an *n-* prefix used in the common nomenclature of some organic compounds, as in the following example:

$$CH_3CH_2CH_2CH_2CH_3$$

pentane
(sometimes called *n*-pentane)

$$CH_3CH_2CH_2CH_2\text{---}Br$$

1-bromobutane (substitutive nomenclature)
butyl bromide (common nomenclature)
(sometimes called *n*-butyl bromide)

The *n-* prefix stands for "normal." At one time, the prefix *n-* and the word normal were used to indicate an isomer containing a functional group at the end of an unbranched carbon chain. However, this prefix is superfluous and unnecessary. For example, the name butyl bromide itself means the structure shown on the right above; an additional prefix is unnecessary. Branched-chain isomers have other names, such as isobutyl bromide, *sec*-butyl bromide, or *tert*-butyl bromide. Because common names are unambiguous without the prefix *n-*, the IUPAC recommended abandoning it. Despite this recommendation, the prefix continues to be used.

Solutions

Solutions to In-Text Problems

8.1   (a) Isobutyl fluoride is a primary alkyl halide.   (c) Cyclopentyl bromide is a secondary alkyl halide.

8.2   (a)

$$CH_3CCH_2CH_2CHCH_3$$

with Cl, CH₃ groups

2,2-dichloro-5-methylhexane

(c)

H_3C—   —Br, Cl

6-bromo-1-chloro-3-methylcyclohexene

8.3   (a) 3-Bromo-3-chloro-1-methylcyclopropene   (c) 3-Bromo-1,1,1-trichloro-2-fluorobutane
(e) 2-Chloro-5-methylhexane   (g) 1-Chloro-3-isopropyl-1-methylcyclohexane

8.4   (a)

$$CH_3CHCH_2CH_3$$ with OH

sec-butyl alcohol

(c)

CH_3CH_2 ... —OH

3-ethylcyclopentanol

(e)

Cl, CHCH₃, C=C structure

CH_3CHCH_2 with OH

(*E*)-6-chloro-4-hepten-2-ol

8.5   (a) 3-Bromo-1-butanol   (c) 2-Chloro-5-methyl-2-cyclopentenol
(e) 3-Butyl-2,4-pentanediol   (g) 2,5-Cyclohexadienol
(h) 1-Mercapto-2-pentanol

8.6   (a)

$$CH_3CH_2OCH_2CH_2CH_3$$

ethyl propyl ether

(c)

—S—

dicyclopentyl sulfide

(e)

$$H_2C=CHCH_2OCH_2Ph$$

allyl benzyl ether

(g)

O
H'''' ⌂ ''''CH₃
CH₃   H

(2*R*,3*R*)-2,3-dimethyloxirane

(h)

$$CH_3CHCH_2CH_2CHCH_2CH_3$$
with CH₃ and SCH₂CH₃

5-(ethylthio)-2-methylheptane

8.7   (a) 2-ethoxyethanol   (c) (*E*)-5-methoxy-3-penten-1-ol   (d) 1-(isobutylthio)-2-methylpropane

8.8   Notice from Table 8.1 that the bond lengths in a given period are about 0.04 Å greater in the column containing oxygen and sulfur than in the column containing the halogens. Assuming that this trend holds true for CH₃SeH, then the predicted carbon-selenium bond length in this compound is about 1.98 Å. The prediction for (CH₃)₂Se can be refined by noting in Fig. 8.1, text p. 345, that the carbon-sulfur bond length in dimethyl sulfide is about 0.02 Å smaller than that in methanethiol. Assuming a similar trend in the corresponding selenium derivatives, the predicted bond length for $(CH_3)_2Se$ is 1.96 Å.

 The point of this problem is that more often than not, molecular properties vary in a regular way throughout the periodic table, and that reasonably good predictions can be made within one group from trends in a nearby group.

8.9 (a) Because of the molecular geometry of *trans*-2-butene, the C—H bonds are oriented in opposite directions, as are the C—CH_3 bonds. Consequently, their bond dipoles cancel, and the molecular dipole moment of *trans*-2-butene is zero. (The cancellation of the C—CH_3 bond dipoles is shown in the diagram below; the C—H bond dipoles also cancel for the same reason.) The corresponding bond dipoles of *cis*-2-butene do not cancel; consequently it is the stereoisomer with the dipole moment of 0.25 D.

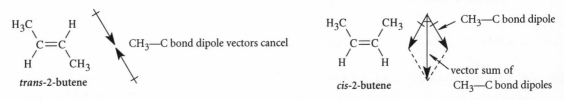

(b) A molecular dipole moment tends to increase boiling point because it is a source of intramolecular attractions. (See text p. 346.) Consequently, *cis*-2-butene has the higher boiling point.

8.11 (a) Recall that a bromine contributes about the same molecular mass (80 units) as a pentyl group (73 units), and that alkyl bromides have boiling points *lower* than alkanes of about the same molecular mass; then 2-bromopropane should have a boiling point that is lower than that of an eight-carbon alkane; because the two alkanes listed have more than eight carbons, their boiling points are correspondingly higher. Because the two alkanes have the same type of branching, their boiling points should be in the same order as their molecular masses. These considerations suggest that the order of increasing boiling points is 2-bromopropane < 4-ethylheptane < 4-ethyloctane. (The actual boiling points are 59°, 141°, and 168°, respectively.)

8.12 (a) HBr is primarily a hydrogen-bond donor because it is a strong acid; —Br is a poor acceptor.
(c) Acetone is a hydrogen-bond acceptor; the oxygen can accept hydrogen bonds at its unshared electron pairs. (See structure below.)
(e) Phenol is both a hydrogen-bond donor and a hydrogen-bond acceptor. The O—H can participate in hydrogen-bond donation, and the oxygen can accept hydrogen bonds at its unshared electron pairs.

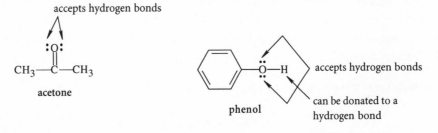

8.13 (a) 2-Methoxyethanol is a polar, protic, donor solvent.
(c) This compound, 2-butanone, is a polar, aprotic, donor solvent.

8.14 Use the "like-dissolves-like" criterion. Hexane should be least soluble in ethanol because hexane is an aprotic, apolar, nondonor substance, whereas ethanol is a polar, protic, donor solvent. 1-Octanol is also protic, but its structure has a rather substantial hydrocarbon portion that would interact more favorably than ethanol with hexane.

8.16   Water can accept a hydrogen bond from the hydrogen of the N—H bond, and can donate a hydrogen bond to the oxygen of the C=O bond as well as to the nitrogen. This hydrogen bonding tends to solubilize acetanilide in water. In contrast, the CH_3 group and the phenyl ring cannot hydrogen-bond with water; consequently, these apolar ("greasy") groups interact poorly with water and tend to make acetanilide insoluble.

> Groups such as alkyl and phenyl groups that interact unfavorably with water are sometimes termed *hydrophobic groups.*

8.17   (a)

$$K^+ \ (CH_3)_3C\!-\!O^-$$

potassium *tert*-butoxide

(c)

$$\left[\begin{array}{c} CH_3 \\ | \\ CH_3CH_2CCH_2O^- \\ | \\ CH_3 \end{array} \right]_2 Mg^{2+}$$

magnesium 2,2-dimethyl-1-butanolate

8.18   (a) Cuprous ethanethiolate [or copper(I) ethanethiolate]

8.19   (a) $Cl(CH_2)_3OH$ < $ClCH_2CH_2OH$ < Cl_2CHCH_2OH

               C   *A*   *B*

Compound *B* is more acidic than compound *A* because it has more chlorines at the same separation from the oxygen; and compound *A* is more acidic than compound *C* because the separation of the electronegative chlorine from the oxygen is greater in *C*. (Recall that polar effects decrease with increasing separation of the two interacting groups; see text p. 110.)

(c) CH_3CH_2OH < $ClCH_2CH_2OH$ < $ClCH_2CH_2SH$

               C   *B*   *A*

Compound *A* is most acidic because of the element effect; that is, thiols are more acidic than alcohols. The polar effect of the chlorine makes alcohol *B* more acidic than alcohol *A*.

8.20   (a) $(CH_3)_2CH\!-\!Br \ + \ Mg \ \xrightarrow{\text{ether}} \ (CH_3)_2CH\!-\!MgBr$

(c) $Ph\!-\!Br \ + \ 2\,Li \ \xrightarrow{\text{hexane}} \ Ph\!-\!Li \ + \ LiBr$

> As this problem illustrates, aryl halides can be used to form aryllithium reagents just as alkyl halides are used to form alkyllithium reagents.

(e) Organoboranes are prepared by hydroboration of alkenes (see text p. 183).

$$3\,(CH_3)_2C\!=\!CH_2 \ + \ BH_3 \ \xrightarrow{\text{THF}} \ [(CH_3)_2CH\!-\!CH_2]_3B$$

8.21   (a)

                         1-cyclohexenylmagnesium bromide

8.22   (a) Whenever a Grignard or organolithium reagent reacts with a protic substance such as water, an alcohol, or an acid, a Brønsted acid-base reaction occurs, as illustrated in Eq. 8.25 of the text.

$$CH_3\ddot{O}{-}H \quad CH_3{-}Li \longrightarrow CH_3\ddot{O}{:}^{-} \quad Li^+ \ + \ CH_4$$

8.23 (a) $(CH_3)_2CH{-}MgBr$ and $CH_3CH_2CH_2{-}MgBr$

8.24 The initiation step is shown in Eq. 8.29 on text p. 372; the first series of propagation steps are shown in Eqs. 8.30 and 8.31. Methylene chloride is formed by the following additional propagation steps:

$$Cl\cdot \quad H{-}CH_2Cl \longrightarrow Cl{-}H \ + \ \cdot CH_2Cl$$

$$Cl{-}Cl \quad \cdot CH_2Cl \longrightarrow Cl\cdot \ + \ Cl{-}CH_2Cl$$

methylene chloride

8.26 Ethane can be formed as a by-product by the recombination reaction of two methyl radicals, which are in turn formed in the propagation reaction shown in Eq. 8.30 of the text.

$$H_3C\cdot \quad \cdot CH_3 \longrightarrow H_3C{-}CH_3 \quad \text{ethane}$$

Solutions to Additional Problems

8.28 (a–d) The alcohols with the formula $C_5H_{11}OH$:

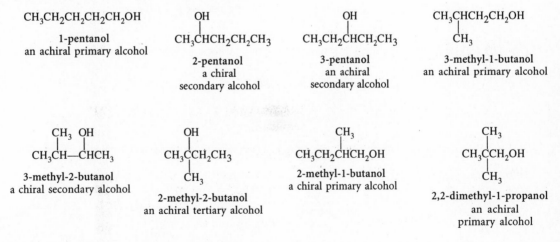

$CH_3CH_2CH_2CH_2CH_2OH$

1-pentanol
an achiral primary alcohol

OH
|
$CH_3CHCH_2CH_2CH_3$

2-pentanol
a chiral
secondary alcohol

OH
|
$CH_3CH_2CHCH_2CH_3$

3-pentanol
an achiral
secondary alcohol

$CH_3CHCH_2CH_2OH$
|
CH_3

3-methyl-1-butanol
an achiral primary alcohol

$CH_3 \quad OH$
|　　|
$CH_3CH{-}CHCH_3$

3-methyl-2-butanol
a chiral secondary alcohol

OH
|
$CH_3CCH_2CH_3$
|
CH_3

2-methyl-2-butanol
an achiral tertiary alcohol

CH_3
|
$CH_3CH_2CHCH_2OH$

2-methyl-1-butanol
a chiral primary alcohol

CH_3
|
CH_3CCH_2OH
|
CH_3

2,2-dimethyl-1-propanol
an achiral
primary alcohol

8.30 (a) The systematic name of halothane is 2-bromo-2-chloro-1,1,1-trifluoroethane.

8.31 (a) 2-butene-1-thiol

8.32 (a) The order of boiling points is *tert*-butyl alcohol < 2-pentanol < 1-hexanol. (The actual boiling points are 82°, 119°, 158°.) This follows the order of molecular masses. The increased branching of 2-pentanol relative to 1-hexanol, and *tert*-butyl alcohol relative to 2-pentanol, makes the differences between these boiling points even greater than they would be for unbranched alcohols of the same molecular masses.

(c) The order of boiling points is 1-hexene < 1-chloropentane < 1-hexanol. (The actual boiling points are 63°, 108°, 158°.) Use the fact that the mass of a chlorine (35 units) is about the same as that of an ethyl group (29 units), and that alkyl chlorides have about the same boiling points as alkanes (and alkenes) of the same molecular mass. Consequently, 1-chloropentane has about the same boiling point as 1-heptene, which, in turn, has a higher boiling point than 1-hexene. 1-Hexanol has about the same molecular mass as heptane, and its boiling point is considerably higher because of hydrogen bonding.

(e) The order of boiling points is propane < diethyl ether < 1,2-propanediol. (The actual boiling points are –42°, 37°, 189°.) Diethyl ether has a higher boiling point than propane because diethyl ether is more polar and because it has a greater molecular mass. 1,2-Propanediol has the highest boiling point because of hydrogen bonding.

8.33 (a) The first compound, acetic acid, has a higher boiling point than the second, ethyl acetate, because hydrogen bonding between acetic acid molecules can take place; the O—H hydrogen on one molecule can be donated to form a hydrogen bond to either of the oxygens on another molecule. Because ethyl acetate has no O—H, molecules of ethyl acetate cannot associate by hydrogen bonding. The intermolecular hydrogen bonds must be broken in order to vaporize acetic acid, and the energy required to break these hydrogen bonds is reflected in a higher boiling point.

8.34 (a) The unsaturation number of the ether is 1; because it has no double bonds, it must contain a ring. Two of the many acceptable possibilities are

(c) The unsaturation number is 2, and both rings and/or multiple bonds are allowed in this case. Two of several possibilities are

(e) The following glycol, although it has asymmetric carbons, is achiral because it is *meso*.

(f) 2,3-Butanediol (see text p. 244) can exist as a pair of enantiomers and a *meso* stereoisomer.

(h)

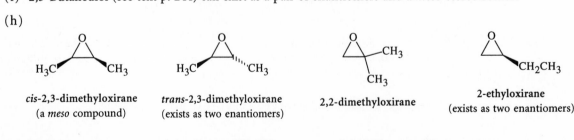

cis-2,3-dimethyloxirane
(a *meso* compound)

trans-2,3-dimethyloxirane
(exists as two enantiomers)

2,2-dimethyloxirane

2-ethyloxirane
(exists as two enantiomers)

8.35 (a) hydrogen, H_2 (c) ethane, CH_3CH_3 (e) hydrogen, H_2

Remember that when a Grignard reagent reacts with a proton source, whether it is water or an alcohol, the *conjugate acid* of the "carbon anion" part of the Grignard reagent is formed. (See Eq. 8.25, text p. 371.)

8.36 (a) The water-miscible compound is propyl alcohol; alcohols of low molecular mass are miscible with water because of their hydrogen-bonding interactions with water. Isomeric ethers are considerably less soluble, because they cannot donate hydrogen bonds.

(b) Allyl methyl ether, $H_2C{=}CH{-}CH_2{-}OCH_3$, decolorizes a solution of bromine in CCl_4 because the bromine adds to the double bond; the alcohol has no double bond to react with bromine.

(c) The compounds must be cyclic because they contain no double bonds; and they must be alcohols because they give off a gas (methane) when treated with the Grignard reagent methylmagnesium iodide. The compounds can only be the four stereoisomers of 2-methylcyclopropanol: the enantiomeric pair of *cis* isomers, and the enantiomeric pair of *trans* isomers.

H_3C OH 2-methylcyclopropanol

8.38 (a) Compound *A* is soluble in hydrocarbon solvents by the "like-dissolves-like" criterion. The alkyl groups, particularly the large cetyl ($C_{16}H_{33}$) group, interact favorably with the hydrocarbon chains of the solvent. In contrast, there is no hydrogen-bond acceptor available in a hydrocarbon solvent to stabilize ammonium chloride.

(b) Although compound *A* is soluble in hydrocarbon solvents, it exists in such solvents as ion pairs and higher aggregates rather than as free ions. The reason is that hydrocarbon solvents have a very low dielectric constant ($\varepsilon \cong 2$), whereas a high dielectric constant is required to separate ionic aggregates into free ions. Furthermore, a hydrocarbon offers no solvation by hydrogen bonding to the bromide counterion. Hence, this anion remains in proximity to its positive partner at all times.

8.39 (a) Neopentane, $(CH_3)_4C$, gives only one monochlorination product, $(CH_3)_3C{-}CH_2Cl$.

8.40 (a) *tert*-butyl alcohol < isopropyl alcohol < propyl alcohol < 1-propanethiol. Alcohols with a greater degree of branching are less acidic because their conjugate bases are more poorly solvated; and thiols are substantially more acidic than alcohols (element effect).

(c) $CH_3NHCH_2CH_2CH_2OH$ < $CH_3NHCH_2CH_2OH$ < $(CH_3)_3\overset{+}{N}CH_2CH_2OH$. The last compound is most acidic because it has a full-fledged positive charge that stabilizes a negative charge in the conjugate-base anion. In the other two compounds, the nitrogen stabilizes the conjugate-base anion by its electron-withdrawing polar effect; in the more acidic of these compounds, the nitrogen is closer to the oxygen, which is the site of negative charge in the conjugate-base alkoxide anion.

(e) The order of increasing acidity is

$$CH_3CH_2\overset{..}{\underset{..}{O}}H \quad < \quad CH_3CH_2\overset{\overset{H}{\overset{+}{|}}}{\underset{..}{O}}CH_2CH_3 \quad < \quad CH_3\overset{\overset{H}{\overset{+}{|}}}{\underset{..}{S}}CH_3$$

The protonated ether is more acidic than the alcohol for the same reason that H_3O^+ is more acidic than H_2O: a positively charged oxygen is much more electronegative than a neutral oxygen. The sulfonium ion is most acidic because of the element effect.

8.41 Water forms a Lewis acid-Lewis base complex with Cu^{2+} ion:

$$Cu^{2+} \quad \overset{..}{\underset{..}{O}}H_2 \quad \longrightarrow \quad Cu^+{-}\overset{..}{O}H \quad \text{more acidic than the hydrogens of water}$$

the species with
$pK_a = 8.3$

Ionization of a hydrogen of the complexed water removes a positive charge from the oxygen and eliminates the charge-charge repulsion in the complex. Ionization of water itself does not have this driving force. Hence, water that is complexed to copper is more acidic.

8.42   In the presence of concentrated acid, dibutyl ether is protonated. The protonated ether is an ionic compound, and ionic compounds are soluble in water:

$$R = \text{butyl}$$

$$R{-}\overset{..}{\underset{..}{O}}{-}R \;+\; H{-}\overset{..}{\underset{..}{O}}NO_2 \longrightarrow R{-}\overset{\overset{H}{|}}{\overset{+}{\underset{..}{O}}}{-}R \quad :\overset{..}{\underset{..}{O}}NO_2$$

not ionic; not   nitric acid   an ionic compound;
soluble in water   soluble in water

8.43   Think of the dissociation reaction of the crown ether-K$^+$ complex as you would any other equilibrium:

$$\text{crown ether–K}^+ \;+\; \text{solvent} \;\rightleftharpoons\; \text{crown ether} \;+\; \text{K}^+(\text{solvated})$$

Anything that stabilizes the species on the right side of the equation will increase the dissociation constant of the complex. Because potassium ion, a typical positive metal ion, is solvated more effectively by water than by ether, the dissociation constant for the complex is greater in water.

8.44   As Eq. 8.18, text p. 369, shows, ethers solubilize Grignard reagents by forming Lewis acid-base complexes with them. Because a tertiary amine is also a Lewis base, it can solubilize Grignard reagents by the same type of interaction:

$$\underset{R_3N:}{\overset{\delta+}{}} \longrightarrow \overset{\overset{R}{|}}{\underset{\underset{Br}{|}}{\overset{\delta-}{Mg}}} \longleftarrow \underset{:NR_3}{\overset{\delta+}{}}$$

8.45   (a) When the concentration of ethanol increases, the concentration of dimer increases.
       (b) Use the usual relationship between $\Delta G°$ and K_{eq}; recall that $2.3RT$ at 298 K is 5.706 kJ/mol.

$$\Delta G° = -2.303RT(\log K_{eq}) = -5.706(\log 11) = -5.94 \text{ kJ/mol}$$

(c) Neglect the volume of ethanol and assume that 1 mole of ethanol is dissolved in 1 L of solution so that the initial concentration of ethanol is 1 M. (This introduces an error of only 2%.) Assume that x M of ethanol react to form dimer. The concentration of ethanol remaining is then $(1.0 - x)$ M, and the concentration of dimer is $x/2$ M. (Remember, it takes two ethanol molecules to form one molecule of dimer.) The equilibrium-constant expression for dimerization becomes

$$K_{eq} = 11 = \frac{x/2}{(1-x)^2}$$

From the quadratic formula, $x = 0.808$ M. Thus, the concentration of dimer is $x/2 = 0.404$ M; that of free ethanol is $(1 - x)$ or 0.192 M. Thus, about 80% of the ethanol is dimerized in a 1 M solution.

8.47   Flick's frangent flailings were fundamentally futile because his ether is also an alcohol. The Grignard reagent was destroyed by the —OH group of the alcohol in a protonolysis reaction:

$$CH_3CH_2{-}MgBr \;+\; H{-}OCH_2CH_2OC_2H_5 \longrightarrow CH_3CH_3 \;+\; BrMg^+ \; {}^-OCH_2CH_2OC_2H_5$$

8.48   (a) The free-radical bromination of 2-methylpropane (isobutane) gives the following two alkyl bromides:

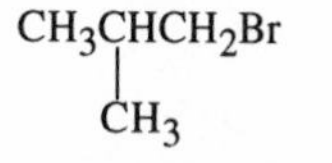

$$CH_3CHCH_2Br \qquad and \qquad CH_3CCH_3$$

(with CH₃ branch below left structure; Br above and CH₃ below right structure)

isobutyl bromide
(1-bromo-2-methylpropane)

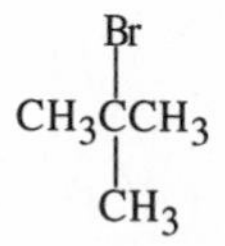

tert-butyl bromide
(2-bromo-2-methylpropane)

(b) The propagation steps in the formation of alkyl halides from alkanes are as follows, with X = Br and R = alkyl. (See Eqs. 8.29–8.31, text p. 372.)

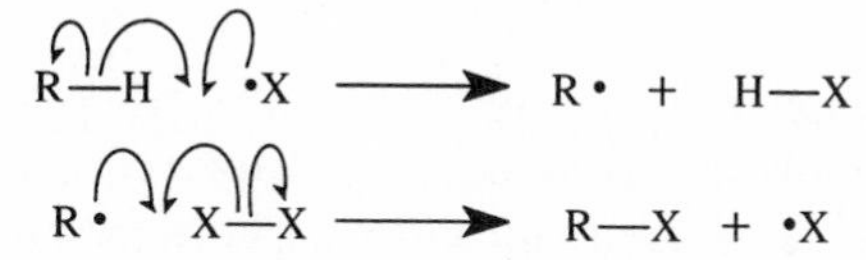

$$R{-}H \quad \cdot X \longrightarrow R\cdot \; + \; H{-}X$$

$$R\cdot \quad X{-}X \longrightarrow R{-}X \; + \; \cdot X$$

When the primary hydrogens of isobutane react, the free-radical intermediate R· is the *isobutyl radical,* $(CH_3)_2CH\overset{\bullet}{C}H_2$, a *primary* free radical. When the tertiary hydrogen of isobutane reacts, the free-radical intermediate R· is the *tert-butyl radical,* $(CH_3)_3\overset{\bullet}{C}$, a *tertiary* free radical. Table 5.3, text p. 205, shows that formation of a primary free radical by dissociation of a C—H bond requires about 16 kJ/mol more energy than formation of a tertiary free radical. Because the *tert*-butyl radical is more stable than the isobutyl radical, its formation should be faster; consequently, the product resulting from its reaction with Br₂, namely, *tert*-butyl bromide, should be (and is) the major product formed.

Notice that there are *nine* primary hydrogens and only *one* tertiary hydrogen in isobutane. Thus, if the reaction were random, there would be nine times as much isobutyl bromide formed as *tert*-butyl bromide. The fact that *tert*-butyl bromide is the major product shows the importance of the relative stabilities of the free-radical intermediates in determining the outcome of the reaction.

8.49   The outcome of the Grignard protonolysis shows that all of the alkyl halides have the same carbon skeleton, that of 2,4-dimethylpentane. The protonolysis in D₂O confirms the fact that the bromines are bound at different places on the carbon skeleton. The only three possibilities for the alkyl halides are

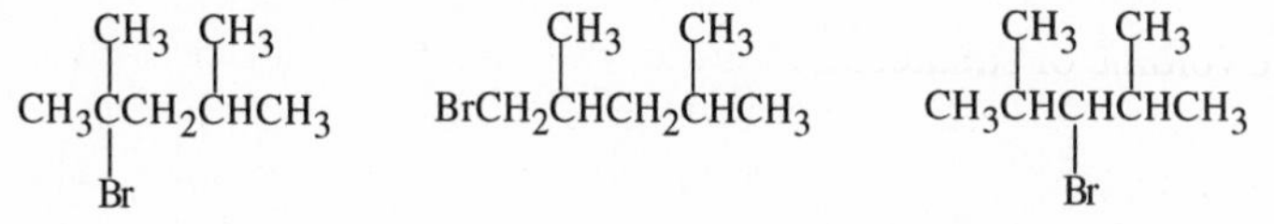

$$CH_3CCH_2CHCH_3 \qquad BrCH_2CHCH_2CHCH_3 \qquad CH_3CHCHCHCH_3$$

A   *B*   *C*

Compound *B* is the chiral alkyl halide, and compounds *A* and *C* are the other two.

8.50   Figure 5.2, text p. 203, shows that free radicals are sp^2-hybridized, and that the unpaired electron resides in a *p* orbital. Combining this fact with the mechanism shown in the solution to Problem 8.48(b) shows that the free-radical intermediate involved in the bromination of *sec*-butylbenzene has the structure shown on the left of the following equation. Because the radical is planar, and therefore achiral, the chirality of the starting *sec*-butylbenzene is irrelevant; the radical can react with Br₂ equally well at either the top lobe *(A)* or bottom lobe *(B)* of the *p* orbital. Because the two reaction pathways are enantiomeric, they occur with equal rates. Consequently, the product of the reaction is the *racemate* of (1-bromo-1-methylpropyl)benzene, and therefore this product should be optically inactive, even if the *sec*-butylbenzene starting material is optically active.

8.51 (a) In compound *A*, the equatorial position of the —OH group is preferred for the usual reason: it avoids 1,3-diaxial interactions with hydrogens on the ring. However, compound *B* prefers to have an axial —OH group because it affords the opportunity for intramolecular hydrogen bonding; this additional bonding is a stabilizing effect, and lowers the energy of the axial conformation below that of the equatorial conformation.

intramolecular hydrogen bonds

axial conformation is
stabilized by intramolecular
hydrogen bonding

equatorial conformation;
intramolecular hydrogen bonding
is not possible

(c) The result in part (b) suggests that intramolecular hydrogen bonding should stabilize the *gauche* conformations of both stereoisomers. Indeed, in either enantiomer of the racemate, such hydrogen bonding can occur in one conformation in which the large *tert*-butyl groups are *anti* to each other. However, in the *meso* stereoisomer, the necessity that the hydroxy groups be *gauche* in order for hydrogen bonding to occur also means that the large *tert*-butyl groups must also be *gauche*. The stabilizing effect of intramolecular hydrogen bonding cannot compensate for the magnitude of the resulting van der Waals repulsions between the *tert*-butyl groups.

one hydrogen-bonding conformation
of (3*R*,4*R*)-2,2,5,5-tetramethyl-3,4-hexanediol;
tert-butyl groups are *anti*

gauche conformation of
meso-2,2,5,5-tetramethyl-3,4-hexanediol

8.52 Refer to the solution to Problem 2.30 on p. 30 of this manual. The energy of unstable conformation *D* is the highest energy that must be attained by the molecule during internal rotation. The energy of this conformation, and thus the height of the energy barrier to rotation, is raised not only by van der Waals repulsions of the two chlorines, but also by unfavorable electrostatic interactions of the two carbon-chlorine bond dipoles. That is, the ends of the bond dipoles with like charges are adjacent. In the corresponding conformation of butane, the van der Waals repulsions between methyl groups are present, but the electrostatic repulsions do not exist because the carbon-carbon bonds have no significant bond dipole.

8.53 (a) In the chair conformation, for every C—O bond dipole in a given direction, there is another C—O bond dipole of the same magnitude pointing in the opposite direction. Thus, in the following diagram, the grey dipoles cancel each other, and the black dipoles cancel each other.

In other words, all bond dipoles cancel and the net dipole moment is zero.

(b) To the extent that the twist-boat form is present, it will contribute a nonzero dipole moment, because the C—O bond dipoles do not cancel in this form.

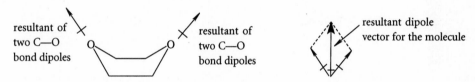

resultant of two C—O bond dipoles

resultant of two C—O bond dipoles

resultant dipole vector for the molecule

In other words, the nonzero dipole moment of 1,4-dioxane results from the presence of a very small amount of a conformation that has a large dipole moment.

8.54 (a) Equilibrium (2) contains more of the conformation with the methyl group in an axial position than equilibrium (1) does; that is, equilibrium (2) lies farther to the right. The reason is that, because C—O bonds are shorter than C—C bonds, the axial methyl group in equilibrium (1) is closer to the axial hydrogens than the axial methyl group in equilibrium (2). Therefore, the 1,3-diaxial interactions in the axial-methyl conformation of equilibrium (1) are somewhat more severe than those in the axial-methyl conformation of equilibrium (2).

smaller distance; greater van der Waals repulsion

larger distance; smaller van der Waals repulsion

8.55 Because BF_3 is electron-deficient, it readily reacts as a Lewis acid with the unshared pairs of the Lewis base ethanol:

$$C_2H_5\ddot{O}H \quad BF_3 \longrightarrow C_2H_5\overset{+}{\underset{\cdot\cdot}{O}}\text{—}\overset{-}{B}F_3$$

Fluoride ion dissociates from the boron in a Lewis acid-base dissociation, and then removes the proton from the oxygen in a Brønsted acid-base reaction:

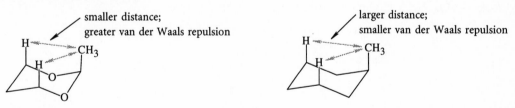

The resulting species contains an electron-deficient boron, which is attacked by ethanol, and a similar sequence of reactions is repeated:

$$C_2H_5O\!-\!BF_2 \longrightarrow C_2H_5O\!-\!\overset{\ddot{\text{:}}\!\underset{|}{\overset{\ddot{F}\text{:}}{\text{B}}}F}{\underset{C_2H_5\overset{+}{O}\!-\!H}{}} \longrightarrow C_2H_5O\!-\!\underset{C_2H_5\overset{+}{\underset{\ddot{\text{:}}}{O}}\!-\!H}{\overset{|}{\text{B}}}F \quad :\!\ddot{F}\!:^- \longrightarrow C_2H_5O\!-\!\underset{C_2H_5\ddot{O}\text{:}}{\overset{|}{\text{B}}}F \;+\; H\!-\!\ddot{F}\text{:}$$

A third, exactly analogous, sequence, which you should write, finishes the mechanism.

· ·

Chemistry of Alkyl Halides

Terms

Concepts

I. General Reaction Issues

A. EQUILIBRIA

1. Knowledge of the equilibrium constant for a reaction provides no information about the rate at which the reaction takes place.
   a. Whether the equilibrium in a nucleophilic substitution reaction is favorable can be predicted from an analysis of the corresponding Brønsted acid-base reaction.
   b. If the Brønsted acid-base reaction strongly favors one side of the equation, then the analogous substitution reaction likewise favors the same side of the equation.
2. The equilibrium in any nucleophilic substitution reaction, as in an acid-base reaction, favors release of the weaker base.

$$I—CH_3 \;+\; Na^+ \;{}^-C\!\!\equiv\!\!N \;\underset{\longleftarrow}{\overset{\longrightarrow}{}}\; CH_3—C\!\!\equiv\!\!N \;+\; Na^+ \; I^-$$

stronger base   weaker base

  a. LeChatelier's principle states that if an equilibrium is disturbed, the components of the equilibrium will react so as to offset the effect of the disturbance.

  b. Some equilibria that are not too unfavorable can be driven to completion by applying LeChatelier's principle, *e.g.*, by precipitation of a by-product.

B. Rate Law

1. The mathematical statement of how a reaction rate depends on concentration is called the rate law.
   a. Each reaction has its own characteristic rate law.
   b. The concentration terms of the rate law indicate what atoms are involved in the rate-limiting transition state.
2. The rate law provides fundamental information about the mechanism of a reaction.
   a. Mechanisms not consistent with the rate law are ruled out.
   b. Of the chemically reasonable mechanisms consistent with the rate law, the simplest one is provisionally adopted.
   c. The mechanism of a reaction is modified or refined if required by subsequent experiments.
   d. Although the rate law indicates what atoms are present in the transition state, it provides no information about how they are arranged.
3. A rate law is determined experimentally by varying the concentration of each reactant (including any catalysts) independently and measuring the resulting effect on the rate.
4. In the rate law, the constant of proportionality, k, between the rate and the concentrations of reactants is called the rate constant.
   a. In general, the rate constant is different for every reaction, and it is a fundamental physical constant for a given reaction under particular conditions of temperature, pressure, solvent, etc.
   b. The rate constant is related to the standard free energy of activation $\Delta G^{o\ddagger}$.
      i. The standard free energy of activation, or energy barrier, determines the rate of a reaction under standard conditions.
      ii. If the rates of two reactions A and B are compared, the relationship between their rate constants and their standard free energies of activation is $-2.3RT \log(k_A/k_B) = \Delta G_A^{o\ddagger} - \Delta G_B^{o\ddagger}$.
   c. The dimensions of the rate constant depend on the kinetic order of the reaction.
      i. The overall kinetic order for a reaction is the sum of the powers of all the concentrations in the rate law.
      ii. The kinetic order in each reactant is the power to which its concentration is raised in the rate law.

C. Nucleophilicity and Solvents

1. Nucleophilic reactivity is termed nucleophilicity.
   a. The nucleophilicity of a base is determined by the rate of its reaction with a standard Lewis acid.
   b. When the attacking atom is the same, the relative nucleophilicities of different nucleophiles correlate roughly with their relative Brønsted basicities.
   c. Nucleophilicity, not Brønsted basicity, governs the rate of an S_N2 reaction.
2. The relationship between nucleophilicity and basicity depends strongly on the solvent.
   a. In protic solvents:
      i. nucleophilicity increases towards greater atomic number within a group of the periodic table;
      ii. hydrogen bonding occurs between the protic solvent molecules and the nucleophilic anions;
      iii. within a group of the periodic table, the most strongly basic anions are most strongly hydrogen-bonded, and are therefore least nucleophilic.
   b. In aprotic solvents:
      i. there is no hydrogen bonding to reduce the nucleophilicity of dissolved anions; consequently, anions dissolved in such solvents are considerably more nucleophilic than they are in protic solvents;
      ii. within a group of the periodic table, the most basic anions are the best nucleophiles;

iii. The nucleophilicity of an anion is much greater in a polar aprotic solvent than in a protic solvent.

3. Solvents can affect the reactivity of nucleophiles:

   a. polar protic solvents best solubilize charged ions, but anionic nucleophiles are least reactive in these solvents;

   b. polar aprotic solvents, in many cases, provide the best combination of solubility and reactivity for anionic nucleophiles; and

   c. the reactivity of uncharged nucleophiles shows less dependence on solvent.

D. Nucleophiles

1. Most strong Brønsted bases are good nucleophiles, but some excellent nucleophiles are relatively weak Brønsted bases.

   a. Many nucleophiles are anions, others are uncharged, or in a few cases, even positively charged.

   b. Uncharged nucleophiles show a much weaker dependence on solvent.

2. The strongest bases are generally the most reactive nucleophiles; however, the reverse is true in protic solvents for anions in which the attacking atoms come from within a group (column) of the periodic table.

3. The nucleophilicities of anionic nucleophiles are much greater in a polar aprotic solvent than in a protic solvent.

E. Primary Deuterium Isotope Effect

1. When a hydrogen is transferred in the rate-limiting step of a reaction, a compound in which that hydrogen is replaced by its isotope deuterium will react more slowly in the same reaction.

   a. This effect of deuterium substitution on reaction rates is called a primary deuterium isotope effect.

   b. The theoretical basis for the primary deuterium isotope effect lies in the comparative strengths of C—H and C—D bonds.

      i. The bond to the heavier isotope D is stronger (and thus requires more energy to break) than the bond to the lighter isotope H.

      ii. The energy barrier, or free energy of activation, for the compound with the C—D bond is greater; as a result, its rate of reaction is smaller.

2. A primary deuterium isotope effect is observed only when the hydrogen that is transferred in the rate-determining step is substituted by deuterium.

II. *Reactions of Alkyl Halides*

A. Nucleophilic Substitution Reactions

1. A nucleophilic substitution reaction, or nucleophilic displacement reaction, is a very general type of reaction:

   a. It is *substitution* because one group is substituted for (or displaces) another group, which departs.

      i. The group that is displaced in a nucleophilic substitution reaction is termed the leaving group.

      ii The best leaving groups give the weakest bases as products.

   b. It is *nucleophilic substitution* because the substituting group acts as a nucleophile, or Lewis base.

      i. Many nucleophiles are anions, others are uncharged, or in a few cases, even positively charged.

      ii. The roles of nucleophile and leaving group are reversed if the reaction can be run in the reverse direction.

   c. Nucleophilic substitution reactions occur by two mechanisms:

      i. The S_N1 mechanism is characterized by a first-order rate law that contains only a term in alkyl halide concentration.

      ii. The S_N2 mechanism occurs in a single step with inversion of stereochemical configuration and is characterized by a second-order rate law.

$$\text{Nuc:}^- \quad C{-}X \longrightarrow \text{Nuc}{-}C{,,,} + X{:}^- \qquad \text{rate} = k\left[\text{Nuc:}^-\right]\left[\,C{-}X\right]$$

2. The reaction of an alkyl halide with a solvent in which no base or nucleophile has been added is termed a solvolysis.
   a. The solvolysis reactions of tertiary alkyl halides are fastest in polar protic donor solvents.
   b. Rearrangements are observed in S_N1 solvolysis reactions that involve rearrangement-prone carbocations as intermediates.
   c. Tertiary and secondary alkyl halides undergo solvolysis reactions by the S_N1 and E1 mechanisms; tertiary alkyl halides are more reactive.
3. An intramolecular substitution reaction is a reaction in which the nucleophile and the leaving group are part of the same molecule.
4. Nucleophilic substitution reactions can be used to transform alkyl halides into a wide variety of other functional groups.
5. For the most part, polar protic solvents, polar aprotic solvents, or mixtures of these are used in substitution reactions.
6. Each nucleophilic substitution reaction is conceptually similar to a Brønsted acid-base reaction.

B. ELIMINATION REACTIONS

1. An elimination reaction is a reaction in which two or more groups are lost from within the same molecule.
   a. In an alkyl halide, the carbon bearing the halogen is often referred to as the α-carbon, and the adjacent carbons are referred to as the β-carbons.

$$\beta\text{-hydrogen} \longrightarrow H \quad H \longleftarrow \alpha\text{-hydrogen}$$
$$-C{-}C{-}X$$
$$\beta\text{-carbon} \nearrow \qquad \nwarrow \alpha\text{-carbon}$$

   b. An elimination that involves loss of two groups from adjacent carbons is termed a β-elimination.
   c. A β-elimination reaction is conceptually the reverse of an addition to an alkene.
2. An α-elimination, as in the formation of dichloromethylene from chloroform, is an elimination of two groups (in this case the elements of HCl) from the same atom.

$$\text{Base:}^- + H{-}\overset{Cl}{\underset{Cl}{C}}{-}Cl \longrightarrow \overset{Cl}{\underset{Cl}{C}}{:} + \text{BaseH} + Cl{:}^-$$

chloroform   dichloromethylene

3. β-Elimination reactions occur by two mechanisms:
   a. The E1 mechanism is an alternative product-determining step of the S_N1 mechanism in which a carbocation intermediate loses a β-hydrogen to form an alkene.

$$-\overset{H}{C}{-}C{-}X \xrightarrow{\text{rate-limiting}} -C{-}C^+ + X{:}^- \longrightarrow C{=}C + HX \qquad \text{rate} = k\left[-\overset{H}{C}{-}C{-}X\right]$$

   b. The E2 mechanism, which competes with the S_N2 mechanism, has a second-order rate law and occurs with *anti* stereochemistry.

$$\text{Base:}^- \quad -\overset{H}{C}{-}\underset{X}{C}{-} \longrightarrow C{=}C + X{:}^- + \text{BaseH} \qquad \text{rate} = k\left[\text{Base:}^-\right]\left[-\overset{H}{C}{-}\underset{X}{C}{-}\right]$$

4. Base-promoted β-elimination reactions typically follow a rate law that is second order overall and first order in each reactant.

    a. A Brønsted base attacks a β-hydrogen of the alkyl halide, not a carbon atom as in a nucleophilic substitution reaction.

    b. If the reacting alkyl halide has more than one type of β-hydrogen atom, then more than one β-elimination reaction is possible.

    c. Strong bases, such as sodium ethoxide [Na^+ $^-OCH_2CH_3$] and potassium *tert*-butoxide [K^+ $^-OC(CH_3)_3$], promote the β-elimination reactions of alkyl halides.

    d. Often the conjugate-acid alcohols of these bases are used as solvents, for example, sodium ethoxide in ethanol or potassium *tert*-butoxide in *tert*-butyl alcohol.

5. *Anti* elimination is preferred for three reasons:

    a. *Syn* elimination occurs through a transition state that has an eclipsed conformation, whereas *anti* elimination occurs through a transition state that has a staggered conformation.

    b. *Syn* elimination requires the base and leaving group to be on the same side of the molecule, where they can interfere sterically with each other, whereas *anti* elimination requires the base and leaving group to be on opposite sides of the molecule, out of each other's way.

    c. *Syn* elimination requires a frontside electronic displacement on the carbon-halogen bond, whereas *anti* elimination is more favorable because *anti* elimination involves all-backside electronic displacements.

6. A *syn* elimination is conceptually the reverse of a *syn* addition, and an *anti* elimination is conceptually the reverse of an *anti* addition.

C. CARBENES AND CARBENOIDS

1. A haloform reacts with base in an α-elimination reaction to give dihalomethylene, a carbene.

2. A carbene is an unstable and highly reactive species having a divalent carbon atom.

    a. The divalent carbon of a carbene can act as a nucleophile and an electrophile at the same time.

       i. An atom with an unshared electron pair reacts as a Lewis base, or nucleophile.

       ii. An atom that lacks an electronic octet is an electron-deficient compound that can accept an electron pair and is thus a powerful electrophile.

    b. The carbon atom of dichloromethylene bears three groups (two chlorines and an unshared pair of electrons) and therefore has approximately trigonal-planar geometry.

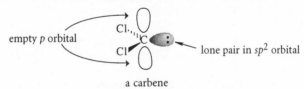

empty *p* orbital   lone pair in sp^2 orbital

a carbene

3. The *syn* addition of carbenes or carbenoids to alkenes yields cyclopropanes in a reaction that forms new carbon-carbon bonds.

    a. Dihalomethylene adds to alkenes to give dihalocyclopropanes.

    b. Methylene iodide ($CH_2 I_2$) reacts with a zinc-copper couple to give a carbenoid organometallic reagent which adds to alkenes to give cyclopropanes.

III. *Unimolecular Reactions*

A. THE S_N1 REACTION

1. A substitution mechanism that involves a carbocation intermediate is called an S_N1 mechanism.

    a. Substitution reactions that take place by the S_N1 mechanism are called S_N1 reactions.

    b. The meaning of the S_N1 "nickname" is $\mathbb{S}$ for substitution, $\mathbb{N}$ for nucleophilic, and $\mathbb{1}$ for unimolecular (meaning that the transition state of the reaction involves one species).

2. The S_N1 mechanism is observed mostly with tertiary and secondary alkyl halides because the reaction involves a carbocation intermediate, the formation of which is promoted by branching at the α-carbon.

3. Th S_N1 mechanism occurs mostly in polar protic solvents.

B. THE E1 REACTION

1. An elimination mechanism that involves carbocation intermediates is termed an E1 mechanism.

    a. Reactions that occur by E1 mechanisms are called E1 reactions.

b. The meaning of the E1 "nickname" is $\mathbb{E}$ for elimination and 1 for unimolecular (meaning that the transition state of the reaction involves one species).

2. A strong base is not required for the E1 reaction as it is for the E2 reaction.

3. When an alkyl halide contains more than one type of β-hydrogen, more than one type of elimination product can be formed.

4. The alkene with the greatest number of alkyl substituents on the double bond predominates.

C. S_N1 vs E1 REACTIONS

1. S_N1-E1 reactions are most rapid with tertiary alkyl halides, they occur more slowly with secondary alkyl halides, and they are almost never observed with primary alkyl halides.

   a. The reactivity order of the alkyl halides is: fluorides $\ll$ chlorides $<$ bromides $<$ iodides.

   b. If an alkyl halide has β-hydrogens, elimination products formed by the E1 reaction accompany substitution products formed by the S_N1 reaction.

2. The S_N1 and E1 reactions of an alkyl halide share a common intermediate, the carbocation.

   a. The first step, ionization of the alkyl halide to the carbocation, is the rate-limiting step and thus has the transition state of highest free energy.

   i. This step, a Lewis acid-base dissociation, is the rate-limiting step of both the substitution and elimination reactions.

   ii. Substitution and elimination products arise from competing reactions of the carbocation.

   b. The different products that can be formed in S_N1-E1 reactions reflect three reactions of carbocation intermediates:

   i. reaction with a nucleophile (the S_N1 reaction);

   ii. loss of a β-proton (the E1 reaction); and

   iii. rearrangement of the initially formed carbocation intermediate to a more stable carbocation followed by (*i*) or (*ii*).

   c. Because the relative rates of these steps determine the ratio of products, they are said to be the product-determining steps.

   d. The rates of the product-determining steps have no effect on the rate at which the alkyl halide reacts.

3. The S_N1 and E1 reactions are accelerated by polar protic donor solvents; ionic dissociation is favored by solvents that separate ions and by solvents that solvate ions.

IV. *Bimolecular Reactions*

A. S_N2 REACTION

1. A mechanism in which attack of a nucleophile on an atom (usually carbon) displaces a leaving group from the same atom in a concerted manner is called an S_N2 mechanism.

   a. The meaning of the "nickname" S_N2 is $\mathbb{S}$ for substitution, $\mathbb{N}$ for nucleophilic, and 2 for bimolecular (meaning that the transition state of the reaction involves two species).

   b. Reactions that occur by S_N2 mechanisms are called S_N2 reactions.

   c. An S_N2 mechanism, because it is concerted, involves no reactive intermediates.

2. In general, secondary alkyl halides undergo S_N2 reactions much more slowly than typical primary alkyl halides, and tertiary alkyl halides are even less reactive.

a. The reaction rate is second order overall: first order in the nucleophile and first order in the alkyl halide.

b. The mechanism involves backside attack of the nucleophile on the alkyl halide and inversion of stereochemical configuration.

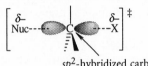

c. The reaction is retarded by branching (a steric effect) at both the α- and β-carbon atoms; neopentyl halides are virtually unreactive.

d. The fastest S_N2 reactions involve leaving groups that have the weakest bonds to carbon; that is, those that give the weakest bases as products.

e. The S_N2 mechanism is especially rapid in polar aprotic solvents.

f. The strongest bases are generally the most reactive nucleophiles; however, the reverse is true in protic solvents for anions with attacking atoms from within a group (column) of the periodic table.

3. In the transition state for an S_N2 reaction on carbon, the nucleophile and the leaving group are partially bonded to opposite lobes of the carbon *p* orbital.

a. The central atoms is turned "inside out," and is approximately sp^2-hybridized in the transition state.

$$\left[\begin{array}{ccc} \delta^- & & \delta^- \\ Nuc--- & C & --X \end{array} \right]^{\ddagger}$$

sp^2-hybridized carbon

b. van der Waals repulsions (steric effects) raise the energy of the transition state and therefore reduce the reaction rate.

B. E2 MECHANISM

1. A mechanism involving concerted removal of a β-proton by a base and loss of a halide ion is called an E2 mechanism.

a. Reactions that occur by the E2 mechanism are called E2 reactions.

b. The meaning of the "nickname" E2 is E for elimination and 2 for bimolecular (meaning that the transition state of the reaction involves two species).

2. The E2 reaction is a β-elimination reaction of alkyl halides that is promoted by strong bases.

a. The rates of E2 reactions are second order overall: first order in base and first order in the alkyl halide.

b. E2 reactions are normally *anti* eliminations.

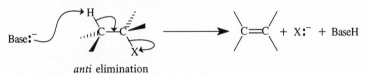

anti elimination

c. The best leaving groups give the weakest bases as products.

3. The rates of E2 reactions show substantial primary deuterium isotope effects at the β-hydrogen atoms because a β-hydrogen is removed in the transition state.

4. When an alkyl halide has more than one type of β-hydrogen, more than one alkene product can be formed.

a. The most stable alkene (the alkene with the greatest numbers of alkyl substituents at their double bonds) are formed in greatest amounts.

b. The standard free energy of the E2 transition state, like that of an alkene, is lowered by branching.

5. The greatest use of the E2 elimination for the preparation of alkenes is when the alkyl halide has only one type of β-hydrogen, and only one alkene product is possible.

C. S$_N$2 vs E2 Reaction

1. The S$_N$2 and E2 reactions are competing processes; the reaction pathway that occurs most rapidly is the one that predominates.
   a. A greater fraction of S$_N$2 reaction is observed when the attacking atom is a good nucleophile yet is a relatively weak base.
   b. Elimination is favored by:
      i. alkyl branches in the alkyl halide at the α- or β-carbon atoms,
      ii. alkyl branches in the base; and
      iii. stronger bases.
2. When simple alkoxide bases such as methoxide and ethoxide are used, the predominant product of an E2 reaction is usually the most stable alkene isomer.
   a. The reaction with the transition state of lower energy (the one with more branching at the developing double bond) is the faster reaction.
   b. The alkene products are stable under the conditions of the reaction.
   c. Because the product mixture, once formed, does not change, the distribution of products reflects the relative energies of the competing transition states.
3. The effect of structure of the alkyl halide:
   a. alkyl halides with greater amounts of branching at the α-carbon give greater amounts of elimination.
   b. Tertiary alkyl halides give more elimination than secondary, which give more than primary.
   c. Alkyl halides with greater amounts of branching at the β-carbon give greater amounts of elimination.
4. The effect of structure of the base:
   a. More highly branched bases give a greater fraction of elimination than unbranched ones.
   b. More highly branched bases give a greater fraction of the alkene product with less branching at the double bond.
   c. Weaker bases that are good nucleophiles give a greater fraction of substitution.

. . . .

Reactions

. . . .

I. Formation of Cyclopropanes

A. Dihalocarbenes

1. Reaction of a haloform with base in the presence of an alkene yields a 1,1-dihalocyclopropane.
2. The addition of a dihalomethylene to an alkene is a concerted *syn* addition reaction.

$$CHCl_3 \;+\; \begin{array}{c} H \\ C \\ \parallel \\ C \\ H \end{array}\begin{array}{c} CH_3 \\ \\ \\ CH_3 \end{array} \xrightarrow[\text{(CH}_3)_3\text{COH}]{K^+ \; ^-OC(CH_3)_3} \quad \underset{Cl}{\overset{Cl}{\diagdown}}\!\!\!\!\triangle\!\!\!\!\begin{array}{c} H \;CH_3 \\ \\ H \; CH_3 \end{array}$$

B. Carbenoids: The Simmons-Smith Reaction

1. Cyclopropane without halogen atoms can be prepared by allowing alkenes to react with methylene iodide (CH_2I_2) in the presence of a zinc-copper couple (a copper-activated zinc preparation) in a reaction called the Simmons-Smith reaction.
   a. The active reagent in the Simmons-Smith reaction is believed to be an α-halo organometallic compound, a compound with halogen and a metal on the same carbon.
   b. The formation of this species is analogous to formation of a Grignard reagent.
   c. The Simmons-Smith reagent can be conceptualized as methylene coordinated to the zinc atom.
   d. Because they show carbenelike reactivity, α-halo organometallic compounds are sometimes termed carbenoids, that is, a reagent that is not a free carbene but has carbenelike reactivity.

2. Addition reactions of methylene from the Simmons-Smith reagent to alkenes, like the reactions of dichloromethylene, are *syn* additions.

$$CH_2I_2 \xrightarrow{\text{Zn–Cu couple}} I—CH_2—ZnI \longrightarrow$$

Study Guide Links

SGL

✓9.1   Deducing Mechanisms from Rate Laws

Suppose you are an expert in traffic safety, and you are asked to determine as much as you can about traffic accidents on a bridge. In fact, your job is to come up with a description for a "typical" accident and formulate a strategy to eliminate the accidents. Of course, you would interview eyewitnesses and read police reports to gather your data. But suppose eyewitness accounts are not available to you for some reason. This is the situation in deducing reaction mechanisms. We cannot directly observe transition states (that is, molecular collisions).

Imagine that the traffic data you have available are the number of accidents as a function of the traffic density on the bridge in the two directions. Suppose you find that the number of accidents is proportional to the number of cars traveling east *and* to the number of cars traveling west. From this information you can construct a mental picture—a mechanism—for a typical collision. Because the number of accidents in a given time is proportional to the traffic flow in *both* directions, you could reasonably conclude that a typical accident is a head-on collision between a car in one lane and a car in the other. An *unreasonable* picture would be two cars in the same lane having a rear-end collision, because, in this case, the number of accidents in a given time would be proportional to the some function of the traffic density in only that lane. The head-on collision is as good a picture of an accident as any other until other data force you to refine it. In fact, it suggests an experiment for reducing the number of accidents: build a wall between the two lanes of traffic. If this "experiment" reduces the number of accidents, it would support your picture of a typical accident. Notice that it would *not prove your picture to be correct*; it would only *show that your picture is not incorrect*. Nothing short of eyewitnesses can prove you correct.

Similarly, a starting point for deducing a chemical mechanism is the rate law. The rate law tells us about "molecular collisions" that lead to a reaction, but does not tell us whether these collisions occur by a head-on mechanism, a rear-end mechanism, or some less direct pathway. The simplest mechanism possible is adopted, and a modification is considered only when required by new data.

✓9.2   Reaction Stereochemistry and Fischer Projections

It may be tempting to attempt to deduce reaction stereochemistry directly from a Fischer projection. For example, a student might look at the starting material in Eq. 9.38a of the text and say, "Obviously, the phenyls are on the same side of the molecule, so that's the way they have to be in the product." The problem with this reasoning is that it ignores the fact that *Fischer projections do not indicate conformation; they are only meant to indicate stereochemical configuration.* Suppose that the same starting material is redrawn in a different Fischer projection:

$$
\begin{array}{c}
\mathrm{H} \\
\mathrm{X}\!-\!\!\!-\!\!\!-\!\mathrm{Ph} \\
\mathrm{Ph}\!-\!\!\!-\!\!\!-\!\mathrm{H} \\
\mathrm{CH_3}
\end{array}
$$

This is exactly the same molecule, but drawn in a Fischer projection in which the phenyls happen to be on opposite sides of the projection. You can see that the reasoning above would not work, because it predicts that the phenyl groups would end up *trans*. An experimental result can hardly depend on how we choose to draw a structure!

The Fischer projection stands for a particular stereoisomer that might exist in many conformations. In order to examine a particular conformation—which is what must be done to understand the elimination stereochemistry—the Fischer projection must be translated into the conformation of interest. This translation process was explained in Sec. 6.11, text p. 256, but let's review it again in the present context.

First, interpret the Fischer projection directly as an eclipsed conformation:

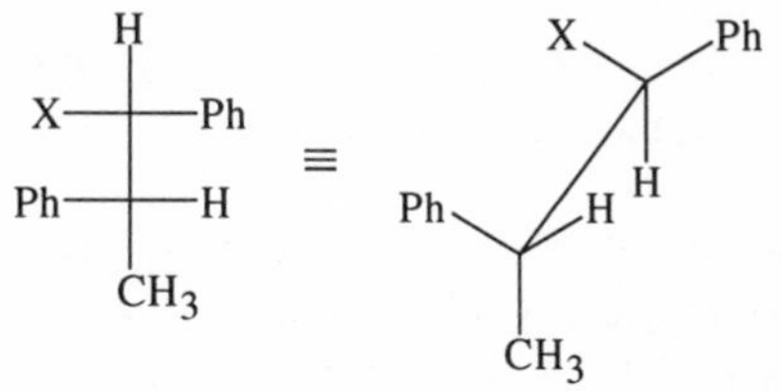

Then change the conformation so that the X on the α-carbon and the H on the β-carbon are *anti:*

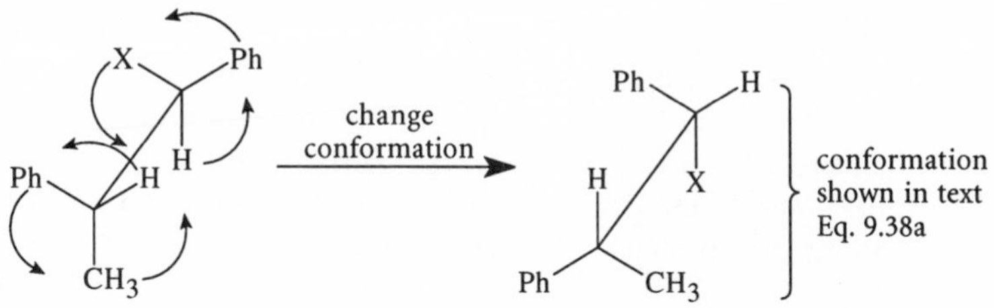

Notice that when the halogen on one carbon and the β-hydrogen on the other are *anti,* the phenyl groups are on the same side of the molecule; that is why they are also on the same side of the molecule in the product—that is, they are *cis.*

The message here is that if you are looking at the stereochemistry of a reaction, and any of the molecules involved are drawn in a Fischer projection, the first thing you should do is interpret the Fischer projections in terms of three-dimensional perspective structures. Don't try to see the reaction stereochemistry directly from the Fischer projection.

You might ask why we just don't draw the appropriate conformation in the text, and forget about Fischer projections. The reason is that the starting material exists in a variety of conformations. The Fischer projection system has nothing to say about conformation. To draw a single conformation as a starting material for Eq. 9.38a might suggest that the molecule exists only in that conformation, and such an implication would be wrong.

✓9.3   **Branching in Cyclic Compounds**

In Study Problem 9.1, the text refers to β-branches in the starting alkyl halide *B* and branching at the double bond of alkene *C*. Some students have difficulty in seeing branches within a cyclic compound, even though they have no problem with acyclic compounds. The reason for this difficulty is that the branches in a cyclic compound are "tied together" into a ring. For example, "branching at the double bond" in compound *C* should be viewed as follows:

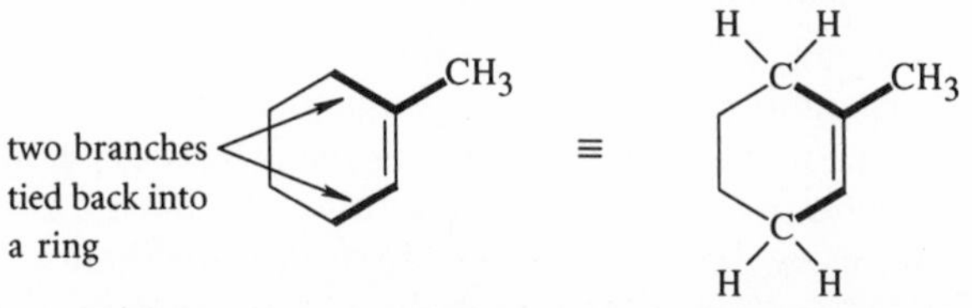

(The methyl group, of course, is a third branch on the double bond, and it is not tied back.) A similar approach is taken to the analysis of β-branching in alkyl halide *B*:

The fact that the branching carbons are linked in a ring doesn't change the fact that they are branches. After all, tying your hands together wouldn't change the fact that you still have two hands, would it?

✓9.4   Diagnosing Reactivity Patterns in Substitution and Elimination Reactions

The last case study in Sample Problem 9.2 points out a danger in using a guide such as Table 9.4: *you can sometimes expect special cases not covered by the table or borderline cases that are hard to predict.* You can't diagnose every possible reaction with a summary table any more than a physician can diagnose a patient with a table of symptoms and diseases! When you encounter borderline cases that are difficult to predict, you should be able to narrow the possibilities of what can happen. The important thing is for you to be able to predict trends and to understand the reasoning behind them. Table 9.4 will prove most useful if you *make the effort to understand each entry in terms of the principles that underlie it,* and then *think* about each problem you are asked to solve in terms of these principles rather than attack it by rote.

A number of professors of organic chemistry, the author included, find that the section of the organic chemistry course on substitution and elimination reactions—the section covered by this chapter—is a point at which a number of students have particular difficulty. We have discerned at least two reasons for this problem. First, this is a point at which the number of reactions and ideas have accumulated to the stage that the student who is not carefully organized is overwhelmed. If you are in this situation, you should read (or re-read) Study Guide Link 5.1, "How to Study Reactions," on p. 88 of this manual, and follow the suggestions there. The second reason for difficulty is the fact that all four mechanisms of substitution and elimination—S_N1, S_N2, E1, and E2—can potentially happen simultaneously, and it can be difficult to keep straight exactly when each process can take place. Again, to use an analogy from the medical world, when a physician diagnoses a problem with a patient, he/she can't focus simply on one system, such as the heart; the physician must understand that the many systems in the body interact, and that there can be many root causes contributing to a given symptom. In some cases, the problems are so complex that they can't be readily solved! Table 9.4 is designed to help you keep all the possibilities in substitution and elimination reactions organized.

A broader philosophical issue is relevant to this discussion. The temptation is for students to believe that there are "recipes" or set formulas for solving every problem. This section on substitutions and eliminations shows that there are *principles* that can guide you into the right "ballpark," but inevitably, uncertainties will arise in some situations. Again: Your goal should be not to memorize every conceivable case, but rather to bring to bear your knowledge of the principles so as to narrow the possibilities.

If you were a laboratory scientist trying to predict the outcome of a reaction, you would make your best prediction using the principles that you know, and then you would go into the laboratory and run an experiment to see whether you are correct. A physician facing such uncertainty runs tests to gather as many facts as possible, adopts a hypothesis, and checks it by administering appropriate medication or carrying out other tests. Two people with difficulties in a relationship adopt a hypothesis as to the nature of the problem and, if they are behaving rationally, change their behavior and see whether the problem is solved. For most problems there aren't magic recipes and tables that allow us to solve problems by rote, nor are there simple formulas into which we "plug" data for an answer. Uncertainty is in the very nature of science, and indeed, of life itself. If this frustrates you—welcome to the real world! *One reason that organic chemistry is valued as a prerequisite course of study by various disciplines is that a student must, in order to succeed, develop a body of principles that can be applied to situations in which there is inherent uncertainty.* The good news is that if you master this skill in organic chemistry, you'll find it easier to do so in your other endeavors as well.

Solutions

Solutions to In-Text Problems

9.1   (a) $CH_3CH_2\overset{+}{N}H_3$ I^- ethylammonium iodide

9.2   (a) Two elimination products are expected because there are two types of β-hydrogens in the starting alkyl halide.

from elimination of a   from elimination of
hydrogen a and bromine   hydrogen b and bromine

9.3   (a) Notice that there are two types of β-hydrogens, and there are correspondingly two possible β-elimination products (three, if you count stereoisomers).

substitution   from elimination of   from elimination of
product   a hydrogen a a hydrogen b
                      and bromine   and bromine

(c) Methyl iodide can only undergo substitution to give dimethyl ether, CH_3—O—CH_3.

9.4   The principle to be applied is that each reaction favors the side with the weaker base.

(a) Because iodide ion is a weaker base than chloride ion, the equilibrium lies to the left.
(c) Because chloride ion is a weaker base than azide ion, the equilibrium lies to the right.

9.5   (a) The reaction is third order overall; the reaction is first order in alkene and second order in bromine. The dimensions of the rate constant are such that the overall rate has the dimensions mol L^{-1} sec^{-1}; thus, the rate constant has the dimensions L^2 mol^{-2} sec^{-1}, or M^{-2} sec^{-1}.

9.6   (a) Use Eq. 9.21 on text p. 392 with $(\Delta G_A^{\circ\ddagger} - \Delta G_B^{\circ\ddagger}) = -14$ kJ/mol:

$$\log (k_A/k_B) = \frac{\Delta G_A^{\circ\ddagger} - \Delta G_B^{\circ\ddagger}}{-2.3RT} = \frac{-14 \text{ kJ/mol}}{-5.706 \text{ kJ/mol}} = 2.4536$$

$$k_A/k_B = 10^{2.4536} = 284$$

Therefore, the rate constant of reaction A is 284 times as large as that of reaction B.

9.7   Because A and B are converted into C, their concentrations decrease with time; hence, the rate also decreases with time. Since the rate is the *slope* of the curves of [reactants] or [products] with time, then the slopes of these curves continually decrease with time, and approach zero at long times. A plot of these concentrations with time is shown in Fig. SG9.1 on the following page.

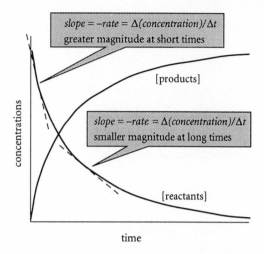

Figure SG9.1 *A plot of reactant and product concentrations to accompany the solution to Problem 9.7. Notice that the rate is the* slope *of the plot at a given time. Because the rate is proportional to concentration, the rate decreases as the concentrations decrease.*

 Note that the *rate* of a reaction is not the same as a *rate constant*. The rate changes with time because concentrations change with time; the rate constant does not change.

9.8 A mechanism consistent with the rate law is a bimolecular process in which ammonia displaces acetate ion from the proton:

$$CH_3\overset{O}{\overset{\|}{C}}\!-\!\overset{..}{\underset{..}{O}}\!-\!H \quad :NH_3 \longrightarrow \left[CH_3\overset{O}{\overset{\|}{C}}\!-\!\overset{\delta-}{\underset{..}{\overset{..}{O}}}\!\text{---}H\text{----}:\overset{\delta+}{NH_3} \right]^{\ddagger} \longrightarrow CH_3\overset{O}{\overset{\|}{C}}\!-\!\overset{..}{\underset{..}{O}}\!:^- + \overset{+}{NH_4}$$
acetate ion

9.10 (a) The reaction of the nucleophile iodide ion with (*S*)-2-bromobutane occurs with inversion of configuration; the resulting product has the *R* configuration.

(*S*)-2-bromobutane → (*R*)-2-iodobutane

9.11 (a) In an S_N2 reaction, the stronger base is the better nucleophile when the same attacking atom is involved. Because ethoxide is a much stronger base than ethanol, it is the better nucleophile.

9.12 (a) Table 8.2, text p. 355, shows that formamide is a polar, *protic* solvent. Consequently, it can donate hydrogen bonds to the acetate ion, thus reducing its nucleophilicity. *N,N*-Dimethylformamide (DMF) is a polar, *aprotic* solvent, and does not reduce the nucleophilicity of acetate by hydrogen bonding. Consequently, sodium acetate is more nucleophilic in DMF, and therefore the reaction of sodium acetate with ethyl iodide is much faster in DMF.

9.13 (a) The discussion of leaving-group effects in Sec. 9.5B indicates that leaving-group effects are about the same in S_N2 and E2 reactions; and Eq. 9.30, text p. 403, indicates that bromide is a better leaving group than chloride; alkyl bromides are about fifty times as reactive as alkyl chlorides. Therefore, compound *A* reacts most rapidly. Because of the primary deuterium isotope effect, compound *C* reacts about 3–8 times as rapidly as compound *B*. (Note that leaving-group effects are more important than isotope effects.) Thus, the order of increasing reactivity is *B* < *C* < *A*.

9.14 (a) The hydration of styrene should be (and is) slower in D_2O, because the transferred proton is a deuterium, and a rate-retarding primary deuterium isotope effect operates when the proton transferred in the rate-limiting step is isotopically substituted.

The products differ by isotopic substitution. In H_2O/H_3O^+, the hydration product is compound *A* below; in D_2O/D_3O^+, the hydration product is compound *B*.

$$\overset{OH}{\underset{}{PhCHCH_3}} \quad A \qquad\qquad \overset{OD}{\underset{}{PhCHCH_2D}} \quad B$$

(b) The hydration rate of the deuterium-substituted styrene should differ very little, if at all, from that of styrene itself, because the deuteriums are not transferred in the rate-limiting step.

> There is a small effect of isotopic substitution in this case because of the differential effect of deuterium and hydrogen on the rehybridization of carbon in the transition state. (The carbon bearing the deuteriums rehybridizes from sp^2 to sp^3 in the rate-limiting step.) However, this effect on rate amounts to only a few percent. Effects of this sort are called *secondary deuterium isotope effects.*

9.15 (a) Analyze one enantiomer of (±)-stilbene dibromide; the result is the same for the other. First, draw the Fischer projection; then translate it into a conformational projection (*Step 1* below). Next, internally rotate about the central carbon-carbon bond until the H and Br that are eliminated are *anti* (*Step 2* below). Finally, draw the alkene that is formed and note its stereochemistry (*Step 3* below). The process starts with the Fischer projection because it is easy to distinguish between the chiral and *meso* stereoisomers of the starting material with a Fischer projection. Of course, you can start at whatever point in the process is easiest.

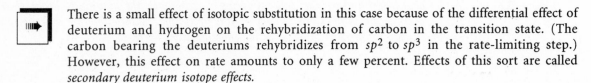

9.16 The diastereomer of the starting alkyl halide in Eq. 9.38a gives the diastereomer of the product. This can be verified by using the process similar to that used in the solution to Problem 9.15.

$$\underset{CH_3}{\overset{X}{\underset{\displaystyle Ph\!-\!\!-\!H \atop H\!-\!\!-\!Ph}{}}} \quad\xrightarrow{Na^+ \, C_2H_5O^-, \, C_2H_5OH}\quad \underset{H_3C \quad Ph}{\overset{Ph \qquad H}{C\!=\!C}}$$

(*E*)-α-methylstilbene

9.17 (a) $(CH_3)_2CHCH_2Br \; + \; (CH_3)_2S \;\xrightarrow{acetone}\; (CH_3)_2CHCH_2\overset{+}{S}(CH_3)_2 \;\; Br^-$

isobutyl bromide   dimethyl sulfide

(c) $(CH_3)_2CHCH_2Br \;\xrightarrow[(CH_3)_3COH]{K^+ \, (CH_3)_3CO^-}\; (CH_3)_2C\!=\!CH_2$

isobutyl bromide

9.18   Analyze the branching pattern in each alkyl halide; branching at both the α- and β-positions increases the ratio of elimination to substitution. Compound D has both an α-branch and two β-branches; compound B has two β-branches; compound C has one β-branch; and compound A, of course, is unbranched. Therefore, the ratio of E2 elimination to S_N2 substitution decreases in the order $D > B > C > A$.

9.20   (a) Products A and B result from attack of ethanol and water, respectively, on the carbocation intermediate that is formed by ionization of the alkyl halide starting material. Products C and D result from attack of ethanol and water, respectively, on the rearranged carbocation intermediate. (Compound C is the product shown in Eq. 9.58.)

Alkenes can form by loss of a β-proton from either of the two carbocation intermediates. (The solvent, indicated below by ROH, acts as the base that removes the proton.)

Alkene F would be formed in greatest amount because it has the greatest amount of branching at the double bond. Alkene G would be formed in smaller amount; alkene E would be formed in least amount for two reasons: it comes from a carbocation that can rearrange, and it has only one branch at the double bond.

9.21   The mechanism for the formation of the rearrangement product in Eq. 9.58:

(equation continues)

$$\underset{\substack{\\ \overset{+}{:}\text{OC}_2\text{H}_5 \\ | \\ \text{H}}}{\underset{\text{HOC}_2\text{H}_5}{\overset{\text{CH}_3 \quad \text{CH}_3}{\text{CH}_3-\text{C}-\text{CH}-\text{CH}_3}}} \quad :\!\ddot{\text{Cl}}\!: \quad \longrightarrow \quad \underset{:\ddot{\text{OC}}_2\text{H}_5}{\overset{\text{CH}_3 \quad \text{CH}_3}{\text{CH}_3-\text{C}-\text{CH}-\text{CH}_3}} \quad + \quad \text{HO}\overset{+}{\text{C}}_2\text{H}_5 \quad :\!\ddot{\text{Cl}}\!:^{-}$$

9.22 (a) Entry 3 of Table 9.4 covers this case; the major product is the substitution product 1-methoxybutane, which is formed by an S$_N$2 mechanism:

$$\text{CH}_3\text{CH}_2\text{CH}_2\text{CH}_2 \!-\! \ddot{\text{Br}}\!:^{-} \quad \longrightarrow \quad \text{CH}_3\text{CH}_2\text{CH}_2\text{CH}_2 \!-\! \ddot{\text{O}}\text{CH}_3 \quad + \quad :\!\ddot{\text{Br}}\!:^{-}$$
$$^{-}\!:\!\ddot{\text{O}}\text{CH}_3$$

In early printing of the text, ethanol was specified as the solvent rather than methanol. If ethanol were used as the solvent, some ethoxide would also be present, and some ethyl ether would also be formed.

A small amount of the alkene 1-butene will also be formed:

$$\underset{\text{CH}_3\ddot{\text{O}}:^{-}}{\text{CH}_3\text{CH}_2\text{CH}-\text{CH}_2 \!-\! \ddot{\text{Br}}\!:^{-}} \quad \longrightarrow \quad \text{CH}_3\text{CH}_2\text{CH}\!=\!\text{CH}_2 \quad + \quad \text{CH}_3\ddot{\text{O}}\!-\!\text{H} \quad + \quad :\!\ddot{\text{Br}}\!:^{-}$$

(c) 2-Bromo-1,1-dimethylcyclopentane is a secondary alkyl halide with significant β-branching; a protic solvent is used without a strong base. Entry 9 in Table 9.4 covers this case. Rearrangement from the initially formed secondary carbocation *a* to the tertiary carbocation *b* is likely to occur.

(Note that rearrangement of the carbocation *a* to another tertiary carbocation *c* is also possible:

This rearrangement will not occur because it produces a carbocation with significantly greater ring strain.)

Each carbocation can react to give S$_N$1 substitution and E1 elimination products. Carbocation *a* gives the substitution product *A* and the elimination product *B*.

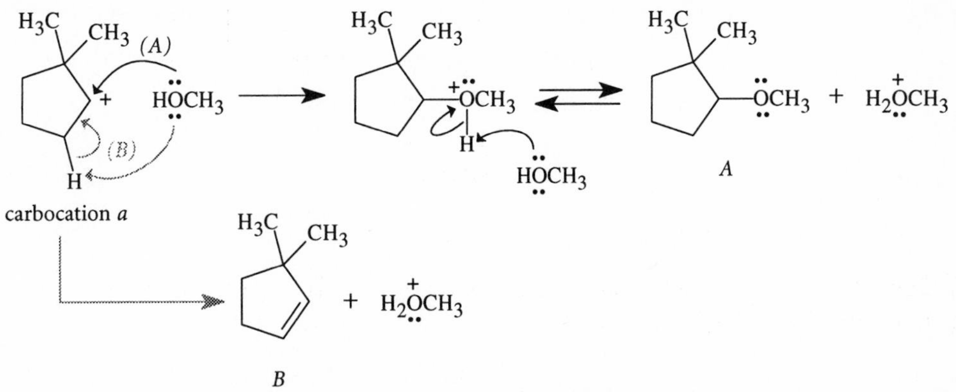

carbocation *a*

B

Carbocation *b* gives the substitution product *C* and the elimination products *D–F*. (The curved-arrow mechanism for formation of *C* is analogous to the mechanism shown above for the formation of *A*; and the curved-arrow mechanisms for the formation of alkenes *D–F* are analogous to the mechanism shown above for the formation of *B*.)

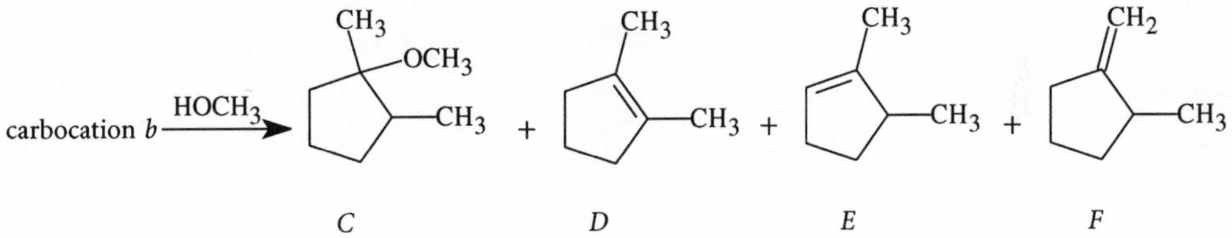

carbocation *b* $\xrightarrow{HOCH_3}$

C   *D*   *E*   *F*

Compounds *C*, *D*, and *E* are expected to be the major products of the reaction.

9.23   (a)

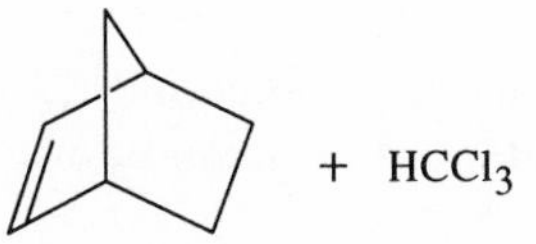

norbornene   chloroform

+ $HCCl_3$

(c)

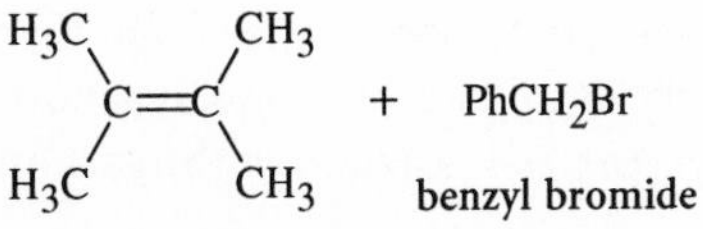

+ $PhCH_2Br$

2,3-dimethyl-2-butene   benzyl bromide

In part (c), why not use $(CH_3)_2CHBr$ (isopropyl bromide) and the alkene $PhCH{=}C(CH_3)_2$? Because isopropyl bromide has β-hydrogens, it will not form a carbene, but instead will form the alkene propene, $CH_3CH{=}CH_2$.

9.24   (a)   (c)

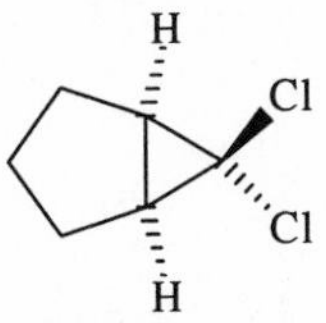

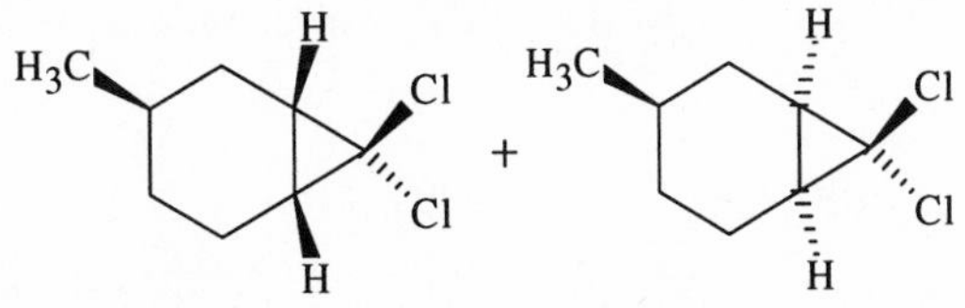

Because the two *syn*-addition products of part (c) are diastereomers, they are formed in different amounts.

9.25   (a)   (c)

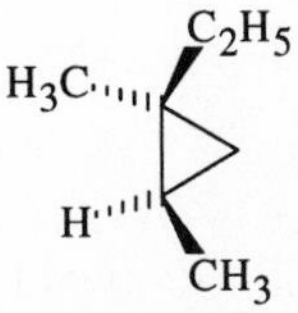

(racemate)

9.26 (a)

$$CH_2$$

structure: 4-methylenecyclohexane (cyclohexane ring with $=CH_2$ groups at top and bottom)

Solutions to Additional Problems

9.27 The first step in any problem that requires structures is to draw the structures:

$$CH_3(CH_2)_4CH_2Br$$

1-bromohexane

(1)

$$CH_3CH_2\overset{\overset{\displaystyle CH_3}{|}}{\underset{\underset{\displaystyle Br}{|}}{C}}CH_2CH_3$$

3-bromo-3-methylpentane

(2)

$$BrCH_2\overset{\overset{\displaystyle CH_3}{|}}{\underset{\underset{\displaystyle CH_3}{|}}{C}}CH_2CH_3$$

1-bromo-2,2-di-methylbutane

(3)

$$CH_3\overset{\overset{\displaystyle CH_3}{|}}{C}H\overset{\overset{}{\underset{\underset{\displaystyle Br}{|}}{C}}}HCH_2CH_3$$

3-bromo-2-methyl-pentane

(4)

$$CH_3\overset{\overset{\displaystyle Br}{|}}{C}H\overset{\overset{}{\underset{\underset{\displaystyle CH_3}{|}}{C}}}HCH_2CH_3$$

2-bromo-3-methyl-pentane

(5)

(a) Compounds (4) and (5) can exist as enantiomers.

(c) Compound (1) gives the fastest S_N2 reaction with sodium methoxide, because it is the only primary alkyl halide.

(e) Compound (1) can give only one alkene in the E2 reaction.

(g) Compounds (4) and (5) undergo the S_N1 reaction with rearrangement. If compound (3) is forced to react, it will also undergo the S_N1 reaction with rearrangement, but under ordinary conditions it will not react.

9.28 The products of the reactions of isopentyl bromide, $(CH_3)_2CHCH_2CH_2Br$: (When the solvent is a by-product it is not shown.)

(a)

$(CH_3)_2CHCH_2CH_2I$

$+ K^+ Br^-$

(b)

$(CH_3)_2CHCH_2CH_2OH + (CH_3)_2CHCH_2CH_2OC_2H_5$

$+ (CH_3)_2CHCH=CH_2$ (small amount) $+ K^+ Br^-$

(c)

$(CH_3)_2CHCH=CH_2$

$+ K^+ Br^-$

(d)

$$(CH_3)_2CHCHCH_3$$
$$\underset{\displaystyle Br}{|}$$

$$+ \ (CH_3)_2\overset{}{C}CH_2CH_3$$
$$\underset{\displaystyle Br}{|}$$

(from rearrangement)

(e)

$(CH_3)_2CH$ — cyclopropane ring with Cl, Cl

$+ (CH_3)_3COH$

(f)

$(CH_3)_2CH$ — cyclopropane ring

$+ ZnI_2$

(g)

$(CH_3)_2CHCH_2CH_3$

$+ Li^+ Br^-$

(h)

$(CH_3)_2CHCH_2CH_2OCH_3$

$+ (CH_3)_2CHCH=CH_2$

(small amount)

$+ Na^+ Br^-$

9.30 The products of the reactions of 2-bromo-2-methylhexane:

$$CH_3\overset{\overset{\displaystyle CH_3}{|}}{\underset{\underset{\displaystyle Br}{|}}{C}}(CH_2)_3CH_3 \quad \text{2-bromo-2-methylhexane}$$

(a)

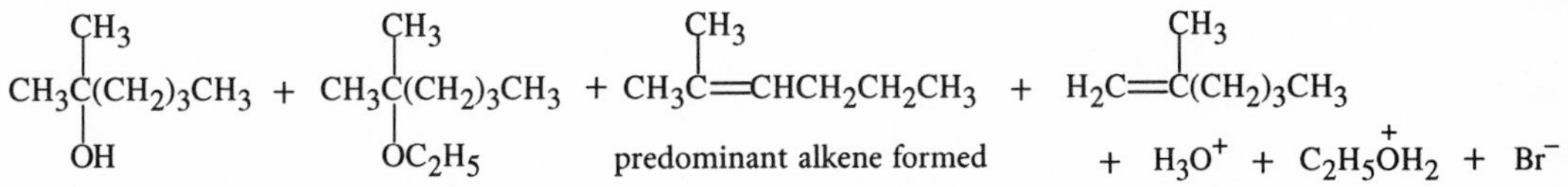

$$CH_3\underset{OH}{\overset{CH_3}{\underset{|}{\overset{|}{C}}}}(CH_2)_3CH_3 \; + \; CH_3\underset{OC_2H_5}{\overset{CH_3}{\underset{|}{\overset{|}{C}}}}(CH_2)_3CH_3 \; + \; CH_3\overset{CH_3}{\overset{|}{C}}\!\!=\!\!CHCH_2CH_2CH_3 \; + \; H_2C\!\!=\!\!\overset{CH_3}{\overset{|}{C}}(CH_2)_3CH_3$$

predominant alkene formed $+ \; H_3O^+ \; + \; C_2H_5\overset{+}{O}H_2 \; + \; Br^-$

(b) 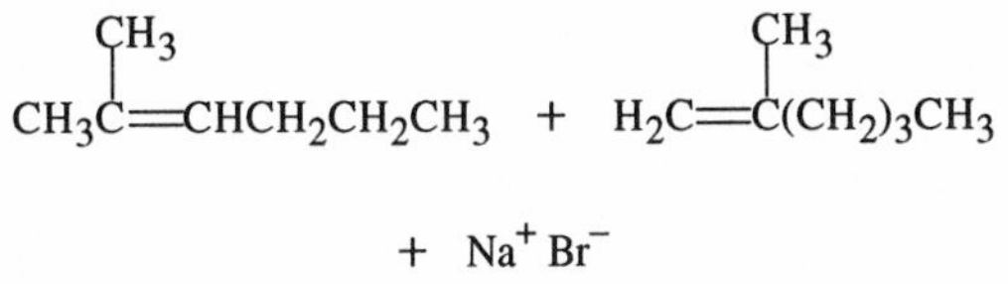   (c)

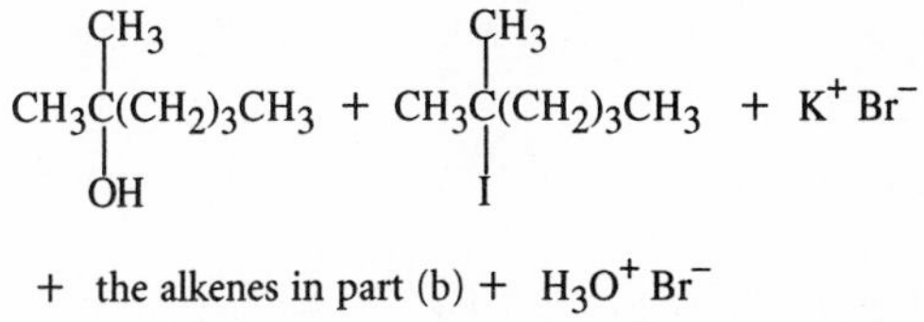

$$CH_3\overset{CH_3}{\overset{|}{C}}\!\!=\!\!CHCH_2CH_2CH_3 \; + \; H_2C\!\!=\!\!\overset{CH_3}{\overset{|}{C}}(CH_2)_3CH_3$$

$$+ \; Na^+Br^-$$

$$CH_3\underset{OH}{\overset{CH_3}{\underset{|}{\overset{|}{C}}}}(CH_2)_3CH_3 \; + \; CH_3\underset{I}{\overset{CH_3}{\underset{|}{\overset{|}{C}}}}(CH_2)_3CH_3 \; + \; K^+Br^-$$

$$+ \; \text{the alkenes in part (b)} \; + \; H_3O^+Br^-$$

(d)

$$CH_3\underset{Br}{\overset{CH_3}{\underset{|}{\overset{|}{C}}H}}CHCH_2CH_2CH_3 \; + \; BrCH_2\overset{CH_3}{\overset{|}{C}H}(CH_2)_3CH_3$$

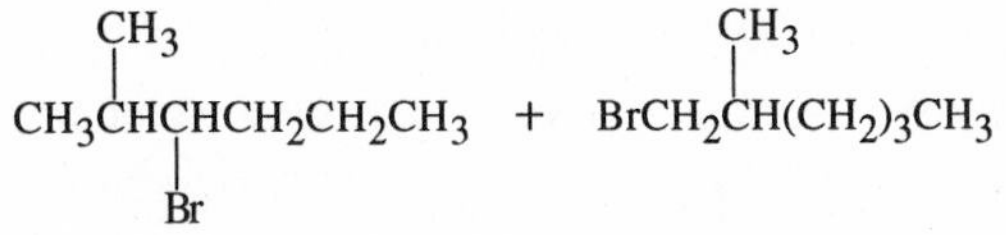

9.32   The order of increasing S_N2 reaction with KI in acetone is $A < B < C < E < D$. Tertiary alkyl halides react slowly, if at all, in S_N2 reactions; secondary alkyl halides react more rapidly; primary alkyl halides with two β-branches react even more rapidly; and unbranched primary alkyl halides react most rapidly. Of the two butyl halides shown, the one with the better leaving group reacts more rapidly.

9.34   To deduce the structure of the nucleophile, disconnect the appropriate atom from the ethyl group, add a pair of electrons and calculate the formal charge, if any.

(a)   (c)   (e)

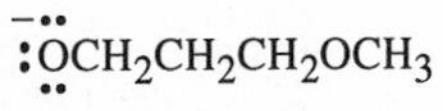

$$:\overset{..}{\underset{..}{O}}CH_2CH_2CH_2OCH_3 \qquad (CH_3)_3N:$$

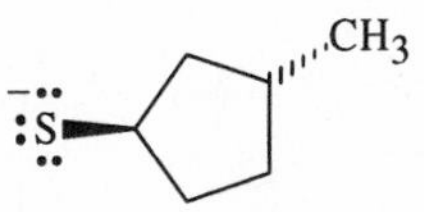

9.35   The substitution reaction occurs by an S_N2 mechanism with inversion of configuration.

(a)

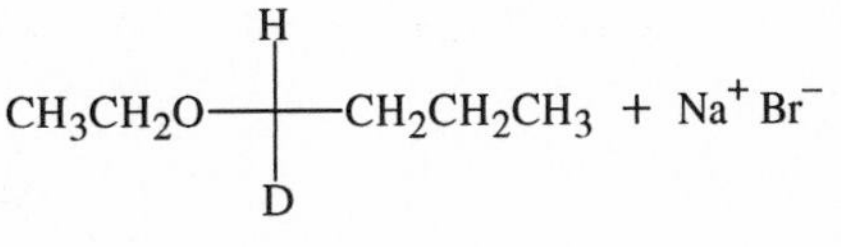

$$CH_3CH_2O\underset{D}{\overset{H}{\underset{|}{\overset{|}{-\!\!-\!\!-}}}}CH_2CH_2CH_3 \; + \; Na^+Br^-$$

9.36   Because the alkyl halide starting material is tertiary, solvolysis proceeds by an S_N1 mechanism, that is, through a carbocation intermediate. The carbocation has a trigonal planar geometry; unlike the starting alkyl halide, it is *achiral*. Water can attack this carbocation at either lobe of the vacant p orbital with equal probability. Abbreviating the $(CH_3)_2CHCH_2CH_2CH_2-$ group as $C_6H_{13}-$, this process is as shown in the diagram on the top of page 212. This analysis shows that the enantiomeric alcohols should be obtained in equal amounts; in other words, the product is predicted to be *racemic*.

Despite this apparently straightforward prediction based on the carbocation hypothesis, it was found experimentally that the alcohol obtained in this reaction was not completely racemic; about 60% of the *S* enantiomer and about 40% of the *R* enantiomer were obtained. In other words, a slight predominance of *stereochemical inversion* takes place. Results such as this led to the hypothesis that there are at least *two* reactive intermediates in S_N1 reactions. The first is an *ion pair,* a carbocation that is closely associated with its counter-ion and solvated by water at only one face—the face opposite to the chloride ion. Attack of this solvent water on the ion pair must necessarily occur with inversion. The other intermediate is the achiral carbocation shown in the solution above, which can be solvated by water at either or both faces; this cation can be attacked from either face with equal probability. To accommodate the experimental results, about 20% of the product must be formed with inversion from the ion pair, and the remaining 80% from the free carbocation.

9.37 (a) This compound can react to give a mixture of alkenes, because it contains more than one type of β-hydrogen:

$$CH_3CH=C(CH_3)_2 \quad + \quad CH_3CH_2\underset{\underset{\displaystyle CH_3}{|}}{C}=CH_2$$

(c) This compound can also react to give a mixture of alkenes:

9.38 (a) Even though it is a weaker base, the thiolate is a better nucleophile in methanol; therefore, the major product is the sulfide $CH_3SCH_2CH_2CH_3$. In a protic solvent, the alkoxide, because it is a stronger base than the thiolate, is involved in much stronger hydrogen bonds from the solvent; this hydrogen bonding diminishes the nucleophilicity of the alkoxide ion relative to that of the thiolate ion in the same sense that hydrogen bonding diminishes the nucleophilicity of fluoride ion relative to that of the chloride ion.

(b) In a polar *aprotic* solvent, the hydrogen bonding discussed in part (a) is not present; consequently, the alkoxide is the better nucleophile because it is the stronger base. Therefore, the major product is the ether $CH_3OCH_2CH_2CH_3$.

9.39   Notso has attempted the S_N2 reaction of a *tertiary* alkyl halide with a strong base. As you no doubt realize, and Notso would have realized, had he consulted a superior organic chemistry text such as yours, that these are precisely the conditions that promote the E2 reaction. Consequently, Notso formed an alkene. Ima, on the other hand, allowed a strong base to react with methyl iodide, which cannot undergo elimination; the only alternative is the desired S_N2 reaction.

9.41   (Crown ethers are discussed in Sec. 8.4C, text pp. 358–360.) Potassium fluoride, an ionic compound, is not soluble in hydrocarbons such as benzene. Consequently, it will not react in such solvents because its dissolved concentration is essentially zero. When the crown ether is added, the crown ether forms a benzene-soluble complex with the potassium ion, which, by electrical neutrality, includes an accompanying "naked" (that is, essentially unsolvated) fluoride ion. This soluble but unsolvated fluoride ion is a very good nucleophile, and reacts with benzyl bromide to yield benzyl fluoride, $PhCH_2F$, plus "naked" bromide ion, which then exchanges with insoluble potassium fluoride to give more dissolved fluoride and solid KBr. The newly dissolved fluoride can react with more benzyl bromide. The cycle is repeated until one of the reactants is exhausted. To summarize:

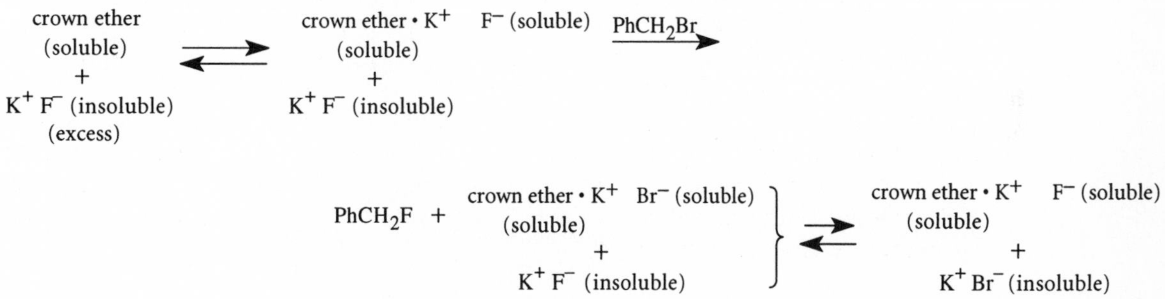

This scheme shows that complexation of the cation by the crown ether is essential for the reaction to occur. Because [18]-crown-6 does not bind the smaller lithium cation—it is selective for the larger potassium cation—lithium fluoride is not solubilized by the crown ether, and is therefore unreactive whether the crown ether is present or not.

9.42   In the transition state of the S_N2 reaction, the carbon at which substitution occurs is sp^2-hybridized; consequently, the ideal bond angles at this carbon are 120°. (See Fig. 9.2, text p. 397.) This requirement for a 120° bond angle means that as cyclopropyl iodide is converted into the S_N2 transition state, the bond angle within the cyclopropane ring should increase. The increasing angle causes additional angle strain in the cyclopropane ring that raises the energy of the transition state. Consequently, S_N2 reactions of cyclopropyl halides are unusually slow.

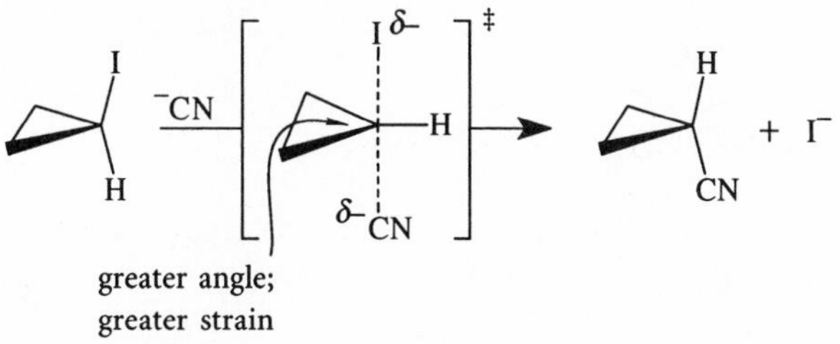

greater angle;
greater strain

Such rate-retarding angle strain is present in neither compound (b), isopropyl iodide, nor its S_N2 transition state. Consequently, the S_N2 reactions of isopropyl iodide are faster than those of cyclopropyl iodide.

9.43   In base, some of the chlorines are eliminated as chloride ion in E2 reactions. One such reaction is as follows;

can you find another? (S_N2 reactions probably do not occur; why?)

9.44   (a) Draw either enantiomer of the molecule in two staggered conformations in which a β-hydrogen or a β-deuterium, respectively, is *anti* to the bromine. These conformations reveal that, when the deuterium is *anti* to the bromine, the methyl groups are on opposite sides of the molecule; hence, elimination of D and Br gives a *trans*-alkene that does not contain deuterium. When the hydrogen is *anti* to the bromine, the methyl groups are on the same side of the molecule; hence, elimination of H and Br gives a *cis*-alkene that contains deuterium.

9.45   The S_N1 solvolysis products result from attack of the respective solvent molecules on the carbocation intermediate:

(a)

(b) The reaction is faster in formic acid, because its higher dielectric constant promotes the separation of the chloride ion leaving group from the carbocation. (This effect is discussed in Sec. 9.6C on text p. 421.)

9.46   (a) The principle to apply in this case is that the right side of the equilibrium is favored by the solvent that better separates and solvates ions. Because ethanol is the more polar solvent—it has the higher dielectric constant—it better separates ions. Because it is a protic, donor solvent, it better solvates ions; consequently the equilibrium lies farther to the right in ethanol.

(b) In this case, the dielectric constants of the two solvents are the same. Consequently, the ability of the solvent to solvate the ions determines the relative position of the equilibrium. Because dimethylacetamide is *aprotic,* it does not solvate anions as well as the aqueous methanol solvent. Consequently, the equilibrium lies farther to the right in aqueous methanol.

9.47   Because it is a tertiary alkyl halide without β-hydrogens, trityl chloride can react only by the S_N1 mechanism. A characteristic feature of this mechanism is that the first step is rate-limiting, whereas attack of the nucleophile(s) occurs on the carbocation intermediate in the second step. Hence, the rate of reaction of trityl chloride is *independent of the nucleophile concentration,* and therefore independent of the nucleophile that reacts with the carbocation.

$$Ph_3C\text{—}Cl \xrightarrow{\text{rate-limiting step}} Ph_3C^+ \; Cl^-$$

trityl chloride   trityl cation

However, the product-determining steps for the various reactions are different. In the product-determining step, a nucleophile attacks the carbocation. The principle to apply is that the better nucleophile reacts more rapidly with the carbocation intermediate; hence, the major product is derived from the better nucleophile.

Thus, in reaction (1), the only nucleophile available is water. After attack of water on the carbocation to give $Ph_3C\text{—}\overset{+}{O}H_2$, a proton is lost to solvent to give trityl alcohol as the solvolysis product. In reaction (2), both water and azide ion are in competition for the trityl cation. The basicities of water and azide are determined by the pK_a values of their respective conjugate acids. The conjugate acid of water, H_3O^+, has a pK_a value of -1.7. The conjugate acid of azide ion, HN_3, has a pK_a value of 4.7. Consequently, azide is the more basic nucleophile, and is evidently the better nucleophile. When, in reaction (2), sodium azide is added to the reaction mixture, trityl chloride ionizes at the same rate, but the carbocation intermediate reacts preferentially with azide ion.

When hydroxide ion is added, the rate of ionization of trityl chloride again remains unchanged; however, because both hydroxide ion and azide are present, both nucleophiles react with the carbocation to give a mixture trityl alcohol and trityl azide.

> One might expect trityl alcohol to be the major product in case (3) because hydroxide is considerably more basic than azide. However, hydroxide ion, because of its basicity, forms very strong hydrogen bonds to water. This hydrogen bonding reduces the nucleophilicity of hydroxide ion to the point that its nucleophilicity is comparable to that of azide ion.

9.48   Remember that the key to working problems of this type is to find a known structure and work from it. (See point 3 in Study Guide Link 4.6, p. 63 of this manual.) Compound *B* is an alkene, because it decolorizes Br_2 in an inert solvent. The structure of the ozonolysis product *C* leads to the identity of *B*.

The elemental analysis of *A* leads to the molecular formula $C_8H_{13}Br$, which is equivalent to that of the alkene *B* plus the elements of HBr. Because the carbon skeleton must be bicyclic, there can be no additional unsaturation in *A*, a point confirmed by the fact that it does not add bromine. Evidently, compound *A* is an alkyl halide that undergoes a reaction with a strong base to give the alkene *B*; this reaction is then an E2 reaction. The alkyl halide *A*, and its E2 reaction, are as follows:

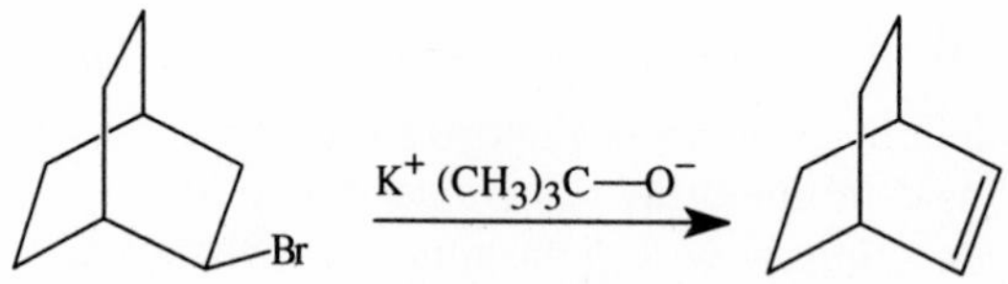

A *B*

9.49 The fact that protonolysis reactions of the corresponding Grignard reagents give the same hydrocarbon indicates that the two compounds have the same carbon skeleton. The conditions of ethanol and no added base are S_N1 conditions. Since compound *A* reacts rapidly to give a solution containing bromide ion, it must be an alkyl halide that readily undergoes an S_N1 reaction, and therefore it is probably a tertiary alkyl bromide. Because the two alkyl halides give the same ether, the product from compound *B* must be formed in a rearrangement. The only tertiary alkyl halide with the formula $C_5H_{11}Br$ is 2-bromo-2-methylbutane, and this is therefore compound *A:*

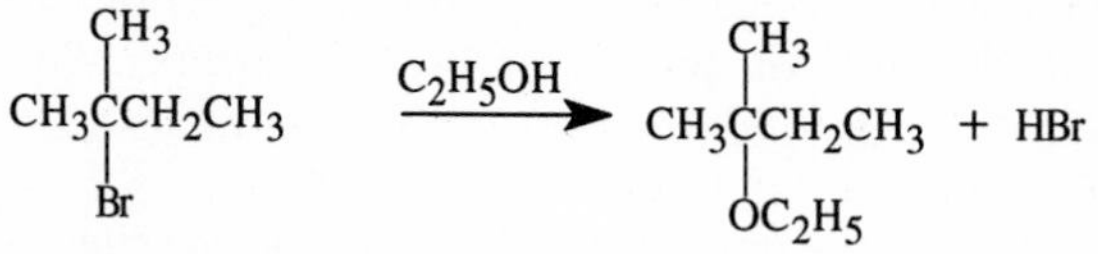

A *C*

2-bromo-2-methylbutane

There are two possible alkyl halides with the same carbon skeleton as *A* that could rearrange in respective S_N1 reactions to give the same carbocation, and hence the same ether product, as *A*; these are labeled *B1* and *B2* below:

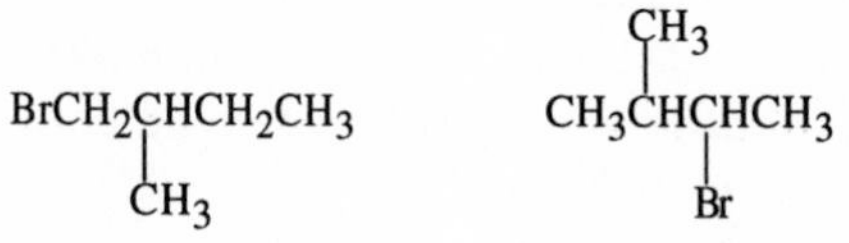

B1 *B2*

However, only *B2* can react in an E2 reaction with sodium ethoxide to give an alkene that furnishes acetone as one of its ozonolysis products:

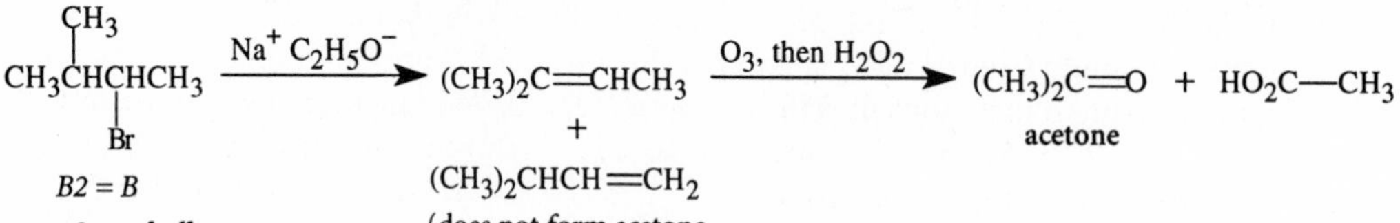

B2 = *B*

2-bromo-3-methylbutane

(does not form acetone on ozonolysis)

acetone

Consequently, compound *B2* is compound *B*, 2-bromo-3-methylbutane.

9.50 Compound *A* is an alkyl halide with two degrees of unsaturation; because one unsaturation is accounted for by the cyclohexane ring found in subsequent products, the other must be a double bond. Subsequent elimination reactions of *A* yield two dienes *B* and *C*. The catalytic hydrogenation of these two compounds establishes that compounds *A–C* and 1-isopropyl-4-methylcyclohexane have the same carbon skeleton. Because ozonolysis leaves the cyclohexane ring intact, the double bond must be external to the ring; and the identity of product *F* shows that both double bonds are in fact attached to the ring. There is only one possibility for compound *B*:

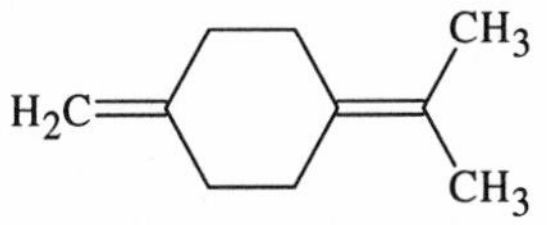

compound B

The following possibilities for A can be enumerated:

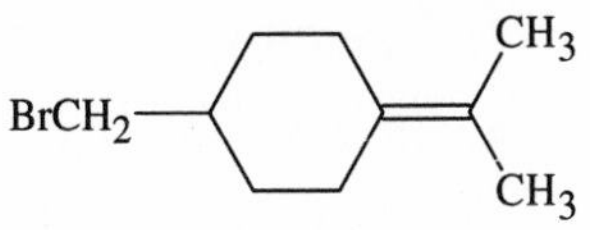

A1 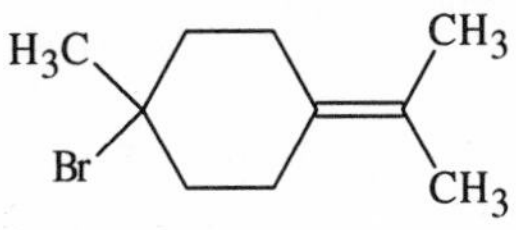   *A2* 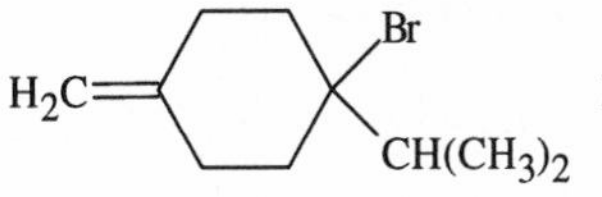   *A3* 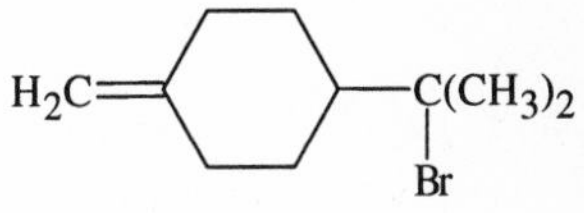   *A4*

Notice that any other possibility for A is ruled out both by the structure of B and by the fact that compound A is achiral. Compound A1 is ruled out by the fact that it can only give one diene in the E2 reaction; yet compound A gives two dienes, compounds B and C. That compound A undergoes S_N1 solvolysis rapidly also rules out compound A1, because it is a primary alkyl halide that would not react under S_N1 conditions. Because compound A undergoes ozonolysis to give acetone—and because structure A1 was ruled out—*only structure A2 fits all the data.* The structure of alkene B is given above; the structure of alkene C, the other E2 reaction product of compound A = A2, is as follows:

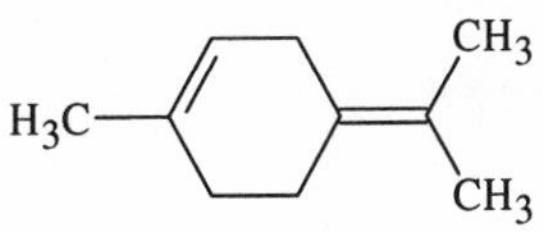

compound C

Finally, the two achiral compounds D and E obtained from the bromine addition reaction of A are the following two diastereomers:

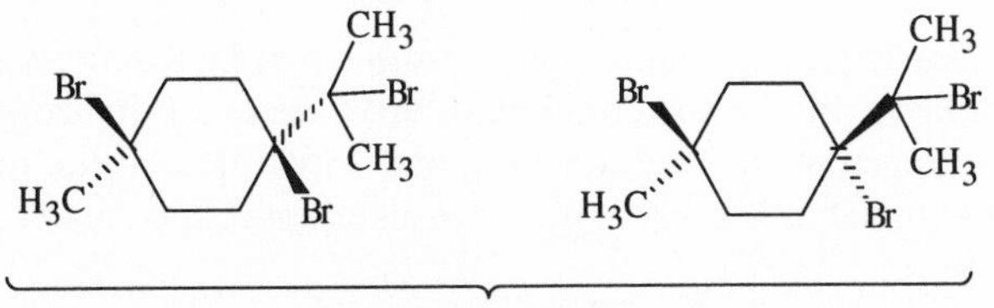

compounds D and E

Of course, which is D and which is E cannot be determined from the data.

9.51   The important keys to solving this problem are (1) that the E2 reaction is fastest when the hydrogen and chlorine are *anti;* and (2) the substituted cyclohexanes can undergo the chair flip rapidly. In order for menthyl chloride to undergo *anti* elimination, it must undergo the chair flip so that the chlorine is axial:

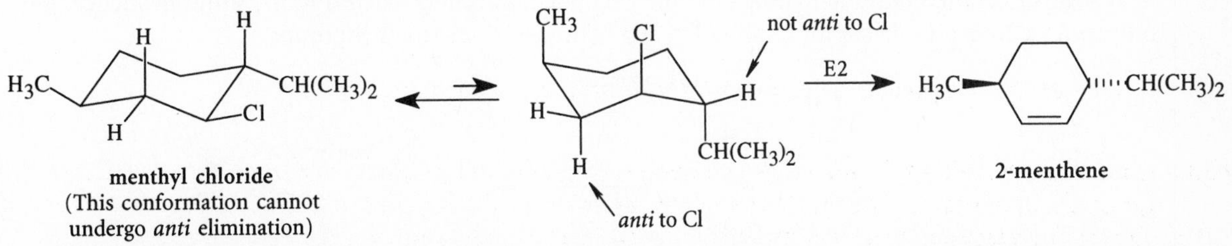

menthyl chloride
(This conformation cannot
undergo *anti* elimination)

2-menthene

The only alkene that can be formed by *anti* elimination is 2-menthene. In contrast, neomenthyl chloride has two β-hydrogens, H^a and H^b, that can be lost in *anti* eliminations along with the chlorine to give 2-menthene and 3-menthene, respectively:

neomenthyl chloride

More 3-menthene is formed because it is the alkene with the greater number of alkyl branches at its double bond.

9.52 (a) Because the alkene product is *cis*, the transition state for the elimination must have the two methyl groups on the same side of the molecule. Drawing the molecule in a conformation with the methyl groups on the same side shows that the bromines are eliminated from opposite sides of the molecule; that is, the elimination is *anti*.

9.53 As in Problem 9.51, the key to the solution is to assume that E2 reactions are faster when they can occur with *anti* stereochemistry. In the first compound, there are two β-hydrogens that can be eliminated with the bromine: the hydrogen at the ring junction, and a hydrogen of the methyl group. The hydrogen at the ring junction is eliminated because it gives the alkene with the greater number of alkyl branches on the double bond.

In the second compound, the hydrogen at the ring junction cannot be *anti* to the bromine; hence, the only *anti* hydrogen available for elimination is a hydrogen of the methyl group:

9.54 (a) Initiation of the reaction by AIBN (see text p. 198) suggests a free-radical chain mechanism. The first initiation step is shown in Eq. 5.44, text p. 198. Let the free-radical product of this reaction be abbreviated as R•, and let the butyl group be abbreviated as Bu. The second initiation step is

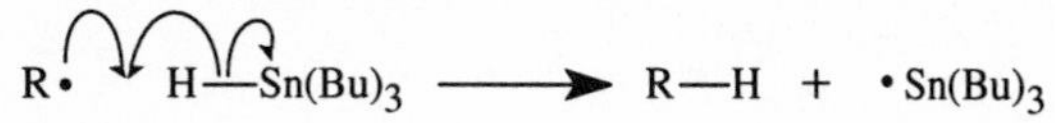
R•
H—Sn(Bu)$_3$
R—H + •Sn(Bu)$_3$

The propagation steps are as follows:

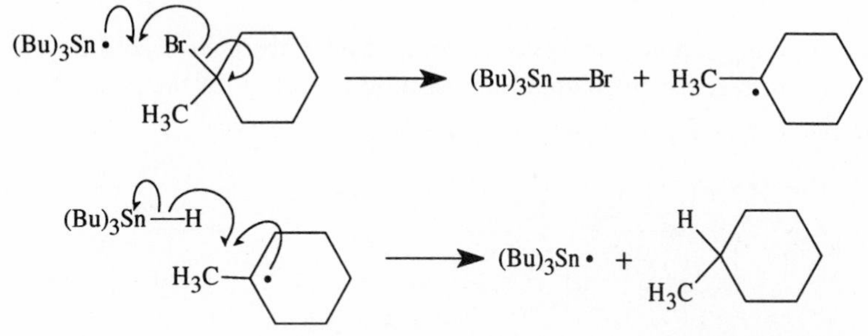
(Bu)$_3$Sn•
Br
H$_3$C
(Bu)$_3$Sn—Br + H$_3$C
(Bu)$_3$Sn—H
H$_3$C
(Bu)$_3$Sn• +
H
H$_3$C

methylcyclohexane

(b) One sequence of reactions is to convert the alkyl halide into a Grignard reagent, which is then treated with water:

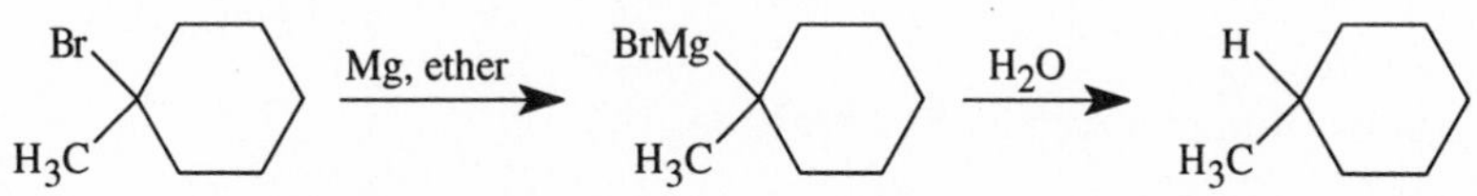
Br
H$_3$C
Mg, ether
BrMg
H$_3$C
H$_2$O
H
H$_3$C

Another is to carry out an E2 reaction to obtain an alkene (or a mixture of alkenes), which is then subjected to catalytic hydrogenation to give methylcyclohexane.

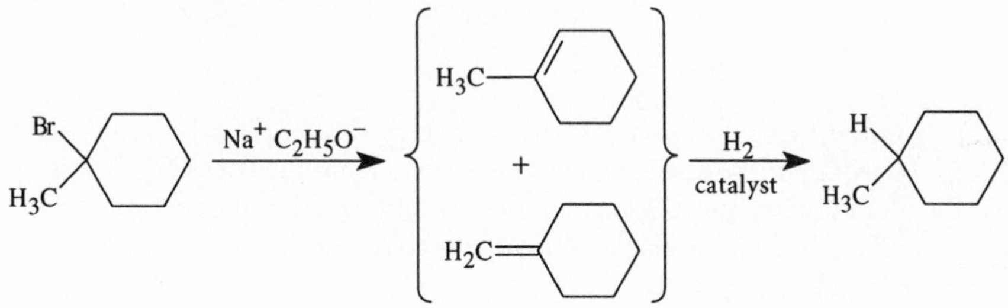
Br
H$_3$C
Na$^+$ C$_2$H$_5$O$^-$
H$_3$C
+
H$_2$C
H$_2$
catalyst
H
H$_3$C

9.55 The reaction of butylamine with 1-bromobutane is a typical S_N2 reaction:

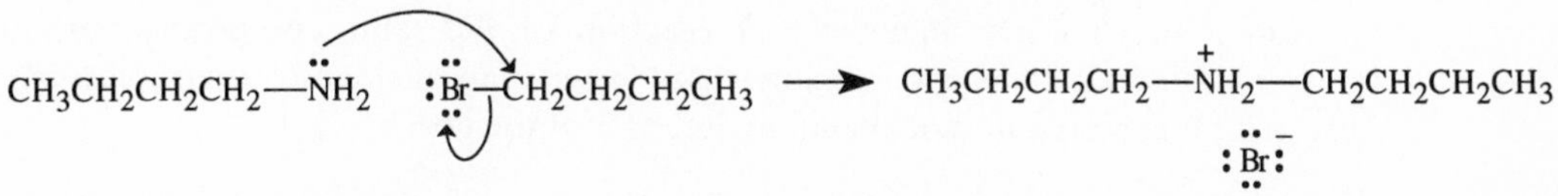
CH$_3$CH$_2$CH$_2$CH$_2$—NH$_2$
:Br—CH$_2$CH$_2$CH$_2$CH$_3$
CH$_3$CH$_2$CH$_2$CH$_2$—NH$_2$—CH$_2$CH$_2$CH$_2$CH$_3$
:Br:

This mechanism is consistent with the second-order rate law, because the rate law requires one molecule of amine and one molecule of alkyl halide in the transition state.

The second reaction is also a nucleophilic substitution reaction, but, because it is intramolecular, the nucleophile and the alkyl halide are part of the same molecule. Hence, the reaction is first-order.

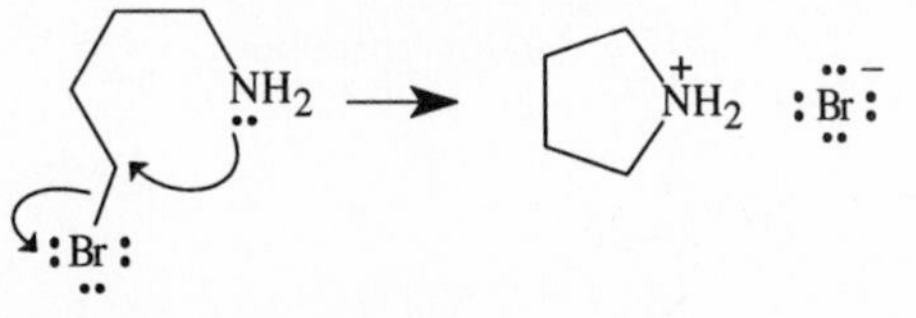
NH$_2$
:Br:
NH$_2$
:Br:

9.56 (a) Product *A* results from a conventional E2 reaction:

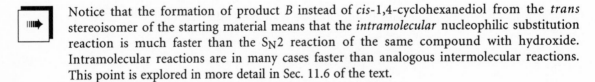

(The same sort of elimination reaction gives compound *A* from *cis*-4-chlorocyclohexanol.) Product *B* is formed by an intramolecular nucleophilic substitution reaction of the alkoxide that occurs in the twist-boat conformation:

Product *C* results from an ordinary S$_N$2 reaction:

(b) The intramolecular substitution that yields the bicyclic compound *B* requires, like all nucleophilic substitutions, approach of the nucleophile to the carbon from the backside. This is only possible in the twist-boat form of the *trans* stereoisomer. Because a conformation that permits backside attack of the intramolecular alkoxide is not available to the *cis* stereoisomer, it reacts instead by an ordinary S$_N$2 reaction with the external nucleophile ⁻OH.

> Notice that the formation of product *B* instead of *cis*-1,4-cyclohexanediol from the *trans* stereoisomer of the starting material means that the *intramolecular* nucleophilic substitution reaction is much faster than the S$_N$2 reaction of the same compound with hydroxide. Intramolecular reactions are in many cases faster than analogous intermolecular reactions. This point is explored in more detail in Sec. 11.6 of the text.

9.57 (a) Bromine addition to alkenes is *anti;* consequently, the stereochemistry of compound *B* is as follows:

(b) Draw the structure of compound *B* in a conformation in which the butyl (Bu) group and the Br that remains after the elimination are on opposite sides of the molecule, because this is the way they are in the alkene product. This shows that the trimethylsilyl group and the bromine are *anti;* consequently, the elimination shown is an *anti* elimination.

9.58   (a) Sodium hydroxide acts as a base to form trichloromethyl anion, which then forms dichlorocarbene. (See Eq. 9.60a–b on text p. 427.) Dichlorocarbene reacts with iodide ion to form a new anion, which is protonated by water.

The fact that the reaction is not observed in the absence of NaOH rules out a simple S_N2 reaction of iodide ion with chloroform.

(b) Butyllithium acts as a base to form an anion that decomposes to phenylcarbene, which then reacts with the alkene to form the cyclopropane.

9.59   This result can be accounted for by a carbocation rearrangement.

10

Chemistry of Alcohols, Glycols, and Thiols

Terms

Concepts

I. Oxidation and Reduction in Organic Chemistry

A. GENERAL

1. In organic chemistry, whether a transformation is an oxidation or a reduction is determined by the oxidation numbers of the reactants and products.

   a. An oxidation is a transformation in which electrons are lost.

      i. The product has a greater (more positive, less negative) oxidation number than the reactant.

      ii. The reagent that brings about the transformation is called an oxidizing agent.

   b. A reduction is a transformation in which electrons are gained.
      i. The product has a smaller (less positive, more negative) oxidation number than the reactant.
      ii. The reagent that effects the transformation is called a reducing agent.
   c. Oxidations and reductions always occur in pairs; whenever something is oxidized, something else is reduced.

B. OXIDATION NUMBERS
   1. Assign an oxidation level to each carbon that undergoes a change between reactant and product by the following method:
      a. Assign a –1 for:
         i. every bond from the carbon to a less electronegative element (including hydrogen)
         ii. every negative charge on the carbon.
      b. Assign a 0 (zero) for:
         i. every bond from the carbon to another carbon atom
         ii. every unpaired electron on the carbon.
      c. Assign a +1 for:
         i. every bond from the carbon to a more electronegative element
         ii. every positive charge on the carbon.
      d. Add the numbers assigned under (a), (b), and (c) to obtain the oxidation level of the carbon under consideration.
   2. Determine the oxidation numbers N_{ox} of both the reactant and product by adding, with each compound, the oxidation levels of all the carbons computed in Step 1.
   3. Compute the difference N_{ox}(product) – N_{ox}(reactant) to determine whether the transformation is an oxidation, reduction, or neither.
      a. If the difference is a positive number, the transformation is an oxidation.
      b. If the difference is a negative number, the transformation is a reduction.
      c. If the difference is zero, the transformation is neither an oxidation nor a reduction.

C. OXIDATIONS AND REDUCTIONS
   1. Whether a reaction is an oxidation or a reduction does not necessarily depend on the introduction or loss of oxygen.
   2. A balanced half-reaction shows the loss or gain of electrons.
      a. Use H_2O to balance missing oxygens.
      b. Use H^+ to balance missing hydrogens.
      c. Use e^- to balance charges.
      d. The change in oxidation number is the same as the number of electrons lost or gained in the corresponding half-reaction.
   3. The change in oxidation number for a transformation indicates whether an oxidizing or reducing agent is required to bring about the reaction.
   4. The oxidation-number concept can be used to organize organic compounds into functional groups with the same oxidation level; carbons with larger numbers of hydrogens have a greater number of possible oxidation states.

D. OXIDATIONS IN BIOLOGICAL SYSTEMS
   1. An understanding of the fundamental types of organic reactions and their mechanisms is useful in the study of biochemical processes.
   2. A naturally occurring oxidation catalyzed by the enzyme alcohol dehydrogenase is the conversion of ethanol into acetaldehyde by NAD^+.
   3. Nicotinamide adenine dinucleotide (NAD^+) is a biological oxidizing agent.
      a. NAD^+ is an example of a coenzyme, a molecule required, along with an enzyme, for certain biological oxidation reactions to occur.
      b. NADH is the reduced form of NAD^+ and is a reducing agent.
      c. The coenzymes NAD^+ and NADH are derived from the vitamin niacin.

II. General Chemical Concepts

A. CHEMICAL EQUIVALENCE AND NONEQUIVALENCE

1. It is sometimes important to know when two groups in a molecule are chemically equivalent, that is, when they behave in exactly the same way towards a chemical reagent.
2. Two groups within a molecule can be classified as constitutionally equivalent or constitutionally nonequivalent according to their connectivity relationship.
   a. Groups within a molecule are constitutionally equivalent when they have the same connectivity relationship to all other atoms in the molecule.
3. Whether two constitutionally equivalent groups are chemically equivalent depends on their stereochemical relationship.
   a. Constitutional nonequivalence is a sufficient but not a necessary condition for chemical nonequivalence.
   b. Constitutionally nonequivalent groups are chemically nonequivalent.
4. Constitutionally equivalent groups are of three types:
   a. Homotopic groups are chemically equivalent and indistinguishable under all circumstances.
   b. Enantiotopic groups are chemically equivalent towards achiral reagents, but are chemically nonequivalent towards chiral reagents.
   c. Diastereotopic groups are chemically nonequivalent under all conditions.

B. DETERMINATION OF CONSTITUTIONAL EQUIVALENCY

1. The stereochemical relationship between constitutionally equivalent groups is revealed by a substitution test.
   a. Substitute each constitutionally equivalent group in turn with a fictitious circled group and compare the resulting molecules.
   b. Their stereochemical relationship determines the relationship of the circled groups.
2. When the substitution test give identical molecules, as in this example, the constitutionally equivalent groups are said to be homotopic.

identical

3. When the substitution test gives enantiomers, the constitutionally equivalent groups are said to be enantiotopic.

enantiomers

4. When a substitution test gives diastereomers, the constitutionally equivalent groups are said to be diastereotopic.

diastereomers

   a. Diastereotopic groups are easily recognized when two constitutionally equivalent groups are present in a molecule that contains an asymmetric carbon (as in the example above).
   b. Diastereotopic groups are easily recognized when two groups on one carbon of a double bond are the same and the two groups on the other carbon are different.

diastereomers

III. *Organic Synthesis and Reactions*

A. ORGANIC SYNTHESIS

1. The preparation of organic compounds from other organic compounds by the use of one or more reactions is called organic synthesis.
2. The molecule to be synthesized is called the target molecule.
3. In order to assess the best route to the target molecule from the starting material, work backward from the target molecule towards the starting material.

B. PRINCIPLE OF MICROSCOPIC REVERSIBILITY

1. Any reaction and its reverse proceed by the forward and reverse of the same mechanism.
   a. This statement is known as the principle of microscopic reversibility.
   b. It follows from this principle that forward and reverse reactions must have the same intermediates and the same rate-limiting transition states.
2. Any reaction catalyzed in one direction is also catalyzed in the other.

C. DEHYDRATION REACTIONS

1. A reaction in which the elements of water are lost from the starting material is called a dehydration reaction.
   a. Lewis acids such as alumina (aluminum oxide, Al_2O_3) and/or heat can also be used to catalyze or promote dehydration reactions.
   b. Strong acids such as H_2SO_4 and H_3PO_4 catalyze a β-elimination reaction in which water is lost from a secondary or tertiary alcohol to give an alkene.
2. Most acid-catalyzed dehydrations of alcohols are reversible reactions and can easily be driven towards the alkene products by applying LeChatelier's principle.

D. ALKYLATING AGENTS

1. In a nucleophilic substitution, an alkyl group is transferred from a leaving group to a nucleophile.
   a. A nucleophile is said to be alkylated by an alkyl halide or a sulfonate ester in the same sense that a Brønsted base is protonated by a strong acid.
   b. Alkyl halides, sulfonate esters, and related compounds are sometimes referred to as alkylating agents.
2. Alkyl esters of strong inorganic acids are typically very potent alkylating agents because they contain leaving groups that are very weak bases.

IV. *Reactions Involving Alcohols and Thiols*

A. SUBSTITUTION AND ELIMINATION REACTIONS OF ALCOHOLS

1. Several reactions of alcohols involve breaking the C—O bond.
   a. In acid-catalyzed dehydration, the —OH group of an alcohol is converted by protonation into a good leaving group and is eliminated as water to give an alkene.
      i. The initial step in the dehydration of secondary and tertiary alcohols is protonation of the alcohol oxygen and formation of a carbocation.
      ii. The relative rates of alcohol dehydration are in the order tertiary > secondary > primary.

    b. In the reaction with hydrogen halides, the —OH group of an alcohol is converted by protonation into a good leaving group and is displaced by halide (in reactions with hydrogen halides) to give an alkyl halide.

    c. In the reaction of an alcohol with $SOCl_2$, the reagent itself converts the —OH into a good leaving group, which in a subsequent reaction is displaced by halide ion.

       i. When an alcohol reacts with thionyl chloride, a chlorosulfite ester intermediate is formed.

      ii. The chlorosulfite ester reacts readily with nucleophiles because the chlorosulfite group, —O—SO—Cl, is a very good leaving group.

     iii. The chlorosulfite ester is usually not isolated, but reacts with the chloride ion formed to give the alkyl chloride.

     iv. The displaced $^-$O—SO—Cl ion is unstable and decomposes to SO_2 and Cl^-.

pentanol   thionyl chloride   pyridine   pentyl chlorosulfite   pyridinium ion

chloropentane

chlorosulfite ion   sulfur dioxide

    d. In the reaction with a sulfonyl chloride, the alcohol is converted into a sulfonate ester, which is an excellent leaving group in substitution or elimination reactions.

  2. Alkyl halide formation and dehydration to alkenes are alternative branches of a common mechanism.

  3. Hydration of alkenes and dehydration of alcohols are the forward and reverse of the same reaction.

    a. Both hydration and alkenes to alcohols and dehydration of alcohols to alkenes are catalyzed by acids.

B. Considerations for the Conversion of Alcohols into Alkyl Halides

  1. A variety of reactions that can be used to convert alcohols into alkyl halides are:

    a. reaction with hydrogen halides;

    b. formation of sulfonate esters and displacement with halide ions; and

    c. reaction with $SOCl_2$.

  2. The method of choice depends on the structure of the alcohol and on the type of alkyl halide to be prepared.

  3. In order to break the carbon-oxygen bond, the —OH group must first be converted into a good leaving group by

    a. protonation: protonated alcohols are intermediates in both dehydration to alkenes and substitution to give alkyl halides.

    b. conversion into sulfonate esters or inorganic esters: these esters, to a useful approximation, react like alkyl halides.

    c. reaction with thionyl chloride: this reagent effects the conversion of alcohols into chlorosulfite esters, which are converted within the reaction mixture into alkyl chlorides.

  4. When a primary alcohol is the starting material, the reaction occurs as a concerted displacement of water from the protonated alcohol by halide ion.

    a. The reaction is an S_N2 reaction in which water is the leaving group.

    b. Primary alcohols require heating for several hours.

  5. Secondary alcohols can react by either the S_N1 or S_N2 mechanism or both.

a. When carbocation intermediates are involved in the reactions of alcohols with hydrogen halides, rearrangements occur in certain cases.

b. Rearrangements are best avoided in the preparation of secondary alkyl halides by the reaction of a halide ion with a sulfonate ester in a polar aprotic solvent.

6. In the reactions of tertiary alcohols, protonation of the alcohol oxygen is followed by carbocation formation.

a. The carbocation reacts with the halide ion which is present in great excess.

b. Once the alcohol is protonated, the reaction is essentially an S_N1 reaction with H_2O as the leaving group.

c. Typically, tertiary alcohols react with hydrogen halides within a few minutes.

7. The thionyl chloride method is most useful with primary alcohols; it can also be used with secondary alcohols, although rearrangements in such cases have been known to occur.

C. Considerations for the Dehydration of Alcohols to Alkenes

1. The basicity of alcohols is important to the success of the dehydration reaction.

2. Acid-catalyzed alcohol dehydration is essentially an E1 reaction.

a. If the alcohol has more than one type of β-hydrogen, a mixture of alkene products can be expected.

b. Alcohols that react to give rearrangement-prone carbocation intermediates yield rearranged alkenes.

c. The most stable alkene (the one with the greatest number of branches at the double bond) is the alkene formed in greatest amount.

D. Oxidation of Alcohols and Thiols

1. Alcohols can be oxidized to carbonyl compounds with Cr(VI).

a. Chromate (CrO_4^{2-}) and dichromate ($Cr_2O_7^{2-}$) reagents are customarily used under strongly acidic conditions.

b. CrO_3 is often used in pyridine.

c. A complex between CrO_3 and two molecules of pyridine in methylene chloride solvent is commonly known as Collins' reagent.

2. The oxidation of alcohols with Cr(VI) reagents involves an acid-catalyzed displacement of water from chromic acid by the alcohol to form a chromate ester.

a. After protonation of the chromate ester, it decomposes in a β-elimination reaction.

b. The last step is much like an E2 reaction, except that it does not involve a strong base.

c. Primary alcohols are oxidized to aldehydes (in the absence of water) or carboxylic acids (in the presence of water).

d. Secondary alcohols are oxidized to ketones.

e. Tertiary alcohols are not oxidized.

3. Primary alcohols can be oxidized to carboxylic acids with $KMnO_4$.

4. Vicinal glycols react with periodic acid to give aldehydes or ketones derived from cleavage of the carbon-carbon bond between the —OH groups.
5. Thiols are oxidized at sulfur rather than at the α-carbon.
6. Disulfides and sulfonic acids are two common oxidation products of thiols.

Reactions

I. Substitution and Elimination Reactions of Alcohols and Their Derivatives

A. SULFONATE ESTER AND INORGANIC ESTERS OF ALCOHOLS

1. Sulfonate esters are derivatives of sulfonic acids, which are compounds of the form R—SO$_3$H.
   a. A sulfonate ester is a compound in which the acidic hydrogen of a sulfonic acid is replaced by an alkyl or aryl group.
   b. Sulfonate esters are prepared from alcohols and other sulfonic acid derivatives called sulfonyl chlorides.

$$H_3C--SO_3H \qquad\qquad H_3C--SO_2Cl$$

p-toluenesulfonic acid   *p*-toluenesulfonyl chloride (tosyl chloride)

   i. This is a nucleophilic substitution reaction in which the oxygen of the alcohol displaces chloride ion from the tosyl chloride.
   ii. The pyridine used as the solvent is a base; besides catalyzing the reaction, it also prevents HCl from forming in the reaction.
2. Sulfonate esters have approximately the same reactivities as the corresponding alkyl bromides in substitution and elimination reactions.
   a. Sulfonate anions, like bromide ions, are good leaving groups.
   b. Sulfonate anions are weak bases; they are the conjugate bases of sulfonic acids, which are strong acids.
3. Sulfonate esters prepared from primary and secondary alcohols undergo S$_N$2 reactions in which a sulfonate ion serves as the leaving group.
4. Secondary and tertiary sulfonate esters also undergo E2 reactions with strong bases, and S$_N$1-E1 solvolysis reactions in polar protic solvents.

B. PREPARATION OF PRIMARY ALKYL HALIDES

1. Primary alkyl chlorides are best prepared from primary alcohols by the thionyl chloride method.

$$CH_3(CH_2)_2CH_2OH \xrightarrow[CH_2Cl_2]{SOCl_2} CH_3(CH_2)_2CH_2Cl$$

2. Primary alkyl bromides are prepared from primary alcohols by the reaction of the alcohol with concentrated HBr.

$$CH_3(CH_2)_2CH_2OH \xrightarrow{HBr} CH_3(CH_2)_2CH_2Br$$

3. Primary alkyl iodides are prepared from primary alcohols by the reaction of the alcohol with concentrated HI, which is often provided by mixing KI and H$_3$PO$_4$.

$$CH_3(CH_2)_2CH_2OH \xrightarrow{KI/H_3PO_4} CH_3(CH_2)_2CH_2I$$

4. The sulfonate ester method works well for primary alkyl halides but requires two separate reactions:
   a. formation of the sulfonate ester
   b. reaction of the ester with halide ion.

$$CH_3(CH_2)_2CH_2OH \xrightarrow[\text{pyridine}]{RSO_2Cl} CH_3(CH_2)_2CH_2OSO_2R \xrightarrow[\text{acetone}]{X^-} CH_3(CH_2)_2CH_2X$$

5. Because all these methods have an S_N2 mechanism as their basis, alcohols with a significant amount of β-branching do not react under the usual conditions.

C. PREPARATION OF SECONDARY ALKYL HALIDES

1. Secondary alkyl chlorides can be prepared from secondary alcohols by the thionyl chloride method if the secondary alcohol has relatively little β-branching.

$$(CH_3)_2CH-OH \xrightarrow{SOCl_2} (CH_3)_2CH-Cl$$

2. Secondary alkyl bromides can be prepared from alcohols by the HBr method if the desired alkyl halide product is derived from a carbocation that does not rearrange.

$$H_3C-CH_2-\overset{\overset{\displaystyle OH}{|}}{CH}-CH_3 \xrightarrow{HBr} H_3C-CH_2-\overset{\overset{\displaystyle Br}{|}}{CH}-CH_3$$

3. Secondary alkyl halides can be prepared from secondary alcohols via the sulfonate ester which should be treated with the appropriate halide ion in a polar aprotic solvent to avoid rearrangements.

4. Specialized methods that have not been discussed are required for primary and secondary alcohols that have significant β-branching.

D. PREPARATION OF TERTIARY ALKYL HALIDES

1. Tertiary alkyl chlorides can be prepared from the corresponding tertiary alcohols by reaction with HCl under mild conditions.

$$HO-\overset{\overset{\displaystyle CH_3}{|}}{\underset{\underset{\displaystyle CH_3}{|}}{C}}-CH_3 \xrightarrow[\text{H}_2\text{O}]{HCl} Cl-\overset{\overset{\displaystyle CH_3}{|}}{\underset{\underset{\displaystyle CH_3}{|}}{C}}-CH_3$$

2. Tertiary alkyl bromides can be prepared from the corresponding tertiary alcohols by reaction with HBr under mild conditions.

$$HO-\overset{\overset{\displaystyle CH_3}{|}}{\underset{\underset{\displaystyle CH_3}{|}}{C}}-CH_3 \xrightarrow[\text{H}_2\text{O}]{HBr} Br-\overset{\overset{\displaystyle CH_3}{|}}{\underset{\underset{\displaystyle CH_3}{|}}{C}}-CH_3$$

E. PREPARATION OF ALKENES

1. The dehydration of alcohols occurs by a three-step mechanism involving a carbocation intermediate (E1 mechanism).
   a. In the first step, the —OH group of the alcohol accepts a proton from the catalyzing acid in a Brønsted acid-base reaction.
   b. In the second step, the carbon-oxygen bond of the alcohol breaks in a Lewis acid-base dissociation to give water and a carbocation.
   c. In the third, water removes a β-proton from the carbocation in another Brønsted acid-base reaction to generate the alkene product and regenerate the catalyzing acid H_3O^+.

$$H—\overset{|}{\underset{|}{C}}—\overset{|}{\underset{|}{C}}—\overset{+}{\underset{..}{O}}H_2 \longrightarrow H—\overset{|}{\underset{|}{C}}—\overset{|}{C}{+} \ + \ H_2\overset{..}{\underset{..}{O}}:$$

$$H—\overset{|}{\underset{|}{C}}—\overset{|}{\underset{|}{C}}—\overset{+}{\underset{..}{O}}H_2 \longrightarrow H_3\overset{+}{\underset{..}{O}} \ + \ \overset{\diagdown}{\diagup}C{=}C\overset{\diagup}{\diagdown} \ + \ H_2\overset{..}{\underset{..}{O}}:$$

$$H_2\overset{..}{\underset{..}{O}}:$$

2. Tertiary sulfonates readily undergo the E2 reaction with alkoxide bases instead of undergoing S_N2 reactions.

II. *Oxidation of Alcohols and Thiols*

A. OXIDATION OF ALCOHOLS TO ALDEHYDES AND KETONES

1. Primary and secondary alcohols are oxidized by reagents containing Cr(VI), that is, chromium in the +6 oxidation state, to give carbonyl compounds (compounds containing the carbonyl group, $>$C=O).

   a. A carbon atom must bear a hydrogen atom in order for oxidation of an alcohol to an aldehyde or ketone to occur.

   b. Primary alcohols react with Cr(VI) reagents to give aldehydes, but if water is present, aldehydes are further oxidized to carboxylic acids.

$$CH_3(CH_2)_4CH_2OH \quad \text{hexanol}$$

$$\xrightarrow[\text{pyridine/CH}_2\text{Cl}_2]{\text{CrO}_3} \quad CH_3(CH_2)_4{-}\overset{\overset{\displaystyle O}{\|}}{C}{-}H \quad \text{hexanal}$$

$$\xrightarrow[\text{H}_2\text{SO}_4/\text{H}_2\text{O}]{\text{K}_2\text{Cr}_2\text{O}_7} \quad CH_3(CH_2)_4{-}\overset{\overset{\displaystyle O}{\|}}{C}{-}OH \quad \text{hexanoic acid}$$

B. OXIDATION OF ALCOHOLS TO CARBOXYLIC ACIDS

1. Primary alcohols can be oxidized to carboxylic acids using aqueous solutions of Cr(VI) such as aqueous potassium dichromate ($K_2Cr_2O_7$) in acid.

$$\text{CH}_2\text{OH} \xrightarrow[\text{H}_2\text{SO}_4/\text{H}_2\text{O}]{\text{K}_2\text{CrO}_7} \text{CO}_2\text{H}$$

2. Primary alcohols can be oxidized to carboxylic acids with potassium permanganate ($KMnO_4$) in basic solution.

   a. Because $KMnO_4$ reacts with alkene double bonds, Cr(VI) is preferred for the oxidation of alcohols that contain double or triple bonds.

   b. Potassium permanganate is not used for the oxidation of secondary alcohols to ketones because many ketones react further with the alkaline permanganate reagent.

C. OXIDATIVE CLEAVAGE OF GLYCOLS

1. The carbon-carbon bond between the —OH groups of a vicinal glycol can be cleaved with periodic acid to give two carbonyl compounds.

   a. The cleavage of glycols with periodic acid takes place through a cyclic periodate ester intermediate that forms when the glycol displaces water from H_5IO_6.

   b. The cyclic ester spontaneously breaks down by a cyclic flow of electrons in which the iodine accepts an electron pair.

a cyclic periodate ester

2. A glycol that cannot form a cyclic ester intermediate is not cleaved by periodic acid.

D. OXIDATION OF THIOLS

1. Oxidation of a thiol takes place not at the carbon, but at the sulfur.

   a. The most commonly occurring oxidation products of thiols are disulfides and sulfonic acids.

   $$CH_3CH_2S—SCH_2CH_3 \qquad CH_3SO_3H$$

   diethyl disulfide   methanesulfonic acid

   b. The Lewis structures of these derivatives require either violation of the octet rule or separation of formal charge.

   c. Sulfur can accommodate more than eight valence electrons because, in addition to its $3s$ and $3p$ orbitals, it has unfilled $3d$ orbitals of relatively low energy.

   d. The same oxidation-number formalism used for carbon can be applied to oxidation at sulfur.

2. Sulfonic acids are formed by vigorous oxidation of thiols or disulfides with $KMnO_4$ or nitric acid (HNO_3).

   $$CH_3SH \xrightarrow{HNO_3} CH_3SO_3H$$

   methanethiol   methanesulfonic acid

3. Many thiols spontaneously oxidize to disulfides merely on standing in air (O_2).

4. Thiols can also be converted into disulfides by mild oxidants such as I_2 in base or Br_2 in CCl_4.

   a. These reactions can be viewed as a series of S_N2 reactions in which halogen and sulfur are attacked by thiolate anion.

   $$H_3C—S—H \xrightarrow[-H_2O]{^-OH} H_3C—S^- \xrightarrow[-I^-]{I_2} H_3C—S—I \xrightarrow[-I^-]{RS^-} H_3C—S—S—CH_3$$

   methanethiol   methyl disulfide

   b. When thiols and disulfides are present together in the same solution, an equilibrium among them is rapidly established and the equilibrium constant is close to 1.0.

   $$CH_3SH + CH_3CH_2S—SCH_2CH_3 \rightleftharpoons CH_3CH_2SH + CH_3S—SCH_2CH_3$$

   methanethiol   diethyl disulfide   ethanethiol   ethyl methyl disulfide

Study Guide Links

✓10.1 Dehydration of Alcohols

The text stresses the close analogy between the E1 reaction and alcohol dehydration. Could the analogy between the chemistry of alcohols and that of alkyl halides be pursued even further? Could an alcohol be dehydrated by an E2 reaction that uses a strong base? Let's consider whether this would be reasonable. First, in order for an elimination to take place, the —OH group must depart as *hydroxide ion,* $^-$OH. Chapter 9 showed that —OH does not act as a leaving group in S_N2 reactions—it is too strong a base. For the same reason, it also cannot act as a leaving group in E2 reactions. (There are a few exceptions in cases in which the β-hydrogen is unusually acidic.)

Notice that in the dehydration mechanism shown in Eq. 10.3, text p. 445, *water* is the base that removes the β-hydrogen, not hydroxide. When H_3O^+ acts as the acid, in most cases its conjugate base H_2O acts as the base. *Solutions that contain substantial amounts of hydronium ion cannot also contain enough hydroxide ion to act as a base in base-catalyzed reactions.* Likewise, in solutions that contain substantial hydroxide ion, water, not hydronium ion, serves as the acid, because hydronium ion is not present to a significant extent in solutions of hydroxide.

✓10.2 Rearrangements Involving Cyclic Carbon Skeletons

The rearrangement shown in Eq. 10.6 of the text occurs within a ring. Following such rearrangements sometimes proves to be tricky even for students who have a good understanding of carbocation rearrangements. The key to following such rearrangements, at least at first, is to *draw out the individual carbon atoms implied by the skeletal formula,* as demonstrated below. As this equation shows, the rearrangement is nothing more than the migration of carbon-5 from carbon-2 to carbon-1.

Notice that the rearranged carbocation is drawn first with all the atoms as they were before rearrangement; then the structure is drawn more conventionally.

As a result of this rearrangement, carbon-2 becomes electron-deficient, and carbon-1 becomes part of a ring that is now one carbon larger. Conceptually, this ring expansion is analogous to the expansion of a noose in a rope:

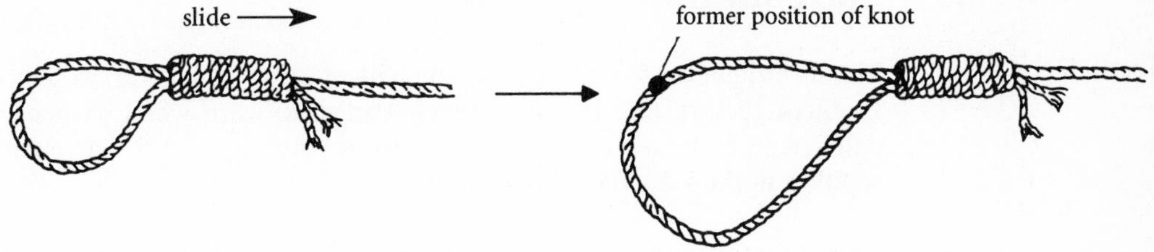

slide ⟶   former position of knot

The product is formed by loss of a β-proton from the rearranged carbocation:

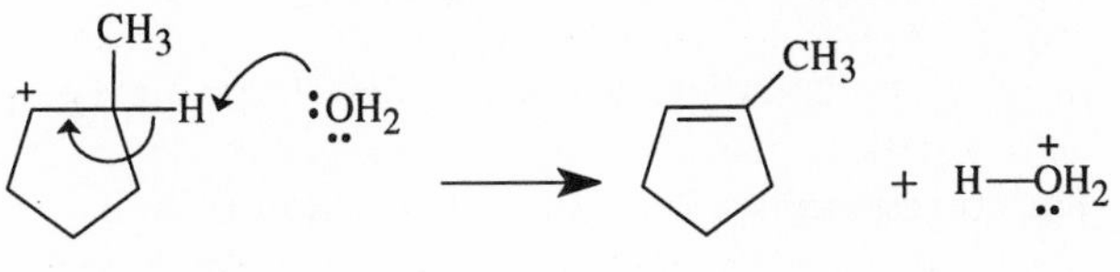

10.3 Mechanism of Sulfonate Ester Formation

As the text suggests, the formation of sulfonate esters is another nucleophilic substitution reaction. The role of the pyridine is to form a small amount of the conjugate base alkoxide ion of the alcohol:

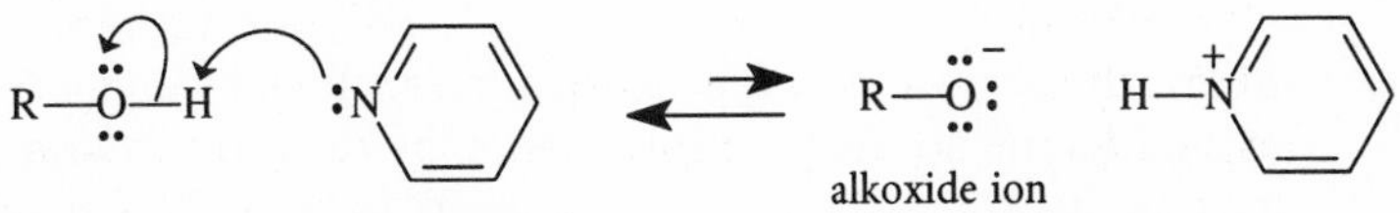

pyridine   alkoxide ion

This alkoxide, acting as a nucleophile, displaces chloride ion from the sulfonyl chloride in a nucleophilic substitution reaction to form the sulfonate ester.

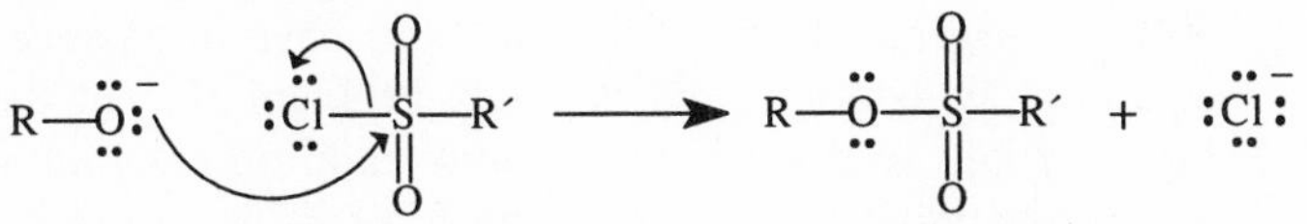

a sulfonyl chloride   a sulfonate ester

Notice that there is really nothing very new in this mechanism; the only difference is that a sulfonyl chloride instead of an alkyl halide undergoes substitution.

✓10.4 Oxidations and Reductions

You may have learned more extensive rules for assigning oxidation numbers in general chemistry. In organic chemistry, the major concern is the oxidation states of carbon atoms, because the vast majority of oxidations in organic chemistry involve oxidation at carbon. Occasionally, the same ideas are applied to other atoms, such as sulfur, but such cases follow essentially the same rules.

✓10.5   More on Half-Reactions

The technique used for balancing the half-reaction in Eq. 10.29a of the text, shown in Study Problem 10.4, text p. 462, is relatively straightforward when oxygens are involved. How can half-reactions be written when oxygens are not involved? For example, what is the half-reaction in the following oxidation?

$$H_2C{=}CH_2 \xrightarrow{\ Br_2\ } Br{-}CH_2{-}CH_2{-}Br$$

Since oxygens are balanced with H_2O, the bromines can be balanced with HBr. Taking this approach results in the following half-reaction:

$$2\,HBr \;+\; H_2C{=}CH_2 \longrightarrow Br{-}CH_2{-}CH_2{-}Br \;+\; 2e^- \;+\; 2H^+$$

The corresponding inorganic half-reaction is

$$2e^- \;+\; 2H^+ \;+\; Br_2 \longrightarrow 2\,HBr$$

If the two half-reactions are added, and if the "dummy electrons" and the HBr molecules on each side of the equation are canceled, the sum is the overall bromine addition reaction:

$$2\,\cancel{HBr} \;+\; H_2C{=}CH_2 \longrightarrow Br{-}CH_2{-}CH_2{-}Br \;+\; \cancel{2e^-} \;+\; \cancel{2H^+}$$
$$\cancel{2e^-} \;+\; \cancel{2H^+} \;+\; Br_2 \longrightarrow 2\,\cancel{HBr}$$

Sum: $\qquad H_2C{=}CH_2 \;+\; Br_2 \longrightarrow Br{-}CH_2{-}CH_2{-}Br$

Although the notion of a half-reaction can thus be applied to bromine addition, it isn't very useful because not only "dummy electrons" but also "dummy HBr" molecules have to be used to make it work; HBr, as you know, is not involved in the bromine addition reaction of alkenes. On the other hand, water and protons *really are* involved in the oxidation of ethanol to acetic acid, so that the idea of a half-reaction is somewhat more realistic.

When you want to determine the number of "electrons lost" and "electrons gained" in an oxidation or reduction reaction, you'll find it much simpler as a rule to calculate the change in oxidation number than to balance a half-reaction. The only time half-reactions are worth the effort in organic chemistry is when you have to balance a complicated oxidation-reduction equation, as demonstrated in Study Problem 10.5 of the text. (One certainly doesn't need half-reactions to balance the bromine-addition reaction above.) Nevertheless, it is important for you to understand the concept of a half-reaction because biochemists frequently discuss redox processes in terms of half-reactions. If you take biochemistry, you'll probably be glad you learned (or re-learned) these concepts here.

One other point about half-reactions: Just as inorganic "half-reactions" can be run by supplying or removing electrons at cathodes or anodes of electrochemical cells, in many cases organic compounds as well can be reduced or oxidized electrochemically as suggested by organic "half-reactions." So don't get the idea that organic half-reactions are fictitious processes that cannot actually occur.

10.6   Symmetry Relationships Among Constitutionally Equivalent Groups

In Sec. 6.1A, you learned that enantiomers are molecules that are noncongruent mirror images. You also learned in Sec. 6.1C that enantiomers lack certain *symmetry elements,* such as internal mirror planes. The purpose of this study guide link is to show that the different types of constitutionally equivalent groups—homotopic, enantiotopic, and diastereotopic groups—can also be classified by their symmetry relationships.

Homotopic groups are interchanged by a rotation of the molecule that gives an indistinguishable structure. For example, rotation of the methylene chloride molecule 180° about an axis bisecting the two C—Cl (or the two C—H) bonds (dashed line) exchanges the hydrogens and the chlorines, yet leaves the molecule looking exactly as it did before the rotations:

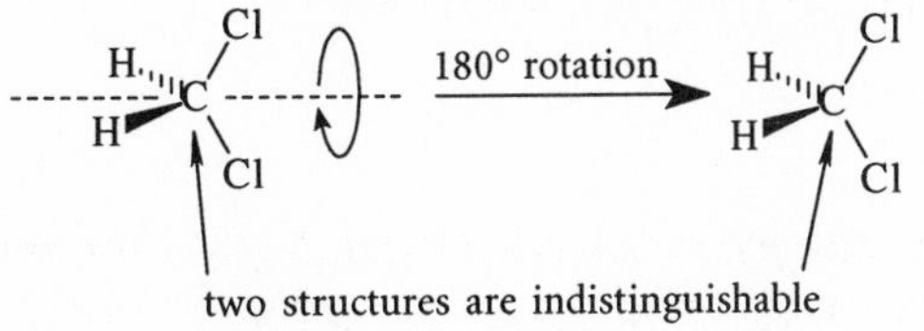

two structures are indistinguishable

Consequently, the hydrogens are homotopic, and the chlorines are also homotopic. (The hydrogens of CH_3Cl are also homotopic, as shown in the text; perhaps you might want to demonstrate to yourself the rotation of this molecule that interchanges the three hydrogens while leaving the molecule invariant.)

Enantiotopic groups cannot be interchanged by such a rotation. For example, a rotation that interchanges the two enantiotopic α-protons of ethanol gives a molecule that looks different:

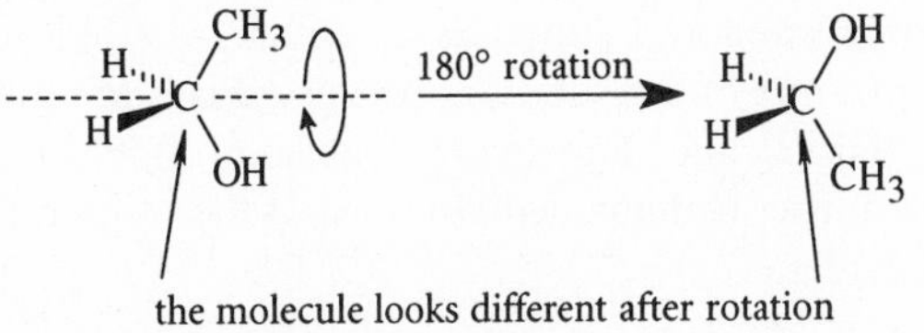

the molecule looks different after rotation

That is, the CH_3 group is "up" before the rotation and the OH group is "down;" after the rotation; these groups have changed positions. The second structure is not congruent to the first without turning it over again.

Enantiotopic groups are related as object and mirror image by an internal mirror plane of the molecule. The internal mirror plane of the ethanol molecule (dashed line in the following diagram) contains the CH_3, the OH, and the central carbon, and it bisects the angle between the two C—H bonds at the α-carbon.

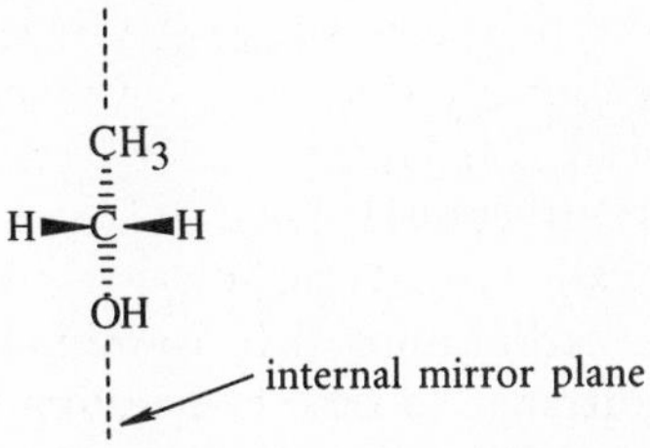

internal mirror plane

The two hydrogens of ethanol are related as object and mirror image by this plane; hence, these two hydrogens are enantiotopic. Notice that two homotopic groups might also be related by an internal mirror plane, but in order to be homotopic, they must *in addition* be related by a rotation of the molecule as described above. Since a chiral molecule by definition cannot contain an internal mirror plane, it follows that a chiral structure *cannot* contain enantiotopic groups (but it can contain homotopic and diastereotopic groups!).

Finally, diastereotopic protons have no symmetry relationship whatsoever.

Solutions

Solutions to In-Text Problems

10.1 (a) (c)

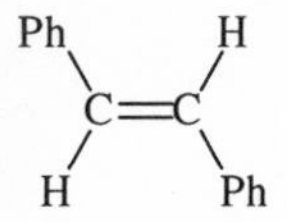
Ph
H
C=C
H
Ph

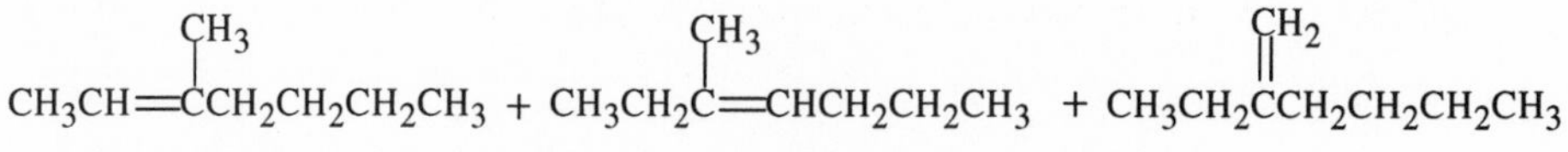
CH3
CH3CH=CCH2CH2CH2CH3 + CH3CH2C=CHCH2CH2CH3 + CH3CH2CCH2CH2CH2CH3
CH3
CH2

(*E* and *Z*) (*E* and *Z*) *C*

A *B*

In part (a), none of the *Z* alkene is formed because the *E* alkene is considerably more stable. (Recall that the E1 reaction, of which dehydration is an example, gives the most stable alkene isomers.)

10.2 (c) Because they have the most branches at their double bonds, alkenes *A* and *B* are the most stable, and are therefore formed in greatest amount. There should not be great differences in stability among these alkenes; consequently, they should all be formed in significant amounts.

10.3 (a) 1-Methylcyclohexanol and either *cis*- or *trans*-2-methylcyclohexanol should give 1-methylcyclohexene as the major product of dehydration, because this is the most stable alkene that could be formed in each case. The tertiary alcohol 1-methylcyclohexanol should react most rapidly because it involves a tertiary carbocation intermediate; dehydrations of the other alcohols involve a less stable secondary carbocation intermediate.

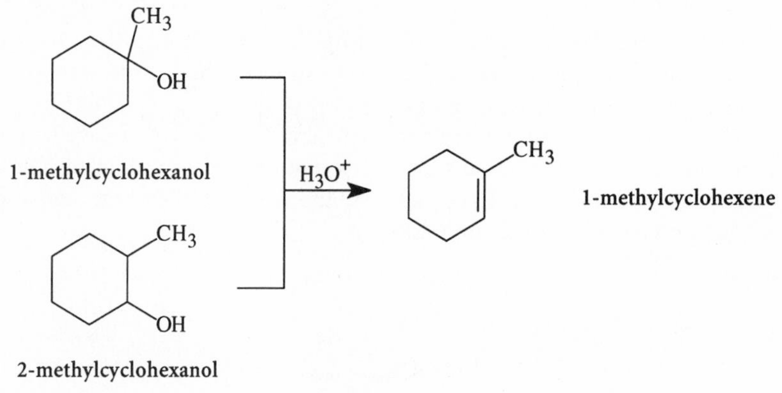
CH3
OH
1-methylcyclohexanol
H3O+
CH3
1-methylcyclohexene
CH3
OH
2-methylcyclohexanol

Note that the carbocation that is initially formed from 2-methylcyclohexanol would undoubtedly rearrange to the same tertiary carbocation involved in the dehydration of 1-methylcyclohexanol; the same product must result from the same intermediate. 1-Methylcyclohexene should also be the major product formed from the unrearranged carbocation because it is the more branched of the two possible alkenes.

10.4 (a) As the text suggests, the mechanism involves a rearrangement of the initially formed carbocation to a more stable carbocation.

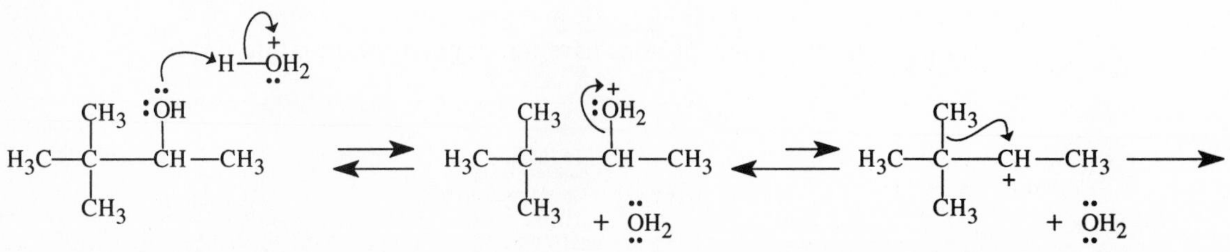
H—OH2
CH3 :OH
H3C—C—CH—CH3
CH3
CH3 :OH2
H3C—C—CH—CH3
CH3 + OH2
CH3
H3C—C—CH—CH3
CH3 + OH2

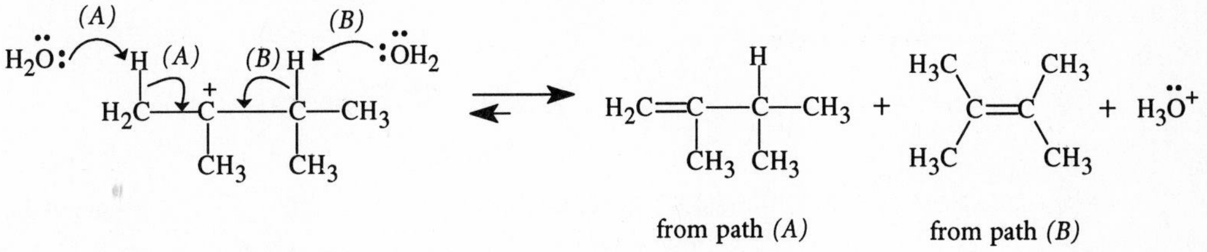

from path *(A)*   from path *(B)*

(b) Rearrangement occurs because a more strained secondary carbocation is converted into a less strained, and therefore more stable, secondary carbocation.

10.5 **(a)** Reaction of either HOCH$_2$CH$_2$CH$_2$CH$_2$CH$_2$I or HOCH$_2$CH$_2$CH$_2$CH$_2$CH$_2$OH with concentrated HI will give the desired 1,5-diiodopentane.

Because concentrated HI readily oxidizes to iodine, it is usually generated *in situ* (that is, in the reaction flask) by using a combination of H$_3$PO$_4$ and KI.

10.6 Aha! Another carbocation rearrangement:

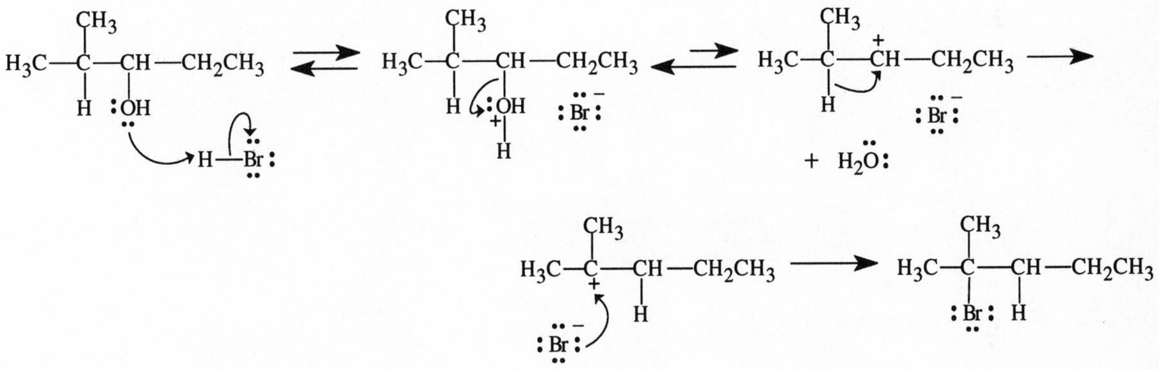

10.7 **(a)**

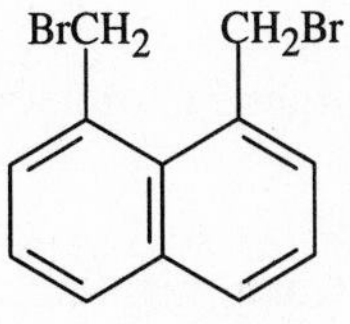

(c)

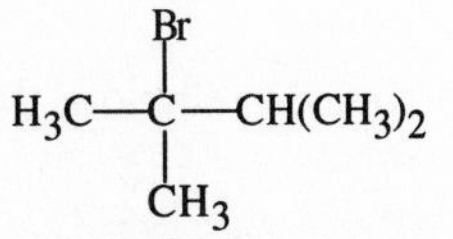

(a rearrangement product)

(d) This compound, neopentyl alcohol, will not react, because it is primary, and therefore cannot undergo the S$_N$1 reaction; and it has too much β-branching to undergo the S$_N$2 reaction.

10.8 **(a)**

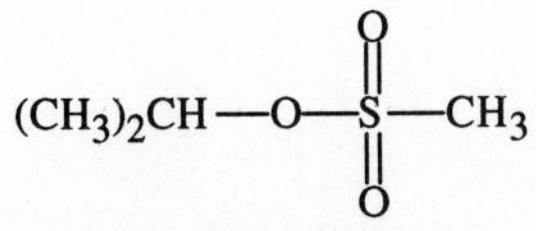

isopropyl methanesulfonate
(isopropyl mesylate)

(c)

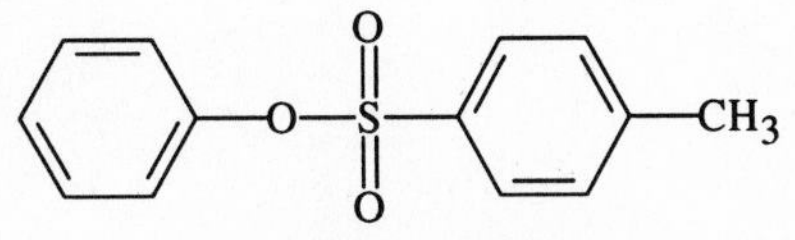

phenyl tosylate
(phenyl *p*-toluenesulfonate)

10.9 (a)

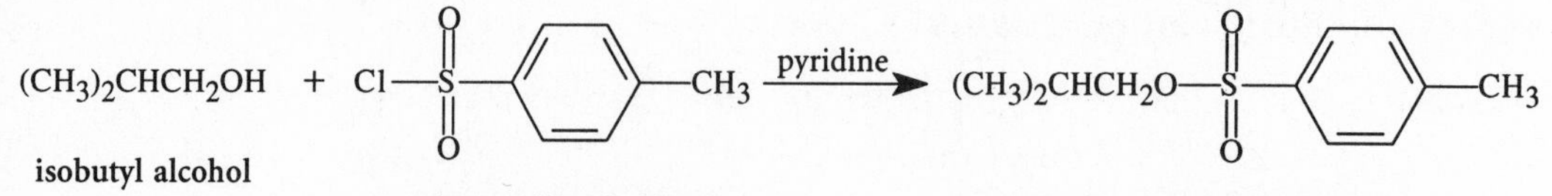

isobutyl alcohol + *p*-toluenesulfonyl chloride (tosyl chloride) → isobutyl tosylate

10.10 (a) There are two possible syntheses. In the first, the alkoxide 3-cyclopentylpropanolate is allowed to react with methyl tosylate:

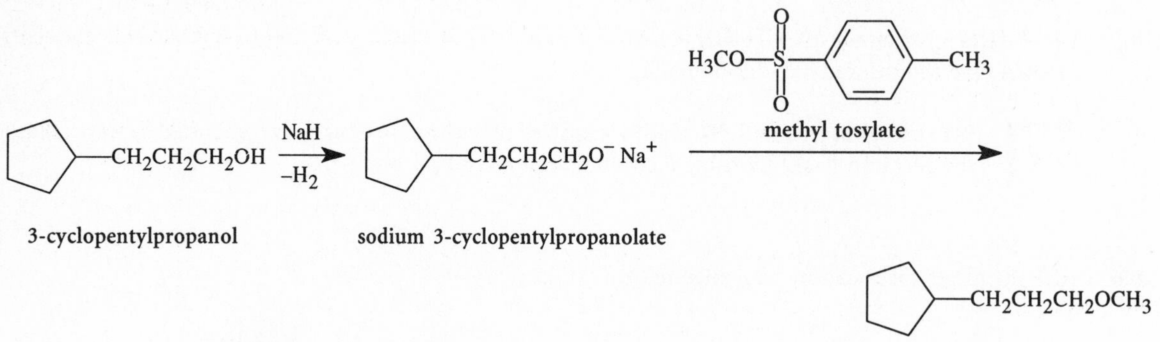

3-cyclopentylpropanol → sodium 3-cyclopentylpropanolate

methyl tosylate

In the second, the tosylate of 3-cyclopentylpropanol is allowed to react with sodium methoxide:

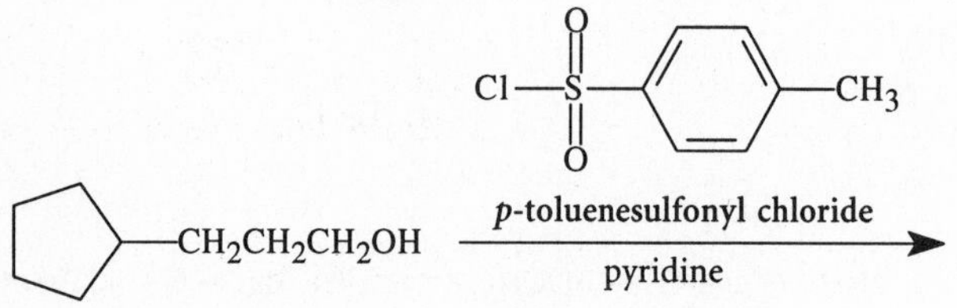

3-cyclopentylpropanol

p-toluenesulfonyl chloride
pyridine

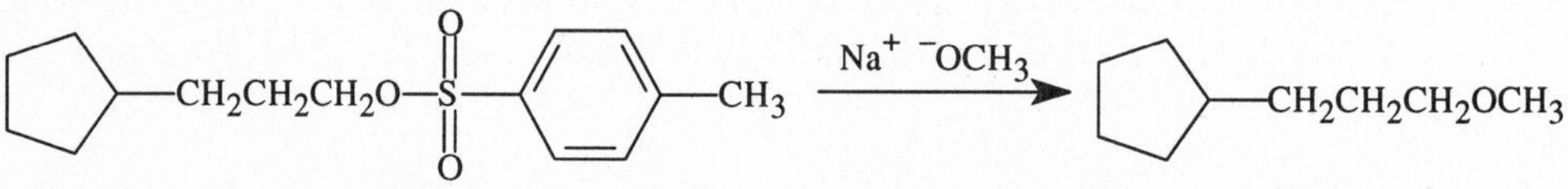

Because methyl tosylate cannot undergo the competing E2 reaction, the first synthesis is probably the better one, although both are reasonable.

10.11 (a) The first reaction forms the methanesulfonate (mesylate) ester. Because a strong, highly branched alkoxide base is used, the major product of the second reaction is the alkene 4-methyl-1-pentene.

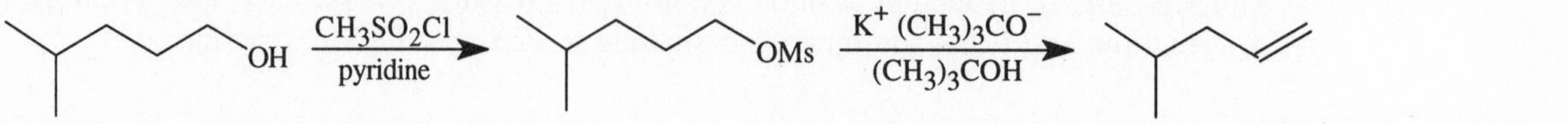

4-methyl-1-pentene

10.12 (a)

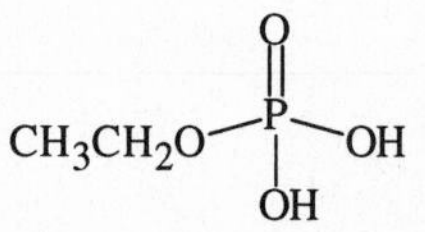

CH_3CH_2O—P—OH   ethyl phosphate

10.13 Because dimethyl sulfate is a superb alkylating agent, it transfers a methyl group to each of the nucleophiles. The products are as follows:

(a)

$$HO—CH_3 \quad + \quad HO—\overset{\displaystyle O}{\underset{\displaystyle O}{S}}—OCH_3$$

methanol

(c)

$$CH_3CH_2CH_2SCH_3 \quad + \quad Na^+ \; {}^-O—\overset{\displaystyle O}{\underset{\displaystyle O}{S}}—OCH_3$$

1-(methylthio)propane
(methyl propyl sulfide)

10.14 (a) 1-Butanol can be converted into 1-chlorobutane by treating it with thionyl chloride and pyridine. (See Eq. 10.21 on text p. 456.) 1-Butanol can also be converted into 1-chlorobutane by first converting it into a sulfonate ester such as butyl tosylate, then treating that ester with sodium chloride in a polar aprotic solvent such as DMSO.

10.16 The first step is the conversion of the alcohol —OH group into a good leaving group. This group is then displaced by bromide ion.

good leaving group

$$R—\overset{\cdot\cdot}{\underset{\cdot\cdot}{O}}H \quad Br_2P—\overset{\cdot\cdot}{\underset{\cdot\cdot}{Br}}: \quad \longrightarrow \quad R\overset{+}{\overset{\cdot\cdot}{—O}}H \atop PBr_2 \quad \longrightarrow \quad R—\overset{\cdot\cdot}{\underset{\cdot\cdot}{Br}}: \; + \; \overset{\cdot\cdot}{HO}—PBr_2$$

$$:\overset{\cdot\cdot}{\underset{\cdot\cdot}{Br}}:{}^-$$

10.17 (a) Because the alcohol groups are primary, concentrated HBr will bring about the desired reaction.
(c) The alcohol reacts with HBr to give a carbocation intermediate that rearranges. However, this is precisely the reaction desired; reaction of the rearranged carbocation with bromide ion gives the desired product. (See Study Guide Link 10.2 on p. 226 of this manual.)

10.18 (a) The bromination of methane to give methyl bromide is a two-electron oxidation.
(c) The transformation shown is a two-electron reduction.
(e) The ozonolysis of 2-methyl-2-butene is a six-electron oxidation. Notice that both carbons of the double bond must be considered, even though the molecule is "split in two" as a result of the reaction.
(g) This is a two-electron reduction; note that the negative charge contributes –1 to the oxidation number of the product.

10.19 (d) Notice that two electrons are "lost" in this half-reaction, and that this is consistent with its classification in the previous problem as a two-electron oxidation:

$$CH_3CH{=}CHPh \quad + \quad 2H_2O \quad \longrightarrow \quad CH_3\overset{\overset{\displaystyle OH}{|}}{C}H{-}\overset{\overset{\displaystyle OH}{|}}{C}HPh \quad + \quad 2e^- \; + \; 2H^+$$

10.20 (a) The organic compound is reduced, and the $^-AlH_4$ is oxidized.
(c) The alkene is oxidized, and the Br_2 is reduced.

10.21 First balance the equation; then, from the balanced equation, determine how many moles of permanganate per mole of toluene are required.

$$2H_2O \; + \; PhCH_3 \quad \longrightarrow \quad PhCO_2H \; + \; 6e^- \; + \; 6H^+$$

$$\overset{8}{\cancel{4}}H^+ \; + \; \overset{2}{}MnO_4^- \; + \; \overset{6}{\cancel{3}}e^- \quad \longrightarrow \quad \overset{2}{}MnO_2 \; + \; \overset{4}{\cancel{2}}H_2O$$

$$PhCH_3 \; + \; 2MnO_4^- \; + \; 2H^+ \quad \longrightarrow \quad PhCO_2H \; + \; 2H_2O \; + \; 2MnO_2$$

This balanced equation shows that two moles of permanganate are required to oxidize one mole of toluene.

 A shorter way to solve a problem of this type is to count the number of electrons lost and gained in each half reaction, and then use enough of each reagent to make these balance. To illustrate, the oxidation of toluene is a six-electron oxidation, and the reduction of permanganate to MnO_2 is a three-electron reduction. Thus two permanganates are required to "accept" the six electrons necessary to oxidize one toluene.

10.23 (a) The primary alcohol is oxidized, but the tertiary alcohol is not.

10.24 (a) The carboxylic acid could be prepared from $(CH_3)_2CHCH_2CH_2CH_2CH_2OH$ (5-methyl-1-hexanol) by oxidation either with aqueous chromic acid [or any other form of aqueous Cr(VI)], or with aqueous potassium permanganate.

(c) The aldehyde can be prepared from 5-methyl-1,5-hexanediol by the Collins oxidation or other variation of nonaqueous oxidation with Cr(VI).

$$CH_3CCH_2CH_2CH_2CH_2OH \quad \text{5-methyl-1,5-hexanediol}$$

10.25 (a)   (c)

10.26 (a) Because the large *tert*-butyl group virtually locks the molecule into the chair conformation in which this group is equatorial, the two —OH groups in the compound *A* are locked into a *trans*-diaxial arrangement. The cyclic periodate ester intermediate cannot be formed from two hydroxy groups in this arrangement without introducing significant ring strain. Because the required intermediate cannot form, the oxidation of compound *A* cannot occur. In contrast, compound *B* can be oxidized with periodic acid because the two —OH groups are *trans*-diequatorial.

10.27 This oxidation occurs in much the same way that NAD^+ oxidations occur: by transfer of a hydride from the alcohol to the carbocation. (Notice that it is a *hydride* (a hydrogen with two electrons), not a proton, that is transferred.)

The reaction scheme shows:

:ÖH
CH₃CCH₂CH₂CH₂CH₂CH₃ → [CH₃CCH₂CH₂CH₂CH₂CH₃ ↔ CH₃CCH₂CH₂CH₂CH₂CH₃] ⁻BF₄ →
|
H ← hydride transfer
Ph₃C+

+ Ph₃CH

:Ö:
‖
HBF₄ + CH₃CCH₂CH₂CH₂CH₂CH₃

10.28 (a) The two methyl groups *a* and *b* are constitutionally equivalent and diastereotopic. The pair of methyl groups *a* and *c* are constitutionally nonequivalent, as are the pair *b* and *c*.

 Notice that when investigating the stereochemical relationships among several groups, the appropriate technique is to consider *each pair*. Thus, there are three relationships that must be categorized within three groups (although it is possible that some relationships might be the same.) Here's an amusing little puzzle that uses the same sort of reasoning. Man *A* looks at a picture and comments about the person *B* in the picture, saying, "Brothers and sisters have I none; this man's father is my father's son." What is the relationship between *A* and *B*?

(c) Hydrogens *a* and *b* within a particular chair conformation are constitutionally equivalent and diastereotopic. However, the chair flip interchanges the positions of these two hydrogens and makes them completely equivalent, that is, homotopic, *over time*.

(e) By turning the Fischer projection 180°, you can verify that hydrogen *a* is indistinguishable from hydrogen *d*, and that hydrogen *b* is indistinguishable from hydrogen *c*. Consequently H^a and H^d are constitutionally equivalent and homotopic, as are H^b and H^c.

180° rotation in the page shows that the two molecules are congruent; hence, the two circled hydrogens are homotopic.

Replacing H^a and H^b in turn with a "circled H" shows that these hydrogens are constitutionally equivalent and enantiotopic, as are H^c and H^d.

The two molecules are enantiomers; hence, the two circled hydrogens are enantiotopic.

The replacement test along with the allowed rotation of Fischer projections 180° in the plane of the page shows that H^a is constitutionally equivalent and enantiotopic to H^c, and that H^b is constitutionally equivalent and enantiotopic to H^d.

180° rotation of either molecule shows that the two molecules are enantiomers; hence, the two circled hydrogens are enantiotopic.

identical

enantiomers

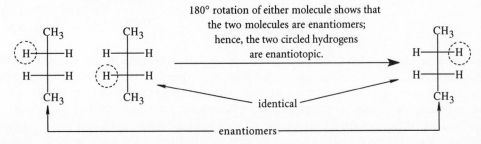

It is particularly easy to see the enantiotopic relationships discussed in the last two cases by applying the symmetry criterion discussed in Study Guide Link 10.6, beginning on p. 228 of this manual. Recall that a molecule containing enantiotopic groups must contain at least one internal mirror plane, and that enantiotopic groups are related as object and mirror image by such an internal mirror plane.

10.29   In every case, the hydrogen (or deuterium) is delivered from the pro-*(R)* position on NADH (the "up" position when NADH is drawn as it is in the problem) so that it ends up in the pro-*(R)* position of deuterated ethanol. In both parts (a) and (c), the resulting isotopically substituted ethanol is chiral.

(a)   (c)

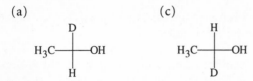

10.30   (a) In a thiol, the oxidation number of sulfur is –1; in a sulfonic acid, it is +5; hence, the change in oxidation number is +6. The oxidation of a thiol to a sulfonic acid is a six-electron oxidation. (This can be verified by writing a balanced half-reaction.)

10.31   The reaction is a nucleophilic substitution in which one sulfur attacks another. Because a thiolate is a better nucleophile than a thiol—thiolates are more basic than neutral thiols—a thiolate serves as the nucleophile in this reaction. A base is required to form the thiolate anion. The reaction is faster in the presence of base because the reaction rate depends on the nucleophile concentration, and base increases the concentration of the nucleophile.

$$CH_3CH_2\ddot{S}\!-\!H \quad :\!\ddot{O}CH_2CH_3 \;\rightleftharpoons\; CH_3CH_2\ddot{S}:^- \;+\; H\!-\!\ddot{O}CH_2CH_3$$

$$CH_3CH_2\ddot{S}:^- \quad \overset{CH_2CH_2CH_3}{\underset{|}{S\!-\!\ddot{S}CH_2CH_2CH_3}} \;\rightleftharpoons\; CH_3CH_2\ddot{S}\!-\!SCH_2CH_2CH_3 \;+\; :\!\ddot{S}CH_2CH_2CH_3$$

10.32   (a) 2-Methyl-3-pentanol can be prepared by hydroboration-oxidation of 2-methyl-2-pentene, which can be prepared by dehydration of 2-methyl-2-pentanol:

$$\underset{\underset{CH_3}{|}}{CH_3\overset{OH}{\underset{|}{C}}CH_2CH_2CH_3} \;\xrightarrow{H_2SO_4}\; \underset{\underset{CH_3}{|}}{CH_3C\!=\!CHCH_2CH_3} \;\xrightarrow[\text{2) }H_2O_2/NaOH]{\text{1) }BH_3/THF}\; \underset{\underset{CH_3}{|}}{CH_3\overset{OH}{\underset{|}{C}}HCHCH_2CH_3}$$

2-methyl-2-pentanol   2-methyl-2-pentene   2-methyl-3-pentanol

(c) The carboxylic acid can be prepared by oxidation of a primary alcohol, which, in turn, can be prepared from the alkene starting material by hydroboration-oxidation.

$$\text{C}_6H_{10}\!=\!CH_2 \;\xrightarrow[\text{2) }H_2O_2/NaOH]{\text{1) }BH_3/THF}\; \text{C}_6H_{11}\!-\!CH_2OH \;\xrightarrow[\text{2) }H_3O^+]{\text{1) }KMnO_4/NaOH}\; \text{C}_6H_{11}\!-\!CO_2H$$

10.33   This reaction is an S_N2 reaction of iodide ion with the alkyl chloride. It occurs with inversion of stereochemistry at the α-carbon because the nucleophile displaces the leaving group by a backside-attack mechanism. (See Fig. 9.2, text p. 397.) An S_N2 reaction can be carried out successfully only on primary and secondary alkyl halides; even in these cases there must not be too much β-branching. The reason is that branches suffer van der Waals repulsions with both the nucleophile and the leaving group; these repulsions raise the energy of the transition state and thus retard the reaction. (See Fig. 9.3, text p. 399.)

Solutions to Additional Problems

10.34 By-products that are the same as the solvent are not shown explicitly.

(a)   (b)   (c)   (d)

$CH_3CH_2CH_2CH_2Br$ $CH_3CH_2CH_2CH_2\overset{+}{O}H_2\ HSO_4^-$ $CH_3CH_2CH_2CH{=}O$ $CH_3CH_2CH_2CH_2O^-\ Na^+\ +\ H_2$
                                                                       $+\ Cr^{3+}$

(e)   (f)   (g)

$CH_3CH_2CH_2CH_2OCH_3$ $CH_3CH_2CH_2CH_2OTs$ $CH_3CH_2CH_2CH_2O^-\ {}^+MgBr\ +\ CH_3CH_2CH_3$
$+\ Na^+\ I^-$ $+\ \langle\!\!\bigcirc\!\!\rangle\overset{+}{N}H\ Cl^-$

(h)   (i)   (j)   (k)

$CH_3CH_2CH_2CH_2Cl$ $CH_3CH_2CH_2CH_2MgBr$ $CH_3CH_2CH{=}CH_2$ $CH_3CH_2CHCH_2OH$
$+\ SO_2\ +\ \langle\!\!\bigcirc\!\!\rangle\overset{+}{N}H\ Cl^-$ $+\ K^+\ Br^-$ $\qquad\quad\ \ |$
$\qquad\qquad\qquad\qquad\qquad\qquad\qquad\qquad\qquad\qquad\qquad\qquad\qquad\qquad\qquad\qquad\qquad\quad\ \ OH$

10.36 (a)   (c)

CH_3CH_2 OH
$\qquad\ |$ $|$
$CH_3CH_2C{-}OH$ $PhCHCH_2Ph$
$\qquad\ |$ 1,2-diphenylethanol
CH_3CH_2

3-ethyl-3-pentanol

(e) Oxidation of 2-methyl-1-butanol gives the same product as ozonolysis of *trans*-3,6-dimethyl-4-octene:

$$\textit{trans-}CH_3CH_2\overset{\overset{\displaystyle CH_3}{|}}{C}HCH{=}CH\overset{\overset{\displaystyle CH_3}{|}}{C}HCH_2CH_3 \xrightarrow[\text{2) }H_2O_2]{\text{1) }O_3} CH_3CH_2\overset{\overset{\displaystyle CH_3}{|}}{C}HCO_2H \xleftarrow{KMnO_4} CH_3CH_2\overset{\overset{\displaystyle CH_3}{|}}{C}HCH_2OH$$

3,6-dimethyl-4-octene   2-methyl-1-butanol

10.37 This is a triester of 1,2,3-propanetriol (glycerol) and nitric acid, known by its traditional name of nitroglycerin. It is derived from glycerol (a triol) and nitric acid:

$$\overset{\overset{\displaystyle OH}{|}}{CH_2}{-}\overset{\overset{\displaystyle OH}{|}}{CH}{-}\overset{\overset{\displaystyle OH}{|}}{CH_2}\ +\ HONO_2\ \longrightarrow\ \overset{\overset{\displaystyle ONO_2}{|}}{CH_2}{-}\overset{\overset{\displaystyle ONO_2}{|}}{CH}{-}\overset{\overset{\displaystyle ONO_2}{|}}{CH_2}$$

1,2,3-propanetriol   nitric acid   glyceryl trinitrate
(glycerol, or glycerin)   (nitroglycerin)

10.38 (a) In this compound, fluorines *a* are constitutionally equivalent and homotopic; fluorines *b* are constitutionally equivalent and enantiotopic; fluorines *a* are constitutionally nonequivalent to fluorines *b*.

$$\overset{\overset{\displaystyle F^a\ \ F^b}{|\quad\ \ |}}{F^a{-}C{-}C{-}H}\atop{\underset{\displaystyle F^a\ \ F^b}{|\quad\ \ |}}$$

(c) Fluorines *a* are constitutionally equivalent and diastereotopic; fluorines *a* are constitutionally nonequivalent to fluorine *b*.

(solution continues)

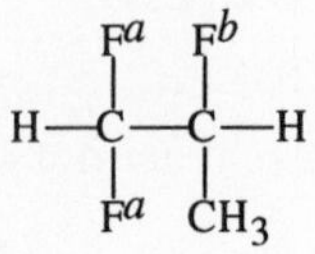

$$H-\overset{\overset{\displaystyle F^a}{|}}{\underset{\underset{\displaystyle F^a}{|}}{C}}-\overset{\overset{\displaystyle F^b}{|}}{\underset{\underset{\displaystyle CH_3}{|}}{C}}-H$$

10.39 Hydrogens that are constitutionally nonequivalent are chemically nonequivalent; hydrogens that are diastereotopic are chemically nonequivalent; hydrogens that are either homotopic or enantiotopic are chemically equivalent. In each structure below, the chemically equivalent hydrogens have the same letter. In structure (c), hydrogens with the same letter are homotopic; and hydrogens b and c are diastereotopic, as are hydrogens d and e. In structure (e), the two hydrogens b are enantiotopic, as are the two hydrogens c; hydrogens b are diastereotopic to hydrogens c. The methyl groups a are enantiotopic, but, within each methyl group, the hydrogens are homotopic.

(a)

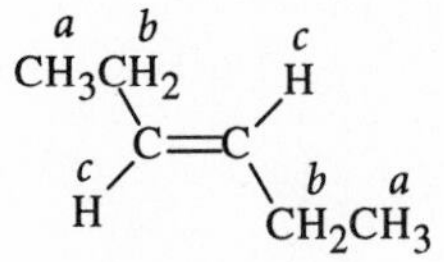

$$\begin{array}{c} \overset{a}{CH_3}\overset{b}{CH_2} \qquad \overset{c}{H} \\ \underset{\underset{\displaystyle H}{c}}{\overset{}{}}C=C\underset{\underset{\displaystyle CH_2CH_3}{b\ a}}{} \end{array}$$

three chemically
nonequivalent
sets of chemically
equivalent hydrogens

(c)

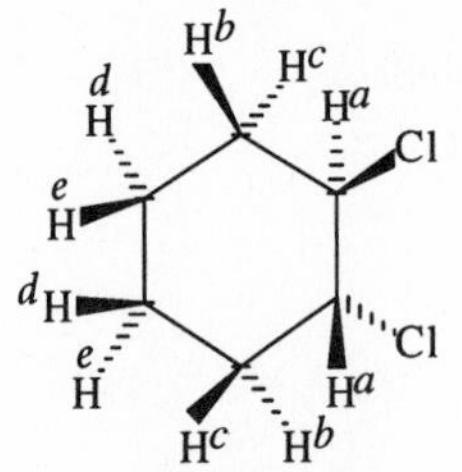

five chemically
nonequivalent
sets of chemically
equivalent hydrogens

(e)

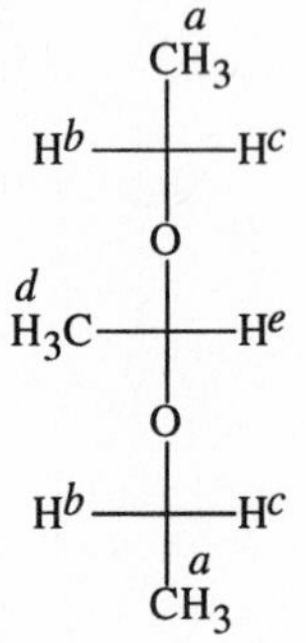

five chemically
nonequivalent
sets of chemically
equivalent hydrogens

10.40 Protonation of the —OH group, formation of a carbocation, and attack of isotopically labeled water on this carbocation are the key steps in this mechanism. (The character Ö represents ^{18}O.)

$(CH_3)_3C-\overset{..}{\underset{..}{O}}H \rightleftharpoons (CH_3)_3C-\overset{+}{\underset{..}{O}}H_2 \rightleftharpoons (CH_3)_3C+ \overset{\overset{\displaystyle \ddot{O}H_2}{}}{\rightleftharpoons}$

$+\ \overset{..}{\underset{..}{O}}H_2 \qquad +\ \overset{..}{\underset{..}{O}}H_2$

$(CH_3)_3C\overset{+}{-}\overset{\overset{\displaystyle H}{|}}{\underset{..}{O}}H \rightleftharpoons (CH_3)_3C-\overset{..}{\underset{..}{O}}H\ +\ \overset{..}{\underset{..}{O}}H_2$

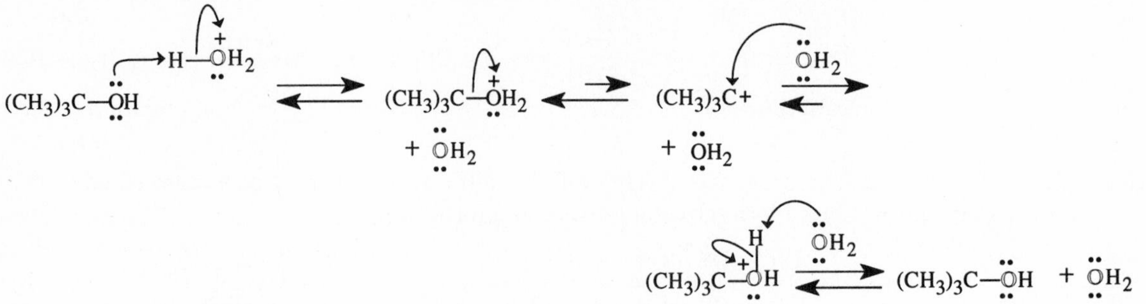

10.41 (a) A two-electron reduction.
(c) A two-electron oxidation.
(e) Neither an oxidation nor a reduction.

10.42 (a) First convert the alcohol into a bromide using a method that involves an inversion of configuration. Then displace the bromide with $^{18}\overset{-}{O}H$ to provide the alcohol with the desired configuration.

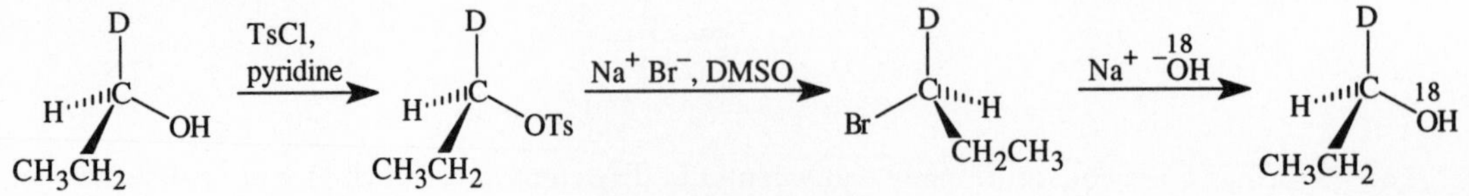

10.43 (a)

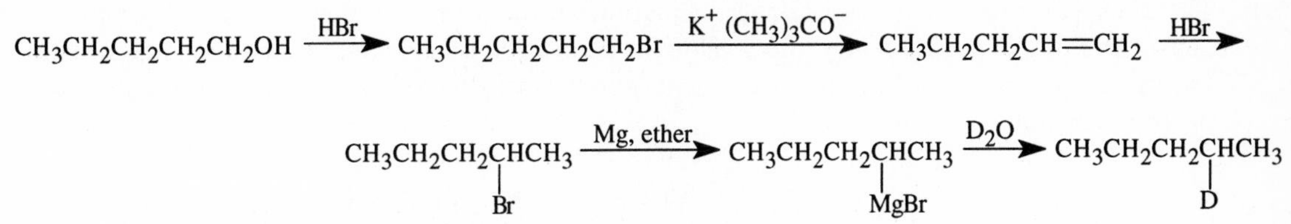

$$CH_3CH_2CH_2CH_2CH_2OH \xrightarrow{HBr} CH_3CH_2CH_2CH_2CH_2Br \xrightarrow{K^+ (CH_3)_3CO^-} CH_3CH_2CH_2CH=CH_2 \xrightarrow{HBr}$$

$$\underset{\overset{|}{Br}}{CH_3CH_2CH_2CHCH_3} \xrightarrow{Mg, \text{ ether}} \underset{\overset{|}{MgBr}}{CH_3CH_2CH_2CHCH_3} \xrightarrow{D_2O} \underset{\overset{|}{D}}{CH_3CH_2CH_2CHCH_3}$$

(c)

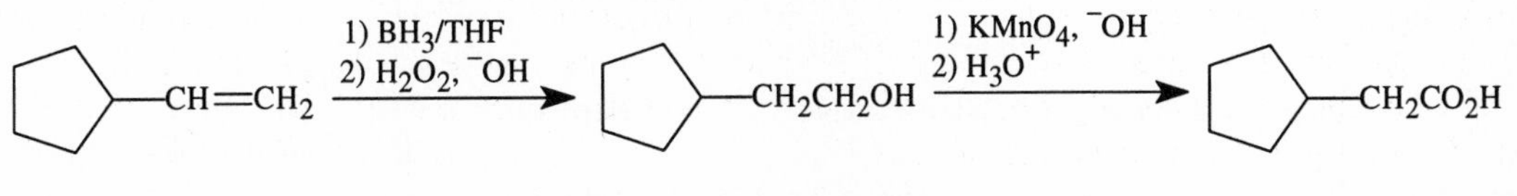

In many cases, there is more than one acceptable synthesis for a given target, or, within a synthesis, there may be more than one acceptable method for accomplishing a particular conversion. For example, in part (a) above, you might have elected to convert the starting alcohol into a sulfonate ester such as a tosylate and carry out the elimination on the tosylate. In part (c), you might have elected to oxidize the primary alcohol with a different oxidizing agent, such as aqueous dichromate. Be sure to ask your teaching assistant or professor if you have a different synthesis and you are not sure whether it is acceptable.

(e)

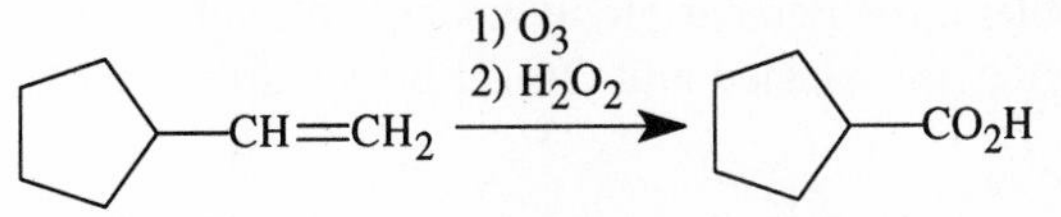

(g) The hint is to remind you that cyanide ion can be used as a nucleophile in S_N2 reactions.

$$CH_3CH_2CH=CH_2 \xrightarrow[\text{2) } H_2O_2,\ ^-OH]{\text{1) } BH_3/THF} CH_3CH_2CH_2CH_2OH \xrightarrow{TsCl, \text{ pyridine}}$$

$$CH_3CH_2CH_2CH_2OTs \xrightarrow{NaCN, DMSO} CH_3CH_2CH_2CH_2CN$$

10.44 (a) The principle is that the best leaving groups are the weakest bases. The question becomes, then, which of the two leaving groups is the weaker base. The 3,5-dibromobenzenesulfonate anion is a weaker base than the benzenesulfonate anion because the *polar effect* of the bromines stabilizes the negative charge in this anion. (See Sec. 3.6B, text p. 108.) Consequently, this group is the better leaving group.

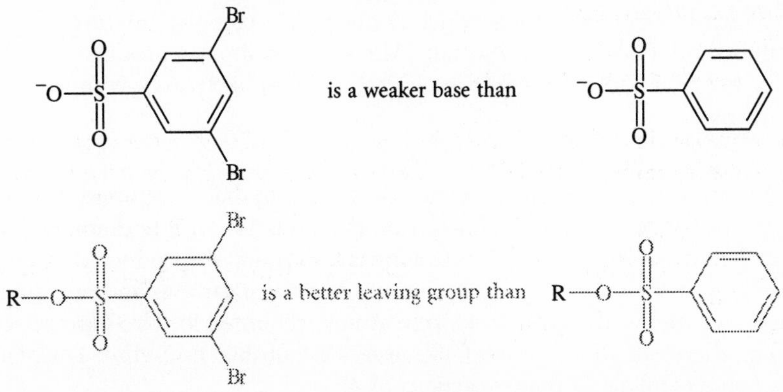

10.45   The essence of this solution is a balanced equation for the oxidation of ethanol by dichromate and conversion of the stoichiometry into a value for percent blood alcohol. Because the dichromate is aqueous, assume that ethanol is oxidized to acetic acid (a four-electron oxidation). A balanced equation for this process is as follows (verify this):

$$16H^+ + 3CH_3CH_2OH + 2Cr_2O_7^{2-} \longrightarrow 3CH_3CO_2H + 11H_2O + 4Cr^{3+}$$

This equation shows that 3/2 mole of ethanol is oxidized for every mole of $Cr_2O_7^{2-}$ ion consumed. Required to solve the problem are the grams of ethanol oxidized and the mL of blood in which that ethanol is contained. The process can be diagrammed as follows:

$$\left. \begin{array}{l} \text{moles of dichromate} \Rightarrow \text{moles of ethanol} \Rightarrow \text{g of ethanol} \\ \text{mL of air} \Rightarrow \text{mL of blood} \end{array} \right\} \text{divide and multiply the result by 100}$$

The amount of ethanol oxidized is (3/2 mol ethanol per mol of dichromate)(0.506×10^{-6} mol of dichromate) = 0.760×10^{-6} mol of ethanol. The molecular mass of ethanol is 46.07 g/mol; hence, the g of ethanol in the sample of blood is (46.07 g/mol)(0.760×10^{-6} mol) = 3.50×10^{-5}.

The 52.5 mL of air collected is equivalent to 0.0250 mL of blood:

$$\text{mL blood} = \frac{52.5 \text{ mL of air}}{2100 \text{ mL of air per mL of blood}} = 0.0250$$

Finally, the grams of ethanol per mL of blood is (3.50×10^{-5} g of ethanol) ÷ (0.0250 mL of blood) = 1.40×10^{-3}, and, using the formula in the problem, the percent blood alcohol content is this number times 100, or 0.14%. Consequently, Bobbin is legally intoxicated and Officer Order should make the arrest.

10.47   First, write a balanced equation:

$$\overset{3}{H_2O} + \overset{3}{CH_3OCH_2CH_2OH} \longrightarrow \overset{3}{CH_3OCH_2CO_2H} + \overset{12}{4e^-} + \overset{12}{4H^+}$$

$$\overset{12}{4e^-} + \overset{12}{4H^+} + \overset{4}{HNO_3} \longrightarrow \overset{4}{NO} + \overset{8}{2H_2O}$$

Sum: $$3CH_3OCH_2CH_2OH + 4HNO_3 \longrightarrow 3CH_3OCH_2CO_2H + 4NO + 5H_2O$$

Because the alcohol oxidation is a four-electron oxidation, and the reduction of nitric acid to NO is a three-electron reduction, it takes four moles of nitric acid to oxidize three moles of the alcohol, that is, 4/3 mole of nitric acid per mole of alcohol. Therefore, it takes (4/3)(0.100) = 0.133 moles of nitric acid to oxidize 0.1 mole of the alcohol.

10.48   Because thiols and disulfides equilibrate to give a mixture of all possible thiols and disulfides (see Eq. 10.59a–b, text p. 485), Stench has evidently obtained a complex mixture of di(2-octyl) disulfide, dibutyl disulfide, and butyl 2-octyl disulfide. Although the desired product is *statistically* favored (it is twice as likely to be formed as the other disulfides), it is not likely to be comprise more than 50% of the reaction mixture.

10.49   The reactivity data and the molecular formula of *A* indicate that compound *A* is an alkene with one double bond. The identity of compound *D* follows from the oxidation of 3-hexanol; it can only be 3-hexanone (see structure below). It is given that 3-hexanone is an ozonolysis product of alkene *A* (along with O=CH_2 [formaldehyde], not shown below). Since alkene *A* has seven carbons and one double bond, and 3-hexanone has six carbons, the carbon of alkene *A* not accounted for by 3-hexanone must be part of a =CH_2 group. Therefore, the identity of alkene *A* is established as 2-ethyl-1-pentene. The identities of compounds *B* and *C* follow from the reactions of *A*.

10.51 Compound (2) is a *meso*-compound and therefore cannot be optically active; hence, *D* is compound (2). Compound (3) would give the same products of periodic acid oxidation as compound (2); hence, compound (3) is *C*. Because compound (1) is not a vicinal glycol, it does not react with periodic acid, and is therefore *A*. Compound (4) is therefore *B*, and indeed would react with periodic acid to give oxidation products that are different from those obtained from compounds (2) and (3).

10.52 Their reactions with NaH, their formulas, and the fact that they can be resolved into enantiomers show that compounds *A* and *B* are chiral isomeric alcohols. Their eventual conversion into methylcyclohexane shows that they have the same carbon skeleton as methylcyclohexane. When *optically active A* and *B* are converted into tosylates and subjected to the E2 reaction with potassium *tert*-butoxide they give *optically active*, and therefore *chiral*, alkenes. The only possible chiral alkenes with the carbon skeleton of methylcyclohexane are 4-methylcyclohexene and 3-methylcyclohexene, which are compounds *C* and *D*; the data do not determine which is which. To summarize:

4-methylcyclohexene   3-methylcyclohexene   methylcyclohexane

(compounds *C* and *D*; the data
do not determine which is which)

The two tosylates that would both give a mixture of compounds *C* and *D* are the tosylates of stereoisomeric 3-methylcyclohexanols. Because alcohols *A* and *B* have different melting points, they must be the *diastereomeric* 3-methylcyclohexanols. The individual identities of these alcohols follow from an analysis of the E2 reactions of their tosylates. Note that *anti* elimination can take place only when the tosylate group is in an axial position. In the tosylate of *cis*-3-methylcyclohexanol, the tosylate group can only assume the axial position required for E2 elimination when the molecule is in a very unstable conformation in which the methyl group is also axial:

tosylate of *A*   severe 1,3-diaxial
                                   interaction

In the tosylate of *trans*-3-methylcyclohexanol, the tosylate group can assume the axial position in a conformation in which the methyl group is equatorial.

(solution continues)

The compound containing the greater amount of the conformation in which the tosylate group is in the axial position—the tosylate of *B*—undergoes elimination more rapidly. Consequently, compound *A* is *cis*-3-methylcyclohexanol, and compound *B* is *trans*-3-methylcyclohexanol.

cis-3-methylcyclohexanol
(compound *A*)

trans-3-methylcyclohexanol
(compound *B*)

10.53 (a)

(c)   (e)

(g) In this reaction, an example of the *Lemieux-Johnson oxidation,* the OsO_4 hydroxylates the alkene to form a vicinal glycol, which is cleaved by the periodic acid. The periodic acid also re-oxidizes the osmium-containing by-products to Os(VIII), which can then react with another alkene. This "recycling" reaction allows the expensive osmium reagent to be used in considerably less than an equivalent amount.

(i) After the reaction comes to equilibrium, all possible thiols and all possible disulfides are present. (See the solution to Problem 10.48 for the rationale.)

$$(CH_3)_2CHCH_2CH_2\!-\!S\!-\!S\!-\!C_2H_5 \quad + \quad (CH_3)_2CHCH_2CH_2\!-\!S\!-\!S\!-\!CH_2CH_2CH(CH_3)_2$$

$$+ \quad C_2H_5\!-\!S\!-\!S\!-\!C_2H_5 \quad + \quad (CH_3)_2CHCH_2CH_2SH \quad + \quad C_2H_5SH$$

10.54 (a) The primary deuterium kinetic isotope effect of 6.6 suggests that the step of the mechanism shown in Eq. 10.38c is rate-limiting, because the isotopically substituted hydrogen is transferred in this step.

(b) Because of the primary isotope effect in part (a), the deuterium is removed more slowly than the hydrogen (about seven times more slowly). Thus, more of the aldehyde containing deuterium is formed. Because $CrO_3(pyridine)_2$ is an achiral reagent, it cannot differentiate between the two enantiotopic α-hydrogens of ethanol; each is removed with equal frequency. Consequently, the isotope effect causes isotopic discrimination *regardless* of the stereochemical positions of the hydrogen and deuterium.

(c) The enzyme alcohol dehydrogenase, which is a chiral catalyst, distinguishes between the enantiotopic α-hydrogens of ethanol so completely that it removes *only* the pro-(R) hydrogen whether this hydrogen is isotopically substituted or not. Thus, when this hydrogen is substituted by deuterium, the deuterium is removed, and, as a result, the aldehyde contains no deuterium. Although there may be an isotope effect associated with removal of this hydrogen, it is not reflected in the relative amounts of H and D removed from a given enantiomer because the stereochemical discrimination (which may be a factor of 1000 or more) is a much larger effect than the primary deuterium isotope effect (which could be a factor only as large as 7 or so).

10.55 The stereochemistry of the addition reaction as shown indicates that the pro-(R) hydrogen at carbon-3 in the product is the one added from solvent. Thus, by the principle of microscopic reversibility, it is the one eliminated in the reverse reaction.

(a) (c) No reaction.

In part (a), the fumarate contains deuterium because the deuterium in the malate starting material is at carbon-2; a hydrogen at carbon-3, not carbon-2, is eliminated. In part (c), there is no reaction because the enzyme forms *only* (S)-malate from fumarate (see Eq. 7.28, text p. 306). Consequently, in the reverse direction, *only* (S)-malate serves as a substrate for the enzyme; a malate with the 2R configuration will not react regardless of the stereochemistry of carbon-3.

10.56 Unlike HBr, HCN is a rather weak acid ($pK_a = 9.4$), and it provides very little cyanide ion because it is essentially unionized. In addition, it is not acidic enough to protonate the alcohol to any reasonable extent. Hence, the —OH group of the alcohol is not converted into a good leaving group by protonation. With a nucleophile concentration of nearly zero and a poor leaving group, a substitution reaction cannot take place.

Addition of H_2SO_4 to the reaction mixture would indeed solve the leaving-group problem, because the H_2SO_4 would protonate the —OH group. However, added protons suppress the ionization of HCN even more than it is suppressed normally. Without a reasonable concentration of nucleophile, the reaction is doomed.

10.57 Water is displaced from a protonated alcohol molecule by another alcohol molecule:

10.58 (a) The carbocation rearrarangement in this solution, in which a tertiary carbocation is converted into a secondary carbocation, is driven by relief of ring strain.

(solution continues)

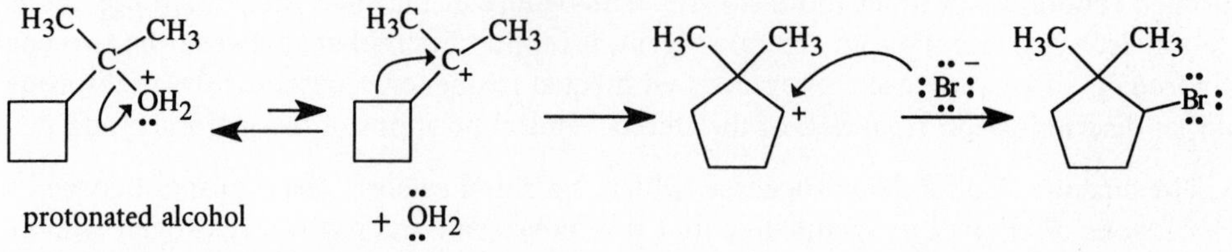

protonated alcohol

$+ \ddot{O}H_2$

11

Chemistry of Ethers, Epoxides, and Sulfides

Terms

Concepts

I. General Synthetic and Mechanistic Concepts

A. NEIGHBORING-GROUP PARTICIPATION

1. Reactions involving the covalent participation of neighboring groups are said to occur with anchimeric assistance, or neighboring-group participation.
   a. Neighboring-group mechanisms involve intramolecular reactions that are in competition with ordinary intermolecular reactions.
   b. Reactions involving neighboring-group participation are generally faster than analogous reactions that do not involve neighboring-group participation.
      i. An intramolecular reaction has a greater probability of occurring.
      ii. Reactions that occur with greater probability have larger rates.
      iii. The formation of a three-membered ring occurs despite the strain in the intermediate.
2. Neighboring-group participation in nucleophilic substitution reactions is common for cases in which three-, five-, and six-membered rings are formed.

B. THE THREE FUNDAMENTAL OPERATIONS OF ORGANIC SYNTHESIS

1. Most reactions used in organic synthesis involve one or more of three fundamental operations:
   a. Functional-group transformation (the most common type of synthetic operation).
   b. Control of stereochemistry (accomplished with stereoselective reactions).

c. Formation of carbon-carbon bonds (a particularly important synthetic operation). Two examples have been presented:
   i. cyclopropane formation from carbenes or carbenoids and alkenes.
   ii. reaction of Grignard reagents with ethylene oxide.

Reactions

I. Synthesis of Ethers and Sulfides

A. WILLIAMSON ETHER SYNTHESIS

1. The Williamson ether synthesis is an S_N2 reaction of the conjugate base of an alcohol (or thiol) as the nucleophile with an unhindered alkyl halide or sulfonate ester to form an ether (or sulfide).

$$(CH_3)_3C-O^- \, Na^+ \quad + \quad CH_3-OTs \quad \longrightarrow \quad (CH_3)_3C-O-CH_3 \quad + \quad TsO^- \, Na^+$$

| sodium 2-methyl-2-propanolate | methyl tosylate | 2-methoxy-2-methylpropane (or *tert*-butyl methyl ether) | sodium tosylate |

$$Ph-S^- \, Na^+ \quad + \quad CH_3CH_2Br \quad \longrightarrow \quad Ph-S-CH_2CH_3 \quad + \quad Na^+ \, Br^-$$

| sodium thiophenolate | bromoethane | (ethylthio)benzene (or ethyl phenyl sulfide) | sodium bromide |

   a. Methyl halides, primary alkyl halides, or the corresponding sulfonate esters can be used in a Williamson synthesis.
   b. Tertiary and many secondary alkyl halides cannot be used in a Williamson synthesis.

$$H_3C-O^- \, Na^+ \quad + \quad (CH_3)_3C-I \quad \longrightarrow \quad H_3C-O-H \quad + \quad Na^+ \, I^- \quad + \quad (CH_3)_2C=CH_2$$

| sodium methanolate | *tert*-butyl iodide (or 2-iodo-2-methylpropane) | methanol | sodium tosylate | 2-methylpropene |

2. In principle, two different Williamson syntheses are possible for any ether with two different alkyl groups.
   a. The preferred synthesis is usually the one that involves the alkyl halide with the greater S_N2 reactivity.
   b. Tertiary and secondary alkyl groups should be derived from the alkoxide.

B. ALKOXYMERCURATION-REDUCTION OF ALKENES

1. An alkoxymercuration-reduction reaction occurs when oxymercuration-reduction is carried out in an alcohol solvent instead of water.

$$(CH_3)_2C=CH_2 \quad \xrightarrow[\text{(CH}_3)_2\text{CHOH}]{\text{Hg(OAc)}_2} \quad \begin{array}{c} (CH_3)_2CHO \\ | \\ (CH_3)_2C-CH_2 \\ | \\ HgOAc \end{array} \quad \xrightarrow{\text{NaBH}_4} \quad \begin{array}{c} (CH_3)_2CHO \\ | \\ (CH_3)_2C-CH_2 \\ | \\ H \end{array}$$

2. The mechanism of the alkoxymercuration reaction is completely analogous to the mechanism of oxymercuration, except that an alcohol instead of water is the nucleophile that attacks the mercurinium ion intermediate.

3. The alkoxymercuration-reduction reaction can be used to prepare ethers that cannot be prepared by a Williamson ether synthesis.

C. ETHERS FROM ALCOHOL DEHYDRATION AND ALKENE ADDITION

1. Two molecules of an alcohol can undergo dehydration to give an ether (a reaction used primarily in industry).

$$2 \, CH_3CH_2OH \quad \xrightarrow[140°]{\text{H}_2\text{SO}_4} \quad CH_3CH_2-O-CH_2CH_3$$

| ethanol | | diethyl ether |

a. This is generally restricted to the preparation of symmetrical ethers derived from primary alcohols.
b. Secondary and tertiary alcohols cannot be used because they undergo dehydration to alkenes.
c. The formation of ethers from primary alcohols is an S_N2 reaction in which one alcohol displaces water from another molecule of protonated alcohol.

2. Tertiary alcohols can be converted into unsymmetrical ethers by treating them with dilute solutions of strong acids in an alcohol solvent.

$$(CH_3)_3C—OH \quad \xrightarrow[CH_3OH]{H_2SO_4 \text{ (trace)}} \quad (CH_3)_3C—O—CH_3 \quad + \quad H_2O$$

tert-butyl alcohol   *tert*-butyl methyl ether

a. One of the alcohol starting materials must readily lose water after protonation to form a relatively stable carbocation.
b. The alcohol that is used in excess must be one that either cannot lose water after protonation to give a carbocation or should form a carbocation much less readily.
c. Any alkene that does form is not removed but is reprotonated to give back a carbocation, which eventually reacts with the alcohol solvent.

3. Treatment of an alkene with a large excess of alcohol in the presence of an acid catalyst gives an ether provided that a relatively stable carbocation intermediate is involved.

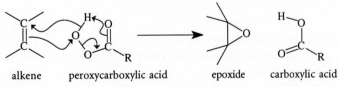

methylcyclohexene   1-ethoxy-1-methylcyclohexane

D. OXIDATION OF ALKENES WITH PEROXYCARBOXYLIC ACIDS

1. One of the best laboratory preparations of epoxides involves the direct oxidation of alkenes with peroxycarboxylic acids.

cyclopentene   a peroxycarboxylic acid   cyclopentene oxide   a carboxylic acid

a. The oxidizing agent, a peroxycarboxylic acid, is a carboxylic acid that contains an —O—O—H (hydroperoxy) group rather than an —OH (hydroxy) group.
b. The general terms peroxyacid or peracid, denoting any acid containing a hydroperoxy group, are sometimes used instead of peroxycarboxylic acid.
c. Many peroxycarboxylic acids are unstable and are formed just prior to use by mixing a carboxylic acid with hydrogen peroxide.

2. The formation of an epoxide from an alkene and a peroxycarboxylic acid is a concerted addition reaction.

alkene   peroxycarboxylic acid   epoxide   carboxylic acid

a. This mechanism is very similar to that for the formation of a bromonium ion in the addition of bromine to alkenes.
b. The formation of epoxides with peroxycarboxylic acids is a stereoselective reaction; it takes place with complete retention of the alkene stereochemistry.

E. CYCLIZATION OF HALOHYDRINS

1. Epoxides can be synthesized by the treatment of halohydrins with base.

   a. This reaction is an intramolecular variation of the Williamson ether synthesis in which the alcohol and the alkyl halide are part of the same molecule.
   b. The alkoxide anion, formed reversibly by reaction of the alcohol with NaOH, displaces halide ion from the neighboring carbon.
2. Like other S_N2 reactions, this reaction takes place by backside attack of the nucleophilic oxygen anion at the halide-bearing carbon.
   a. The attacking oxygen and the leaving halide assume an *anti* relationship in the transition state of the reaction.
   b. Halohydrins derived from cyclic compounds must be able to assume the required *anti* relationship through a conformational change if epoxide formation is to succeed.

II. Reactions of Ethers, Epoxides, and Sulfides

A. CLEAVAGE OF ETHERS

1. Ethers are relatively unreactive compounds; however, the C—O bonds of ethers can be cleaved under acidic conditions to give alcohols and alkyl halides.
   a. Ethers containing only methyl and/or primary alkyl groups require strong acid and relatively harsh conditions to bring about ether cleavage.
   b. Secondary and tertiary ethers cleave more readily than primary or methyl ethers because tertiary carbocation intermediates can be formed.
   c. The alcohol formed in the cleavage of an ether can go on to react with HI or HBr to give a second molecule of alkyl halide.
2. The cleavage mechanism of ethers containing only methyl and/or primary alkyl groups involves
   a. protonation of the ether oxygen, followed by
   b. attack of the protonated ether by the iodide or bromide ion (a good nucleophile) in an S_N2 reaction to form an alkyl halide and liberate an alcohol as a leaving group.

3. The mechanism of cleavage in ethers containing secondary or tertiary alkyl groups involves
   a. protonation of the ether oxygen, followed by
   b. formation of a carbocation by loss of the alcohol leaving group, and finally
   c. attack of the carbocation by halide ion.
4. Because the S_N1 reaction is faster than competing S_N2 processes, none of the primary alkyl halide is formed.

B. REACTIONS OF EPOXIDES WITH ACID OR BASE

1. Because of their ring strain, epoxides undergo ring-opening reactions with ease; the opening of an epoxide relieves the strain of the three-membered ring.

2. The ring opening of epoxides by bases involves backside attack of the nucleophile on the epoxide carbon.
   a. A reaction of this type is essentially an S_N2 reaction in which the epoxide oxygen serves as the leaving group.
   b. Under basic or neutral conditions, nucleophiles typically attack unsymmetrical epoxides at the carbon with fewer branches.

   i. This regioselectivity is expected from the effect of branching on the rates of S_N2 reactions.
   ii. Branching retards the rate of attack; hence, attack at the unbranched carbon is faster and leads to the observed product.
   c. Inversion of configuration is observed if attack occurs at a stereocenter.
3. The regioselectivity of the ring-opening reaction is different under acidic conditions.
   a. The nucleophile reacts at the more branched carbon of the epoxide.
   b. If neither carbon is tertiary, a mixture of products is formed in most cases.

   i. Bonds to tertiary carbon atoms are weaker than bonds to primary carbon atoms and the protonated oxygen is a good leaving group.
   ii. The regioselectivity of acid-catalyzed epoxide ring opening is very similar to attack on bromonium ions.
4. Acid-catalyzed epoxide hydrolysis is generally a useful way to prepare glycols.

   epoxide   glycol

5. Base-catalyzed hydrolysis of epoxides also gives glycols, although, in some cases, polymerization occurs as a side reaction under the basic conditions.

C. REACTION OF ETHYLENE OXIDE WITH GRIGNARD REAGENTS

1. Grignard reagents react with ethylene oxide to give, after a protonation step, primary alcohols.
   a. The carbon of the C—Mg bond of the Grignard reagent has carbanion character and attacks the epoxide as a nucleophile.
   b. The magnesium of the Grignard reagent is a Lewis acid and coordinates with the epoxide oxygen.
   c. Coordination of an oxygen to a Lewis acid makes the oxygen a better leaving group.

   d. After the Grignard reagent has reacted, the alkoxide is converted into the alcohol product in a separate step by the addition of water or dilute acid.

$$Ph{-}CH_2CH_2{-}O^-\ ^+MgBr \xrightarrow{H_3O^+} Ph{-}CH_2CH_2{-}OH\ +\ Mg^{2+}\ +\ Br^-$$

D.   OXIDATION OF ETHERS AND SULFIDES

1. Except for peroxide formation (which occurs on standing in air) and combustion, ethers are relatively inert towards oxidizing conditions.

2. Sulfides oxidize at sulfur rather than at carbon when they react with many common oxidizing agents; sulfides are readily oxidized to sulfoxides and sulfones.

    a. Sulfoxides and sulfones can be prepared by the direct oxidation of sulfides with one and two equivalents, respectively, of hydrogen peroxide, H_2O_2.

| dimethyl sulfide | dimethyl sulfoxide | dimethyl sulfone |

    b. Other common oxidizing agents such as $KMnO_4$, HNO_3, and peroxyacids readily oxidize sulfides.

3. Nonionic Lewis structures for sulfoxides and sulfones cannot be written without violating the octet rule.

$$
\left[\; R\!-\!\underset{\underset{\displaystyle :}{}}{\overset{\overset{\displaystyle :O:}{\parallel}}{S}}\!-\!R \; \longleftrightarrow \; R\!-\!\overset{\displaystyle :\ddot{O}:^-}{\underset{\displaystyle ..}{S^+}}\!-\!R \; \right]
\qquad
\left[\; R\!-\!\overset{\overset{\displaystyle :O:}{\parallel}}{\underset{\underset{\displaystyle :O:}{\parallel}}{S}}\!-\!R \; \longleftrightarrow \; R\!-\!\overset{\overset{\displaystyle :\ddot{O}:^-}{}}{\underset{\underset{\displaystyle :\ddot{O}:^-}{}}{S^{2+}}}\!-\!R \; \right]
$$

sulfoxide   sulfone

E.   REACTIONS OF OXONIUM AND SULFONIUM SALTS

1. If the acidic hydrogen of a protonated ether is replaced with an alkyl group, the resulting compound is called an oxonium salt; the sulfur analog of an oxonium salt is a sulfonium salt.

$$(CH_3)_3O^+ \; {}^-BF_4 \qquad\qquad (C_2H_5)_3S^+ \; {}^-BF_4$$

trimethyloxonium tetrafluoroborate   triethylsulfonium tetrafluoroborate

2. Oxonium and sulfonium salts react with nucleophiles in substitution and elimination reactions.

$$R\!-\!OH \; + \; (CH_3)_3O^+ \; {}^-BF_4 \; \longrightarrow \; R\!-\!O\!-\!CH_3 \; + \; CH_3\!-\!O\!-\!CH_3 \; + \; HBF_4$$

| an alcohol | trimethyloxonium tetrafluoroborate (an oxonium salt) | | a methyl ether | dimethyl ether | tetrafluoroboric acid |

    a. Oxonium salts are among the most reactive alkylating agents known (they react very rapidly with most nucleophiles). They are usually isolated as $^-BF_4$ salts, and must be stored in the absence of moisture.

    b. Sulfonium salts are somewhat less reactive than the corresponding alkyl chlorides in S_N2 reactions and therefore are handled more easily than oxonium salts.

3. The important biological sulfonium salt *S*-adenosylmethionine (SAM) is a methylating agent for biological nucleophiles.

Study Guide Links

✓11.1 Learning New Reactions from Earlier Reactions

Quite often the text points out the close relationship between two reactions—a new one, and one you have already studied. Such connections will help you to understand new reactions in terms of what you already know and will substantially reduce the amount of *really new* material that you must master. Alkoxymercuration-reduction is such a case; this reaction is closely related to oxymercuration-reduction. In order to make the connection between the two reactions, it almost goes without saying that you have to know what oxymercuration-reduction is! If you did not know the meaning of this term when you read this section, did you go back and find out?

The question at the end of Study Problem 11.1 on text p. 498—why *tert*-butyl bromide and sodium methoxide will not work in a Williamson ether synthesis—is another effort to get you to think about earlier material, in this case the reactivities of alkyl halides in the presence of a strong base. Did you answer this question? (If you can't answer it, review Sec 9.5F on text p. 411, as suggested in the problem.)

Remember that *continued review* is one of the keys to successful study in organic chemistry. You must try constantly to remain active rather than passive when you study. Answering questions such as the ones posed in the text will assist you. If you can't answer one of these questions, write it down and get help at the next available opportunity!

✓11.2 Common Intermediates from Different Starting Materials

An important thing to notice about this section is that the same reactive intermediate—in this case, a carbocation—in some cases can be generated in several different ways. For example, consider the many different ways that a *tert*-butyl cation can be formed as a reactive intermediate:

$$CH_3$$
$$H_3C-C-OH$$
$$CH_3$$

"H$^+$", $-H_2O$

$$H_3C$$
$$\;\;\;\;C=CH_2 \quad \xrightarrow{\text{"H}^+\text{"}}$$
$$H_3C$$

$$H_3C$$
$$\;\;\;\;C^+-CH_3$$
$$H_3C$$
tert-butyl cation

$$\xleftarrow[\text{(solvolysis)}]{-Br^-}$$

$$CH_3$$
$$H_3C-C-Br$$
$$CH_3$$

Whether a carbocation is formed by protonation of an alkene, by dehydration of an alcohol, or by solvolysis of an alkyl halide, the same product is obtained if a large excess of the same nucleophile is present in each case. Thus, if methanol is used as the solvent, *tert*-butyl methyl ether is the product regardless of the origin of the carbocation. On the other hand, if concentrated HBr is used, *tert*-butyl bromide is the product.

It is natural at first to think of hydrogen halide addition to alkenes, alcohol dehydration, alkene hydration, and solvolysis of tertiary alkyl halides as different reactions. What you now should be able to see is that these reactions are all linked through a common intermediate, the carbocation. Which product is obtained depends on the conditions used.

✓11.3 Mechanism of Epoxide Formation

The curved-arrow mechanism of epoxide formation incorporates elements of several steps which, when viewed separately, may make the mechanism easier to follow. First, imagine that the alkene π-electrons attack the oxygen of a peroxycarboxylic acid and that a carbocation is formed:

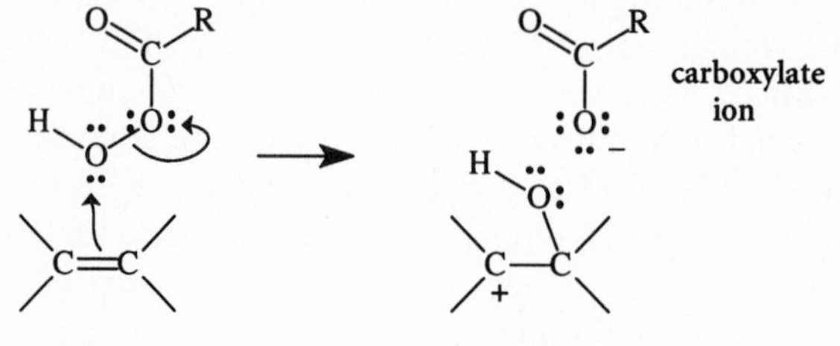

This step breaks a weak O—O bond and liberates a good leaving group, the carboxylate ion, which is a weak base. Next, the oxygen of the —OH group captures the carbocation with an unshared electron pair in an internal Lewis acid-base association reaction:

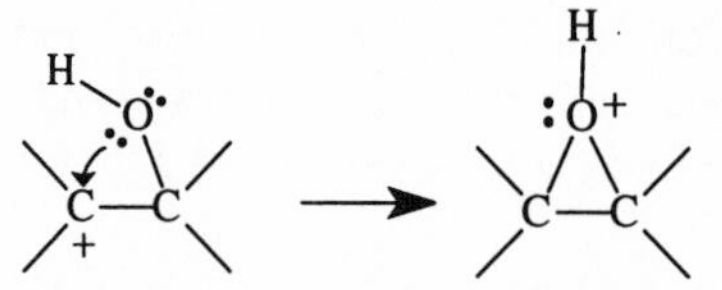

Finally, the carboxylate ion expelled in the first step removes the proton from the protonated epoxide:

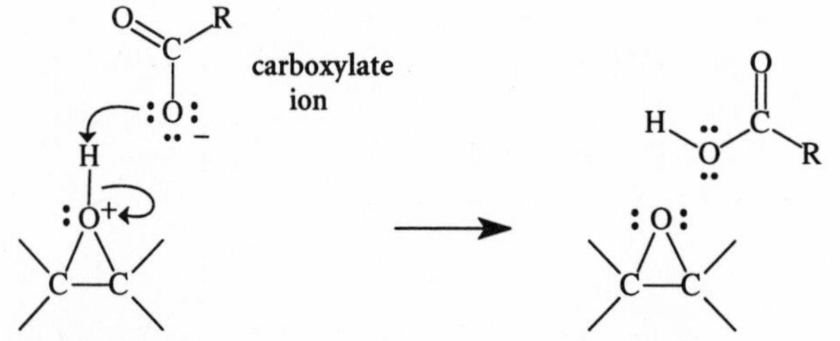

Although the actual mechanism of epoxide formation is *not* stepwise, and does *not* involve carbocations, thinking of it this way helps to see how it is related to other mechanisms that do involve carbocations. A similar approach has been used to help you understand bromine addition (Eq. 5.4a, text p. 176) and ozonolysis (Study Guide Link 5.2, p. 91 of this manual).

11.4 Stereospecific Reactions

Study Guide Link 7.6 on p. 145 of this manual introduced the term *stereospecific reaction:* a stereoselective reaction in which the stereochemistry of the product depends on the stereochemistry of the starting material. Equations 11.17a–b in the text show that epoxidation with peroxycarboxylic acids, like bromine addition to alkenes, is a stereospecific reaction, because the stereochemistry of the epoxide formed depends on the stereochemistry of the alkene starting material.

✓11.5 Neighboring-Group Participation

Probably without realizing it you've studied another reaction in this chapter that owes its success to neighboring-group participation: the formation of epoxides from bromohydrins (Sec. 11.2B, text p. 505). The alkoxide of a bromohydrin has two competing possibilities for reaction. First, it can undergo an *intramolecular reaction* to form an epoxide:

Intramolecular reaction:

epoxide

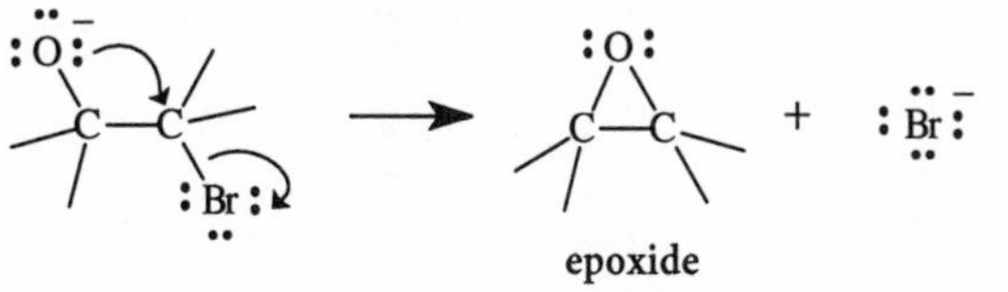

This is the observed reaction (Sec. 11.2B). However, another possible reaction is for the alkoxide to react as a nucleophile in an *intermolecular reaction* with a *second* molecule of bromohydrin:

Intermolecular reaction:

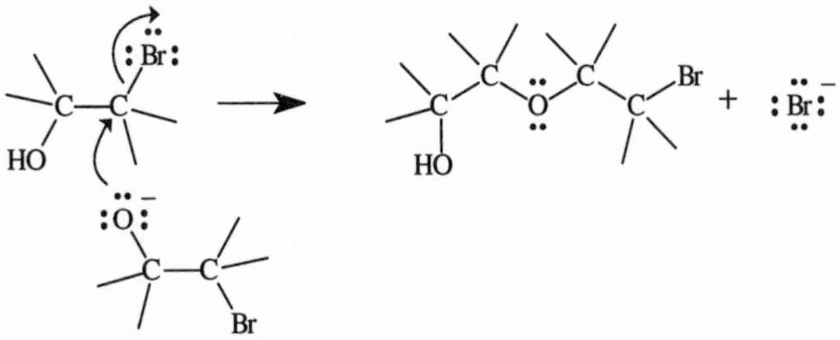

The intramolecular reaction involves neighboring-group participation, and is the observed reaction because it is much faster. This, then, is another reaction in which an *intramolecular* reaction is accelerated over a conceptually similar *intermolecular* counterpart.

11.6   Reaction Probability and Entropy

Recall that a standard free-energy change consists of a contribution from both an enthalpy change and an entropy change (Study Guide Link 4.4 on p. 61 of this manual):

$$\Delta G° = \Delta H° - T\Delta S°$$

Likewise, the standard free energy of activation $\Delta G°^{\ddagger}$ consists of contributions from the *enthalpy of activation* $\Delta H°^{\ddagger}$ and the *entropy of activation* $\Delta S°^{\ddagger}$ according to the equation

$$\Delta G°^{\ddagger} = \Delta H°^{\ddagger} - T\Delta S°^{\ddagger} \qquad \text{(SG11.1)}$$

The "reaction probability" discussed in Sec. 11.6 is roughly equivalent to the entropy of activation; the ring strain contributes to the enthalpy of activation. Study Guide Link 4.4 discussed the relationship of entropy and probability in chemical terms. A loss of *randomness* corresponds to a negative entropy change, which, according to Eq. SG11.1, raises the free-energy barrier for a reaction, and therefore decreases the rate. When two molecules collide to form one molecule, the translational motion associated with two molecules flitting about randomly is converted into the translational motion of a single molecule flitting about randomly; hence, an *intermolecular* reaction causes a loss of randomness, or a decrease in *translational entropy.* When a molecule reacts *intramolecularly,* there is no change in the number of molecules; to a useful approximation, the translational entropy change is zero. Hence, the "reaction probability," or entropy of activation, is less negative (or more positive) for an intramolecular reaction. Counterbalancing this is the loss of internal rotations associated with ring formation; consequently, an intramolecular reaction that results in ring formation also suffers a loss of randomness as well, but the entropy (randomness) associated with internal rotations is much less than that associated with translation.

Forming a three-membered ring introduces strain, which is reflected in an increased $\Delta H^{\circ\ddagger}$. However, as suggested in the text, when intramolecular and intermolecular reactions are compared, the increase in $\Delta H^{\circ\ddagger}$ resulting from an intramolecular reaction that forms a three-membered ring is more than offset by the decrease in $\Delta S^{\circ\ddagger}$ associated with an intermolecular reaction. The balance of $\Delta H^{\circ\ddagger}$ and $\Delta S^{\circ\ddagger}$ is such that three, five, and six-membered rings seem to be most often involved in cases of neighboring-group participation.

Solutions

Solutions to In-Text Problems

11.1 (a) In this reaction, the alkoxide formed by reaction of the alcohol with sodium is alkylated by methyl iodide to give isopropyl methyl ether:

$$(CH_3)_2CHOH \xrightarrow{\text{Na}} (CH_3)_2CHO^- \, Na^+ \xrightarrow{\text{CH}_3\text{I}} (CH_3)_2CHO\text{—}CH_3 \; + \; Na^+ \, I^-$$
$$+ \; H_2 \qquad\qquad \text{isopropyl methyl ether}$$

(c) Alkoxide bases react with tertiary alkyl halides to give elimination products rather than substitution products. (See Sec. 9.5F and Problem 9.39 in the text.) Consequently, the products are $(CH_3)_2C{=}CH_2$ (2-methylpropene), $Na^+ \, Br^-$ (sodium bromide), and CH_3OH (methanol).

11.2 (a)

A second Williamson synthesis, the reaction of sodium ethoxide with (bromoethyl)cyclohexane, is in principle possible; however, because the alkyl halide has a β-branch, this reaction would give, in addition to the desired product, some alkene resulting from elimination.

(c) Di-*tert*-butyl ether cannot be prepared by a Williamson synthesis, because its preparation would require an S_N2 reaction of *tert*-butoxide with a *tert*-butyl halide. With this combination of reagents, an E2 reaction would occur instead, and 2-methylpropene and *tert*-butyl alcohol would be the only organic products formed.

11.3 (a) Let $\text{—R} = \text{—}CH_2CH_2CH_2CH_3$. The mechanism is essentially identical to that shown in Eq. 5.12a–b on text p. 181, except that the nucleophile is isopropyl alcohol rather than water.

Attack of the alcohol on the mercurinium ion intermediate occurs at the more branched carbon because this carbon-mercury bond is weaker (that is, easier to break) than the bond to primary carbon. Another way to state the same point is to say that the mercurinium ion is described by three resonance structures, of which structure *B* is more important than structure *C*, because in structure *B* electron deficiency resides on the more branched carbon:

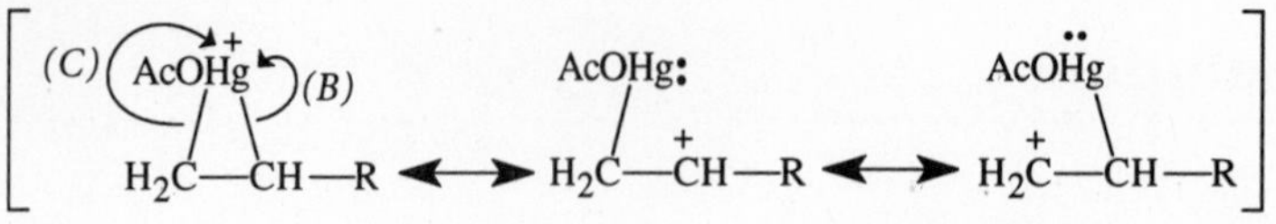

(b) Synthesis of the ether in Eq. 11.6a would require a secondary alkyl halide and an alkoxide with α-branching. A significant, if not predominant, amount of alkene by-product would be obtained by the E2 mechanism.

11.4 (a) In this case the two carbons of the double bond (and hence, the corresponding carbons of the mercurinium ion) have the same number of alkyl branches. Consequently, there is no basis for significant regioselectivity, and a mixture of 2-methoxypentane and 3-methoxypentane would be obtained:

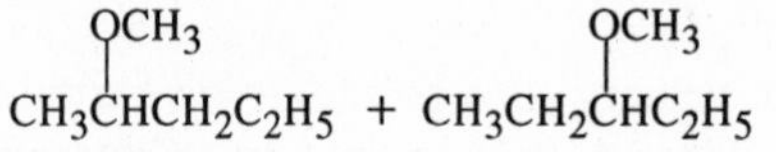

$$\underset{\text{2-methoxypentane}}{\underset{OCH_3}{CH_3\overset{|}{C}HCH_2C_2H_5}} \; + \; \underset{\text{3-methoxypentane}}{\underset{OCH_3}{CH_3CH_2\overset{|}{C}HC_2H_5}}$$

11.5 (a) Cyclohexene is subjected to oxymercuration in cyclohexanol, and the product is then reduced with NaBH₄:

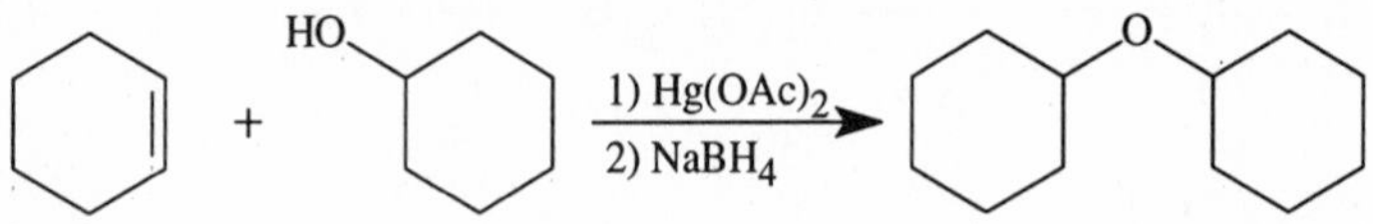

11.6 In each case, the carbocation derived from the tertiary alcohol reacts with the methyl or primary alcohol present in excess.

(a)

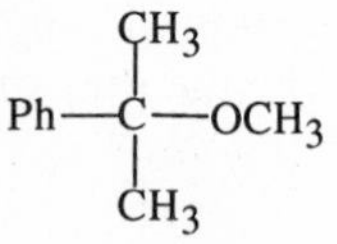

11.7 (a)

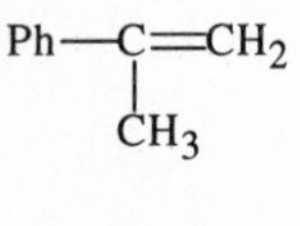

11.8 (a) Because this ether is symmetrical, and because both alkyl groups are primary, alcohol dehydration is the appropriate method. This ether can be prepared by treating $ClCH_2CH_2OH$ (2-chloroethanol) with H_2SO_4 at high temperature.

(c) Because this ether has a tertiary alkyl group, it can be derived by acid-catalyzed addition of an alcohol to an alkene:

$$\underset{\substack{\text{2-methylpropene}}}{(CH_3)_2C{=}CH_2} \; + \; \underset{\substack{\text{2-propanol}\\\text{(solvent)}}}{HOCH(CH_3)_2} \xrightarrow{H_2SO_4} \underset{\substack{\textit{tert}\text{-butyl isopropyl ether}}}{(CH_3)_3C{-}O{-}CH(CH_3)_2}$$

2-Methylpropene and 2-propanol are the reagents of choice because protonation of 2-methylpropene gives a tertiary carbocation. The use of propene and *tert*-butyl alcohol, in contrast, would require the formation of a secondary carbocation by alkene protonation in the presence of an alcohol that could itself form a tertiary carbocation.

11.9   (a)   (c)

11.10   (a)

      =CH₂ →(MMPP)→

11.11   The key to solving this problem is to realize that, in order for epoxide formation to occur, the oxygen of the conjugate-base alkoxide must be *anti* to the bromine in the transition state. Draw each of the stereoisomeric alkoxides in such a conformation and evaluate the relative energies of the resulting transition states.

      In the transition state for the reaction of the alkoxide conjugate base of *A*, the methyl groups are *anti*:

conjugate-base alkoxide
of compound *A*

transition-state conformation for reaction of *A*: methyl groups are *anti*.

In contrast, the methyl groups are *gauche* in the transition state for the reaction of the alkoxide derived from *B*:

conjugate-base alkoxide
of compound *B*

transition-state conformation for reaction of *B*: methyl groups are *gauche*.

The unfavorable *gauche* interaction between the two methyl groups makes the transition state derived from *B* more energetic and therefore more difficult to achieve. Consequently, the reaction of *A* is faster.

11.13   (a) In the protonated ether (see Eq. 11.23a, text p. 507, for a typical protonation reaction) the iodide ion can attack either the carbon of the methyl group or the α-carbon of the butyl group. Because this is an S_N2 reaction, and because S_N2 reactions of methyl compounds are faster than those of butyl compounds (why?), attack occurs at the methyl carbon to give methyl iodide and 1-butanol.

(b) The discussion at the bottom of text p. 507 indicates that the cleavage reactions of tertiary ethers, which are S_N1 reactions, occur under milder conditions than the S_N2 cleavage reactions of methyl and primary ethers. Hence, cleavage of the protonated ether occurs so as to give the *tert*-butyl cation and methanol. (See Eq. 11.24b in the text for the mechanism of a similar reaction.) Attack of iodide on the *tert*-butyl cation gives *tert*-butyl iodide.

(c) As indicated in part (b), acidic cleavage of tertiary ethers occurs by the S_N1 mechanism. As in part (b), the protonated ether reacts to give the *tert*-butyl cation and methanol. Loss of a β-proton from the cation gives 2-methylpropene; although this can protonate to regenerate the *tert*-butyl cation, the conditions of the reaction (distillation of low-boiling compounds) drives the volatile alkene from the reaction mixture as it is formed.

(d) Tertiary ethers and sulfides cleave by an S_N1 mechanism. The rate-limiting step in S_N1 reactions is formation of the carbocation by dissociation. The rate of this dissociation reaction is proportional to the concentration of the reacting species—in this case, the conjugate acid of the ether or sulfide. Because sulfides are much less basic than ethers (Sec. 8.6, text p. 366), the concentration of the reacting species is much smaller in the case of sulfides. Consequently, sulfides are less reactive than ethers toward acid-promoted cleavage.

11.14 (a) The cleavage occurs by the S_N1 mechanism at the secondary alkyl group. The products are therefore isopropyl iodide and ethanol.

$$(CH_3)_2CH-\!\!\xi\!\!-O-CH_2CH_3 \xrightarrow{HI,\ H_2O} (CH_3)_2CH-I\ +\ HO-CH_2CH_3$$

2-ethoxypropane   2-iodopropane   ethanol

(c) The cleavage occurs at the secondary alkyl group by the S_N1 mechanism to give the tertiary iodide 2-iodo-2,3-dimethylbutane and ethanol.

$$\underset{\underset{CH_3}{|}}{\overset{\overset{OCH_2CH_3}{\wr}}{CH_3CCH(CH_3)_2}} \xrightarrow{HI,\ H_2O} \underset{\underset{CH_3}{|}}{\overset{\overset{I}{|}}{CH_3CCH(CH_3)_2}}\ +\ HO-CH_2CH_3$$

2-ethoxy-2,3-dimethylbutane   2-iodo-2,3-dimethylbutane   ethanol

11.15 (a) Ammonia attacks the epoxide at the carbon with fewer branches:

$$\underset{(CH_3)_2CH}{\overset{H_3C\ \ \ O}{\underset{}{C-CH_2}}} +\ NH_3 \longrightarrow \underset{(CH_3)_2CH}{\overset{O^-}{\underset{}{H_3C-C-CH_2-\overset{+}{N}H_3}}} \underset{\xleftarrow{\ \ \ \ \ \ \ }}{\overset{proton}{\overset{transfers}{\xrightarrow{\ \ \ \ \ \ \ }}}} \underset{(CH_3)_2CH}{\overset{OH}{\underset{}{H_3C-C-CH_2-NH_2}}}$$

11.16 The strategy in this problem is to let the —OH group originate from the epoxide oxygen, which, in the starting material, must be attached to the same carbon as the —OH group in the product as well as to the adjacent carbon; the nucleophile is one of the groups attached to the adjacent carbon. To summarize:

$$R-\overset{\overset{OH}{|}}{CH}-\overset{|}{\underset{|}{C}}-X\ \Rightarrow\ R-CH-\overset{O}{C} \diagdown\ +\ X\!:^-$$

(a) The strategy outlined above suggests that the desired alcohol could be obtained by the reaction of cyanide ion with 2-pentyloxirane:

$$\overset{O}{CH_3(CH_2)_4CH-CH_2}\ +\ ^-CN \longrightarrow CH_3(CH_2)_4\overset{\overset{O^-}{|}}{CH}CH_2CN \xrightarrow{H_3O^+} CH_3(CH_2)_4\overset{\overset{OH}{|}}{CH}CH_2CN$$

2-pentyloxirane

(solution continues)

The strategy outlined above suggests another possibility:

$$CH_3(CH_2)_4\overset{O}{\overset{\triangle}{CH}}\!\!-\!\!CHCN \quad + \quad H\!:^-$$

hydride ion

If you came up with this idea, you are reasoning correctly. However, this strategy will not work in this case because the most common source of nucleophilic hydride, $LiAlH_4$ (lithium aluminum hydride), also reacts with the —CN group.

11.17 (a) Attack of methanol on the protonated epoxide occurs with inversion to give $(3S,4R)$-4-methoxy-3-hexanol:

(3S,4R)-4-methoxy-3-hexanol

Attack at the other carbon of the epoxide is equally likely and gives the same stereoisomer of the same compound. (Be sure to verify this point.)

11.18 (a) Under acidic conditions, attack of water occurs at the more branched carbon of the epoxide ring; consequently, the —OH group at that carbon is enriched in ^{18}O. In basic solution, attack of hydroxide occurs at the less branched carbon (the CH_2 carbon) of the epoxide, and the —OH group at that carbon is enriched in ^{18}O. ($^*O = {}^{18}O$.)

attack of ¯OH on the epoxide occurs here

attack of H_2O on the protonated epoxide occurs here

product of acid-catalyzed hydrolysis

product of base-catalyzed hydrolysis

(b) Cleavage of the glycol with periodic acid would yield two compounds, acetone and formaldehyde. In principle, the acetone should contain the ^{18}O in acid-catalyzed hydrolysis, and the formaldehyde should be devoid of the isotope; the formaldehyde should contain the ^{18}O in base-catalyzed hydrolysis, and the acetone should be devoid of the isotope.

product of acid-catalyzed hydrolysis

product of base-catalyzed hydrolysis

11.19 (a) Subtract the group added to the Grignard reagent, namely, the —CH_2CH_2OH group, from the product, and add back the —MgBr to what remains. (See color coding in Eq. 11.39, text p. 515.)

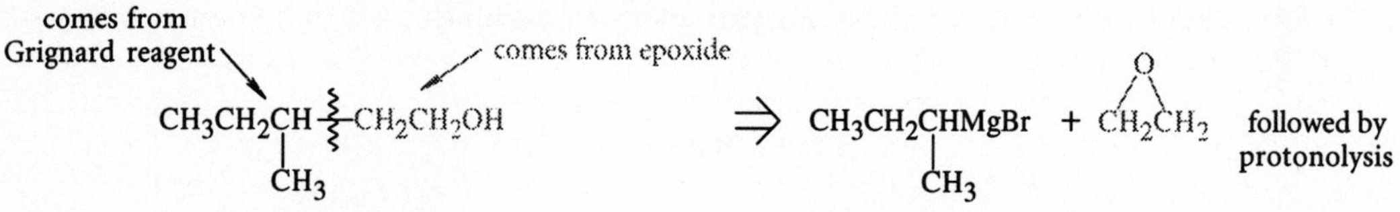

comes from
Grignard reagent ⟶   ⟵ comes from epoxide

$$CH_3CH_2CH\text{—}CH_2CH_2OH \Longrightarrow CH_3CH_2CHMgBr + CH_2CH_2 \quad \text{followed by}$$
$$\underset{CH_3}{|} \qquad\qquad\qquad\qquad \underset{CH_3}{|} \qquad \overset{O}{\diagdown} \quad \text{protonolysis}$$

3-methyl-1-pentanol   *sec*-**butylmagnesium bromide**

11.21 Because iodide ion is a good nucleophile and trimethyloxonium fluoroborate is an excellent alkylating agent, alkylation of the iodide ion to give methyl iodide and dimethyl ether occurs.

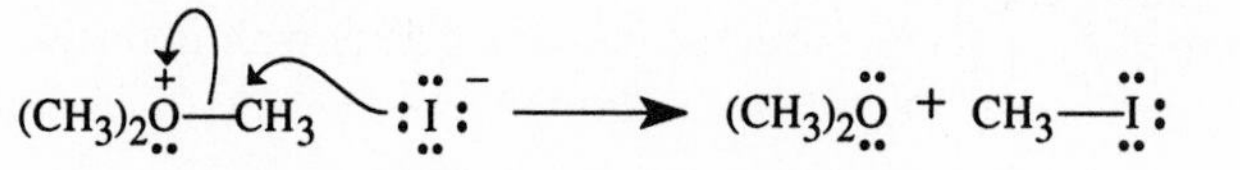

$$(CH_3)_2\overset{+}{\underset{\cdot\cdot}{O}}\text{—}CH_3 \quad :\overset{\cdot\cdot}{\underset{\cdot\cdot}{I}}:^- \longrightarrow (CH_3)_2\overset{\cdot\cdot}{\underset{\cdot\cdot}{O}} + CH_3\text{—}\overset{\cdot\cdot}{\underset{\cdot\cdot}{I}}:$$

11.22 (a) In each case, alkylation of the Lewis base by trimethyloxonium fluoroborate occurs.

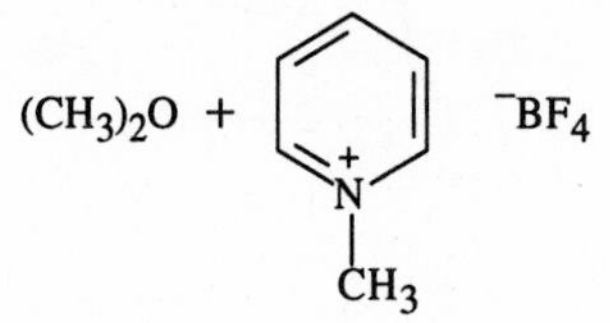

$$(CH_3)_2O + \underset{\underset{CH_3}{|}}{\overset{+}{N}} \qquad ^-BF_4$$

11.23 (a) If neighboring-group participation does not occur, only the alcohol C_2H_5S—$CH_2\overset{*}{C}H_2$—OH (compound *C* in the diagram below) will form (by direct attack of water on the carbon α to the chlorine).

(b) If neighboring-group participation occurs, the cyclic intermediate *A* will be formed. Because this intermediate is symmetrical (except for the isotope), it can be attacked at either carbon to give a mixture consisting of equal amounts of the two labeled alcohols *B* and *C*.

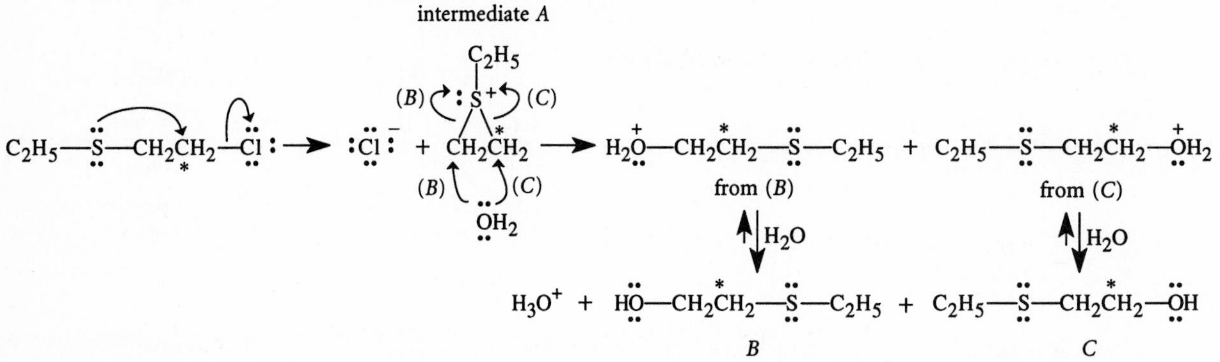

intermediate *A*

$$C_2H_5\text{—}\overset{\cdot\cdot}{\underset{\cdot\cdot}{S}}\text{—}CH_2CH_2\text{—}\overset{\cdot\cdot}{\underset{\cdot\cdot}{Cl}}: \longrightarrow :\overset{\cdot\cdot}{\underset{\cdot\cdot}{Cl}}:^- + CH_2CH_2 \longrightarrow H_2\overset{+}{\underset{\cdot\cdot}{O}}\text{—}CH_2CH_2\text{—}\overset{\cdot\cdot}{\underset{\cdot\cdot}{S}}\text{—}C_2H_5 + C_2H_5\text{—}\overset{\cdot\cdot}{\underset{\cdot\cdot}{S}}\text{—}CH_2CH_2\text{—}\overset{+}{\underset{\cdot\cdot}{O}}H_2$$

from (*B*)   from (*C*)

$$\updownarrow H_2O \qquad\qquad \updownarrow H_2O$$

$$H_3O^+ + \overset{\cdot\cdot}{H\underset{\cdot\cdot}{O}}\text{—}CH_2CH_2\text{—}\overset{\cdot\cdot}{\underset{\cdot\cdot}{S}}\text{—}C_2H_5 + C_2H_5\text{—}\overset{\cdot\cdot}{\underset{\cdot\cdot}{S}}\text{—}CH_2CH_2\text{—}\overset{\cdot\cdot}{\underset{\cdot\cdot}{O}}H$$

               B   *C*

11.25 (a) The intramolecular product tetrahydrofuran results from an internal nucleophilic substitution reaction of the alkoxide on the alkyl halide. The intermolecular product 1,4-butanediol results from the S_N2 reaction of hydroxide ion with the alkyl halide.

HO—$CH_2CH_2CH_2CH_2$—OH

1,4-butanediol

tetrahydrofuran

(b)

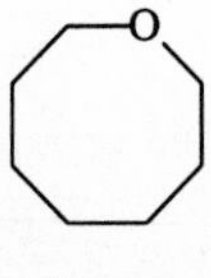

HO—$CH_2CH_2CH_2CH_2CH_2CH_2CH_2$—OH

intermolecular product

intramolecular product

11.26 The bromo alcohol in part (b) of the previous problem should give the least amount of intramolecular substitution product and therefore the greatest amount of intermolecular substitution product because neighboring-group participation (which gives rise to the cyclic ether) is most likely to occur when it results in the formation of three-, five-, or six-membered rings. Because the ether in part (b) has an eight-membered ring, its formation is likely to be slower under ordinary conditions than attack of hydroxide ion.

11.27 (a) Lengthening a carbon chain by two carbons is required; what better way than the reaction of a Grignard reagent with ethylene oxide?

$(CH_3)_2C=CH_2 \xrightarrow[\text{2) } H_2O_2,\ {}^-OH]{\text{1) } BH_3/THF} (CH_3)_2CHCH_2OH \xrightarrow[\text{H}_2SO_4]{\text{conc. HBr,}} (CH_3)_2CHCH_2Br \xrightarrow{\text{Mg, ether}}$

$(CH_3)_2CHCH_2MgBr \xrightarrow[\text{2) } H_3O^+]{\text{1) } \triangle\!\!\!O} (CH_3)_2CHCH_2CH_2CH_2OH \xrightarrow[\text{2) } H_3O^+]{\text{1) } KMnO_4,\ {}^-OH} (CH_3)_2CHCH_2CH_2CO_2H$

(c) The sulfone must be obtained by oxidation of dibutyl sulfide; and the sulfide is obtained from the S_N2 reaction of the thiolate conjugate base of 1-butanethiol with 1-bromobutane.

$CH_3CH_2CH_2CH_2SH \xrightarrow[C_2H_5OH]{Na^+\ C_2H_5O^-} \xrightarrow{CH_3CH_2CH_2CH_2Br} CH_3CH_2CH_2CH_2SCH_2CH_2CH_2CH_3 \xrightarrow{\text{2 equiv. } H_2O_2}$

$$CH_3CH_2CH_2CH_2\overset{\overset{\displaystyle O}{\|}}{\underset{\underset{\displaystyle O}{\|}}{S}}CH_2CH_2CH_2CH_3$$

dibutylsulfone

Solutions to Additional Problems

11.28 (a) Two of the many nine-carbon ethers that cannot be prepared by the Williamson ether synthesis:

2-methyl-2-*tert*-butoxybutane   *tert*-butyl neopentyl ether

In neither case can any of the alkyl groups be introduced by the S_N2 reaction of an alkyl halide with an alkoxide.

(c)   (e)

(g) Since the alkene has an unsaturation number of 3 and forms diepoxides, it must also have a ring.

Notice that 1,3-cyclohexadiene is not a correct answer because it gives two stereoisomeric mono-epoxides and three stereoisomeric di-epoxides.

(g) Since the alkene has an unsaturation number of 3 and forms diepoxides, it must also have a ring.

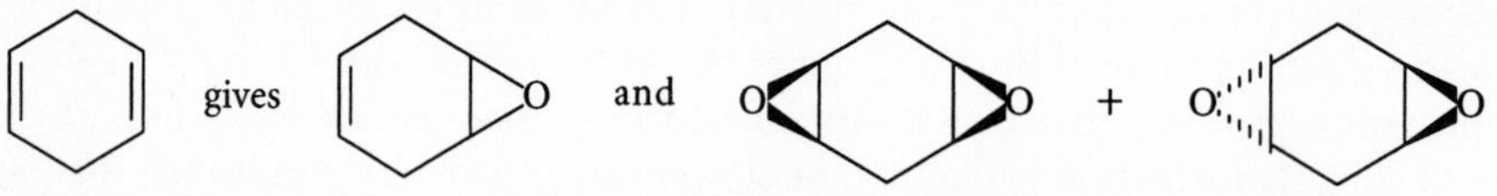

Notice that 1,3-cyclohexadiene is not a correct answer because it gives two stereoisomeric mono-epoxides and three stereoisomeric di-epoxides.

11.29 (a)

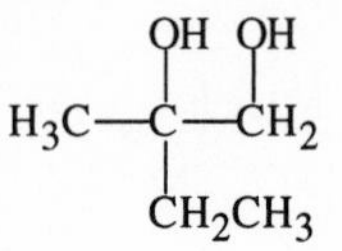

(c)

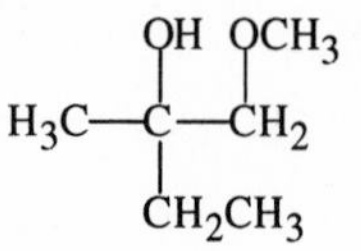

(e)

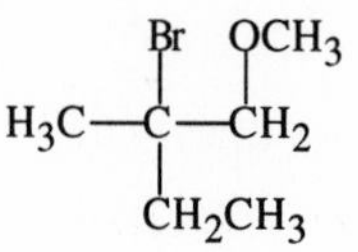

(g)

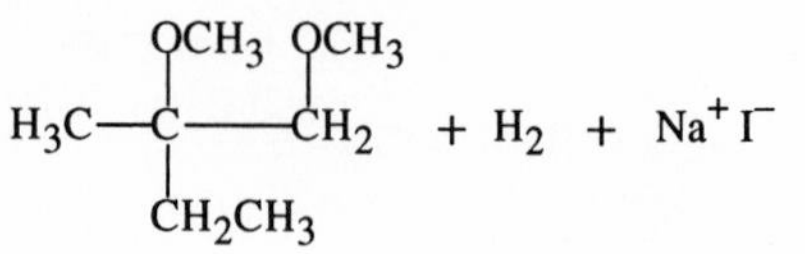

(i)

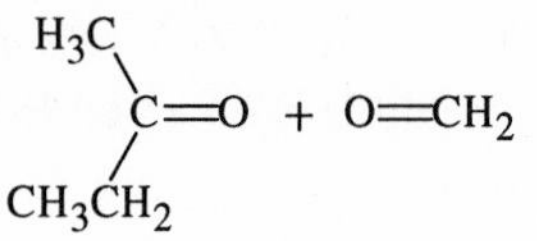

(j)

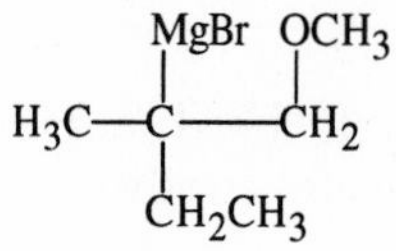

(k)

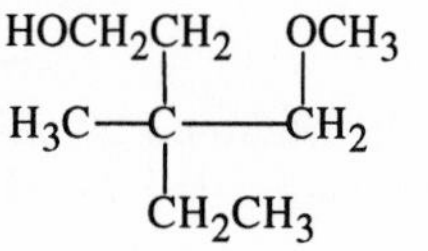

11.30 Because the C—S bond is weaker than the C—O and C—C bonds, and because a thiolate ion is less basic than an alkoxide ion or carbanion, the thiirane (the sulfur compound) reacts most rapidly.

11.32 (a) 3-Ethoxypropene undergoes bromine addition, and therefore decolorizes a solution of Br_2 in an inert solvent; 1-ethoxypropane does not.

(c) 1-Methoxy-2-chloro-2-methylpropane, a tertiary alkyl halide, undergoes the S_N1 reaction on gentle heating in a protic solvent such as aqueous acetone. This results in an acidic solution of aqueous HCl, the formation of which can be detected with litmus paper. Alternatively, the chloride ion produced can be precipitated with silver ion.

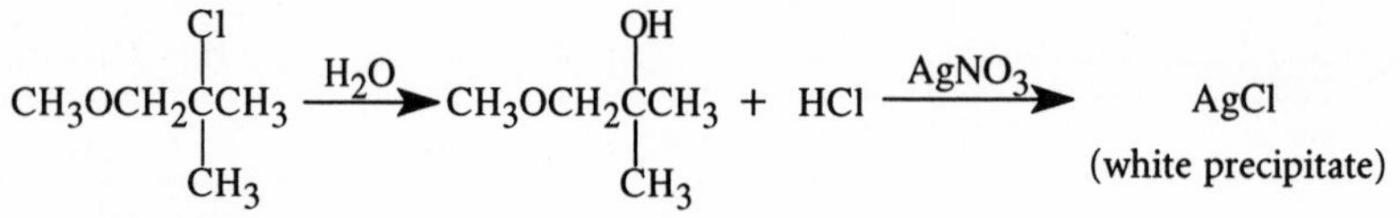

1-methoxy-2-chloro-2-methylpropane + E1 products

11.33 Epoxides add HCl to give neutral chloro alcohols. This reaction is so fast and so favorable that HCl can be quantitatively (that is, completely) removed by adding 2-methyloxirane to a reaction mixture.

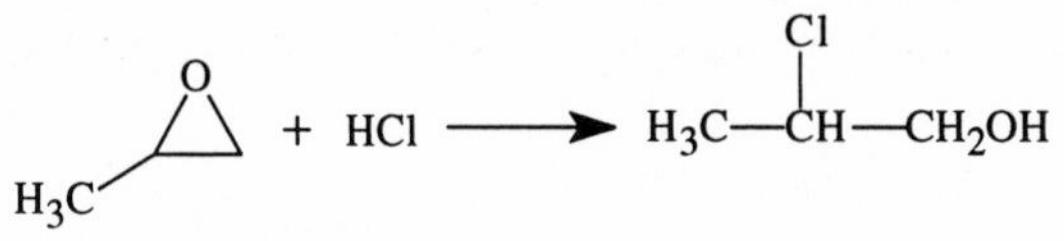

2-methyloxirane

11.34 $(CH_3)_3CCH_2MgBr$ + 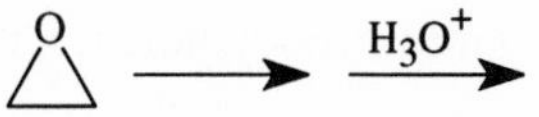 $(CH_3)_3CCH_2CH_2CH_2OH$

4,4-dimethyl-1-pentanol

11.36 (a) Sodium ethoxide reacts with water to give ethanol and sodium hydroxide. Although the pK_a values of water and ethanol are similar, water is present in excess because it is the solvent, and the equilibrium therefore favors sodium hydroxide. Consequently, the alcohol $(CH_3)_2CHCH_2CH_2OH$ rather than the ether will be formed as the major substitution product.

$$Na^{+}\ {}^{-}OC_2H_5 + H_2O \rightleftarrows HOC_2H_5 + Na^{+}\ {}^{-}OH$$

(solvent) — reacts with the alkyl halide

11.37 Because the hydroxide ion reacts with an epoxide at the less branched carbon (the CH_2 group), the configuration at the asymmetric carbon is unaffected. Consequently, (+)-2-methyloxirane has the *R* configuration.

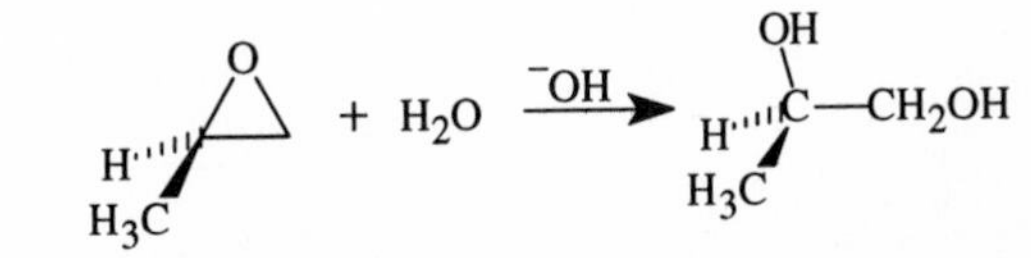

(*R*)-(+)-2-methyloxirane (*R*)-(−)-1,2-propanediol

11.39 The formula of the product indicates one degree of unsaturation. Because the double bond undergoes addition with mercuric acetate, this unsaturation is accounted for by a ring. As the hint and the molecular formula of the product suggest, this is an alkoxymercuration reaction in which the —OH group from within the same molecule serves as the nucleophile:

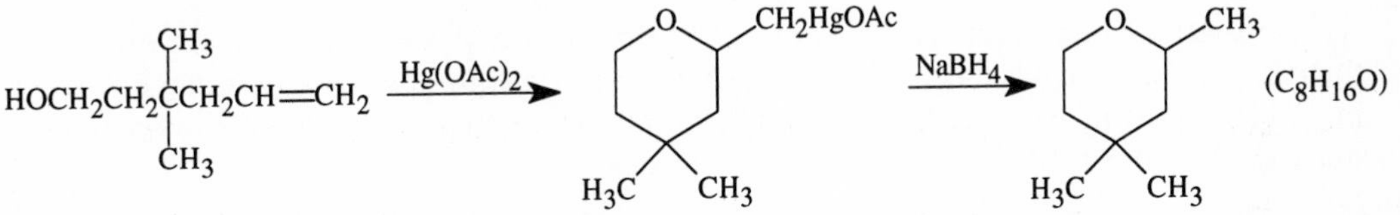

11.41 In both parts the mCPBA reacts at the side of the alkene π bond that involves the less severe van der Waals repulsions.

(a)

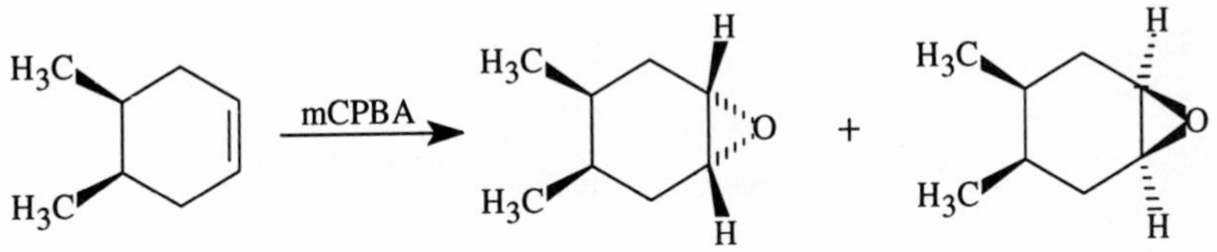

A (major product) *B*

11.42 Synthesis (1) will not accomplish the desired objective, because the —OH group of the alcohol would be lost as water. (See the mechanism in Eqs. 11.10–11.11, text p. 501.) In contrast, synthesis (2) would result in complete incorporation of the isotopic oxygen into the ether. Because this is a conventional Williamson synthesis, the carbon-oxygen bond to the ring is not broken. (See Eq. 11.3, text p. 498.)

11.43 (a) $CH_3CH_2CH_2OC_2H_5$

(b)

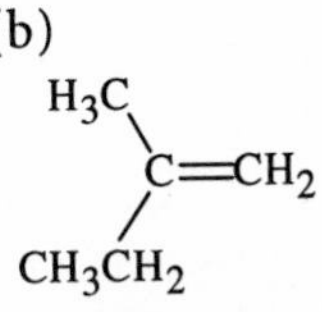

+ $(CH_3)_3COH$

(c)

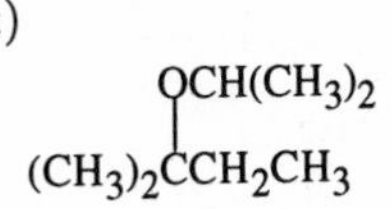

(d)

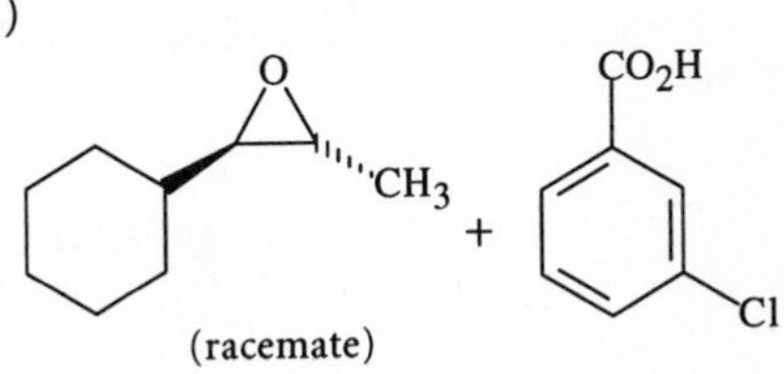

(racemate)

(e) The bromohydrin formed in the first step is oxidized to an α-bromo ketone.

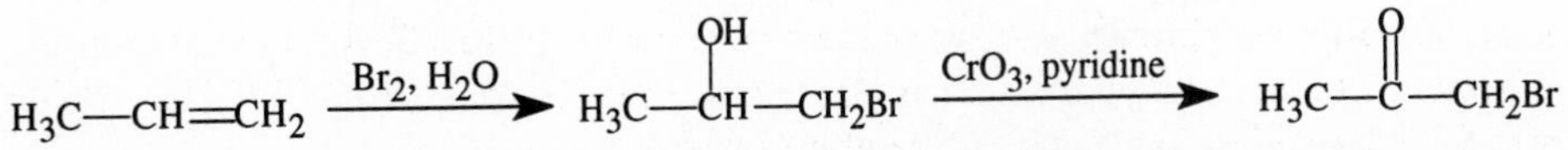

(f)   The epoxide is opened to a glycol by the acidic aqueous reaction conditions, and the glycol is cleaved as it is formed by the periodic acid.

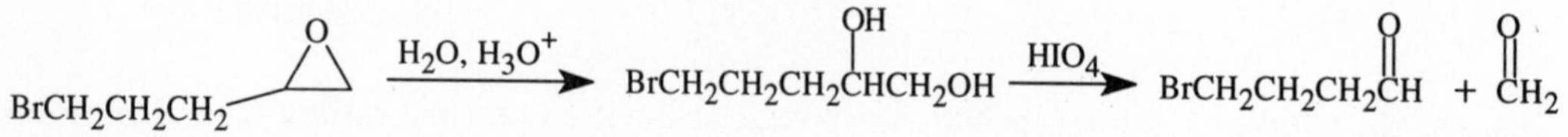

$$BrCH_2CH_2CH_2-\overset{O}{\triangle} \xrightarrow{H_2O, H_3O^+} BrCH_2CH_2CH_2\overset{\overset{OH}{|}}{C}HCH_2OH \xrightarrow{HIO_4} BrCH_2CH_2CH_2\overset{O}{\overset{\|}{C}}H + \overset{O}{\overset{\|}{C}}H_2$$

(g)

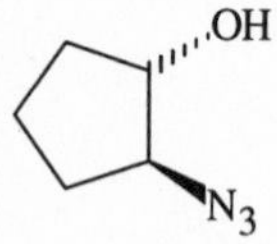

(racemate)

(h)   Sulfide ion displaces one chloride to give a thiolate ion, which then reacts in an internal nucleophilic substitution to give the cyclic sulfide tetrahydrothiophene.

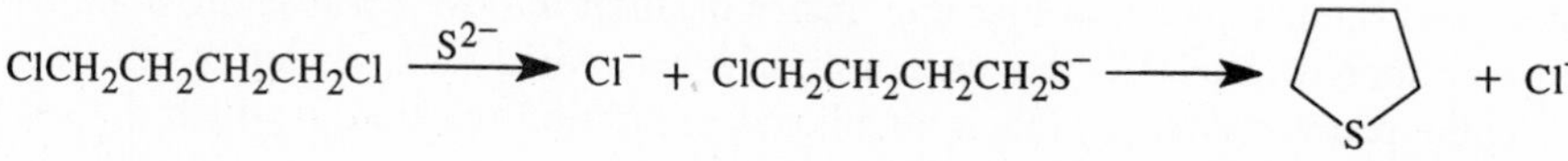

$$ClCH_2CH_2CH_2CH_2Cl \xrightarrow{S^{2-}} Cl^- + ClCH_2CH_2CH_2CH_2S^- \longrightarrow$$

$+ Cl^-$

tetrahydrothiophene

(i)   A tosylate of one —OH group is formed. (Introduction of a second tosylate is much slower because of the van der Waals repulsions between the two large tosylate groups.) The resulting hydroxy tosylate reacts like a bromohydrin. Thus, in base, the —OH group ionizes to an alkoxide, which undergoes an internal nucleophilic substitution reaction to give an epoxide.

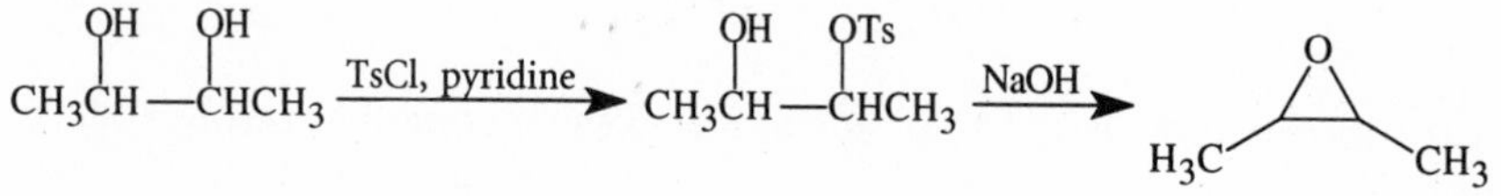

$$CH_3\overset{\overset{OH}{|}}{C}H-\overset{\overset{OH}{|}}{C}HCH_3 \xrightarrow{TsCl, pyridine} CH_3\overset{\overset{OH}{|}}{C}H-\overset{\overset{OTs}{|}}{C}HCH_3 \xrightarrow{NaOH} \underset{H_3C}{\overset{O}{\triangle}}CH_3$$

11.44 (a)

$$(CH_3)_2CHCH=CH_2 \xrightarrow[\text{2) NaBH}_4]{\text{1) Hg(OAc)}_2/C_2H_5OH} (CH_3)_2CH\overset{\overset{OC_2H_5}{|}}{C}HCH_3$$

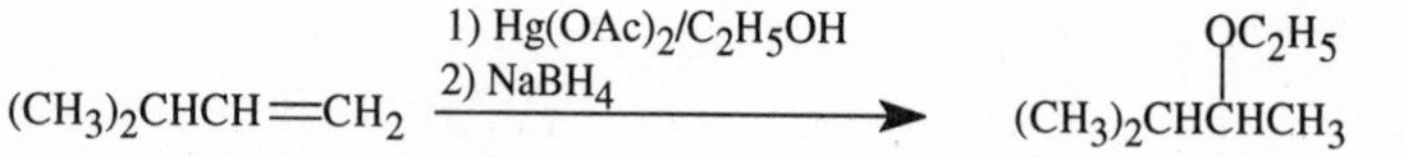

3-methyl-1-butene

2-ethoxy-3-methylbutane

(b)

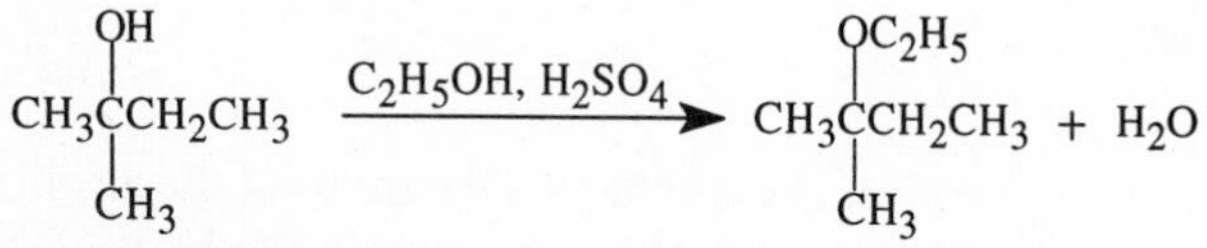

$$CH_3\overset{\overset{OH}{|}}{\underset{\overset{|}{CH_3}}{C}}CH_2CH_3 \xrightarrow{C_2H_5OH, H_2SO_4} CH_3\overset{\overset{OC_2H_5}{|}}{\underset{\overset{|}{CH_3}}{C}}CH_2CH_3 + H_2O$$

2-methyl-2-butanol

2-ethoxy-2-methylbutane

(c)

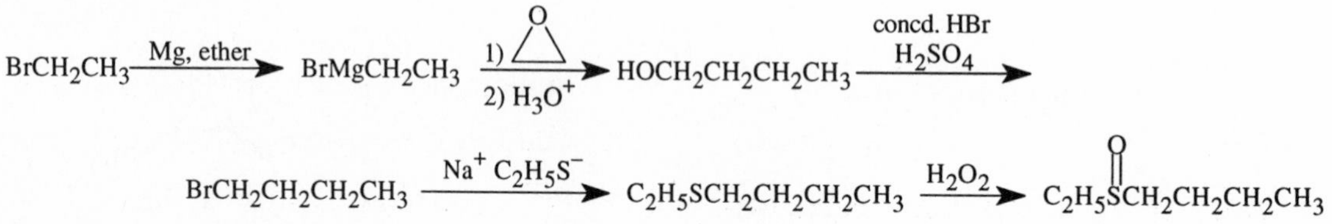

$$BrCH_2CH_3 \xrightarrow{Mg, ether} BrMgCH_2CH_3 \xrightarrow[\text{2) H}_3O^+]{\text{1) }\triangle} HOCH_2CH_2CH_2CH_3 \xrightarrow[H_2SO_4]{\text{concd. HBr}}$$

$$BrCH_2CH_2CH_2CH_3 \xrightarrow{Na^+ C_2H_5S^-} C_2H_5SCH_2CH_2CH_2CH_3 \xrightarrow{H_2O_2} C_2H_5\overset{O}{\overset{\|}{S}}CH_2CH_2CH_2CH_3$$

(d)

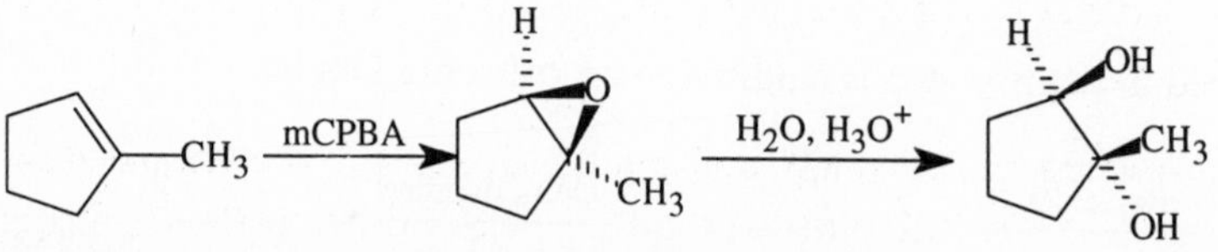

(e)

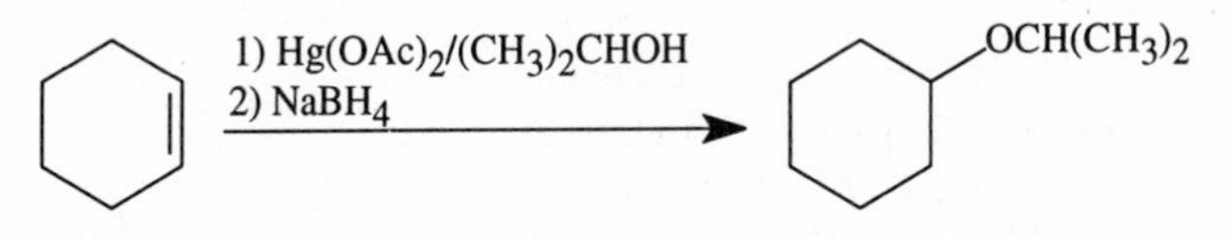

cyclohexene   cyclohexyl isopropyl ether

(f)

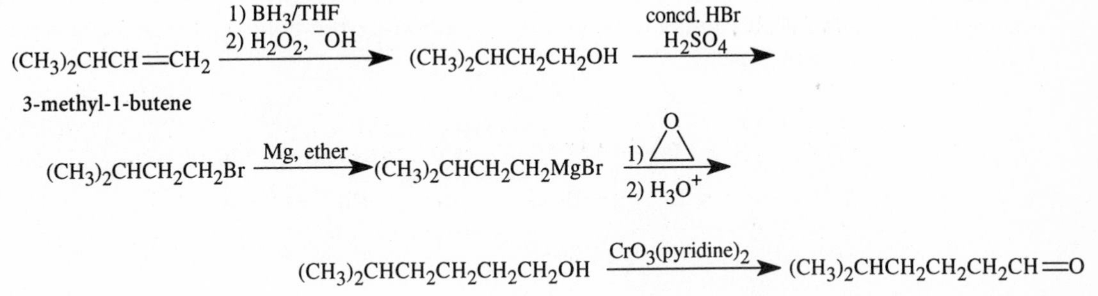

$(CH_3)_2CHCH=CH_2$ $\xrightarrow[\text{2) } H_2O_2, \text{ }^-OH]{\text{1) } BH_3/THF}$ $(CH_3)_2CHCH_2CH_2OH$ $\xrightarrow[H_2SO_4]{\text{concd. HBr}}$

3-methyl-1-butene

$(CH_3)_2CHCH_2CH_2Br$ $\xrightarrow{\text{Mg, ether}}$ $(CH_3)_2CHCH_2CH_2MgBr$ $\xrightarrow[\text{2) } H_3O^+]{\text{1) } \triangle}$

$(CH_3)_2CHCH_2CH_2CH_2CH_2OH$ $\xrightarrow{CrO_3(\text{pyridine})_2}$ $(CH_3)_2CHCH_2CH_2CH_2CH=O$

(g)

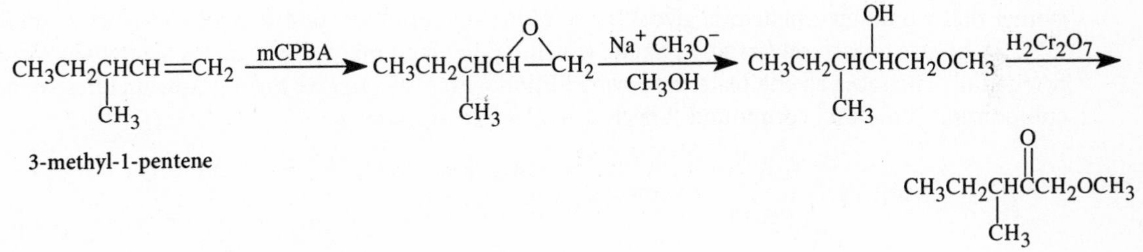

$CH_3CH_2\underset{\underset{CH_3}{|}}{C}HCH=CH_2$ $\xrightarrow{mCPBA}$ $CH_3CH_2\underset{\underset{CH_3}{|}}{C}H\overset{O}{\overset{\triangle}{CH}}-CH_2$ $\xrightarrow[CH_3OH]{Na^+ CH_3O^-}$ $CH_3CH_2\underset{\underset{CH_3}{|}}{C}H\overset{\overset{OH}{|}}{C}HCH_2OCH_3$ $\xrightarrow{H_2Cr_2O_7}$

3-methyl-1-pentene

$CH_3CH_2\underset{\underset{CH_3}{|}}{C}H\overset{O}{\overset{||}{C}}CH_2OCH_3$

(h)

$(CH_3)_2C=CH_2$ $\xrightarrow{mCPBA}$ $(CH_3)_2\overset{O}{\overset{\triangle}{C}}-CH_2$ $\xrightarrow[H_2SO_4]{CH_3CH_2OH,}$ $(CH_3)_2\underset{\underset{OCH_2CH_3}{|}}{C}-CH_2OH$ $\xrightarrow{CrO_3(\text{pyridine})_2}$

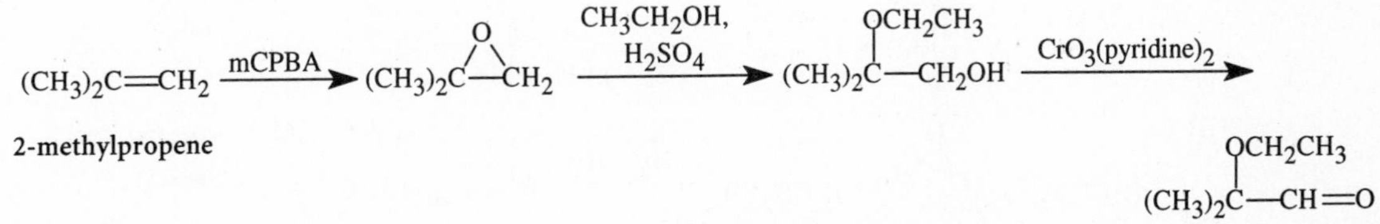

2-methylpropene

$(CH_3)_2\underset{\underset{OCH_2CH_3}{|}}{C}-CH=O$

11.45 (a) The desired compound can be prepared by a ring-opening reaction of $(2R,3R)$-2,3-dimethyloxirane with acidic methanol, because epoxide ring-opening occurs with inversion of configuration:

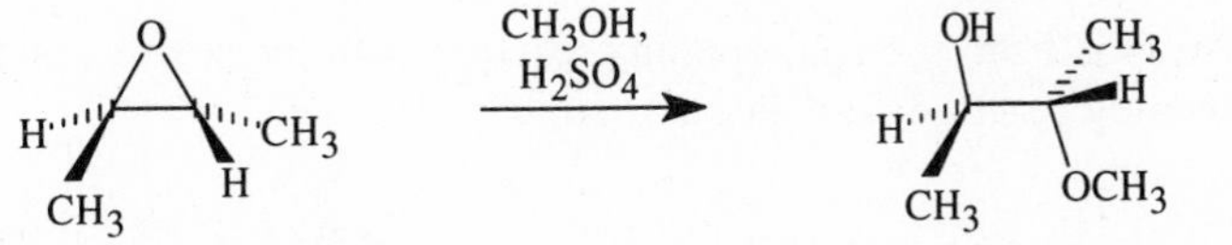

$(2R,3R)$-2,3-dimethyloxirane $(2R,3S)$-3-methoxy-2-butanol

(Be sure to convince yourself that attack of methanol at either carbon of the epoxide ring gives the same stereoisomer of the product.)

(c) Analyze the configurations of the desired compound and the alcohol prepared in part (a):

(solution continues)

$$
\begin{array}{ccc}
\text{(2R,3S)-3-methoxy-2-butanol} & \xrightarrow{\text{identical configurations}} & \text{target compound}
\end{array}
$$

This analysis shows that if the —OH group can be converted into an ethyl ether without changing the configuration of either carbon, the problem will be solved. The Williamson synthesis is a good way to accomplish this objective, because it does not break a carbon-oxygen bond. (The reasoning is similar to that used in solving Problem 11.42.)

$$
\begin{array}{c}
\text{(Fischer projection)} \xrightarrow[\text{2) } C_2H_5I]{\text{1) NaH}} \text{(Fischer projection)}
\end{array}
$$

11.46 Compound *A* is an octene because it undergoes typical alkene reactions and it gives octane on catalytic hydrogenation. Compounds *C* and *D* are stereoisomeric glycols, and glycol *D* is achiral. The only octene isomer that would give an achiral glycol is one of the stereoisomers of 4-octene. *Cis*-4-octene reacts with mCPBA to give an *achiral (meso)* epoxide, which, upon ring opening in aqueous acid, gives *chiral* 4,5-octanediol. The same alkene reacts in a *syn* addition with OsO$_4$ to give *meso*-4,5-octanediol, an achiral compound. Therefore, compound *A* is *cis*-4-octene. To summarize:

cis-4-octene
(compound *A*)

$$\xrightarrow[\text{H}_2\text{O, NaHSO}_3]{\text{OsO}_4,\text{ then}}$$

meso-4,5-octanediol
(compound *D*)

$\downarrow$ mCPBA

meso-2,3-dipropyloxirane
(compound *B*)

$$\xrightarrow{\text{H}_2\text{O, H}_3\text{O}^+}$$

(±)-4,5-octanediol
(compound *D*)

(See the note below the solution to Problem 7.47 on page 157 of this manual.)

11.48 The anion that results from the ring-opening reaction with hydroxide reacts as a nucleophile in another epoxide ring-opening reaction; and so on.

$$
\text{CH}_3\text{CH}\!-\!\text{CH}_2 \quad {}^-\!:\!\ddot{\text{O}}\text{H} \longrightarrow \text{CH}_3\text{CHCH}_2\ddot{\text{O}}\text{H} \longrightarrow \text{CH}_3\dot{\text{C}}\text{HCH}_2\ddot{\text{O}}:
$$

this alkoxide can attack another epoxide

$$
\text{CH}_3\text{CHCH}_2\ddot{\text{O}}\text{H}
$$

11.49 Notice that the transformation of compound *A* to compound *B* occurs with *retention of configuration*. A retention of configuration can be the result of *two successive inversions of configuration*. Fact (3) suggests that neighboring-group participation is occurring. Indeed, a neighboring-group mechanism involves a *meso* and therefore *achiral* cyclic sulfonium-ion intermediate *D*, formed by an intramolecular substitution with inversion of configuration:

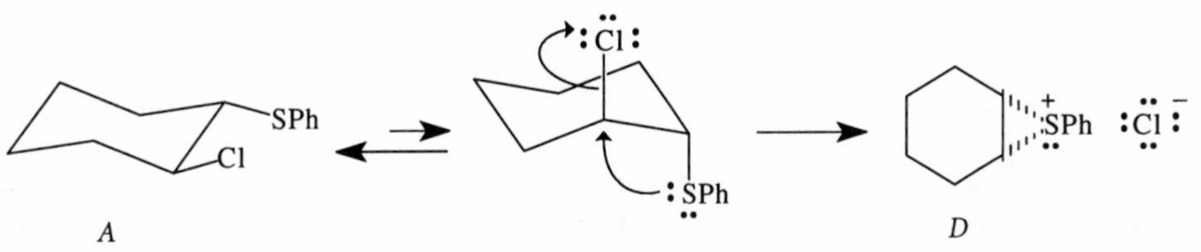

Because intermediate *D* is achiral, any chiral products that result from it must be formed as racemates (Sec. 7.8A, text p. 305); this accounts for fact (2). Reactions of intermediate *D* with ethanol at the two carbons of the three-membered ring give respectively the two enantiomers of product *B*. (The formation of one enantiomer is shown below; you should show the other.) This substitution reaction occurs with inversion of configuration; coupled with the first inversion that takes place in the formation of *D*, this accounts for the overall retention noted in fact (1).

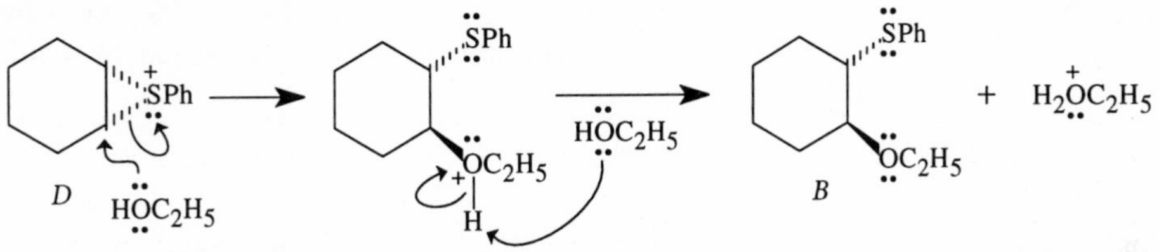

11.51 Reaction (1) is an S_N2 reaction in which oppositely charged species react. This reaction should be promoted by conditions that increase the interaction between charges, that is, by a smaller dielectric constant. Therefore, reaction (1) is faster in ethanol. Reaction (2) is an S_N1 reaction in which the rate-limiting step involves dissociation of an alkyl halide to a carbocation and a chloride ion; this reaction is promoted by conditions that foster separation of charged groups, that is, by a greater dielectric constant. Consequently, reaction (2) is faster in water. Reaction (3) is an S_N1 reaction, but, because the starting sulfonium ion is positively charged, formation of the carbocation intermediate does not involve any change in the number of charges. Consequently, this reaction is not affected very much by a change of solvent.

11.52 (a) In this reaction, the initially formed bromonium ion is attacked intramolecularly by the oxygen of the —OH group. (The mechanism for the formation of a bromonium ion is shown in Eq. 5.4a, text p. 176.)

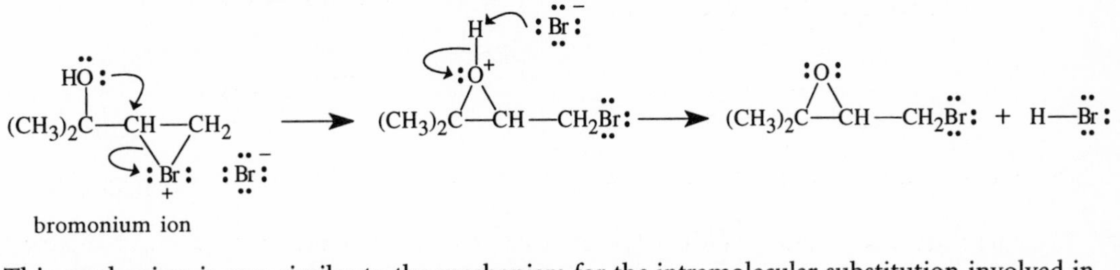

bromonium ion

(c) This mechanism is very similar to the mechanism for the intramolecular substitution involved in Problem 9.56(a); see p. 214 of this manual. The thiolate anion, formed by ionization of the thiol, displaces the tosylate group within a twist-boat conformation.

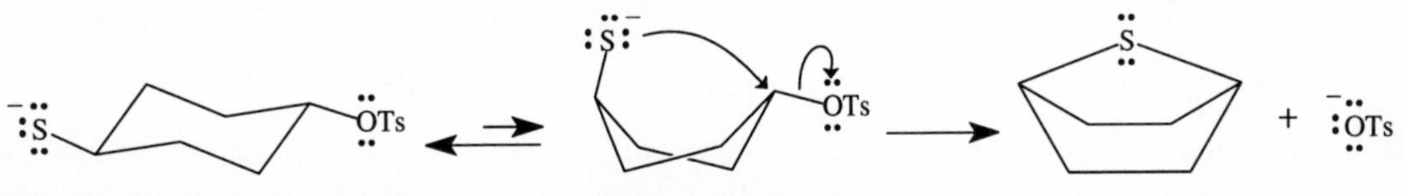

thiolate ion formed by ionization
of thiol starting material

twist-boat conformation
places thiolate ion in position
for backside attack

(e) The carbocation intermediate formed by protonation of the double bond is attacked intramolecularly by the sulfur nucleophile:

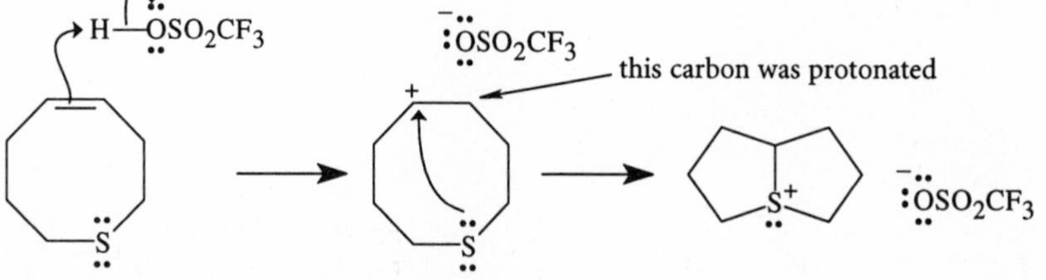

11.53 The principles to use in solving this problem are that (*a*) the *tert*-butyl group, because it is so large, must assume the equatorial position; and (*b*) the —OH and —Cl groups must be able to achieve an *anti* relationship in order for backside attack (and thus epoxide formation) to occur. The required *anti* relationship is impossible for compound (3) because the —OH and —Cl groups are *cis* and are thus fixed in either chair conformation at a dihedral angle of about 60°. Therefore, compound (3) is *A*. In the chair conformation in which the *tert*-butyl group is equatorial, the —OH and —Cl groups in compound (1) are *trans*-diaxial, and are ideally set up for the formation of an epoxide. Consequently, compound (1) is *B*, and the epoxide formed from it is *D*.

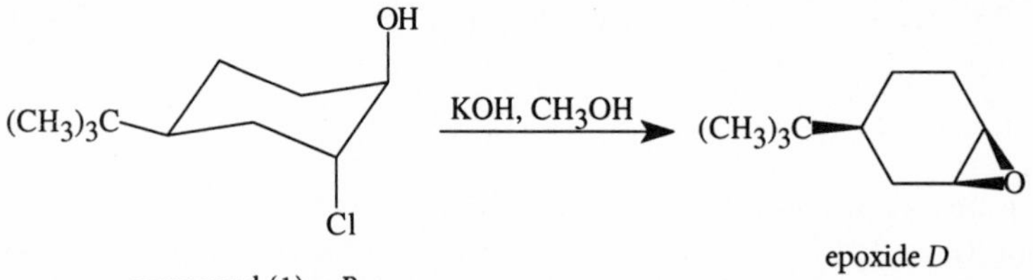

Finally, consider compound (2): the —OH and —Cl groups are *trans*, but they are diequatorial. The only way that they can achieve an *anti* relationship is either for the ring to undergo a chair flip, or for it to assume a twist-boat conformation. In either case, the conformation required for epoxide formation has a very high energy, and very little of it will be present. Because the rate of epoxide formation is proportional to the concentration of the reactive conformation, epoxide formation from compound (2), although possible, is very slow. Therefore, compound (2) = *C* and the epoxide formed from it is *E*. Note that epoxides *D* and *E* are stereoisomers, as required by the problem.

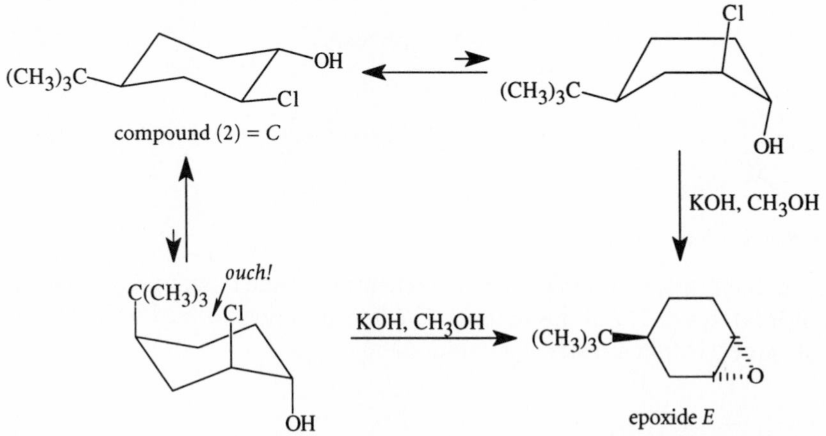

11.54 In both cases, the bromine assists the departure of the protonated —OH group by a backside-attack mechanism, forming a bromonium ion in the process with inversion of configuration. The bromonium ion is opened with a second inversion of configuration by bromide ion from the ionization of HBr.

(a) In this part, the bromonium ion is a *meso* compound; opening by bromide ion gives chiral 2,3-dibromobutane; but, because the bromonium ion is achiral, the 2,3-dibromobutane must be formed as the racemate. The formation of one enantiomer of (±)-2,3-dibromobutane is shown below; you should show formation of the other.

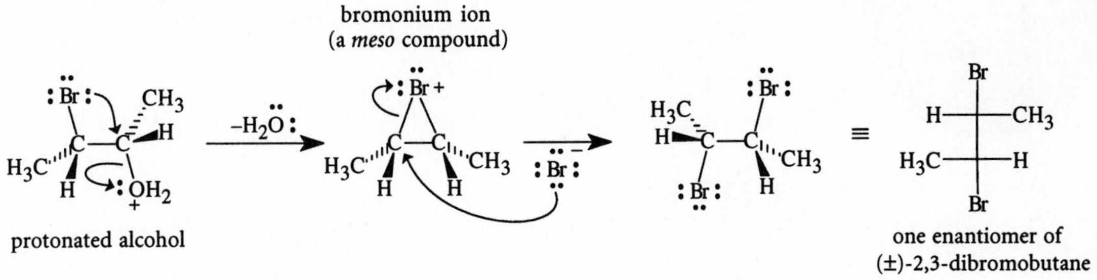

bromonium ion
(a *meso* compound)

protonated alcohol

one enantiomer of
(±)-2,3-dibromobutane

12

Infrared
Spectroscopy and
Mass Spectrometry

Terms

C Concepts

....

I. Introduction to Spectroscopy

A. ELECTROMAGNETIC RADIATION

1. The total range of electromagnetic radiation is called the electromagnetic spectrum.
   a. Visible light as well as X-rays, ultraviolet radiation, infrared radiation, microwaves, and radio waves are various forms of electromagnetic radiation.
   b. Electromagnetic radiation is characterized by its energy, wavelength, and frequency, which are interrelated.
2. Electromagnetic radiation has wave characteristics and can be characterized by its wavelength λ.
   a. The wavelength is the distance between successive peaks or successive troughs in the wave.
   b. The frequency v of a wave is the number of wavelengths that pass a point per unit time when the wave is propagated through space.
      i. The frequency of a wave is given by the equation

      $$v = \frac{c}{\lambda}$$

      in which c = the velocity of light (3×10^8 m/sec).
      ii. The frequency has the dimensions of $\sec^{-1}$ (cycles per second, cps) or hertz, Hz.
3. Light can also show particlelike behavior.
   a. The light particle is called a photon.
   b. The relationship between the energy of a photon and the wavelength or frequency of light is a fundamental law of physics:

   $$E = hv = \frac{hc}{\lambda}$$

   where h is Planck's constant, a universal constant (6.625×10^{-27} erg sec or 6.625×10^{-34} J sec).
   c. The energy and frequency increase as the wavelength decreases.

B. ABSORPTION SPECTROSCOPY

1. Spectroscopy is the study of the interactions of matter and light (or other electromagnetic radiation).
   a. Spectroscopy can be used to determine unknown molecular structures.
   b. The most common type of spectroscopy used for structure determination is absorption spectroscopy.
      i. Matter can absorb energy from certain wavelengths of electromagnetic radiation.
      ii. This absorption is determined as a function of wavelength, frequency, or energy in an instrument called a spectrophotometer or spectrometer.
2. An absorption spectroscopy experiment requires
   a. a source of electromagnetic radiation
   b. the sample to be examined
   c. a detector that measures the intensity of the unabsorbed radiation that passes through the sample
   d. a recorder to graph the result as radiation transmitted (or radiation absorbed) *vs.* wavelength (or frequency).
      i. The graph is commonly called a spectrum of the sample.
      ii. The spectrum of a compound is determined by its structure.
3. The three types of spectroscopy of greatest use (which differ conceptually only in the frequency of radiation used), and the general type of information each provides, are
   a. infrared (IR) spectroscopy—the functional groups present
   b. nuclear magnetic resonance (NMR) spectroscopy—the connectivity of carbons and hydrogens
   c. ultraviolet-visible (UV-VIS) or simply ultraviolet (UV) spectroscopy—the types of π-electron systems present.

II. *Infrared Spectroscopy*

A. THE INFRARED SPECTRUM

1. Infrared spectroscopy deals with the absorption of infrared radiation by molecular vibrations.
2. The infrared (IR) spectrum is measured in an instrument called an infrared spectrophotometer.
3. An IR spectrum is a plot of the infrared radiation transmitted through a sample as a function of the wavenumber or wavelength of the radiation.
   a. The quantity plotted on the lower horizontal axis is the wavenumber $\tilde{v}$ of the light.
      i. The wavenumber, in units of reciprocal centimeters or inverse centimeters (cm^{-1}), is simply another way to express the wavelength or frequency of the radiation.
      ii. Wavenumber is inversely proportional to the wavelength λ.

$$\left. \begin{array}{l} \tilde{v} = \dfrac{10^4}{\lambda} \\[2mm] \lambda = \dfrac{10^4}{\tilde{v}} \end{array} \right\} \quad (\lambda \text{ in micrometers, } \tilde{v} \text{ in } \text{cm}^{-1})$$

   b. The quantity plotted on the vertical axis is percent transmittance (the percent of the irradiation falling on the sample that is transmitted to the detector).
      i. The intensity of an absorption increases with the number of absorbing groups in the sample and the size of the dipole moment change that occurs in the molecule when the vibration occurs.
      ii. Intensities are often expressed qualitatively using the designations vs (very strong), s (strong), m (moderate), or w (weak).
      iii. Some peaks are sharp (narrow), whereas others are broad (wide).
   c. Sometimes an IR spectrum is not presented in graphical form, but is summarized completely or in part using descriptions of peak positions, intensities, and shapes.
4. The absorptions observed in an IR spectrum are the result of vibrations within a molecule.
   a. Absorption of energy from infrared radiation can occur only when there is a match between the frequency of the radiation and the frequency of the bond vibration.
   b. Energy is absorbed and the intensity of the bond vibration increases when radiation of the matched frequency interacts with a vibrating bond.

B. INFRARED ABSORPTION AND CHEMICAL STRUCTURE

1. Each peak in the IR spectrum of a molecule corresponds to absorption of energy by the vibration of a particular bond or group of bonds.
   a. Certain absorptions are diagnostic; that is, they indicate with reasonable certainty that a particular functional group is present.
      i. The =C—H stretching and bending absorptions and the C≡C stretching absorptions are very useful for the identification of alkenes.
      ii. The O—H stretching absorption is diagnostic for alcohols.
   b. Other peaks are confirmatory; that is, similar peaks can be found in other types of molecules, but their presence confirms a structural diagnosis made in other ways.
      i. The absorptions in the 1000 cm^{-1} to 1600 cm^{-1} region of the spectrum, known as the "fingerprint region," are generally not interpreted in detail, but they serve as a valuable "molecular fingerprint."
   c. The absence of absorptions in an IR spectrum restricts the possible structures under consideration.
2. Two aspects of IR absorption peaks are particularly important:
   a. the position of the peak (the wavenumber or wavelength at which it occurs); and
   b. the intensity of the peak.

C. FACTORS THAT DETERMINE IR ABSORPTION POSITION

1. The IR absorptions of stronger bonds—bonds with greater bond dissociation energy—occur at higher wavenumber.
2. The stretching frequency is governed primarily by the mass of the lighter atom in a bond between two atoms that differ significantly in mass; vibrations of lighter atoms occur at higher wavenumbers than vibrations of heavier atoms.

3. The type of vibration being observed affects the position of the absorption in the IR spectrum.
   a. The allowed vibrations of a molecule are termed its normal vibrational modes.
   b. A stretching vibration occurs along the line of the chemical bond.
   c. A bending vibration is any vibration that does not occur along the line of the chemical bond.
      i. Bending vibrations occur at lower frequencies (higher wavelengths) than stretching vibrations of the same groups.
      ii. Bending vibrations can be such that the hydrogens move in the plane of the group, or out of the plane of the group.
   d. Stretching and bending vibrations can be symmetrical or unsymmetrical with respect to a plane between the vibrating group.

D. FACTORS THAT DETERMINE IR ABSORPTION INTENSITY

1. A greater number of molecules in the sample (that is, a more concentrated sample or a greater optical path) and more absorbing groups within a molecule give a more intense spectrum.
2. Absorptions can be expected only for vibrations that cause a change in the molecular dipole moment.
   a. Any vibration that gives rise to an IR absorption is said to be infrared active.
   b. Certain symmetrical (or nearly symmetrical) molecules lack IR absorptions that otherwise might be expected to be observed.
      i. Absorptions that result in no dipole moment change are infrared inactive.
      ii. The symmetrical vibrations occur, but they are simply not observed in the IR spectrum.
      iii. IR-inactive vibrations can be observed by a less common type of spectroscopy called Raman spectroscopy.
   c. Because the intensity of an IR absorption depends on the size of the dipole moment change that accompanies the corresponding vibration, IR absorptions differ widely in intensity.

E. THE INFRARED SPECTROMETER

1. In a conventional IR spectrometer, the slow scanning of wavelengths through the range of interest takes several minutes.
2. In a newer type of IR spectrometer, called a Fourier-transform infrared spectrometer (FTIR spectrometer), the IR spectrum can be obtained in just a few seconds.

III. *Mass Spectrometry*

A. GENERAL

1. The utility of mass spectrometry is that:
   a. it can be used to determine the molecular mass of an unknown compound, and
   b. it can be used to determine the structure (or a partial structure) of an unknown compound by an analysis of the fragment ions in the spectrum.
2. A mass spectrum is a graph of the relative amount of each ion (called the relative abundance) as a function of the ionic mass (or mass-to-charge ratio m/z).
   a. The mass spectrum can be determined for any molecule that can be vaporized in a high vacuum, and this includes most organic compounds.
   b. Only ions are detected by the mass spectrometer—neutral molecules and radicals do not appear as peaks in the mass spectrum.
3. The instrument used to obtain a mass spectrum is called a mass spectrometer.
   a. In one type of mass spectrometer, a compound is vaporized in a vacuum and bombarded with an electron beam of high energy in which an electron is ejected from the molecule.
      i. A species formed in this process is both a radical and a cation—a radical cation.
      ii. Radical cations decompose in a series of reactions called fragmentation reactions to give positively charged fragment ions of differing mass.
   b. In the mass spectrometer, these fragment ions are separated according to their mass-to-charge ratio, m/z (m = mass, z = the charge of the fragment).
4. The ion derived from electron ejection before any fragmentation takes place is known as the molecular ion, and abbreviated M.
   a. The molecular ion occurs at an m/z value equal to the molecular mass of the sample molecule.

    b. Except for peaks due to isotopes, the molecular ion peak is the peak of highest *m/z* in any ordinary mass spectrum.

  5. The base peak is the ion of greatest relative abundance and is arbitrarily assigned a relative abundance of 100% to which the other peaks in the mass spectrum are scaled.

    a. The relative abundances of the fragment ions are recorded as a function of their mass-to-charge ratios *m/z*, which, for most ions, equal their masses.

    b. Both molecular masses and partial structures can be derived from the masses of these ionic fragments.

B. FRAGMENTATION MECHANISMS

  1. The molecular ion is formed by loss of an electron.

    a. A stable molecular ion decomposes slowly and is detected by the mass spectrometer as a peak of large relative abundance.

    b. A less stable molecular ion decomposes, in some cases completely, into smaller pieces detected as smaller ions, called fragment ions.

      i. A type of cleavage called inductive cleavage, involving the heterolytic cleavage of a bond to a positively charged electronegative atom, is very common in ethers, alkyl halides, and other compounds containing a very electronegative element.

inductive cleavage

      ii. A type of cleavage called α-cleavage or β-scission, involving the homolytic cleavage of a bond to a carbon atom attached to a positively charged electronegative atom, is important in secondary and tertiary ethers, alcohols, alkenes, and several other types of compounds.

α-cleavage (β-scission)

  2. Molecular ions derived from compounds of even mass, as well as some fragment ions produced in other ways, have even masses.

    a. When such a compound undergoes fragmentation to give separate radical and cation species, each of these must have an odd mass.

    b. When a compound contains only C, H, and O, its fragment ions of odd mass must be even-electron ions and its fragment ions of even mass must be odd-electron ions.

      i. Most of the fragment ions observed in the mass spectra of molecules containing only C, H, and O have odd molecular masses.

      ii. Since the unpaired electron is carried off in the radical fragment, the carbocation fragment detected by the mass spectrometer must have no unpaired electrons.

  3. Analyze a mass spectrum by thinking of fragment ions as coherent pieces that result from the breaking of chemical bonds in the molecular ion.

    a. Every possible fragment may not be seen in the mass spectrum of a compound.

    b. The relative abundances of the various fragments in a mass spectrum depend on their relative lifetimes, that is, the relative rates at which they break apart into smaller fragments.

  4. The masses of fragments lost as well as the masses of the fragments observed can be used to postulate the structures of fragment ions and thus convey information about their electronic structures.

       a. Ions containing no unpaired electrons are called even-electron ions and are formed by such processes as α-cleavage, inductive cleavage, and direct fragmentation at a σ bond.

       b. Ions containing an unpaired electron (a radical cation) are called odd-electron ions.

         i. Hydrogen-atom transfer followed by loss of a stable neutral molecule is a very common mechanism for the formation of odd-electron ions in primary alcohols, alkyl halides, and other compounds.

C. IDENTIFYING THE MOLECULAR ION

    1. The molecular ion is the most important peak in the mass spectrum because it provides the molecular mass of the molecule under study and thus is the basis for calculating the losses involved in fragmentation.

    2. Whether the ion of highest mass is the molecular ion can be determined by:

       a. knowing that all compounds containing only the elements C, H, and O have even molecular masses and therefore must have a molecular ion of even mass;

       b. calculating the losses of fragments observed from the peak that is assumed to be the molecular ion and determining whether these losses correspond to reasonable combinations of atoms;

       c. noting the mass change that results from converting the compound into a derivative by known reactions; or

       d. using a different ionization technique.

D. ISOTOPIC PEAKS

    1. Associated with each peak in a mass spectrum are other peaks at higher mass that arise from the presence of isotopes at their natural abundance.

       a. A peak that occurs one mass unit higher than the molecular ion (M) is termed an M + 1 peak.

       b. Every peak in a mass spectrum has isotopic peaks.

    2. The relative abundance of each isotopic peak is directly related to the natural abundance of the isotope involved.

       a. Isotopic peaks are particularly useful for diagnosing the presence of elements that consist of more than one isotope of high natural abundance such as chlorine and bromine.

       b. The naturally occurring isotopes of chlorine (^{35}Cl and ^{37}Cl) exist in a ratio of 3:1.

       c. The naturally occurring isotopes of bromine (^{79}Br and ^{81}Br) exist in about equal amounts.

    3. Isotopes are especially useful because they provide specific labels at particular atoms without changing their chemical properties.

       a. The fate of specifically labeled atoms can be used to decide between two mechanisms.

       b. Nonradioactive isotopes can be used for biological metabolic studies (studies that deal with the fates of chemical compounds when they react in biological systems).

E. THE MASS SPECTROMETER

    1. Various techniques are available to obtain the ions necessary for observation in a mass spectrometer:

       a. Electron-impact (EI) mass spectrometry involves bombardment of the vaporized sample with a beam of electrons.

       b. Chemical ionization (CI) mass spectrometry involves treating the vaporized sample with a proton source.

       c. Fast-atom bombardment (FAB) involves compounds in solution that are converted directly into gas-phase ions by subjecting them to a beam of heavy atoms (such as xenon, argon, or cesium) that have been accelerated to high velocities.

       d. Electrospray mass spectrometry involves atomizing a solution of the sample within highly charged droplets.

    2. More complex mass spectrometers, called high-resolution mass spectrometers, can resolve ions that are separated in mass by only a few thousandths of a mass unit.

       a. Nominal mass is the mass of the ion to the nearest whole number.

       b. Exact mass is the mass of the ion to four or more decimal places.

       c. The exact mass provides an elemental analysis of the molecular ion as well as an elemental analysis of each fragment in the mass spectrum.

Study Guide Links

12.1   IR Absorptions and the Vibrating Spring

The classical law of physics governing the vibrating spring is Hooke's law. Perhaps you studied Hooke's law in general physics. Hooke's law says that the restoring force of a stretched spring depends on how far you stretch it. Put another way, the more you stretch a spring, the tougher it gets. Hooke's law applied to bond vibrations permits the estimation of IR absorption wavenumbers with a remarkable degree of accuracy. Hooke's law for the wavenumber $\tilde{v}$ of the vibration of a heavier atom of mass M connected to a lighter atom of mass m is

$$\tilde{v} = \frac{1}{2\pi c}\sqrt{\frac{\kappa(m + M)}{mM}} \qquad \text{(SG12.1)}$$

In this equation, κ (the Greek letter *kappa*) is the **force constant**, a quantity that measures the tightness of the spring. If frequencies are in cm^{-1} (to correspond to the units used in IR spectra), then the cgs unit of force, the dyne, is required. (One dyne = one $g\ cm\ sec^{-2}$) Likewise, c, the velocity of light, is 3×10^{10} cm/sec. The units of the force constant are dyne/cm.

Stronger bonds—or tighter springs—have greater force constants. In fact, the force constants for single, double, and triple bonds are about 5×10^5, 10×10^5, and 15×10^5 dyne/cm, respectively—essentially in the ratio 1:2:3, as intuition would dictate. Eq. SG12.1 shows that the wavenumber of an absorption increases with the square root of the force constant. Thus we have the conclusion that *stronger bonds absorb at higher wavenumber.*

The mass effect is also nicely illustrated by Eq. SG12.1. If $M >> m$, the smaller mass can be ignored in the numerator of Eq. SG12.1, and the larger mass M then cancels and vanishes from the equation, leaving only the smaller mass m in the denominator. Eq. SG12.1 then becomes

$$\tilde{v} = \frac{1}{2\pi c}\sqrt{\frac{\kappa}{m}} \qquad \text{(for } m << M\text{)} \qquad \text{(SG12.2)}$$

(Be sure you see why this is true.) In other words, if the mass of the heavier partner is much greater than that of the lighter partner, the heavier mass simply doesn't matter any more. This is the quantitative basis for the "cannonball and Empire State Building" analogy used in the text.

12.2   FTIR Spectroscopy

This section will give you a better idea how an FTIR spectrometer works. In an FTIR spectrometer, radiation to be passed through the sample is split and sent over two paths simultaneously by a system consisting of two mirrors, one moving and one fixed (a Michelson interferometer). The difference between the lengths of the two light paths at various times is precisely calibrated with a laser. The superimposed beams of light from the two paths is passed through the sample. The interference patterns, or interferograms, that result can be thought of as superimposed plots of absorption *vs.* time from both light beams in which all wavelengths in the spectrum contribute simultaneously. The interferograms are stored in a small computer and analyzed by *Fourier transformation.* Fourier transformation is a mathematical technique for decomposing time-based wave motions into their contributing frequency components. (For example, the unique sound of a bell as it dies out over time might be analyzed in terms of the different contributing pitches by the same technique.) Fourier analysis converts the interferograms into a plot of absorption *vs.* wavenumber—the IR spectrum. The entire process takes just a few seconds.

Besides saving time, FTIR has much better wavenumber resolution than conventional IR spectroscopy. (This means that two closely-spaced peaks would appear as separate peaks rather than as a single broad peak.) In addition, multiple spectra from the same sample can be digitally added; because "noise" is random, it decreases when multiple spectra are summed, and the signal due to the IR spectrum grows. Consequently, many spectra can be accumulated from the same sample in a very short time, and thus strong spectra can be obtained from very dilute samples. The principles of FTIR have been known for many years, but only in relatively recent times has the technique become practical because of the availability of inexpensive, high-powered computers. The digital revolution has come to spectroscopy just as it has to sound and video technology!

12.3　The Mass Spectrometer

This section will give you a more quantitative understanding of how a mass spectrometer operates. When an ion in a mass spectrometer is accelerated to a voltage V its kinetic energy is $mv^2/2$, where m is its mass, and v is its velocity. Since all ions have, to a good approximation, the same velocity as a result of the accelerating voltage, then it is the *ionic mass* that gives one ion a different kinetic energy from another. The basic equation describing the bending of an ion by a magnetic field H is

$$m/z = k\frac{\text{H}^2 r^2}{V}$$

where r is the radius of the circular path over which the ion is bent, z is the charge on the ion, and k is a proportionality constant. This equation shows that the paths of ions of a larger mass are bent over a circle of greater radius. Similarly, for an ion of a given mass, increasing the magnetic field H decreases the radius of its path.

In the mass spectrometer, a collector is positioned to sense only the ions traversing a path of a particular radius r. By increasing the magnetic field H in a regular way, ions of progressively increasing mass are focussed onto the collector. The relative intensity of the ion beam of a given mass is recorded as an ion current. Thus is obtained a plot of ion current *vs.* mass—the mass spectrum.

Solutions

Solutions to In-Text Problems

12.1 (a) Apply Eq. 12.1 and include the conversion factor 100 cm/m.

$$v = \frac{(3.00 \times 10^{10}\text{ cm/sec})}{(9 \times 10^{-6}\text{ m})(100\text{ cm/m})} = 3.33 \times 10^{13}\text{ sec}^{-1}$$

12.2 (a) Multiply the frequency obtained in the solution to Problem 12.1 times Planck's constant:

$$E = hv = (3.99 \times 10^{-13}\text{ kJ sec mol}^{-1})(3.33 \times 10^{13}\text{ sec}^{-1}) = 13.3\text{ kJ/mol}$$

12.3 (a) The energy of X-rays is greater than that of any visible light, including blue light. In fact, it is so much greater that prolonged exposure to X-rays is harmful.

12.4 (a) Apply Eq. 12.6a:

$$\tilde{v} = \text{wavenumber} = (1 \times 10^4\ \mu\text{m/cm}) \div 6.0\ \mu\text{m} = 1667\text{ cm}^{-1}$$

12.5 Because $v = c/\lambda = c(1/\lambda)$ and $\tilde{v} = 1/\lambda$, then $v = c\tilde{v}$. (Notice that the factor of 10^4 in Eqs. 12.6a–b is a conversion factor between cm and μm, and is not necessary if consistent units are used.)

 There is an error in early printings of the text. The problem should read, "...follows from Eq. 12.6a and 12.1."

12.6 Convert the frequency into wavenumber and then consult Fig. 12.4.

$$\tilde{v} = v/c = (9 \times 10^{13}\text{ sec}^{-1})/(3 \times 10^{10}\text{ cm sec}^{-1}) = 3000\text{ cm}^{-1}$$

The group of peaks around 3000 cm^{-1} are due to the C—H stretching vibrations in nonane.

12.7 Bond strength is the major determining factor. If atomic mass were the major determining factor, H—F would have the smallest vibrational frequency, not the largest, and the vibrational frequencies would not be very different, because the mass of the smaller atom in each pair (hydrogen) affects the frequency the most. In contrast, Table 5.3 on text p. 205 shows that bond strength varies significantly for the three types of bonds; a related variation in force constant, and thus in stretching frequency, would also be expected and is observed.

12.8 (a) Active
(c) Inactive. The zero dipole moment of cyclohexane is unaffected by this symmetrical vibration.
(e) Active. Because the two N—O bonds have bond dipoles that do not cancel, changing the lengths of these bonds at the same time changes the magnitudes of the bond dipoles and therefore changes the magnitude of their resultant. Consequently, the dipole moment of the molecule is changed by the vibration.
(g) Inactive. The zero dipole moment of this alkene is unchanged by the C=C stretching vibration.

12.9 The moderately strong C=C stretching absorption at about 1650 cm^{-1} and the C—H bending vibration at 890 cm^{-1} suggest that (a) is the spectrum of 2-methyl-1-hexene. The C—H bending absorption at about 970 cm^{-1} and the virtual absence of a C=C stretching absorption in spectrum (b) confirms that this is the spectrum of *trans*-2-heptene.

12.10 The absence of C=C stretching absorption near 1650 cm^{-1} and an O—H stretching absorption near 3300 cm^{-1}, and the presence of strong absorption near 1100 cm^{-1} (C—O stretching absorption), show that this is the spectrum of the ether.

12.11 The two C—O bonds of an ether can undergo both symmetrical and unsymmetrical stretching vibrations; each of these normal vibrational modes has an associated infrared absorption. (These vibrations are described on the first two lines of Fig. 12.7 on text p. 547.)

12.12 Think of molecular association by hydrogen bonding as a reaction like any other and apply LeChatelier's principle.

$$2 \text{ RO—H} \rightleftharpoons \text{RO—H---OR}$$

unassociated
alcohols

H

hydrogen-bonded
alcohols

Increasing the concentration of alcohol drives this equilibrium to the right, and decreasing the concentration of the alcohol drives this equilibrium to the left. At sufficiently low concentrations of alcohol, the spectrum is that of the unassociated alcohol.

12.13 Table 12.2 shows that Si has the following isotopic abundances: ^{28}Si, 92.21; ^{29}Si, 4.67; and ^{30}Si, 3.10. These abundances result in the following relative intensities for the $m/z = 74$ and 75 peaks:

$$m/z = 74: \text{ relative intensity } (4.67)/(92.21) = 0.0506, \text{ or } 5.06\%$$
$$m/z = 75: \text{ relative intensity } (3.10)/(92.21) = 0.0336, \text{ or } 3.36\%$$

To this must be added the contribution due to ^{13}C. The molecular mass of $(CH_3)_4Si$ is 88; therefore, the base peak at $m/z = 73$ results from loss of 15 mass units (CH_3). Hence, the $m/z = 74$ peak will have a contribution from three isotopic carbons equal to $3(1.11)/(92.21) = 0.0361$ (3.61%). Add this contribution to that calculated above to get the final value for the $m/z = 74$ peak:

$$m/z = 74: \text{ relative intensity } = 5.06\% + 3.61\% = 8.67\%$$

The contribution of ^{13}C to the $m/z = 75$ peak is negligible. Hence, the $m/z = 74$ peak has a relative intensity of 8.67%, and the $m/z = 75$ peak has a relative intensity of 3.36%.

12.15 (a) 2-Methyl-2-pentanol has molecular mass = 102. The fragment at $m/z = 59$ corresponds to the loss of 43 mass units (a propyl group). This loss can occur by an α-cleavage mechanism:

$$(CH_3)_2\overset{:\ddot{O}H}{C}\text{—}CH_2CH_2CH_3 \xrightarrow{-e^-} (CH_3)_2\overset{+\cdot\cdot\ddot{O}H}{C}\text{—}CH_2CH_2CH_3 \longrightarrow (CH_3)_2\overset{+\ddot{O}H}{C} + {}^\bullet CH_2CH_2CH_3$$

2-methyl-2-pentanol   molecular ion M $m/z = 59$

12.16 One fragmentation of *sec*-butyl isopropyl ether by inductive cleavage is shown in Eq. 12.17(b) of the text, and this is followed by the comment that a similar mechanism accounts for the formation of the $m/z = 43$ fragment. That inductive cleavage mechanism is as follows:

$$CH_3CH\text{—}\overset{+}{\underset{\cdot\cdot}{O}}\text{—}CH(CH_3)_2 \longrightarrow CH_3CH\text{—}\overset{\cdot}{\underset{\cdot\cdot}{O}}: + {}^+CH(CH_3)_2$$

C_2H_5 C_2H_5 $m/z = 43$

molecular ion of
sec-butyl isopropyl ether

12.17 (a) Because even-electron ions of compounds containing only C, H, and Cl have odd mass, the ion with $m/z = 57$ is the even-electron ion and the ion with $m/z = 56$ is the odd-electron ion.
(b) HCl can be lost to give the odd-electron ion.

(c) The molecular mass of 2-chlorobutane is 92 (for the ^{35}Cl isotope). The $m/z = 57$ peak is due to loss of a chlorine atom from the molecular ion by an inductive-cleavage mechanism to give the *sec*-butyl cation:

$$CH_3\overset{\overset{\displaystyle :\overset{\cdot\cdot}{Cl}\cdot\,+}{|}}{C}HCH_2CH_3 \longrightarrow CH_3\overset{+}{C}HCH_2CH_3 \;+\; :\overset{\cdot\cdot}{\underset{\cdot\cdot}{Cl}}\cdot$$

molecular ion   *sec*-butyl cation
of 2-chlorobutane $m/z = 57$

The $m/z = 56$ peak is due to hydrogen-atom transfer within the molecular ion followed by loss of HCl in a process analogous to that shown in Eq. 12.21 of the text:

$$H_3C\!-\!CH\!-\!CHCH_3 \longrightarrow H_3C\!-\!CH\!-\!\overset{\cdot}{C}HCH_3 \longrightarrow :\overset{\cdot\cdot}{\underset{\cdot\cdot}{Cl}}\!-\!H \;+\; H_3C\!-\!\overset{+}{C}H\!-\!\overset{\cdot}{C}HCH_3$$

molecular ion $m/z = 56$
of 2-chlorobutane

12.18   Formation of a methyl ether adds 14 mass units to the molecular mass of an alcohol. Postulate, then, that the molecular mass of alcohol *A* is $(116 - 14) = 102$. The prominent peaks at $m/z = 87$ and $m/z = 73$ in the spectrum of the alcohol, then, represent mass losses of 15 units (a methyl group) and 29 units (an ethyl group), respectively. Now, subtract the masses of the known fragments in the alcohol from its molecular mass, and thus determine what fragment is missing. A C—OH unit contributes 29 mass units; the ethyl and methyl groups noted above contribute a total of 44 mass units. A group or groups totalling 29 mass units is (are) required to complete the structure; another ethyl group is consistent with this requirement. A tertiary alcohol containing one methyl and two ethyl groups at the α-carbon, that is, 3-methyl-3-pent-anol, fits the data. Notice that there are two ways in which an ethyl group could be lost by α-cleavage; consequently, a peak corresponding to a mass loss of 29 units, that is, a peak at $m/z = 73$, should be particularly strong, as observed.

$$H_3C\!-\!\overset{\overset{\displaystyle OH}{|}}{\underset{\underset{\displaystyle CH_2CH_3}{|}}{C}}\!-\!CH_2CH_3 \qquad \text{3-methyl-3-pentanol}$$

Solutions to Additional Problems

12.19   The three factors that determine the wavenumber, or position, of an infrared absorption are: strength of the bonds involved, masses of the atoms involved, and the type of vibration.

12.21   (a) Use a branched alkoxide base such as K^+ $(CH_3)_3C\!-\!O^-$ to bring about an E2 elimination. Confirm the reaction by observing appearance of the alkene C=C stretch at about 1655 cm^{-1}.

(c) Carry out a Williamson ether synthesis:

$$CH_3(CH_2)_4CH_2OH \xrightarrow{\text{NaH}} \xrightarrow{\text{CH}_3\text{I}} CH_3(CH_2)_4CH_2OCH_3$$

1-hexanol   hexyl methyl ether
(1-methoxyhexane)

Confirm the reaction by observing loss of the O—H stretch of the alcohol in the 3200–3400 cm^{-1} region of the IR spectrum.

12.22 The principle is that compounds which have different physical properties have different IR spectra.

(a) Because 3-pentanol and racemic 2-pentanol are constitutional isomers, they have different IR spectra.
(b) Because enantiomers have identical physical properties, the IR spectra of (*R*)- and (*S*)-2-pentanol are identical.
(c) This one is a little tricky! The two chair conformations of cyclohexanol are diastereomers and, because diastereomers have different physical properties, the two conformations also have different IR spectra. The problem is *observing* the two conformations independently. Any sample of cyclohexanol at room temperature is a mixture of these diastereomeric conformations, and the IR spectrum of cyclohexanol is therefore a spectrum of the mixture.

12.23 Spectrum (1) has no peaks characteristic of alkenes or alcohols, but it does have a strong peak near 1100 cm^{-1} that is a C—O stretching absorption. This is the spectrum of compound (d), dipropyl ether.

Spectrum (2) shows the typical absorptions of a —CH=CH$_2$ group: the double C—H bending absorption at 910 and 990 cm^{-1}, and the strong C=C stretching absorption at 1640 cm^{-1}. This is the spectrum of compound (a), 1,5-hexadiene.

Spectrum (3) is virtually featureless; the only strong absorptions are the C—H stretching absorptions in the 2800–3000 cm^{-1} region. This is the spectrum of compound (f), cyclohexane.

Spectrum (4) shows both O—H stretching and C=C stretching absorptions, and is the spectrum of compound (c), 1-hexen-3-ol. The double peaks at 910 and 990 cm^{-1} confirm the presence of a —CH=CH$_2$ group. Compound (g), 3-hexanol, is ruled out because it should not have alkene absorptions in its IR spectrum.

Spectrum (5) has C=C stretching absorption at 1675 cm^{-1} and and shows the typical high-wavenumber shoulder for the alkene C—H stretching absorption. The position of the C=C stretching absorption along with the C—H bending absorption near 800 cm^{-1} indicates an alkene with three alkyl substituents on its double bond. This is the spectrum of compound (b), 1-methylcyclopentene. The definite absence of a C—H bending absorption near 970 cm^{-1} rules out compound (e), *trans*-4-octene.

12.25 Because stronger bonds absorb at higher wavenumber (text p. 545), the order of increasing C—H bond strength is alkane < alkene < alkyne. This deduction can be verified by consulting the relevant C—H bond dissociation energies in Table 5.3, text p. 205; these bond energies are alkane (418 kJ/mol), alkene (460 kJ/mol), and alkyne (548 kJ/mol).

12.27 Table 5.3 on text p. 205 indicates that the C—F bond is considerably stronger than the C—O bond; consequently, the C—F absorption would be expected to occur at a greater wavenumber than the C—O absorption. However, this absorption would not be in the region of carbon double bonds, which are considerably stronger than carbon single bonds. Therefore, this absorption should occur somewhere between the C—O and the C=C absorptions, that is, somewhere between 1050 cm^{-1} and 1600 cm^{-1}, probably closer to the lower wavenumber than the higher. (In fact, a typical C—F absorption is at 1100 cm^{-1}.)

12.28 (a) The bond dipole of the S—H bond is much less than that of the O—H bond because the electronegativities of sulfur and hydrogen differ less than do the electronegativities of oxygen and hydrogen (Table 1.1, text p. 10). As a result, the S—H vibration is less active in the infrared, and the corresponding absorption is less intense. The S—H absorption occurs at lower frequency mostly because the S—H bond is much weaker than the O—H bond; compare the bond dissociation energies for H—SH and H—OH in Table 5.3, text p. 205. (The difference between the masses of sulfur and oxygen has little effect on the vibration frequency because the mass of the smaller atom, hydrogen, is the major determinant of the vibration frequency.)

(b) Compound *B* has a typical strong, broad O—H stretching absorption in the 3200–3400 cm^{-1} region, and is therefore the alcohol. Compound *A*, then, is the thiol. Note the considerably weaker S—H stretching absorption in the spectrum of compound *A* at about 2530 cm⁻¹, as suggested by part (a) of the problem.

12.29 (a) Both compounds have the same absorptions, except that the absorptions of compound *C* in the spectrum at the top of text page 575 are displaced to lower frequency, an observation that implies a higher mass for the absorbing group; see the discussion of the mass effect on text p. 545. In particular, a peak at 3000 cm⁻¹ in compound *D*, undoubtedly a C—H stretching absorption, is displaced to 2250 cm⁻¹ in compound *C*. Hence, compound *C* is CDCl₃.

Eq. SG12.1 in Study Guide Link 12.1 (p. 276 of this manual) gives the quantitative basis of the mass effect. If the force constants of C—H and C—D bonds are nearly the same (and they are), then the ratio of stretching frequencies of these two bonds should be equal to the square root of the ratios of the mass terms. Eq. SG12.1 predicts a ratio of 1.36; the actual ratio, (3000/2250) = 1.33, is very close to this prediction.

(b) The two compounds could be distinguished by mass spectrometry by the masses of their parent ions. Both CHCl₃ and CDCl₃ have four molecular ions (why four?). Each molecular ion peak of CDCl₃ lies at one unit higher mass than the corresponding molecular ion peak of CHCl₃.

12.31 (a) Loss of a propyl radical from the molecular ion of 3-methyl-3-hexanol by α-cleavage gives a fragment with *m/z* = 73.

$$CH_3CH_2-\overset{\overset{\overset{..}{HO}:}{|}}{\underset{\underset{CH_3}{|}}{C}}-CH_2CH_2CH_3 \longrightarrow CH_3CH_2-\overset{\overset{..}{HO}+}{\underset{\underset{CH_3}{|}}{C}} + \cdot CH_2CH_2CH_3$$

molecular ion of 3-methyl-3-hexanol   *m/z* = 73

(c) Fragmentation of neopentane at any one of its four carbon-carbon bonds gives a methyl radical and the *tert*-butyl cation, which has the correct mass:

$$H_3C-C(CH_3)_3 \xrightarrow{-e^-} H_3C \overset{+}{\overset{|}{\cdot}} C(CH_3)_3 \longrightarrow H_3C\cdot + \overset{+}{C}(CH_3)_3$$

neopentane   *tert*-butyl cation
                                          m/z = 57

(d) 1-Pentanol has molecular mass = 88; therefore, a fragment with *m/z* = 70, which must be an odd-electron ion, could result from the loss of H₂O (18 mass units.) A hydrogen-transfer mechanism followed by loss of H₂O, exactly as shown for 1-heptanol in Eq. 12.21 on text p. 566, accounts for this ion.

$$CH_3(CH_2)_2\overset{H}{\underset{|}{C}}H-CH_2 \longrightarrow CH_3(CH_2)_2\overset{\cdot}{C}H-\overset{\overset{+}{OH}}{\underset{}{C}}H_2 \longrightarrow CH_3(CH_2)_2\overset{\cdot}{C}H-\overset{+}{C}H_2 + H-\overset{..}{\underset{..}{O}}H$$

molecular ion of 1-pentanol   *m/z* = 70

12.32 (a) A mass of 28 units could be accounted for by a molecule of ethylene, H₂C=CH₂.

(c) The data indicate the presence of chlorine, which has a mass of 35; loss of 36 mass units, then, could correspond to loss of HCl. (See the solution to Problem 12.17(c) for an example of such a process.)

12.33 First ask what products are expected from the reaction. Two constitutional isomers, both with molecular mass = 88, are anticipated:

trans-2-pentene

1) BH₃/THF
2) H₂O₂/NaOH

$CH_3CHCH_2CH_2CH_3$ + $CH_3CH_2CHCH_2CH_3$

2-pentanol   3-pentanol

Spectrum (2) is consistent with 3-pentanol; the base peak at $m/z = 59$ corresponds to loss of C_2H_5, and there are two ethyl groups in this compound that could be lost by α-cleavage. Spectrum (1) has a base peak at $m/z = 45$ that corresponds to loss of 43 units (a propyl group). 2-Pentanol has a propyl branch that could be lost as a radical by α-cleavage.

When unknown compounds come from a chemical reaction, use what you know about the reaction as a starting point for postulating structures.

12.35 The three peaks correspond to dichloromethane molecules that have, respectively, two atoms of ^{35}Cl, one atom each of ^{35}Cl and ^{37}Cl, and two atoms of ^{37}Cl. The probability of any combination of two isotopes is the product of their separate probabilities. Make a table of the possibilities:

	Cl #1	Cl #2	
$m/z = 84$	35	35	relative probability = $(0.7577)^2 = 0.574$
$m/z = 86$}	35	37	relative probability = $(0.7577)(0.2423) = 0.184$
	37	35	relative probability = $(0.7577)(0.2423) = 0.184$
			total relative probability 0.368
$m/z = 88$	37	37	relative probability = $(0.2423)^2 = 0.059$

Taking the peak at $m/z = 84$ as 100%, the ratios of the peaks are 100%, 64.1%, and 10.3%, respectively.

The middle entry of the table above shows that two *different* isotopes can occur within the same compound in two ways. An analogy is the combinations that can be rolled with a pair of six-sided dice. There is only one way to roll a "2" (a "1" on each of the dice), but there are two ways to roll a "3" (a "1" on one of the dice and a "2" on the other, and vice-versa).

12.37 (a) The ratio of the M + 1 to the M peak is 1.06/19 = 0.056. If each carbon atom contributes 0.011 to this ratio, then this ratio indicates a compound containing five carbon atoms. Five carbon atoms account for 60 mass units. One oxygen accounts for 16 mass units. (A compound with molecular mass = 86 and five carbons can have no more than one oxygen.) Ten hydrogens are required to fulfill the molecular mass of the compound. The formula is $C_5H_{10}O$.

12.38 (a) The molecular mass of ethyl bromide (CH_3CH_2Br) is 108 (for ^{79}Br) and 110 (for ^{81}Br). Consequently, the ion with $m/z = 110$ is the molecular ion of the molecule containing the heavier isotope. It is formed by ejection of an electron from one of the bromine unshared pairs:

$$CH_3-CH_2-\overset{..}{\underset{..}{Br}}: \xrightarrow{-e^-} CH_3-CH_2-\overset{+}{\underset{..}{Br}}:$$

$$m/z = 108, 110$$

(c) The fragment at $m/z = 81$ corresponds to the formation of a bromine cation and an ethyl radical from the compound containing the heavier bromine isotope:

$$CH_3-CH_2 + \overset{..}{\underset{..}{Br}}:^+$$

$$m/z = 79, 81$$

(e) The fragment at $m/z = 29$ is the ethyl cation formed by inductive cleavage at the carbon-bromine bond:

$$CH_3—CH_2—\overset{+}{\underset{..}{\ddot{B}r}}: \longrightarrow CH_3—\overset{+}{C}H_2 \ + \ :\overset{.}{\underset{..}{\ddot{B}r}}:$$

$$m/z = 29$$

(g) The ion at $m/z = 27$ could arise by elimination of a hydrogen atom by β-scission (α-cleavage) from the ion with $m/z = 28$; see part (f).

$$H_2\overset{.}{C}—\overset{+}{C}H \longrightarrow H_2C{=}\overset{+}{C}H \ + \ \cdot H$$

$$m/z = 28 \qquad\qquad m/z = 27$$

12.39 The presence of an odd number of nitrogens in a molecule containing, as the other atoms, any combination of C, H, O, and halogen reverses the odd-electron/even-electron mass correlations in Sec. 12.6D on text p. 565 because such a molecule must have an odd molecular mass. (Molecules containing no nitrogen or an even number of nitrogens have even molecular masses,)

(a) The molecular ion of any molecule, because it is formed by ejection of a single electron, must be an odd-electron ion.

(b) A fragment of even mass containing one nitrogen is an even-electron ion. (The ion in Problem 12.31(b) is an example.)

12.40 (a) Sigma (σ) electrons are held closer to the nucleus than π electrons; by the electrostatic law, attraction between particles of opposite charge (such as an electron and a nucleus) is greater when the particles are closer. Hence, there is a greater energy of attraction of the nucleus for σ electrons. This means that more energy must be used to remove a σ electron; hence, a π electron is removed more easily. An equivalent explanation is that the bond dissociation energy of a typical π bond is considerably less than that of a σ bond (see Table 5.3, text p. 205). It should take less energy to eject an electron from a weaker bond—a π bond—than it does from a stronger bond—a σ bond.

(b) The fragment with $m/z = 41$ is the allyl cation, which is formed by α-cleavage of the molecular ion.

$$H_2C{=}CH—CH_2—CH_2CH_2CH_2CH_3 \ \xrightarrow{-e^-}$$

1-heptene

$$\left[H_2\overset{.}{C}—\overset{+}{C}H—CH_2—CH_2CH_2CH_2CH_3 \longleftrightarrow H_2\overset{+}{C}—\overset{.}{C}H—CH_2—CH_2CH_2CH_2CH_3 \right] \longrightarrow$$

molecular ion

$$H_2\overset{+}{C}—CH{=}CH_2 \ + \ \cdot CH_2CH_2CH_2CH_3$$

allyl cation
$m/z = 41$

The resonance structures of the molecular ion may seem strange because they are interconverted by movement of a single electron between adjacent carbon atoms. That this is reasonable follows from a consideration of the bonding π molecular orbital of the alkene. The molecular ion is formed by removal of a single electron from that molecular orbital, but the molecular orbital remains with its one electron in the product radical-cation. The single remaining electron is shared between the two carbons involved in the π orbital just as two electrons are shared between the same two carbons in the starting alkene.

(c) The ion product of Eq. 12.21 in the text is identical to the molecular ion in part (b) above. Because identical ionic intermediates are involved, mass spectra containing the same peaks resulting from the fragmentation of this ion are observed.

13

Nuclear Magnetic Resonance Spectroscopy

Terms

Concepts

I. NMR Spectroscopy

A. INTRODUCTION

1. Some nuclei have the property of spin, which is analogous to the spin of electrons.
   a. The hydrogen nucleus ^1H, that is, the proton, has a nuclear spin that can assume either of two values, designated by quantum numbers $+\frac{1}{2}$ and $-\frac{1}{2}$.
   b. The nucleus act like a tiny magnet; hydrogen nuclei, whether alone or in a compound, respond to a magnetic field.
      i. In the absence of a magnetic field, the nuclear magnetic poles are oriented randomly.
      ii. After a magnetic field is applied, the magnetic poles of nuclei with spin of $+\frac{1}{2}$ are oriented parallel to the magnetic field.
      iii. The magnetic poles of nuclei with spin of $-\frac{1}{2}$ are oriented antiparallel to the field.
2. Any nucleus with a spin can be detected by NMR spectroscopy.

B. PHYSICAL BASIS OF NMR SPECTROSCOPY

1. For nuclei to absorb energy, they must have a nuclear spin and must be situated in a magnetic field.
2. The presence of a magnetic field causes the two spin states in a proton to have different energies; the energy difference ΔE between the two spin states for a given proton depends on the intensity of the magnetic field H_p at the proton; the larger the field, the greater is ΔE.
   a. In the absence of the field, the two spin states have the same energy.
   b. The populations of the two spin states are in rapid equilibrium, and the equilibrium favors the state with lower energy.
3. When the nuclei in a magnetic field are subjected to electromagnetic radiation of energy E_0 exactly equal to ΔE, this energy is absorbed by some of the nuclei in the $+\frac{1}{2}$ state.
   a. The absorbed energy causes these nuclei to flip their spins to the more energetic state with $-\frac{1}{2}$ spin.
   b. The electromagnetic radiation with this energy lies in the FM area of the radiofrequency, or "rf," region of the electromagnetic spectrum.
4. This energy absorption by nuclei in a magnetic field is termed nuclear magnetic resonance (NMR).
   a. NMR can be detected in a type of absorption spectrophotometer called a nuclear magnetic resonance spectrometer (NMR spectrometer).
   b. The study of NMR absorption is called NMR spectroscopy.
5. Because almost all organic compounds contain hydrogens, proton NMR spectroscopy is especially useful to the organic chemist.

C. OBTAINING THE NMR SPECTRUM

1. In a typical NMR experiment, radiation at a frequency of 60 MHz, called the operating frequency or applied frequency v_0, is applied to the sample.
   a. The applied magnetic field, H_0, is slowly increased over time.
   b. When the energy separation between the spin energy levels exactly corresponds to the operating frequency, energy is absorbed.
2. When a chemical compound is subjected to a high magnetic field in an NMR spectrometer, the field H_p at a given nucleus within the compound is not the same as the field provided by the spectrometer, H_0.
   a. Electrons circulating in the vicinity of a nucleus provide their own magnetic fields that oppose the external field.
   b. The external field H_0 has to be strong enough to overcome the fields of the circulating electrons in order to make H_p large enough for absorption to occur.
   c. The field provided by the circulating electrons is different for nuclei in different chemical environments.
3. The magnetic field strength required for NMR absorption depends on the chemical environment of the nucleus.

4. An NMR spectrum is a plot of energy absorption *vs.* magnetic field strength.

II. The NMR Spectrum

A. CHEMICAL SHIFT SCALES

1. The strength of the applied field H_0 is plotted on the horizontal axis of an NMR spectrum, and increases to the right, or to the upfield direction.
2. Absorption is plotted on the vertical axis in arbitrary units, increasing from bottom to top; that is, absorption peaks are registered as upward deflections in the spectrum.
    a. Absorption peaks are termed absorptions, lines, or resonances.
    b. Absorption positions are always cited by their chemical shifts.
3. The chemical shift of a proton is defined as the difference between its absorption position and that of TMS [tetramethylsilane, $(CH_3)_4Si$].

$$\text{chemical shift (in Hz)} = \nu - \nu_{TMS}$$

   a. The chemical shift of TMS is set to zero by definition.
      i. TMS was chosen as a standard not only because it is volatile and inert, but also because its NMR absorption occurs at a higher field than the absorption of most organic compounds.
      ii. In most organic molecules, protons give NMR absorptions over a chemical shift range of about 0–10 ppm downfield from TMS.
   b. The chemical shift scale is shown on the upper horizontal axis of the spectrum in units of hertz (Hz).
   c. The chemical shift in Hz is proportional to the operating frequency.

$$\text{chemical shift (in Hz)} = \nu - \nu_{TMS} = \delta\nu_0$$

   where the frequencies of ν and ν_{TMS} are in Hz, and ν_0 is in MHz.
4. Chemical shifts are presented in a form that is independent of the operating frequency

$$\delta = \frac{\nu - \nu_{TMS}}{\nu_0} = \frac{\text{chemical shift in Hz}}{\text{operating frequency in MHz}}$$

   where the units of δ are parts per million, or ppm.
   a. Each nucleus has a characteristic δ value that is independent of the operating frequency.
   b. The ppm scale is shown on the lower horizontal axis of the spectrum.
   c. The ppm scale increases in the downfield direction, toward lower magnetic field.
   d. NMR spectra taken at different operating frequencies have the same chemical shift scale in ppm but a different chemical shift scale in Hz.

B. RELATIONSHIP OF CHEMICAL SHIFT TO STRUCTURE

1. Each chemically nonequivalent set of protons in a molecule gives (in principle) a different resonance in the NMR spectrum.
    a. The chemical shift of a proton provides information about nearby groups.
    b. Factors that increase the proton chemical shift are:
       i. Increasing electronegativity of nearby groups.
       ii. Increasing number of nearby electronegative groups.
       iii. Decreasing distance between the proton and nearby electronegative groups.
2. Electronegative groups affect the chemical shift of a proton through their effects on the magnetic fields caused by the nearby electrons.
    a. Circulation of electrons in the vicinity of a proton causes local fields that shield the proton from the effects of an external field; a proton is deshielded by electronegative groups.
    b. The external magnetic field required to bring a deshielded proton into resonance is smaller than the field required for a proton that is not deshielded.
3. The chemical-shift contributions of different groups are approximately additive.
4. Chemical-shift contributions of various groups can be found in Table 13.2, text page 589, or in Appendix III, text page A-4.

C. THE NUMBER OF ABSORPTIONS IN AN NMR SPECTRUM

1. Chemically equivalent protons have identical chemical shifts.

a. For two protons to be chemically equivalent, they first must have the same connectivity relationship to the rest of the molecule, that is, they must be constitutionally equivalent.
   i. It is possible for chemical-shift differences to be so small that they are not detectable.
   ii. It is possible for the resonances of chemically different groups to accidentally overlap.
b. Homotopic protons are chemically equivalent and have identical chemical shifts under all circumstances.
c. Enantiotopic protons are chemically equivalent as long as they are in an achiral environment.
2. Chemically nonequivalent protons in principle have different chemical shifts.
   a. Constitutionally nonequivalent groups are chemically nonequivalent.
   b. Diastereotopic groups are constitutionally equivalent but are chemically nonequivalent; diastereotopic protons in principle have different chemical shifts.
   c. In an optically active chiral solvent, enantiotopic protons in principle have different chemical shifts.
3. The minimum number of chemically nonequivalent sets of protons in a compound of unknown structure can be determined by counting the number of different groups of resonances in its NMR spectrum.

D. COUNTING PROTONS WITH THE INTEGRAL

1. The size of an NMR absorption is governed essentially by the number of protons contributing to it.
2. The exact intensity of an NMR absorption is given not by its peak height, but rather by the total area under the peak.
   a. NMR instruments are equipped with an integrating device that can be used to display the integral on the spectrum.
   b. The relative height of the integral (in any convenient units, such as chart spaces or millimeters) is proportional to the number of protons contributing to the peak.
   c. The number of absorptions in principle indicates the number of chemically nonequivalent sets of protons.
   d. If the total number of protons is known, the absolute number of protons in each set can be calculated.

E. SPLITTING AND THE $n + 1$ RULE

1. Splitting occurs because the magnetic field due to the spin of a neighboring proton adds to or subtracts from the applied magnetic field and affects the total field experienced by an observed proton.
   a. No splitting is observed between protons with identical chemical shifts.
   b. One of the most important ways that proton spins interact is through the electrons in the intervening chemical bonds.
2. The relationship between the number of lines in the splitting pattern for an observed proton and the number of adjacent protons is given by a rule called the $n + 1$ rule.
   a. If there are n adjacent nonequivalent protons, the resonance of an observed proton is split into $n + 1$ lines.
   b. Protons that split each other are said to be coupled.
   c. With saturated carbon atoms, splitting is observed only between protons on adjacent carbon atoms.
      i. The spacing between adjacent peaks of a splitting pattern (in Hz) is called the coupling constant, J.
      ii. Two coupled protons must have the same J value.
3. In most cases, a splitting pattern can be discerned by the relative intensities of its component lines which have well-defined ratios (see Table 13.3, text page 596).
   a. The chemical shift of a split resonance in most cases occurs at or near the midpoint of the splitting pattern.
   b. The departure from the normal intensity ratios is called leaning.
      i. Leaning is most severe when the chemical-shift difference between two absorptions that split each other are small.
      ii. Leaning is less pronounced when the chemical-shift difference between the two signals is large.
4. A set of protons can be split by protons on more than one adjacent carbon.

5. NMR spectra are often recorded in abbreviated form:
   a. the chemical shift of each resonance
   b. the integral
   c. the splitting (and coupling constant if known) using the following abbreviations:
      i. s—singlet.
      ii. d—doublet.
      iii. t—triplet.
      iv. q—quartet.
      v. m—multiplet.

F. COMPLEX NMR SPECTRA

1. The NMR spectra of some compounds contain splitting patterns that do not appear to be the simple ones predicted by the $n + 1$ rule. This behavior can be due to:
   a. multiplicative splitting—the splitting of one set of protons by more than one other set of protons. (When all the coupling constants are different, the splitting is in general multiplicative.)
   b. a breakdown of the $n + 1$ rule in certain cases.
      i. Spectra that conform to the $n + 1$ rule are called first-order spectra; some compounds contain splitting patterns that are more complex than predicted by the $n + 1$ rule.
      ii. First-order NMR spectra are generally observed when the chemical shift difference (in Hz) between coupled protons is greater than their coupling constants by a factor of about ten or more. (When the chemical shift difference is less than this, non-first-order splitting patterns usually can be anticipated.)
2. When multiplicative splitting occurs and the coupling constants are equal, the splitting is as predicted by the $n + 1$ rule.
3. The conditions for first-order behavior are more likely to be met using an instrument of higher operating field.
   a. Chemical shifts (in Hz) increase with increasing v_0.
   b. Coupling constants do not vary with the operating frequency.
   c. Intensity ratios within splitting patterns more closely approach the idealized values.

III. *Functional-Group NMR Absorption*

A. NMR SPECTRA OF ALKANES AND CYCLOALKANES

1. The protons in a typical alkane are in very similar chemical environments.
   a. The NMR spectra of alkanes and cycloalkanes cover a very narrow range of chemical shifts, typically $\delta\, 0.7$ to $\delta\, 1.5$.
   b. The splitting in such spectra shows extensive non-first-order behavior.
2. The chemical shifts of protons on a cyclopropane ring are unusual for alkanes; they absorb at rather high field, typically $\delta\, 0$ to $\delta\, 0.5$.
   a. The unusual chemical shifts are caused by an induced electron current present in the cyclopropane ring and the resulting magnetic field.
   b. The induced field is oriented so that the chemical shifts of the cyclopropane protons are decreased.

B. NMR SPECTRA OF ALKENES

1. A characteristic proton NMR absorption for alkenes is the absorption for the protons on the double bond, called vinylic protons.
   a. The chemical shifts of vinylic protons ($\delta\, 4.5$–6.0) are much greater than would be predicted from the electronegativity of the alkene functional group.
   b. The applied field H_0 induces a circulation of the π electrons in closed loops above and below the plane of the alkene. (See Fig. 13.15, text p. 613.)
   c. A magnetic field is induced that opposes the applied field H_0 at the center of the loop.
      i. The induced field augments the applied field at the vinyl protons.
      ii. The vinyl protons are thus deshielded and absorb at lower field.
   d. The absorption of internal vinylic protons occur at somewhat lower field than that of terminal vinylic protons.

2. Splitting between vinylic protons in alkenes depends strongly on the geometrical relationship of the coupled protons.
   a. Vinyl protons of a *cis*-alkene have smaller coupling constants than those of their *trans*-isomers.
   b. The very weak splitting (called geminal splitting) between vinylic protons on the same carbon stands in contrast to the much larger *cis* and *trans* splitting.
3. Another characteristic proton NMR absorption for alkenes is the absorption for the protons on carbons adjacent to the double bond, called allylic protons.
   a. Allylic protons have greater chemical shifts (δ 1.5–2.5) than ordinary alkyl protons.
   b. Allylic protons have considerably smaller chemical shifts than vinylic protons.
4. Splitting in alkenes is sometimes observed between protons separated by more than three bonds; these long-distance interactions between protons are transmitted by the π electrons.

C. NMR Spectra of Alkyl Halides and Ethers

1. The chemical shifts caused by the halogens are usually in proportion to their electronegativities; chloro groups and ether oxygens have about the same chemical-shift effect on neighboring protons.
2. Epoxides have considerably smaller chemical shifts than acyclic ethers.
3. Proton resonances are split by neighboring fluorine in the same general way that they are split by neighboring protons; the same $n + 1$ splitting rule applies.
   a. Values of H-F coupling constants are larger than H-H coupling constants.
   b. Coupling between protons and fluorines can sometimes be observed over as many as four single bonds.
4. The common isotopes of chlorine, bromine, and iodine do not cause detectable proton splittings.

D. NMR Spectra of Alcohols

1. Protons on the α-carbons of primary and secondary alcohols generally have chemical shifts in the same range as ethers.
2. Since tertiary alcohols have no α-protons, the observation of an OH stretching absorption in the IR spectrum accompanied by the absence of the CH—O absorption in the NMR is good evidence for a tertiary alcohol (or a phenol).
3. The chemical shift of the OH proton in an alcohol depends on the degree to which the alcohol is involved in hydrogen bonding under the conditions that the spectrum is determined.
   a. The presence of water, acid, or base causes collapse of the OH resonance to a single line and obliterates all splitting associated with this proton.
      i. This effect is caused by a phenomenon called chemical exchange—an equilibrium involving chemical reactions that take place very rapidly as the NMR spectrum is being determined.
      ii. This type of behavior is quite general for alcohols, amines, and other compounds with a proton bonded to an electronegative atom.
   b. Rapidly exchanging protons do not show spin-spin splitting with neighboring protons.
      i. Acid and base catalyze this exchange reaction, accelerating it enough that splitting is no longer observed.
      ii. In the absence of acid or base, this exchange is much slower, and splitting of the OH protons and neighboring protons is observed.
4. The assignment of the OH proton can be confirmed by what is called "the D_2O shake." (When a drop of D_2O is added to the NMR sample tube and the tube shaken, the OH protons rapidly exchange with the protons of D_2O to form OD groups on the alcohol and thus become invisible to NMR.)

IV. *Other Aspects of NMR Spectroscopy*

A. Use of Deuterium in Proton NMR

1. Although deuterium has a nuclear spin, deuterium NMR and proton NMR require different operating frequencies.
   a. Deuterium NMR absorptions are not detected under the conditions used for proton NMR.
   b. Deuterium is effectively "silent" in proton NMR.
   c. The coupling constants for proton-deuterium splitting are very small.
   d. Deuterium substitution can be used to simplify NMR spectra and assign resonances.

2. Solvents used in NMR spectroscopy must either be devoid of protons, or their protons must not have NMR absorptions that obscure the sample absorptions.
   a. $CDCl_3$ (chloroform-*d*, or "deuterochloroform") has no proton-NMR absorption, but it has all the desirable solvent properties of chloroform.
   b. Carbon tetrachloride (CCl_4) is a useful solvent because it has no protons, and therefore has no NMR absorption.

B. Carbon-13 NMR

1. Any nucleus with a nuclear spin can be studied by NMR spectroscopy; for a given magnetic field strength, different nuclei absorb energy in different frequency ranges.
2. The NMR spectroscopy of carbon is called carbon NMR or CMR; the only isotope of carbon that has a nuclear spin is ^{13}C.
   a. The relative abundance of ^{13}C suggests that carbon NMR spectra should be about 1.1% as intense as proton NMR spectra.
   b. The resonance of a ^{13}C nucleus is also intrinsically weaker than that of a proton because of the magnetic properties of the carbon nucleus.
   c. Because of these two factors, ^{13}C spectra are about 0.0002 times as intense as ^1H spectra.
   d. Special instrumental techniques have been devised for obtaining such weak spectra.
3. Coupling (splitting) between carbons generally is not observed.
   a. The reason is the low natural abundance of carbon-13.
   b. Two adjacent ^{13}C atoms almost never occur together within the same molecule.
4. The range of chemical shifts is large compared to that in proton NMR; trends in carbon chemical shifts parallel those for proton chemical shifts.
   a. Chemical shifts in CMR are more sensitive to small changes in chemical environment.
   b. It is often possible to observe distinct resonances for two carbons in very similar chemical environments.
5. Splitting of ^{13}C resonances by protons (^{13}C-^1H splitting) is so large that it adds unwanted complexity to CMR spectra.
   a. Splitting is eliminated by a special technique called proton spin decoupling; a single unsplit line is observed for each chemically nonequivalent set of carbon atoms.
   b. Spectra in which proton coupling has been eliminated are called proton-decoupled carbon NMR spectra.
      i. The DEPT technique yields separate spectra for methyl, methylene, and methine carbons, and each line in these spectra corresponds to a line in the complete CMR spectrum; lines in the complete CMR spectrum that do not appear in the DEPT spectra are assumed to arise from quaternary carbons.
6. CMR spectra are generally not integrated because the instrumental technique used for taking the spectra gives relative peak integrals that are governed by factors other than the number of carbons.
   a. The decoupling technique enhances the peaks of carbons that bear hydrogens.
   b. Peaks for carbons that bear no hydrogens are usually smaller than those for other carbons.

C. NMR Spectroscopy of Dynamic Systems

1. The spectrum of a compound involved in a rapid equilibrium is a single spectrum that is the time-average of all species involved in the equilibrium.
2. The time-averaging effect of NMR is not limited to simple conformational equilibria; the spectra of molecules undergoing any rapid process, such as a chemical reaction, are also averaged by NMR spectroscopy.

D. Other Uses of NMR

1. In pulse-Fourier transform NMR (FT-NMR), a large number of spectra are recorded and stored in a computer.
   a. The random background noise is reduced by adding together large numbers of spectra.
   b. The signals from the individual spectra are mathematically added to give a much stronger spectrum than could be obtained from a single spectrum.
2. Solid-state NMR is used to study the properties of important solid substances.
3. Phosphorus-31 NMR is used to study biological processes.

4. NMR tomography, or magnetic-resonance imaging (MRI), monitors the proton magnetic resonances in signals from water in various parts of the body; clinical scientists can achieve organ imaging without using X-rays or other potentially harmful types of radiation.

E. SOLVING STRUCTURE PROBLEMS WITH SPECTROSCOPY

1. From the mass spectrum determine, if possible, the molecular mass.
2. If an elemental analysis is given, calculate the molecular formula and determine the unsaturation number.
3. Look for evidence in both the IR and NMR spectra for any functional groups that are consistent with the molecular formula; write down any structural fragments indicated by the spectra.
4. Use the CMR spectrum and, if possible, the proton NMR spectrum, to determine the number of nonequivalent sets of carbons and/or protons or to set some limits.
   a. Use the total integral of the entire spectrum and the molecular formula to determine how many chart spaces per proton are in the integral.
      i. Count the chart spaces in the integral of each absorption.
      ii. Use the chart spaces per proton value from above to determine the number of protons in each set.
   b. Determine from the chemical shift of each set which set must be closest to each of the functional groups that are present.
   c. Write down partial structures that are consistent with each piece of evidence, and then write down all possible structures that are consistent with all the evidence.
   d. Estimate the chemical shifts of the protons in each structure, and, if possible, choose the structure that best reconciles the predicted and observed chemical shifts.
5. Rationalize all spectra for consistency with the proposed structure.
6. Most spectra contain redundant structural information.

. . . .

Study Guide Links

13.1   Quantitative Aspects of NMR

The purpose of this section is to give you a more quantitative insight into the fundamentals of the NMR experiment. Figure 13.1 on text p. 580 shows that the separation between proton spin energy levels, ΔE, increases with the magnitude of the field at the proton, H_p. The equation describing this behavior is

$$\Delta E = \frac{h\gamma_H}{2\pi} H_p \tag{SG13.1}$$

in which h is Planck's constant (text p. 537), and γ_H is a constant, called the **magnetogyric ratio,** which is related to the magnetic properties of the hydrogen nucleus. The "H" subscript on the magnetogyric ratio refers to the proton, not to the magnetic field. That is, each nucleus with a spin has a characteristic γ value, and γ_H is the value for the proton, which is 26,753 radians gauss^{-1} sec^{-1}. (The *gauss*—rhymes with *house*—is a unit of magnetic field strength.) The energy units chosen for Planck's constant determine the units of ΔE. *This equation shows that the separation between the spin energy levels is a simple linear function of the magnetic field strength.*

Let's use Eq. SG13.1 to calculate the energy separation ΔE between the +½ and −½ spin energy levels. A typical magnetic field used in proton NMR is 14,092 gauss. If ΔE is in kJ/mol, use $h = 3.99 \times 10^{-13}$ kJ sec mol^{-1}. Then ΔE is given by

$$\Delta E = \frac{(3.99 \times 10^{-13} \text{ kJ sec mol}^{-1})(26{,}753 \text{ rad gauss}^{-1} \text{ sec}^{-1})}{2(3.14 \text{ rad})} \times (14{,}092 \text{ gauss})$$

$$= 2.40 \times 10^{-5} \text{ kJ/mol} = 0.0240 \text{ J/mol} \tag{SG13.2}$$

Recall now from the text that the rf energy E_0 needed for the absorption experiment must match this energy separation. This means that the energy needed to "flip" a mole of proton spins from one state to the other is only about 0.02 J. That this corresponds to the radiofrequency region is shown by calculating the frequency from the relation $E_0 = h\nu_0$, or $\nu_0 = E_0/h$. Dividing the quantity in Eq. SG13.2 by h gives the frequency ν_0 as 60×10^6 sec^{-1}, or 60 million Hz, or 60 MHz. (Remember that the unit sec^{-1} is also called the *hertz*, Hz.) A 60 MHz frequency is in the FM radio band, and is the frequency required for NMR absorption at a field strength of 14,092 gauss. This frequency is the *applied frequency* referred to in the text.

Most of the spectra in the text were taken at 60 MHz. As you'll learn in Sec. 13.4B, modern research spectrometers operate with five to ten times stronger fields, and correspondingly higher frequencies of 300–600 MHz.

✓13.2   Other Chemical Shift Tables

The method used in the text is not the only one available for estimating chemical shifts. A number of books on spectroscopy contain charts or tables from which chemical shift estimates can be read directly.

A very useful tabulation of chemical-shift data was assembled by Tom Curphey (now a professor in the Dartmouth School of Medicine) while a graduate student at Harvard in 1961, and subsequently modified by Harry Morrison (now Professor of Chemistry and Dean of Science at Purdue). The Curphey-Morrison data are given in Table SG 13.1 on the following two pages.

(text continues on p. 295)

Table SG13.1   The Curphey-Morrison Table for Estimating Chemical Shifts[a]

Effect of a Substituent G on the Chemical Shifts of H^α and H^β

$$G\!-\!\overset{|}{\underset{H^\alpha}{C}}\!-\!\overset{|}{\underset{H^\beta}{C}}\!-$$

Base Values of Shift Positions: $-CH_3$ δ 0.90; $-CH_2-$ δ 1.20; $-\overset{|}{C}H-$ δ 1.55

Functional group G	Type of hydrogen	Alpha shift	Beta shift
—Cl	CH₃	2.43	0.63
	CH₂	2.30	0.53
	CH	2.55	0.03
—Br	CH₃	1.80	0.83
	CH₂	2.18	0.60
	CH	2.68	0.25
—I	CH₃	1.28	1.23
	CH₂	1.95	0.58
	CH	2.75	0.00
—Aryl	CH₃	1.40	0.35
	CH₂	1.45	0.53
	CH	1.33	0.22
—CH=O and —CR=O	CH₃	1.23	0.25
	CH₂	1.10	0.30
	CH	0.95	--
—CO₂H and —CO₂R	CH₃	1.23	0.25
	CH₂	1.05	0.30
	CH	1.05	--
—C≡C—	CH₃	0.78	0.05
	CH₂	0.75	0.10
	CH	1.25	0.00
—OH and —OR	CH₃	2.45	0.35
	CH₂	2.30	0.15
	CH	2.20	--
—O—Aryl	CH₃	2.95	0.40
	CH₂	3.00	0.45
	CH	3.30	--

(table continues)

Table SG13.1, continued

Functional group G	Type of hydrogen	Alpha shift	Beta shift
—OCO₂R	CH₃	2.88	0.38
	CH₂	2.98	0.43
	CH	3.45	0.36
—NR₃	CH₃	1.25	0.13
	CH₂	1.40	0.13
	CH	1.35	0.00
—C≡N	CH₃	1.10	0.45
	CH₂	1.10	0.40
	CH	1.05	0.47
—NO₂	CH₃	3.50	0.65
	CH₂	3.15	0.85
	CH	3.05	--
—C≡C—	CH₃	0.90	0.15
	CH₂	0.80	0.05
	CH	0.35	--

[a]Reprinted by permission from the Ph. D. dissertation of T. J. Curphey, Harvard University, 1961, as modified by T. J. Curphey and H. A. Morrison

The use of this table can be illustrated by a calculation of the chemical shifts of the various protons of the following compound.

$$\overset{a}{CH_3}-\overset{\overset{\displaystyle Cl}{|}}{\underset{b}{CH}}-\overset{c}{CH_2}-\overset{d}{CH_2}-Cl$$

Protons a are methyl protons; therefore begin the calculation with a base shift of δ 0.9. To this add δ 0.63 for a chlorine β to a methyl group (first line, last column of the chlorine entry). The predicted chemical shift of protons a is δ 1.53. (The observed shift is δ 1.60.)

Proton b is a methine proton; begin with a base shift of δ 1.55. To this add δ 2.55 for an α-chlorine contribution (third line for chlorine under "Alpha shift") for a predicted chemical shift of δ 4.10. (The observed shift is δ 4.27.)

Protons c are methylene protons with *two* β-chlorines. The predicted shift is δ 1.20 + δ 0.53 + δ 0.53 = δ 2.26. (The observed shift is δ 2.15.)

You should try to predict the chemical shift of the protons d. (The observed shift is δ 3.72.)

You should notice from the comparisons given above between calculated and observed shifts that the chemical shifts calculated by this (or any other) method are estimates; do not rely on them for accuracy to more than a few tenths of a part per million.

The Curphey-Morrison table gives explicitly the contributions for β shifts as well as the shifts for methine protons. Its advantage over the table in the text is that it can be used to calculate the shifts of methine protons. However, it does not predict very well the shifts of

CH_2 or CH protons bound to two or more electronegative groups. (For example, the predicted shift of CH_2Cl_2 is δ 5.8; the method in the text predicts δ 5.2; the observed shift is δ 5.3.)

A review of the approach used in the text can be found in two articles in the *Journal of Chemical Education,* which you should be able to find in your college library:

E. C. Friedrich and K. G. Runkle, *J. Chem. Educ.,* **1984,** *61,* 830–832; *ibid.,* **1986,** *63,* 127–129.

The second article extends the method discussed in the text to the chemical shifts of methine protons.

✓13.3   Approaches to Problem Solving

The approach given in the text for solving NMR structure problems is not necessarily the only approach. For example, you might take step 5 first. The point is that *you should be systematic and extract all the information you can from each piece of evidence in the spectrum before starting to write complete structures.*

A good analogy (suggested to the author by a reviewer) is the various approaches to working a jigsaw puzzle. Some people start by working around the borders, and others by looking for unique patterns or colors in the picture on the box in which the puzzle is packaged. The important point is that fitting together pieces of the puzzle at random is not likely to yield a solution in a reasonable amount of time; *your approach must be systematic.*

The dangers of using "recipes" for problem solving were discussed in Study Guide Link 9.4 on p. 196 of this manual. Remember that exceptions to any recipe will occur, and you will be best equipped to deal with such situations by developing your own approach and understanding the basis of each step. Do not let any "recipe" inhibit you from trying your own style of problem solving. Ultimately the style that works for you is the right one.

Needless to say, learning to work any type of problem requires practice. The text provides plenty of practice problems for you; you have to use them to become proficient.

✓13.4   More NMR Problem-Solving Hints

When you solve structure problems with NMR (or any sort of spectroscopy), it is very important that you be *intellectually honest* with yourself. Do *not* force a structure to fit the data. That is, write down each observation from the spectrum and then write down what it tells you. This point can be illustrated with the jigsaw-puzzle analogy in Study Guide Link 13.3. Anyone who has worked a jigsaw puzzle has encountered pieces that look as if they should go together, but do not quite fit; the temptation to force-fit the pieces is often great—but force-fitting never solves the puzzle. When you propose a structure, be able to defend it on the basis of *evidence that you can cite.* If you must guess a structure, then you haven't really solved the problem. If you are stymied by a structural problem, write down *all possible structures* that seem to fit the data. Ask yourself carefully whether each of these structures fits all the information available: chemical shift, integration, and splitting. Sometimes the act of writing down and analyzing the information in this way will lead to a solution. If you must consult the solution without having completely solved the problem, then at least you have defined the points of uncertainty. If, after looking at the solution, you still don't see why your structures are ruled out, seek assistance! You can sometimes learn more from a situation like this, as frustrating as it is, as you can from a correctly worked problem!

Now to some specific tips. The author has observed a tendency among some students to assume that protons which are physically adjacent in a structure must also have absorptions that are adjacent in their NMR spectra. Don't be caught in this trap. Although this may be true in specific cases, in general it is not something that you can rely on.

The last hint is to remind you about common splitting patterns in NMR spectra that appear over and over again. These patterns are so easy to recognize that they *immediately* suggest the presence of certain groups. Let's consider a few of these.

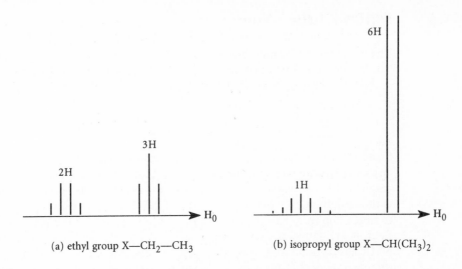

Figure SG13.1   *Some splitting patterns commonly observed in proton NMR spectra. In this diagram, absorption peaks are represented as lines.*

A three-proton triplet at high field along with a two-proton quartet at lower field [Fig. SG13.1(a)] is the earmark of an ethyl group in a structure CH_3CH_2—X, in which the group X has no coupled protons. Ethyl esters and ethyl ethers show this type of pattern. The chemical shift of the quartet provides information about the nature of the group X.

A one-proton septet or multiplet at low field along with a six-proton doublet at high field [Fig. SG13.1(b)] is characteristic of an isopropyl group. Sometimes the one-proton pattern, because of its complexity, is buried among other absorptions, or is otherwise difficult to distinguish. In such cases, the six-proton doublet alone can be used to diagnose an isopropyl group. Likewise, a three-proton doublet at relatively high field is characteristic of the methyl group in the following partial structure:

CH —CH₃   ← three-proton doublet

A three-proton singlet almost always suggests a methyl group, and a nine-proton singlet a *tert*-butyl group. Be on the lookout for these common patterns when you solve unknowns with NMR.

Several collections of known spectra are available in many libraries. Two excellent collections are the *Aldrich Library of NMR Spectra,* and a newer volume, the *Aldrich Library of Carbon-13 and 300 MHz Proton NMR Spectra.* Although these volumes contain a limited number of spectra, they do encompass a large number of simpler compounds. A more extensive collection is the *Sadtler Standard NMR Spectra,* which includes literally thousands of spectra, most of which are 60 MHz spectra. (Most of the spectra in the text are from this collection.) There is a corresponding collection of carbon-13 spectra. (Carbon NMR is discussed in Sec. 13.8 of the text.) All of these collections contain good indexes, including formula indexes, that allow you to find spectra quickly. A good way to improve your ability to interpret NMR spectra is to look through the formula indexes of these collections for compounds with relatively few carbons—say, eight carbons or fewer for aliphatic compounds—and try to write the structure from the name. Then predict the spectrum, and, finally, check your prediction against the actual spectrum. If your prediction doesn't match the actual spectrum, try to explain why.

13.5   NMR of Other Nuclei

The theory of NMR presented in Study Guide Link 13.1 can be generalized to any nucleus. The key point is that each magnetically active nucleus n has a characteristic magnetogyric ratio γ_n. Eq. SG 13.1 generalizes to the following:

$$\Delta E = \frac{h\gamma_n}{2\pi}\,\mathrm{H_p} \qquad\qquad\qquad \text{(SG13.3)}$$

Since the frequency required for energy absorption is $\Delta E/h$, then it follows that there is a characteristic frequency for each nucleus that can be calculated from a knowledge of the magnetic field strength of the NMR instrument being used and the magnetogyric ratio. This is the radiofrequency that must be "tuned in" in order to observe the nucleus of interest. For example, the magnetogyric ratio of the proton, γ_H, is 26,753 rad gauss^{-1} sec^{-1}. In a magnetic field of 117,400 gauss, for example, a frequency of 500 MHz is required to observe proton NMR absorption. Since the magnetogyric ratio for ^{13}C is 6,728 rad gauss^{-1} sec^{-1}, or about one-quarter of that for a proton, then the frequency required for observation of ^{13}C at the same field strength is also one-quarter of that for a proton, or about 125 MHz.

The intrinsic intensity of the NMR signal from each nucleus is proportional to the *cube* of its magnetogyric ratio. Thus, the relative intensity of a proton signal versus that from the same number of ^{13}C atoms is $(\gamma_H/\gamma_C)^3 = (26,753/6,728)^3$, or 62.9. In other words, a carbon NMR signal is about $1/62.9 = 0.0159$ times as intense as a proton signal. Because ^{13}C has a natural abundance of 1.1%, then carbon resonances are only $(0.0159)(0.011) = 0.000175$ times as intense as proton resonances. The only way to achieve ^{13}C spectra of reasonable intensity is to take many carbon spectra, store them digitally in a computer, and add them together. Because electronic noise is random, it is reduced when many spectra are added together, whereas the resonances themselves are enhanced. More than 5000 CMR spectra must be added together in this manner to get the same intensity as a single proton NMR spectrum of the same compound.

13.6   Fourier-Transform NMR

With Fourier-transform NMR a complete NMR spectrum can be taken every second or so. The "5000 spectra" referred to at the end of Study Guide Link 13.5 require about two hours to obtain—about the same amount of time as 5–10 spectra obtained by the older technique.

In FT-NMR, the sample is pulsed with a very intense, short pulse of rf radiation that causes all the nuclear spins to absorb energy simultaneously. The decay of absorption is then monitored for a second or two. During this time, all the nuclear spins in the sample transmit absorption information simultaneously. An analogy is the ringing of a bell. When we "pulse" (strike) the bell, all frequencies (pitches) in the bell are activated simultaneously. As the bell rings, the pitches add together to give the characteristic sound of the bell, which dies out over time. A mathematical technique called *Fourier transformation* can be used to translate the ringing of the bell over time into a plot of sound intensity *vs.* pitch (frequency). Such a plot would be a spectrum showing the different pitches of which the sound is composed and their relative intensities. The same technique can be applied to the decaying NMR signal over time to convert it into a plot of absorption intensity *vs.* frequency (or field), which is the conventional NMR spectrum.

Solutions

Solutions to In-Text Problems

13.1 (a) Apply Eq. 13.2b, with $\delta = 4.40$ and $v_0 = 90$ MHz.

$$\text{chemical shift in Hz} = \delta v_0 = (4.40)(90) = 396 \text{ Hz}$$

13.2 (a) Subtract two equations like Eq. 13.2c for two different frequencies, $v(1)$ and $v(2)$. The difference $\delta(1) - \delta(2)$ is the desired chemical-shift difference.

$$\delta(1) - \delta(2) = \frac{v(1) - v(\text{TMS})}{v_0} - \frac{v(2) - v(\text{TMS})}{v_0} = \frac{v(1) - v(2)}{v_0} = \frac{45}{60} = 0.75 \text{ ppm}$$

13.3 The principle to use is that constitutionally nonequivalent protons and diastereotopic protons in general have different chemical shifts; enantiotopic and homotopic protons have identical chemical shifts.

(a) Because protons H^a and H^b are diastereotopic, their chemical shifts are different.
(c) Because protons H^a and H^b are diastereotopic, their chemical shifts are different. Because the chair flip interconverts the positions of these two protons, their chemical shifts are the same when averaged out over time. (Sec. 13.7 on text p. 619 addresses this point.)

13.4 Parts (a)–(c) are answered in the text discussion that follows the problem.
(d) A derivative $(CH_3)_x M$, in which M is an element more electropositive than Si, should have a negative chemical shift. $(CH_3)_2 Mg$ and $(CH_3)_4 Sn$ are two of several possible correct answers.

13.5 (a) The proton in $Cl_2 CH{-}CHCl_2$ (1,1,2,2-tetrachloroethane) has the greatest chemical shift because it is adjacent to two electronegative atoms.
(c) $(CH_3)_4 Si$ has the greater chemical shift. (See the solution to Problem 13.4(d).)

13.6 To make the required estimates, apply Eq. 13.3, the β-halogen correction, and the data in Table 13.2.

(a) $G_1 = G_2 = {-}CH_3$. Eq. 13.3 gives $\delta (0.2 + 0.6 + 0.6) = \delta 1.4$.

(c) $G_1 = {-}OCH_2 CH_3$, and $G_2 = {-}CH_3$. Eq. 13.3 gives $\delta (0.2 + 2.4 + 0.6) = \delta 3.2$.

(e) $G_1 = BrCH_2{-}$, which is treated as an alkyl group with a β-halogen correction, and $G_2 = {-}CH{=}CH_2$. Eq. 13.3 gives $\delta (0.2 + 0.6 + 0.5 + 1.3) = \delta 2.6$.

(g) $G_1 = {-}CH_3$, and $G_2 = {-}CH_2 Br$, which is treated as an alkyl group with a β-halogen correction. Eq. 13.3 gives $\delta (0.2 + 0.6 + 0.6 + 0.5) = \delta 1.9$.

13.7 (a) The question asks essentially how many chemically nonequivalent sets of protons there are in each case. The compound in this part has two chemically nonequivalent sets of protons; hence, its NMR spectrum consists of two resonances.
(c) *Cis*-2-butene has two resonances, one for the vinylic hydrogens (the hydrogens attached to the double bond) and one for the methyl groups.

13.8 (a)

$$\underset{\delta 6.55}{Cl_2 CH}{-}\underset{\underset{CH_2 Cl \; \longleftarrow \; \delta 4.31}{|}}{\overset{\overset{CH_3 \; \longrightarrow \; \delta 1.99}{|}}{C}}{-}CH_3$$

1,1,3-dichloro-2,2-dimethylpropane

13.9 (a) This is a quantitative analysis problem that can be solved by assigning the resonances and using their integrals. The resonance at δ 1.8 is that of *tert*-butyl bromide, and the resonance at δ 2.20 is that of methyl iodide. (These assignments follow from Table 13.2 and Eq. 13.3.) The ratio of methyl iodide to *tert*-butyl bromide is 15:1. This follows from the fact that there are three times as many protons in the *tert*-butyl bromide molecule as in the methyl iodide molecule; thus, for a given concentration, the *tert*-butyl bromide gives a resonance that is three times as intense as that of methyl iodide.

$$\frac{5 \text{ protons } CH_3I \times (1 \text{ molecule}/3 \text{ protons})}{1 \text{ proton } (CH_3)_3CBr \times (1 \text{ molecule}/9 \text{ protons})} = \frac{15 \text{ molecules of } CH_3I}{1 \text{ molecule of } (CH_3)_3CBr}$$

The mole percent of CH_3I is the number of moles of CH_3I (15) divided by the total moles of both compounds (16) expressed as a percent. Therefore, the mole percent of CH_3I is (15/16) × 100, or 94 mole percent; the mole percent of $(CH_3)_3CBr$ is (1/16) × 100, or 6 mole percent.

(b) The $(CH_3)_3CBr$ impurity in CH_3I is more easily detected, because a given mole fraction of $(CH_3)_3CBr$ gives a resonance that is three times as strong as the resonance for the same amount of CH_3I, as the solution to part (a) of this problem demonstrated.

13.10 (a) Apply Eq. 13.3; the chemical shift for the resonance of the methyl protons should be δ (0.2 + 0.6 + 0.5 + 0.5) = δ 1.8. (Notice that *two* β-halogen corrections are applied.) The resonance of these protons thus should be a doublet (one neighboring proton) at about δ 1.8. The —CHCl$_2$ proton absorbs at about δ (5.3 + 0.7), or about δ 6.0. (The base value of δ 5.3 comes from the chemical shift of CH_2Cl_2 in Table 13.1 on text p. 587, and the δ 0.7 correction is for a methine proton; see the discussion below and at the bottom of text p. 590.) The resonance of this proton is therefore a quartet (three neighboring protons) at about δ 6. Leaning would make the high-field lines of the quartet and the low-field line of the doublet, that is, the inner lines of both patterns, somewhat larger than predicted by the idealized ratios in Table 13.3 of the text.

Throughout these solutions, δ 0.7 is used to convert a chemical-shift estimate for methylene protons into an estimate for a methine hydrogen in a similar chemical environment:

$$\delta(G_1\!-\!\underset{\underset{\text{alkyl}}{|}}{CH}\!-\!G_2) \;=\; \delta(G_1\!-\!CH_2\!-\!G_2) \;+\; 0.7$$

This correction approximates actual corrections that vary from δ 0.5 to δ 1.0 in particular cases.

Do not be concerned if your chemical-shift estimates are not exactly the same as those in these answers; your goal should be to get them in the correct general region of the spectrum, within 0.3–0.5 ppm.

13.11 (a) The resonance for the methyl groups is a six-proton doublet at about δ 1.4, and the resonance for the methine proton is a one-proton septet at about δ 3.1. Here is how these estimates were obtained. The δ 1.4 estimate starts with δ 0.9 for a methyl group bound to an alkyl group, to which is added δ 0.5 for a β-halogen. The δ 3.1 estimate starts with δ 2.4 for an alkyl—CH_2—I group (δ 0.6 for the alkyl group, δ 1.8 for the iodine) to which is added δ 0.7 for the additional alkyl substituent. (See the discussion following the solution to part (a) of the previous problem for an explanation of this correction.)

(c) The resonance for the protons α to the oxygen should be a four-proton triplet at about δ 3.2, and the resonance for the remaining protons should be a two-proton quintet at about δ 1.9.

13.12 There are four possibilities for the spin of three neighboring equivalent protons b; these are shown in Fig. SG 13.2 at the top of the following page. This diagram shows that the resonance for protons a would be split by the three protons b into a quartet whose lines are in the intensity ratio 1:3:3:1.

all spins +½
(one way; relative probability 1)

two spins +½, one spin –½
(three ways; relative probability 3)

two spins –½, one spin +½
(three ways; relative probability 3)

all spins –½
(one way; relative probability 1)

Figure SG13.2   *Spin possibilities for three neighboring protons to accompany the solution to Problem 13.12.*

13.13 (a)

δ 1.03 δ 1.88 δ 3.40

CH_3—CH_2—CH_2—Br

(c)

CH_2Br
|
$BrCH_2CCH_2Br$
|
CH_2Br

13.14 (a) The reaction is the peroxide-promoted HBr addition to the alkene (Sec. 5.6 in the text):

$BrCH_2CH$═CH_2 + HBr $\xrightarrow{\text{peroxides}}$ $BrCH_2CH_2CH_2Br$

3-bromopropene   1,3-dibromopropane

(b) The chemical shifts and splittings are indeed consistent with this structure. The resonance of the central methylene group is predicted to be a quintet at δ 2.4; and the resonance for the two methylenes α to the bromines is predicted to be a triplet at δ 3.1. The splittings agree exactly, and the observed chemical shifts are acceptably close.

 Early printings of the text have an error in the chemical shift of the quintet; it should be δ 2.38, not δ 1.38.

13.15 The resonances for protons H^c and H^a will be doublets at their respective chemical shifts, and the resonance of proton H^b will be split into a quartet ($J = 6$ Hz) by the three protons H^c, and each line of this quartet will be split into a doublet ($J = 3.5$ Hz) by the single proton H^a. The resonance for H^b can be analyzed with a splitting diagram as shown in Fig. SG13.3(a) at the top of the following page. The complete spectrum is sketched in Fig. SG13.3(b)

13.16 (a) Because protons H^b and H^c are diastereotopic, they are chemically nonequivalent; therefore, counting the methyl group, there are four chemically nonequivalent sets of protons. The resonance for proton H^a is split into a quartet by the methyl group; each line of that quartet is split into a doublet by proton H^b, for a subtotal of eight lines; and each line of the eight is split into a doublet by proton H^c, for a grand total of sixteen lines. All three coupling constants are evidently different enough for all sixteen lines to be discernible in the spectrum.

13.17 (a) The alkyl halide *A* with the formula C_4H_9Cl that gives a single line in its NMR spectrum is *tert*-butyl chloride, $(CH_3)_3C$—Cl. For compound *B*, the total integral is 37 spaces, and the low-field sextet integrates for 4 of these spaces; this signal is therefore caused by 4/37, or about 1/9, of the protons in *B*, or one proton. This defines the following partial structure:

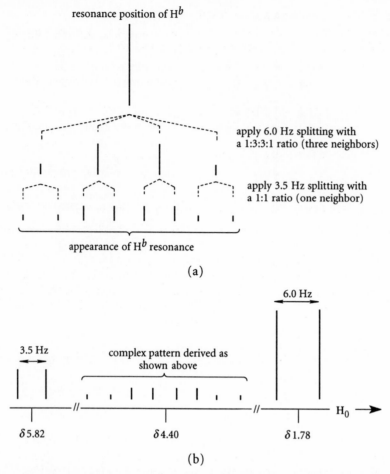

Figure SG13.3   *(a) Splitting diagram for proton H^b to accompany the solution to Problem 13.15. (b) The complete NMR spectrum for 1,1,2-trichloropropane. Compared to the diagram in part (a), horizontal and vertical scales have been compressed.*

The only alkyl halide with this partial structure and the formula C$_4$H$_9$Cl is *sec*-butyl chloride (2-chlorobutane):

H$_3$C—CH$_2$—CH—CH$_3$ *sec*-butyl chloride (2-chlorobutane)
                |
                Cl

13.18   First, prepare CH$_3$CD$_2$CH$_2$CD$_2$Cl. The resonances for the protons of the CH$_3$— and —CH$_2$— groups would be singlets, whose positions would give their chemical shifts directly. Then prepare the compound CD$_3$CH$_2$CD$_2$CH$_2$Cl; its spectrum would show two other singlets, whose chemical shifts could be readily assigned. In each pair of singlets, the one farther downfield would be assigned to the protons nearer the chlorine.

13.19 The integral is 38 spaces (about 2.71 spaces/proton). The resonance at $\delta\,4.6$, in the vinylic proton region, accounts for about 2 protons; the resonance at $\delta\,1.8$, in the allylic proton region, accounts for about 3 protons; and the singlet at $\delta\,1.05$ accounts for 9 protons (a *tert*-butyl group). Structure *A* is the correct one:

$$(CH_3)_3C,\ H_3C \diagdown C{=}CH_2$$

2,3,3-trimethyl-1-butene

A

$$(CH_3)_3C{-}CH{=}CH{-}CH_3$$

4,4-dimethyl-2-pentene

B

Structure *B* is ruled out by several pieces of evidence. First, the chemical shift of the vinylic proton resonance is more consistent with structure *A*. (See Table 13.4 on text p. 612.) Second, the resonance of the —CH₃ group in structure *B* would be a doublet with a typical splitting of about 6–8 Hz, and the resonance of the vinylic protons would be very complex (*cis*- or *trans*-splitting superimposed on a quartet for the splitting by the —CH₃ group). This very complex splitting is not observed. On the other hand, the very small splittings in the spectrum are consistent with structure *A*. The vinylic protons are weakly split by each other (geminal splitting) and by the —CH₃ group (allylic splitting); the resonance of the vinylic —CH₃ group shows the same type of weak allylic splitting by the vinylic protons.

13.20 (a) The large splittings suggest nearby fluorines; there seem to be no smaller splittings, a fact suggesting that the two sets of protons are separated by at least one carbon. A structure that meets all these requirements is $H_3C{-}CF_2{-}CH_2{-}Cl$. Notice that the two H-F coupling constants are similar; this is reasonable, since each set of protons is separated from the fluorine by the same number of bonds.

13.21 The spectrum of ethyl fluoride would look very much like that of ethyl chloride, which, in turn, would resemble that of ethyl bromide (Fig. 13.4, text p. 594), except that the quartet would be at lower field. In ethyl fluoride, the resonance of the methyl group would be a *triplet of doublets*. (Splitting by the —CH₂— group gives three lines; each of these are split into two by the fluorine for a total of six.) The resonance of the —CH₂— group would be a *doublet of quartets*. (Splitting by the fluorine gives a widely spaced doublet; each line of the doublet is split into a quartet by the —CH₃ group.)

13.22 (a) The exchangeable proton suggests the presence of an alcohol. The resonance at $\delta\,5.41$ is a vinylic proton adjacent to two other protons, and the resonance at $\delta\,4.41$ arises from two protons adjacent to both the oxygen and the vinyl group. (Their chemical shift is too great to be caused by either the oxygen or the vinyl group alone.) Evidently the sample is wet enough that the splitting between the —OH proton and the α-proton is not observed. 3-Methyl-2-buten-1-ol, $(CH_3)_2C{=}CHCH_2OH$, is consistent with the data.

13.23 A greater concentration of ethanol pushes the hydrogen-bonding equilibrium toward the hydrogen-bonded form. (See the solution to Problem 12.12 on p. 279 of this manual for a discussion of this point.) Hence, there is more of the hydrogen-bonded form of ethanol in the 2 *M* solution. Since the hydrogen-bonded form of ethanol has a greater O—H chemical shift (see text p. 617), the 2 *M* solution of ethanol has the greater chemical shift.

13.24 At room temperature, the three conformations of 1-bromo-1,1,2-trichloroethane are in rapid equilibrium. Hence, the NMR spectrum of this compound is a singlet whose chemical shift reflects the environment of the proton averaged over time across all conformations. Any changes observed at low temperature would be related to the fact that the interconversion between the different conformations is retarded, and thus certain conformations can be individually observed at the lower temperature. There are three such conformations, two of which are enantiomers:

![Newman projection structures showing three conformations labeled A, B, and C with Cl, Br, Hᵃ, Hᵇ substituents; B and C are bracketed as enantiomeric conformations]

enantiomeric conformations

In conformation A, protons H^a and H^b are equivalent and have the same chemical shift. In either one of the enantiomeric conformations, H^a and H^b are chemically nonequivalent because they are diastereotopic. In either of these conformations, the resonance of H^a is split by H^b and is therefore a doublet; likewise, the resonance of H^a is split by H^b and is also a doublet. Hence, the spectrum of conformation B should consist of two doublets, that is, four lines of equal intensity, assuming no overlap. Because conformations B and C are enantiomeric, their spectra will be identical. Hence, the total spectrum should consist of one line for conformation A, and two doublets for conformations B and C, for a total of five lines at low temperature, assuming no overlaps. The relative integrals depend on the amount of form A relative to the amounts of forms B and C; the latter two forms will be present in equal amounts.

13.26 Because of its symmetry, 4-heptanol has only four chemically nonequivalent sets of carbons, and hence its spectrum consists of four lines, and is therefore spectrum 1. All of the carbons of 3-heptanol are chemically nonequivalent; hence, the CMR spectrum of this compound consists of seven lines, and is therefore spectrum 2.

13.27 (a) This structure almost fits the data; however, it predicts two resonances for carbons attached to a single oxygen atom, that is, two resonances in the δ 50 region of the spectrum. This structure is not consistent with the data because the data indicate the presence of only one such carbon. Because methyl and methylene carbons have significantly different chemical shifts, it is unlikely that two such resonances would overlap.

(c) This structure contains five chemically nonequivalent sets of carbons and therefore would have a CMR spectrum consisting of five lines, not the three observed. The chemical environments of all protons are different enough that accidental overlaps are unlikely.

13.28 (a) Two of the most obvious reasons that *trans*-1-chloro-2-butene is not consistent with the data are that this alkene would have a spectrum in which the integral indicates a total of *two*, not three, vinylic protons, and *two*, not one, protons on the carbon α to the chlorine.

 When considering possible structures for consistency with NMR data, there is a tendency among beginning students to try to wrestle with splitting data before they have considered other aspects of the data that are more straightforward and at least as powerful. Notice in the foregoing solution that the alternative structure can be clearly ruled out on the basis of relative integral.

13.29 (a) A compound of this structure would have only two resonances in its CMR spectrum, not the three observed.

Solutions to Additional Problems

· ·

13.30 The *chemical shift* is used to determine what functional groups are near the proton(s) under observation. The *integral* is used to determine how many protons are being observed relative to the total number in the molecule. The *splitting* is used to determine how many protons are on carbons adjacent to the carbon that bears the proton of interest.

13.31 (a) The spectrum of cyclohexane is a singlet. The spectrum of its constitutional isomer *trans*-2-hexene has many resonances.

(c) The spectrum of 1,1-dichlorohexane would show a one-proton triplet at lowest field; this resonance would be at a chemical shift of at least δ 5.9. 1,6-Dichlorohexane would show a four-proton triplet at lowest field, but this would be at a chemical shift of about δ 3.3. 1,2-Dichlorohexane would show a more complex spectrum at lowest field: a two-proton doublet at about δ 3.8, and a one-proton multiplet (or quintet) at a chemical shift of at least δ 4.3.

(e) The first compound, $Cl_3CCH_2CH_2CHF_2$, has the absorption at lowest field and the more complex spectrum; furthermore, this low-field absorption will show the typically large H-F splitting. Only the second compound, $CH_3CH_2CCl_2CClF_2$, will show a three-proton triplet at about δ 1.

(g) The *cis* and *trans* isomers will have the absorptions at lowest field because they contain protons that are both vinylic and α to the halogens. The *cis* isomer will have smaller proton-proton splitting than the *trans* isomer. The remaining compound will show very small (geminal) splitting. (See Table 13.5 on text p. 614.)

13.32 (a) Chemical shift in Hz is proportional to operating frequency.

(b) Coupling constant J is independent of operating frequency.

(c) The chemical shift in ppm (δ) does not change with operating frequency because it is *defined* as the proportionality constant that relates the chemical shift in Hz to the operating frequency. (See Eq. 13.2c on text p. 584.) Because the latter two quantities are proportional, δ for a given resonance is the same at any operating frequency.

(d) NMR spectroscopy requires that the sample be situated in a magnetic field before absorption of electromagnetic radiation can occur. Other forms of absorption spectroscopy do not require the presence of a magnetic field. The field establishes the energy difference between nuclear spins. NMR spectroscopy also differs conceptually in the phenomenon responsible for absorption: the "flipping" of nuclear spins.

(e) The differences in chemical shift between all signals, in Hz, must be much greater than their coupling constants.

(f) Some of the protons change from a lower-energy spin to a higher-energy spin.

13.33 (a)

(c) $(CH_3)_3C$—O—$C(CH_3)_3$

di-*tert*-butyl ether
(2-*tert*-butoxy-2-methylpropane)

(e)

H_3C $\diagdown$ CH_3

H_3C $\diagup$ CH_3

cyclohexane

1,1,2,2-tetramethylcyclopropane

13.34 (a)

CH_3
|
H_3C—C—CH_3
|
CH_3

2,2-dimethylpropane
(neopentane)

(c) $(CH_3)_2C$=$CHC(CH_3)_3$

2,4,4-trimethyl-2-pentene

(e) $(CH_3)_3CCH_2CH$=CCl_2

1,1-dichloro-4,4-dimethyl-1-pentene

(g) The two protons are evidently equivalent (a single chemical shift), and they must be very close to the fluorines (large splitting). The compound is 1,2-dibromo-1,1-difluoroethane, $BrCH_2CF_2Br$.

(i) Since there are sixteen protons, the integrals account for two, twelve, and two protons, respectively. The compound is 1,1,3,3-tetramethoxypropane, better known to chemists as malonaldehyde dimethyl acetal, $(CH_3O)_2CHCH_2CH(OCH_3)_2$. What makes this problem tricky—and the whole point of the problem—is that not all equivalent protons of a given type are on the same carbon. For example, the two CH protons are equivalent and therefore have exactly the same chemical shift; they are split into a triplet by the adjacent CH_2 protons. Likewise, the CH_2 protons are split by the two adjacent protons into a triplet. We could just as well view this splitting as a doublet of doublets (2×2) in which the coupling constants for both splittings are identical so that, as a result, the inner lines exactly overlap. (See Fig. 13.11 on text pp. 606–607 and the accompanying discussion.)

(k) The integral is 7 spaces/proton; the doublet at δ 3.9 accounts for a —CH_2— group bound to an electronegative group, probably the bromine, and adjacent to a carbon bearing a single hydrogen. The remaining resonances consist of a complex pattern in the vinylic region that integrates for three protons. The compound is 3-bromopropene (allyl bromide, $CH_2\text{=}CH\text{—}CH_2Br$).

13.35 Compound *A* is an alkene with twelve equivalent allylic protons and no vinylic protons. Compound *A* is 2,3-dimethyl-2-butene, $(CH_3)_2C\text{=}C(CH_3)_2$.

     The IR spectrum of compound *B* indicates a $H_2C\text{=}C$ group. The NMR spectrum shows a typical isopropyl pattern: a septet/doublet pattern (δ 2.20, δ 1.07, respectively); from its chemical shift it appears that the isopropyl CH group is allylic. The isopropyl group must be attached to a carbon that bears no protons, since the CH of the isopropyl group is split only by the adjacent methyls. The presence of two vinylic hydrogens (δ 4.60) is in agreement with the IR spectrum, and the three equivalent hydrogens remaining probably correspond to a methyl group. Compound *B* is 2,3-dimethyl-1-butene.

Although the two vinylic hydrogens are in principle chemically nonequivalent, they evidently have the same chemical shift. This is reasonable, since both are surrounded by very similar functional groups. Because they have the same chemical shift, they do not split each other. The methyl groups and the vinylic hydrogens show allylic splitting of 1.5 Hz.

     Compound *C* shows a nine-proton singlet characteristic of a *tert*-butyl group. The remaining three protons are vinylic. The IR spectrum shows C=C stretching and C—H bending absorptions that confirm the presence of a —$CH\text{=}CH_2$ group. Although it is not necessary to interpret the splitting to define the structure, it is very similar to that in Figs. 13.9–13.10 on text pp. 604–605. Compound *C* is 3,3-dimethyl-1-butene: $(CH_3)_3C\text{—}CH\text{=}CH_2$.

     That compound *D* is not an alkene follows from both the chemical data and the NMR absorption. A six-carbon hydrocarbon that is a singlet in the NMR, that has one degree of unsaturation, and that is not an alkene can be only cyclohexane.

cyclohexane
(compound *D*)

The structures of compound *A* and *D* would be difficult to distinguish on the basis of their proton NMR spectra alone; note that additional reactivity data were provided in this problem. However, distinguishing these two compounds would be trivial with CMR; can you see why?

13.36 (a) The NMR spectrum of the starting material has two sets of resonances, one of which is in the vinylic-proton region (δ 4.5–5.0). The NMR of the product is a singlet near δ 1.7.
(b) The NMR spectrum of the starting material is a singlet at δ 1.7; see compound *A* in the solution to Problem 13.35. The NMR spectrum of the product should contain an isopropyl doublet-septet pattern at about δ 0.9 and δ 2, respectively, as well as a six-proton methyl singlet near δ 1.4.

13.38 The doublet near δ 1 indicates a $CH_3\text{—}CH$ partial structure. Both possible structures have the formula C_7H_{12}. With this formula, the total integral of 39.5 chart spaces corresponds to about 3.3 spaces per proton. Therefore, the resonances near δ 5.6, which are the vinylic-proton resonances, integrate for two protons. Only 3-methylcyclohexene is consistent with both of these observations.

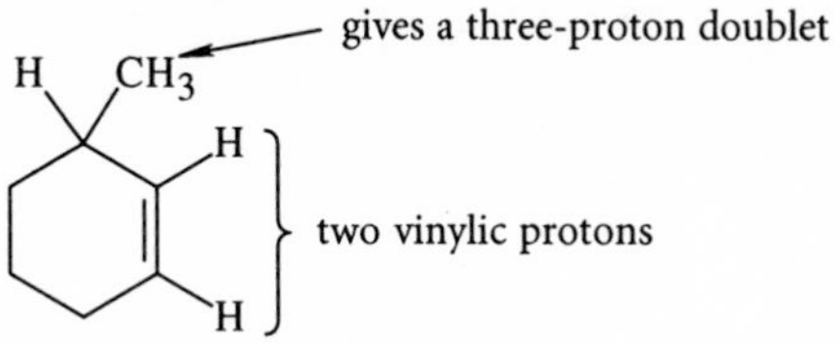

3-methylcyclohexene 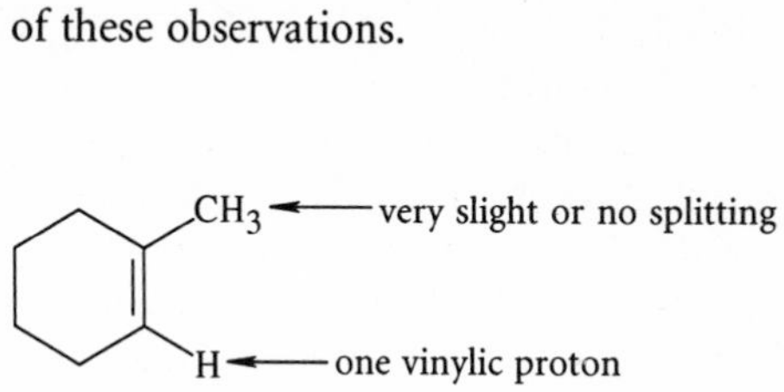   1-methylcyclohexene

13.40 Compound *B* is ruled out because it should have only three carbon resonances. The key to deciding between the remaining two compounds is the resonance with the greatest chemical shift. Only in compound *A* is the methine carbon bound to two oxygens; only a carbon bound to two oxygens can account for a chemical shift of δ 99.6. Therefore the spectrum is of compound *A*.

13.41 (a) Because the chair flip is rapid, the relationships of the carbons in these compounds can be decided from the planar projections shown. The compound shown in this part should have six CMR absorptions because there are six chemically nonequivalent sets of carbons (unless there are accidental overlaps between some absorptions). In the structure of this compound shown below, carbons with the same number are enantiotopic, and therefore chemically equivalent. The following pairs of carbons are diastereotopic, and therefore chemically nonequivalent: 3 and 4; 5 and 6; 1 and 2. All other pairs are constitutionally nonequivalent.

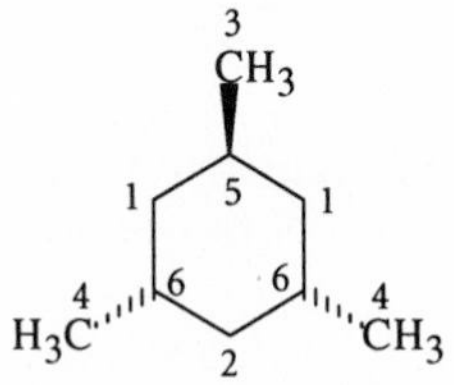

13.42 (a) The NMR spectrum of the compound containing deuterium should consist of two triplets of equal intensity. (The coupling constant between H and D nuclei on adjacent carbons is nearly zero.) The NMR spectrum of the other compound should consist of a triplet at low field and a quintet at higher field; the triplet should have twice the integrated intensity of the quintet. Each set of resonances in one spectrum should have the same chemical shift as one set of resonances in the other.

(c) To decide on the spectra for the first two compounds, first analyze the third compound. The resonance of the methyl group will be a three-proton singlet. The two methylene hydrogens are diastereotopic and are therefore chemically nonequivalent. Hence, they should have different (if only slightly different) chemical shifts, and each will split the other into a doublet. Thus, the resonances for these two protons should consist of a closely spaced leaning doublet of doublets, perhaps resembling the pattern in Fig. 13.6(c) on text p. 597.

    Because the first two compounds are diastereomers, they should have different NMR spectra. To begin with, neglect the splitting by deuterium. (Deuterium splitting, which is considered in Problem 13.43, can be eliminated by an instrumental technique.) Each spectrum should consist of a three-proton singlet for the methyl group and a one-proton singlet for the methylene hydrogen. Although the methyl singlets will likely have identical chemical shifts in the two compounds, the chemical shifts of the other hydrogens will differ by essentially the same amount as the chemical shifts of the two methylene hydrogens in the third compound. The resonance of the methylene hydrogen in the first compound will have the same chemical shift as the resonance of the pro-*S* hydrogen in the third compound; and the resonance of the methylene hydrogen in the second compound will have the same chemical shift as the resonance of the pro-*R* hydrogen in the third compound. (This analysis neglects very slight chemical-shift differences that may be caused by the α-deuterium in each case.) A summary of these spectra is shown in Fig. SG13.4 on the top of the following page.

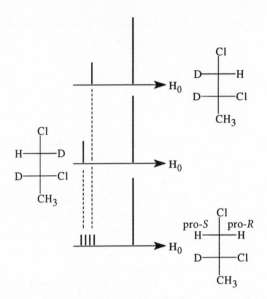

Figure SG13.4 *Schematic spectra to illustrate the differences expected in the NMR spectra of the three compounds in Problem 13.42(c). In these spectra, the pro-S hydrogen is arbitrarily assigned the greater chemical shift, but the chemical shifts of the two diastereotopic hydrogens might be reversed. Splitting by deuterium is neglected.*

13.43 (a) The nitrogen can take on three values of spin with equal probability: +1, 0, and –1. Hence, the *proton* NMR spectrum should (and does) consist of three lines of equal intensity.

(c) The two deuteriums can have the following values of spin:

Total spin = +2:	+1, +1 (one way)
Total spin = +1:	0, +1 and +1, 0 (two ways)
Total spin = 0:	0, 0; +1, –1; and –1, +1 (three ways)
Total spin = –1:	0, –1 and –1, 0 (two ways)
Total spin = –2:	–1, –1 (one way)

Each value of the total deuterium spin makes a different contribution to the position of the proton resonance; thus, there are five lines, with relative intensities proportional to relative probabilities of deuterium spin (that is, to the number of different ways that each spin can be achieved): 1:2:3:2:1.

13.44 (a) The IR spectrum indicates that the compound is both an alcohol (broad absorption near 3400 cm^{-1}) and an alkene (absorption near 1640 cm^{-1}). The alcohol functional group is confirmed in the NMR spectrum by the signal for the exchangeable hydrogen at δ 1.8. The C—H bending region of the IR spectrum is unfortunately obscured by other absorptions, but the integration of the vinylic proton region in the NMR spectrum is for three protons; hence, there is a $H_2C{=}CH{-}$ (terminal vinyl) group. From the formula, the molecule can contain no other unsaturation. The compound is 4-penten-2-ol (structure *A* below). Structure *B*, 1-penten-3-ol, is ruled out by the chemical shift of the α-proton resonance (δ 3.8) and by the splitting of the methyl resonance (the doublet at δ 1.25). In structure *B*, the α-proton resonance would be observed near δ 4.8, and the methyl group resonance would be a triplet.

$$H_2C{=}CHCH_2\overset{\displaystyle OH}{\underset{\displaystyle }{C}}HCH_3 \qquad H_2C{=}CH\overset{\displaystyle OH}{\underset{\displaystyle }{C}}HCH_2CH_3$$

A   *B*

4-penten-2-ol
(correct structure)

(b) The base peak at $m/z = 45$ corresponds to loss of 41 mass units (an allyl radical):

$$\text{molecular ion of} \atop \text{4-penten-2-ol}$$ $$\text{allyl radical} \qquad m/z = 45$$

13.46 The odd mass and the loss of 12 mass units to give the peak at $m/z = 87$ cast serious doubt that the peak at $m/z = 99$ is the molecular ion of compound *A*. Treatment of compound *A* with acid gives a compound with a peak of highest mass at $m/z = 98$. The absorption near 3400 cm^{-1} in the IR spectrum of compound *A* indicates that *A* is an alcohol. If so, it must be a *tertiary alcohol*, because there are no α-proton resonances in the proton NMR spectrum near δ 3.5. The CMR spectrum suggests a high degree of symmetry (only three carbon resonances). Treatment of tertiary alcohols with acid gives alkenes; hence, it seems reasonable that compound *B* is an alkene. If, as its mass spectrum suggests, its molecular mass is 98, then the molecular mass of the tertiary alcohol *A* is 98 + one water molecule, or (98 + 18) = 116. Indeed, the mass spectrum of compound *A* contains peaks for 116 – OH and 116 – H_2O ($m/z = 99$ and 98, respectively). A reasonable formula for an alcohol with molecular mass = 116 is $C_7H_{16}O$, and the only possible alcohol of this formula with only three chemically nonequivalent sets of carbons is 3-ethyl-3-pentanol. Compound *B*, the dehydration product, is then 3-ethyl-2-pentene.

$$\text{3-ethyl-3-pentanol} \atop (\text{compound } A)$$ $$\text{3-ethyl-2-pentene} \atop (\text{compound } B)$$

Notice that the base peak at $m/z = 87$ in the mass spectrum of the alcohol corresponds to a loss of 29 mass units (an ethyl group). Three different ethyl groups can be lost from the molecular ion of compound *A* by α-cleavage.

13.48 The sample of 2,5-hexanediol is a mixture of diastereomers, the racemate and the *meso* compound. Each of these stereoisomers has a CMR spectrum that consists of three resonances, but the two spectra are different.

$$(\pm)\text{-2,5-hexanediol} \atop (\text{one of the two enantiomers})$$ $$meso\text{-2,5-hexanediol}$$

13.49 (a) The proton NMR spectrum of 4-methyl-1-penten-3-ol should contain eight sets of absorptions because the compound contains eight chemically nonequivalent sets of protons (numbered 1–8 below).

4-methyl-1-penten-3-ol

Notice that protons 1 and 2 are diastereotopic and therefore chemically nonequivalent, as are protons 7 and 8.

(b) Because all carbons are chemically nonequivalent, 4-methyl-1-penten-3-ol should have six carbon resonances in its CMR spectrum. (Notice that the carbons of methyl groups 7 and 8 in the structure shown in part (a), like the protons of these groups, are diastereotopic.)

13.51 The ^{17}O NMR resonance is split into a triplet by the two protons of water. The splitting pattern suggests that, in the presence of acid, the ^{17}O resonance is split by *three* protons. Evidently, this splitting is due to the hydronium ion, H_3O^+.

13.52 (a) First, label the protons for discussion:

$$\overset{3}{Cl}-\overset{}{CH_2}-\overset{2}{CH_2}-\overset{1}{OH} \qquad \text{2-chloroethanol}$$

The NMR spectrum of a very dry sample would show splitting between protons 1 and 2. Hence, the resonance of proton 1 should be a triplet; the resonance of protons 2 should be either a quartet (if $J_{12} = J_{23}$) or a doublet of triplets (if $J_{12} \neq J_{23}$). In a wet sample, the splitting associated with proton 1 is obliterated. In that case, the resonance of proton 1 should be a singlet; the resonance of protons 2, which is split only by protons 3, should be a triplet. The resonance of protons 3 should be a triplet in both the wet and the dry sample.

13.53 Note that the ^{13}C resonances from *unenriched* compounds will be much weaker than those from the isotopically enriched compounds and can be ignored. The proton-decoupled CMR spectrum will be a composite spectrum that consists of the spectra of the individual species present. These species, and their relative abundances, will be as follows ($\overset{*}{C} = {}^{13}C$):

$\overset{*}{C}H_3-\overset{*}{C}H_2-Br$ Relative probability $= (0.5)(0.5) = 0.25$ two doublets

$CH_3-\overset{*}{C}H_2-Br$ Relative probability $= (0.5)(0.5) = 0.25$ one singlet

$\overset{*}{C}H_3-CH_2-Br$ Relative probability $= (0.5)(0.5) = 0.25$ one singlet

CH_3-CH_2-Br Relative probability $= (0.5)(0.5) = 0.25$ no ^{13}C signal

Each resonance will show an apparent "triplet" pattern: the line from the singly-enriched species will be in the center of a doublet for the doubly-enriched species. The singly-enriched species are singlets because there is so little ^{13}C resulting from the natural abundance of ^{13}C at each carbon that it can be ignored. Each line of the doublet will have half the total intensity of the doublet, and therefore half the intensity of the singlet. A diagram of the spectrum is as follows:

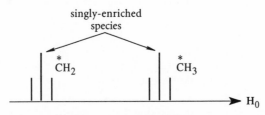

13.54 The resonance of the unpaired electron is split into four lines by the three neighboring protons according to the same $n + 1$ rule that applies to splitting between nuclei.

13.55 (a) At 25 °C the conformations equilibrate too rapidly to be separately observed in the NMR spectrum; a single resonance is observed that is the "time-average" resonance of all conformations. At low temperature, however, the rates of the internal rotations are considerably reduced, and certain conformations can be separately observed:

enantiomeric conformations have
identical NMR spectra

The first of these conformations is the conformational diastereomer of the other two and hence has a different NMR spectrum. The other two conformations are enantiomeric and therefore have identical NMR spectra. Consequently, two resonances are observed: one for the first conformation, and the second for the other two.

(b) The two lines at low temperature have different intensities because the diastereomers are present in different amounts. The intensity of each resonance is proportional to the amount of the conformation of which it is characteristic.

Chemistry of Alkynes

Terms

Concepts

I. Introduction to Alkynes

A. GENERAL

1. Alkynes (acetylenes) are compounds containing carbon–carbon triple bonds; the simplest member of this family is acetylene (ethyne), H—C≡C—H.
   a. The chemistry of the carbon-carbon triple bond is similar in many respects to that of the carbon-carbon double bond.
   b. The unique chemistry of alkynes is mostly associated with the bond between hydrogen and the triply bonded carbon, the ≡C—H bond.
2. Naturally occurring alkynes are relatively rare and do not occur as constituents of petroleum.
   a. Acetylene can be produced by heating coke (carbon from coal) with calcium oxide in an electric furnace to yield calcium carbide, CaC_2.
      i. Calcium carbide is an organometallic compound that can be conceptually regarded as the calcium salt of the acetylene dianion.
      ii. Calcium carbide reacts vigorously with water to yield acetylene and calcium oxide.
   b. Acetylene can also be made by thermal cracking of ethylene.

B. COMMON NOMENCLATURE OF ALKYNES

1. In common nomenclature, simple alkynes are named as derivatives of the parent compound acetylene.

$$CH_3CH_2—C≡C—CH(CH_3)_2 \qquad \text{ethyl isopropyl acetylene}$$

2. Certain compounds are named as derivatives of the propargyl group, H—C≡C—CH_2— (the triple-bond analog of the allyl group).

$$H—C≡C—CH_2—OH \qquad \text{propargyl alcohol}$$

C. SYSTEMATIC NOMENCLATURE OF ALKYNES

1. The systematic nomenclature of alkynes is very much like that of alkenes.
   a. The suffix *ane* in the name of the corresponding alkane is replaced by the suffix *yne*.
   b. The triple bond is given the lowest possible number.
2. When double bonds and triple bonds are present in the same molecule, the principal chain is the carbon chain containing the greatest number of double and triple bonds.
3. The numerical precedence of a double bond or triple bond within the principal chain is decided by the "first point of difference" rule.
   a. Precedence is given to the bond that gives the lowest number in the name of the compound, whether it is a double or triple bond.
   b. The double bond is always cited first in the name by dropping the terminal *e* from the *ene* suffix.

$$\overset{6}{C}H_3—\overset{5}{C}=\overset{4}{C}—\overset{3}{C}H_2—\overset{2}{C}≡\overset{1}{C}—H \qquad \text{5-methyl-4-hexen-1-yne}$$
$$\underset{CH_3}{|}$$

4. Substituent groups that contain a triple bond (called alkynyl groups) are named by replacing the final *e* in the name of the corresponding alkyne with the suffix *yl*, and they are numbered from the point of attachment to the main chain.

$$H_3C—\text{(cyclohexene ring)}—C≡C—CH_3 \qquad \text{4-methyl-1-(1-propynyl)cyclohexene}$$

5. Groups that can be cited as principal groups are given numerical precedence over the triple bond.

$$\overset{OH}{\underset{CH_3}{\overset{|}{\underset{|}{H_3\overset{1}{C}—\overset{2}{C}—\overset{3}{C}H_2—\overset{4}{C}≡\overset{5}{C}—\overset{6}{C}H_3}}}} \qquad \text{2-methyl-4-hexyn-2-ol}$$

II. Physical Properties of Alkynes

A. STRUCTURE AND BONDING IN ALKYNES

1. The bonding in alkynes involves the hybridization of the 2*s* orbital with one 2*p* orbital to form two *sp* hybrid orbitals.
   a. An *sp*-hybridized carbon atom has two *sp* orbitals at a relative orientation of 180°.
      i. An *sp* orbital has much the same shape as an *sp*^2 or *sp*^3 orbital (one small lobe, one large lobe).
      ii. Electrons in an *sp* orbital are somewhat closer to the carbon nucleus than they are in *sp*^2 or *sp*^3 orbitals.
   b. The two remaining unhybridized *p* orbitals lie along axes that are at right angles both to each other and to the *sp* orbitals.

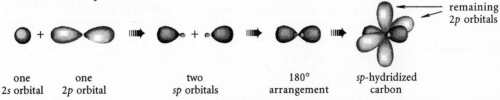

| one 2*s* orbital | one 2*p* orbital | two *sp* orbitals | 180° arrangement | *sp*-hybridized carbon |

   c. The *sp* hybridization state is inherently less stable than the *sp²* hybridization state.

  2. The acetylene molecule is linear.

   a. One bond between the carbon atoms is a two-electron σ bond resulting from the overlap of two *sp* hybrid orbitals.

   b. Two π bonds are formed by the side-to-side overlap of the *p* orbitals.

    i. These bonding π molecular orbitals are mutually perpendicular, and each contains two electrons.

    ii. The total electron density from all the π electrons taken together forms a cylinder, or barrel, about the axis of the molecule.

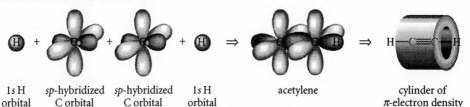

1s H   *sp*-hybridized   *sp*-hybridized   1s H   acetylene   cylinder of
orbital   C orbital   C orbital   orbital π-electron density

   c. *Cis-trans* isomerism cannot occur in alkynes because the bond angles between the C≡C bond and the attached atoms are 180°.

   d. Cycloalkynes smaller than cyclooctyne cannot be isolated under ordinary conditions.

  3. The C≡C bond is shorter than the C=C and C—C bonds.

  4. Heats of formation show that:

   a. alkynes are less stable than isomeric dienes;

   b. the *sp* hybridization state is inherently less stable than the *sp²* hybridization state, other things being equal;

   c. alkynes with internal triple bonds are more stable than isomeric alkynes with terminal triple bonds.

B. BOILING POINTS AND SOLUBILITIES

  1. The boiling points of most alkynes are not very different from those of analogous alkenes and alkanes.

  2. Alkynes have much lower densities than water.

  3. Alkynes are insoluble in water.

III. Spectroscopy of Alkynes

A. IR SPECTROSCOPY OF ALKYNES

  1. Many alkynes have a C≡C stretching absorption in the 2100-2200 cm⁻¹ region of the infrared spectrum; this absorption is very weak or absent in the IR spectra of many symmetrical or nearly symmetrical alkynes because of the dipole moment effect.

  2. A useful absorption of 1-alkynes is the ≡C—H stretching absorption, which occurs at about 3300 cm⁻¹. (The narrow peak for this absorption distinguishes it from the broad peak for the O—H absorption, which occurs at about the same position.)

B. NMR SPECTROSCOPY OF ALKYNES

  1. The proton NMR chemical shifts of allylic and propargylic protons are very similar.

  2. The proton NMR chemical shifts of acetylenic protons and the CMR chemical shifts of carbons involved in triple bonds absorb at considerably higher field than those involved in double bonds.

   a. An induced electron circulation is set up in the cylinder of π electrons that encircles the alkyne molecule.

   b. The resulting induced field H_i opposes the applied field along the axis of this cylinder.

    i. Acetylenic protons and carbons of triple bonds are shielded from the applied field H_0.

    ii. A larger H_0 is required for resonance.

IV. *Acidity of 1-Alkynes*

A. ACETYLENIC ANIONS

1. Carbanions are extremely strong bases (that is, hydrocarbons are very weak acids), and 1-alkynes are the most acidic of all the aliphatic hydrocarbons.
   a. An alkyl anion, the conjugate base of an alkane, has an unshared electron pair in an sp^3 orbital.
   b. A vinylic anion, the conjugate base of an alkene, has an unshared electron pair in an sp^2 orbital.
   c. An acetylenic anion, the conjugate base of an alkyne, has an unshared electron pair in an sp orbital.
2. Alkynes are sufficiently acidic that their conjugate-base acetylide anions can be formed with strong bases.
   a. The sodium salt of an alkyne can be formed from a 1-alkyne quantitatively with sodium amide (or sodamide), $Na^+ \; ^-:\ddot{N}H_2$, dissolved in its conjugate acid, liquid ammonia.
   b. The reaction of an alkylmagnesium halide with an alkyne gives an acetylenic Grignard reagent.
      i. This reaction (transmetallation) is another Brønsted acid-base reaction.
      ii. The release of ethane gas in the reaction of a 1-alkyne with ethylmagnesium bromide is a useful test for 1-alkynes.
   c. Alkynes with internal triple bonds do not react because they lack an acidic alkyne hydrogen.
3. The electronegativity of the sp-hybridized carbon is responsible for the enhanced acidity of alkynes.
   a. The electrons in sp-hybridized orbitals are closer to the nucleus, on the average, than sp^2 electrons, which in turn are closer than sp^3 electrons; the proximity of sp electrons to the nucleus is a stabilizing effect.
   b. Unshared electrons pairs have lower energy when they are in orbitals of greater s-character.

. . . .

Reactions

. . . .

IV. *Addition Reactions of Alkynes*

A. ADDITION OF HYDROGEN HALIDES AND HALOGENS

1. Additions of halogens or hydrogen halides to the triple bond are somewhat slower in most cases than the same reactions of comparably substituted alkenes.
   a. The first addition to an alkyne gives a substituted alkene.
   b. The second addition of a hydrogen halide or a halogen to an alkyne is usually slower than the first because a halo alkene is less reactive than an alkyne.
   c. It is possible to isolate the substituted alkene resulting from addition of only one equivalent of halogen or hydrogen halide.

$$-C{\equiv}C- \xrightarrow{\;Br_2\;} \underset{Br}{\overset{Br}{>C=C<}} \xrightarrow{\;Br_2\;} \underset{Br\;Br}{\overset{Br\;Br}{-C-C-}}$$

2. HBr can be added to the triple bond in a regioselective addition that is analogous to the addition of HBr to alkenes.
   a. The bromine adds to the carbon of the triple bond that bears the alkyl substituent group.
   b. The regioselectivity is reversed in the presence of peroxides because free-radical intermediates are involved.

$$R-C{\equiv}C-H \xrightarrow{\;HBr\;} \underset{R\;\;\;\;H}{\overset{Br\;\;\;\;H}{>C=C<}} \xrightarrow{\;HBr\;} R-\underset{Br\;H}{\overset{Br\;H}{C-C}}-H$$

B. HYDRATION OF ALKYNES

1. The addition of water to the triple bond of an alkyne is catalyzed by strong acid or by a combination of dilute acid and mercuric ion, Hg^{2+} (a faster reaction).

   a. The addition of water to a triple bond is called hydration.

   b. Enols (alcohols containing an —OH group on a carbon of a double bond) are formed in the hydration of alkynes.

      i. Most enols are unstable and are rapidly converted into the corresponding ketones or aldehydes.

      ii. Most aldehydes and ketones are in equilibrium with small amounts of the corresponding enols.

      iii. Because most enols are unstable, any synthesis designed to give an enol gives instead the corresponding aldehyde or ketone.

an enol   an aldehyde or a ketone

2. The mechanism of alkyne hydration is very similar to that of oxymercuration of alkenes.

   a. In the first part the mechanism, a mercuric ion and an —OH group from the solvent water add to the triple bond.

      i. The —OH group of water adds to the carbon of the triple bond that bears the alkyl substituent.

      ii. The bond between the mercury and this carbon in the cyclic ion intermediate is weaker and thus more readily broken.

   b. The second part of the mechanism involves protonolysis, which occurs under the conditions of hydration.

      i. This protonation of the double bond occurs at the carbon bearing the mercury because the resulting carbocation is resonance-stabilized.

      ii. Dissociation of mercury from this carbocation liberates the catalyst Hg^{2+} along with the enol.

   c. Conversion of the enol into the ketone is a rapid, acid-catalyzed process.

      i. Protonation of the double bond gives another resonance-stabilized carbocation, which is also the conjugate acid of a ketone.

      ii. Loss of a proton from the carbocation gives the ketone product.

3. The hydration of alkynes is a useful way to prepare ketones provided that the starting material is a 1-alkyne or a symmetrical alkyne; most other alkynes give difficult-to-separate mixtures of isomers.

C. HYDROBORATION OF ALKYNES

1. When alkynes react with diborane (B_2H_6), the elements of BH_3 are added to the triple bond.
   a. Oxidation of the organoborane with alkaline hydrogen peroxide yields the corresponding "alcohol," an enol.
   b. The enol reacts further to give the corresponding aldehyde or ketone.

2. The reaction conditions can be controlled so that only one addition takes place provided that the alkyne is not a 1-alkyne.
3. The hydroboration of 1-alkynes can be stopped after a single addition provided that organoborane containing highly branched groups is used instead of BH_3.
   a. One reagent developed for this purpose is called disiamylborane.
   b Boron adds to the unbranched carbon atom of the triple bond, and hydrogen adds to the branched carbon.
   c. The disiamylborane molecule is so large and highly branched that only one equivalent can react with a 1-alkyne.

D. CATALYTIC HYDROGENATION OF ALKYNES

1. Alkynes undergo catalytic hydrogenation.
   a. Addition of one molar equivalent of H_2 yields an alkene.
   b. Addition of a second molar equivalent of H_2 gives an alkane.
2. Hydrogenation of an alkyne may be stopped at the alkene stage if the reaction mixture contains a catalyst poison.
   a. A catalyst poison is a compound that disrupts the action of a catalyst; in this case, it selectively blocks the hydrogenation of alkenes without preventing the hydrogenation of alkynes to alkenes.
   b. Useful catalyst poisons are salts of Pb^{2+} and certain nitrogen compounds such as pyridine, quinoline, or other amines.
   c. Lindlar catalyst ($Pd/CaCO_3$) that has been washed with $Pb(OAc)_2$ is a commonly used poisoned catalyst.

3. Hydrogenation of alkynes is a stereoselective *syn*-addition and, when carried out with a poisoned catalyst, is one of the best methods to prepare *cis*-alkenes.

E.   Reduction of Alkynes with Alkali Metals in Ammonia

1. Reaction of an alkyne with a solution of an alkali metal (usually sodium) in liquid ammonia gives an alkene. If the alkyne is not a 1-alkyne the product is a *trans*-alkene.
2. The mechanism of this reaction involves a two-electron reduction:
   a. The deep blue solution of alkali metals dissolved in pure liquid ammonia are sources of electrons complexed to ammonia (solvated electrons).

$$Na \xrightarrow{\text{liquid } NH_3} Na^+ + e^-$$

   b. The addition of an electron to the triple bond results in a species (a radical anion) that has both an unpaired electron and a negative charge.

   c. The radical anion is such a strong base that it removes a proton from ammonia to give a vinylic radical (a radical in which the unpaired electron is associated with one carbon of a double bond).

   d. The resulting vinylic radical rapidly undergoes inversion.
      i. The equilibrium between the *cis* and *trans* radicals favors the *trans* radical for steric reasons.
      ii. Since there is much more of the *trans* radical, the ultimate product of the reaction is the *trans*-alkene.

   e. The vinylic radical accepts an electron to form an anion.

   f. This anion is more basic than the solvent and removes a proton from ammonia to complete the addition.

F.   Acetylenic Anions as Nucleophiles

1. Acetylenic anions are strong bases; they are much stronger than hydroxide or alkoxides.
2. Acetylenic anions can be used as nucleophiles in S_N2 reactions to prepare other alkynes.

$$H_3C-C\equiv C:^- Na^+ + I-CH_2CH_3 \longrightarrow H_3C-C\equiv C-CH_2CH_3 + Na^+ I^-$$

   a. The alkyl halide or sulfonate used in this reaction must be a primary or unbranched secondary compound.
   b. This is another method of carbon-carbon bond formation.

Study Guide Links

✓14.1   Functional Group Preparations

Section 14.5 in the text discusses two ways to prepare aldehydes and ketones. At this point, you should be asking yourself, "What *other* ways do I know to make aldehydes and ketones?" You should recall, for example, that oxidation of alcohols is an important method for preparation of aldehydes and ketones. Keep a separate review list that groups reactions by the *type of product formed.* Thus, you'll have a list of reactions for preparing aldehydes and ketones, a list for preparing alcohols, etc. You should write out each reaction in detail, noting the limitations on each. For example, what are the limitations of alkyne hydration as a method of ketone preparation? (Can *any* ketone be prepared with this method? Look at Problems 14.10–14.11.) You can check how complete your list is by comparing it to the abbreviated list in Appendix IV of the text.

Use your list when working problems in synthesis. When your synthesis requires an aldehyde or ketone, consult your list; review the reactions if they start to look unfamiliar. Remember that putting together such lists keeps your studying *active* rather than passive.

✓14.2   Ammonia, Solvated Electrons, and Amide Ion

Sections 14.6 and 14.7 of the text have discussed several different uses of ammonia. The purpose of this Study Guide Link is to review and contrast these so that there is no confusion.

The first reagent is *liquid ammonia.* Ammonia can be condensed at –33° to a liquid. Once condensed, ammonia is easy to maintain in the liquid state because of its relatively high heat capacity. Liquid ammonia should not be confused with *aqueous* ammonia, which is a solution of ammonia in water. ("Household ammonia" is an aqueous ammonia solution.) If an alkali metal (such as sodium) is dissolved in liquid ammonia, a blue solution is obtained that is an excellent source of electrons—that is, an excellent reducing agent. This solution contains "solvated electrons"—electrons solvated by the hydrogens of the ammonia. The solution of solvated electrons is used in the reduction of alkynes to *trans*-alkenes (Sec. 14.6B).

If a trace of Fe^{3+} ion is added to a solution of sodium in liquid ammonia, a *reaction* occurs between the sodium and the ammonia:

$$2\,Na\cdot \; + \; 2\,\ddot{N}H_3 \; \xrightarrow{\;FeCl_3\;} \; 2\,Na^+ \; \; ^-\ddot{N}H_2 \; + \; H_2$$

sodium amide
(sodamide)

Just as sodium reacts with water to give hydrogen gas and sodium hydroxide, sodium also reacts with ammonia to give hydrogen gas and sodium amide. The difference is that the reaction of sodium with water requires no catalyst, but the reaction with ammonia does. It is this solution—sodium amide in liquid ammonia—that is used in forming acetylenic anions from alkynes. Solid sodium amide can also be purchased as a gray powder. Solutions of sodium amide can also be formed by adding this material to liquid ammonia or other solvents.

Do not confuse the basicity of amide ion with that of ammonia. Amide ion is the *conjugate base* of ammonia, which has a pK_a of about 35. Ammonia is also a base, but is the conjugate base of the *ammonium ion*, NH_4^+, which has a pK_a of 9.2. Most people normally think of ammonia as a base, but in the present context, ammonia is the *acid*, and the amide ion is the conjugate base. (See the discussion of amphoteric substances in Sec. 3.4D on text p. 102.)

Solutions

Solutions to In-Text Problems

14.1 (a)

$(CH_3)_2CHC\equiv CH$

isopropylacetylene

(c)

cyclononyne

(e)

1-ethynylcyclohexanol

(g)

$HC\equiv C-C\equiv C-CH_2CH_3$

1,3-hexadiyne

14.2 (a) The common name is dibutylacetylene; the substitutive name is 5-decyne.

(c) The substitutive name is 3-butyl-8-methoxy-6-octen-1-yne.

14.3 The *sp* hybridization of the carbon atoms of the triple bond requires that these two atoms and the two attached carbon atoms lie on a straight line. With a bond length of 1.2 Å for the triple bond and bond lengths of 1.47 Å for the attached carbon-carbon single bonds, the distance to be bridged by two remaining carbons of cyclohexyne is 4.14 Å. This distance is too great to be connected by two carbon atoms that lie *on the same line,* let alone at reasonable bond angles.

14.5 (a) The proton NMR spectrum of propyne should consist of two resonances. Evidently, the two happen to have the same chemical shift of δ 1.8, which is in fact a reasonable chemical shift for both acetylenic and propargylic protons.

14.6 A nine-proton singlet cries out, "*tert*-butyl group." The other resonance is in the position expected for an acetylenic proton. *Tert*-butylacetylene (3,3-dimethyl-1-butyne, $(CH_3)_3C-C\equiv CH$), which has a molecular mass of 82, is consistent with all the data.

14.7 (a) Because of its symmetry, 3-hexyne should have three chemically nonequivalent sets of carbons; in contrast, all six carbons of 2-hexyne are chemically nonequivalent. Thus, spectrum *A* is that of 2-hexyne, and spectrum *B* is that of 3-hexyne.

14.8 On the assumption that the stereochemistry of bromine addition to alkynes is *anti* as it is in alkenes, the product should be (*E*)-3,4-dibromo-3-hexene:

$$CH_3CH_2C\equiv CCH_2CH_3 \xrightarrow{Br_2} $$

(*E*)-3,4-dibromo-3-hexene

14.10 (a) Only 1-pentyne, $HC\equiv CCH_2CH_2CH_3$, would undergo hydration to give 2-pentanone free of constitutional isomers. (2-Pentyne would give a difficult-to-separate mixture of 2-pentanone and 3-pentanone.)

(c) Only 5-decyne, $CH_3(CH_2)_3C\equiv C(CH_2)_3CH_3$, would give the indicated ketone (5-decanone) free of constitutional isomers.

14.11 (a) The three-carbon alkyne propyne hydrates to give acetone, not propanal (the compound shown in the problem).

(solution continues)

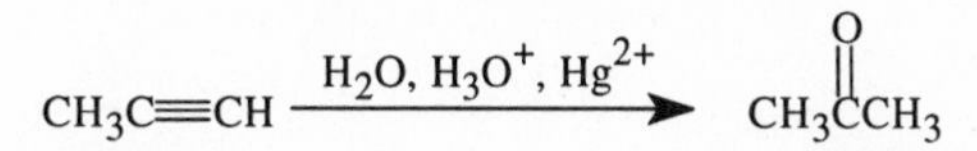

$$CH_3C \equiv CH \xrightarrow{H_2O,\ H_3O^+,\ Hg^{2+}} CH_3\overset{\displaystyle O}{\overset{\|}{C}}CH_3$$

propyne   acetone

(c) The alkyne that would hydrate to cyclohexanone is cyclohexyne. Since cyclohexyne is too unstable to isolate, such a hydration reaction is impractical. (See the solution to Problem 14.3.)

14.12 (a) Two constitutionally isomeric enols are possible, one of which can exist as *E,Z*-stereoisomers.

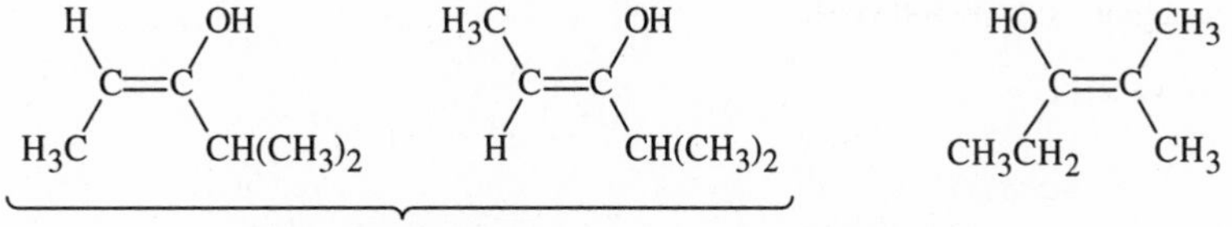

E,Z-stereoisomers

(b) Hydration would not be a good preparative method for this compound. The only alkyne that might give this compound as a product is 4-methyl-2-pentyne; but, because both carbons of its triple bond have one alkyl substituent, that alkyne would give a mixture of two constitutionally isomeric products of which the desired compound is only one component.

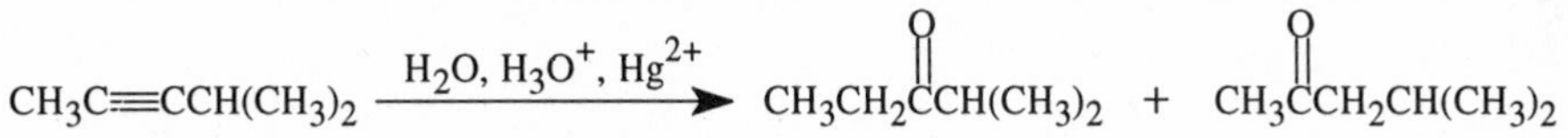

$$CH_3C \equiv CCH(CH_3)_2 \xrightarrow{H_2O,\ H_3O^+,\ Hg^{2+}} CH_3CH_2\overset{\displaystyle O}{\overset{\|}{C}}CH(CH_3)_2 \;+\; CH_3\overset{\displaystyle O}{\overset{\|}{C}}CH_2CH(CH_3)_2$$

4-methyl-2-pentyne

14.13 (a) Cyclohexylacetylene gives different products in the two reactions:

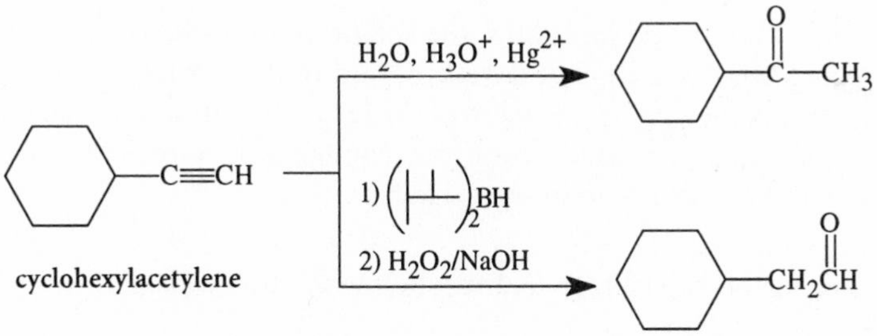

cyclohexylacetylene

14.14 (a) The poisoned catalyst causes hydrogenation to stop at the alkene stage, and the double bond in the starting material is unaffected.

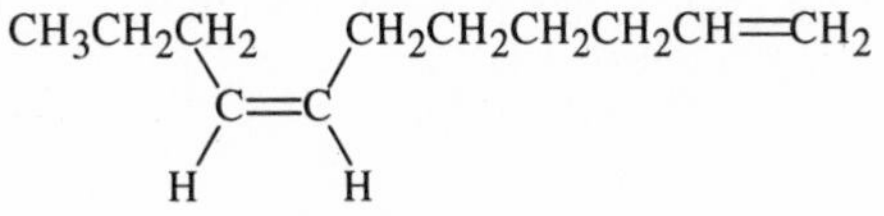

(*Z*)-1,7-undecadiene

(c) Because the molecule contains its own catalyst poison, hydrogenation stops at the alkene.

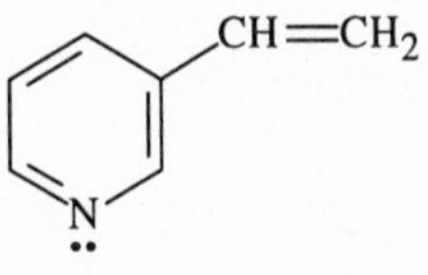

14.15 (a) The first reaction gives *trans*-3-hexene, and the second reaction, a *syn* addition, gives (±)-hexane-3,4-d_2.

(solution continues)

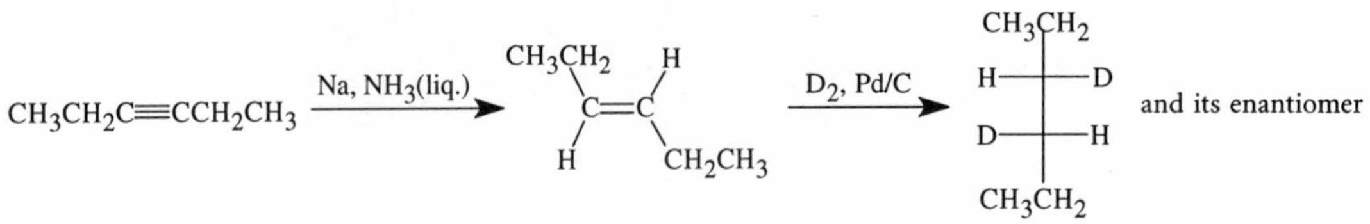

$$CH_3CH_2C\equiv CCH_2CH_3 \xrightarrow{\text{Na, NH}_3\text{(liq.)}}$$

[structure: trans-alkene with CH₃CH₂ and H on left carbon, H and CH₂CH₃ on right carbon]

$$\xrightarrow{\text{D}_2,\text{ Pd/C}}$$

[Fischer projection: CH₃CH₂ at top, H—D, D—H, CH₃CH₂ at bottom] and its enantiomer

(±)-hexane-3,4-d_2

14.16 The conjugate acids are as follows:

$$CH_3{-}CH{=}\overset{+}{N}H_2 \qquad CH_3{-}C{\equiv}\overset{+}{N}H \qquad CH_3{-}CH_2{-}\overset{+}{N}H_3$$

conjugate acid of   *A*   *B*   *C*

The order of increasing basicity:

$$B < A < C$$

This order follows from the principle developed in this section: the hydrogen with the greatest amount of *s* character in its C—H bond is most acidic. This problem demonstrates that it doesn't matter whether the atom to which the hydrogen is attached is charged or neutral, or whether it is a carbon or some other atom.

14.18 (a, Part 1)   The logarithm of the equilibrium constant K_{eq} for an acid-base reaction equals the pK_a value of the acid on the right minus that of the acid on the left. Assuming that the pK_a of ammonia is about 35, and that the pK_a of an alkyne is about 25, log K_{eq} for the reaction in Eq. 14.21 is $(35 - 25) = 10$. Consequently, the equilibrium constant for this reaction is 10^{10}. This equilibrium lies far to the right.

(b) As the previous two calculations show, the equilibrium for the reaction of the base sodium amide with an alkyne strongly favors ionization of the alkyne and protonation of the base; the equilibrium for the reaction of sodium amide with an alkane strongly favors the un-ionized alkane and unprotonated sodium amide. In other words, sodium amide is a strong enough base to remove a proton from an alkyne, but is far too weak a base to remove a proton from an alkane.

14.19 (a) The acetylenic anion reacts in an S$_N$2 reaction with butyl tosylate:

$$CH_3CH_2CH_2CH_2{-}OTs \;+\; {}^-{:}C{\equiv}CPh \;\longrightarrow\; CH_3CH_2CH_2CH_2C{\equiv}CPh \;+\; {}^-OTs$$

butyl tosylate

(c) The acetylenic Grignard reagent, like any other Grignard reagent, opens the epoxide to give an alcohol:

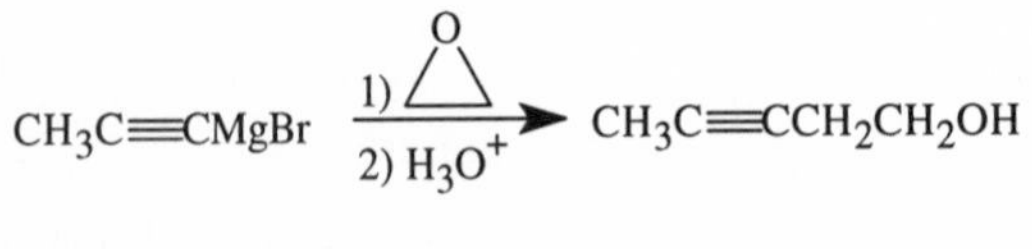

$$CH_3C{\equiv}CMgBr \xrightarrow[\text{2) H}_3\text{O}^+]{\text{1) }\triangle\text{O}} CH_3C{\equiv}CCH_2CH_2OH$$

14.20 Choke forgot the lessons of Sec. 9.5F on text p. 411: tertiary alkyl halides react with strong bases to give elimination products, not substitution products. Choke's products were 2-methylpropene and propyne.

14.21 Either alkyl group can be introduced in an S$_N$2 reaction of an acetylenic anion with an alkyl halide. The two preparations of 2-pentyne ($CH_3C{\equiv}CCH_2CH_3$):

(1)

$$HC{\equiv}CCH_2CH_3 \xrightarrow{\text{NaNH}_2} Na^+ \; {}^-{:}C{\equiv}CCH_2CH_3 \xrightarrow{\text{I}-CH_3} CH_3C{\equiv}CCH_2CH_3$$

(2)

$$CH_3C\equiv CH \xrightarrow{NaNH_2} CH_3C\equiv \overset{-}{C}:Na^+ \xrightarrow{\text{I—CH}_2CH_3} CH_3C\equiv CCH_2CH_3$$

14.23 (a) $CH_3CH_2CH_2C\equiv CH \xrightarrow{NaNH_2} CH_3CH_2CH_2C\equiv \overset{-}{C}:Na^+ \xrightarrow{ICH_2CH_3} CH_3CH_2CH_2C\equiv CCH_2CH_3$

Another satisfactory preparation is to begin with 1-butyne, form its conjugate base with sodium amide, and alkylate with 1-bromopropane.

14.24 (a) An excess of acetylene is used in the first step so that only the *monoanion* is formed. (See the bottom of text p. 667 for the explanation.)

$$HC\equiv CH \xrightarrow[\text{(excess)}]{Na,\ NH_3\ (liq.)} HC\equiv \overset{-}{C}:Na^+ \xrightarrow{BrCH_2CH_2CH_2CH_2CH_3} HC\equiv CCH_2CH_2CH_2CH_2CH_3 \xrightarrow[\substack{\text{Lindlar} \\ \text{catalyst}}]{H_2}$$

$$H_2C=CHCH_2CH_2CH_2CH_2CH_3 \xrightarrow[\substack{1)\ BH_3/THF \\ 2)\ H_2O_2/NaOH}]{} HOCH_2CH_2CH_2CH_2CH_2CH_3$$

14.25 (a)

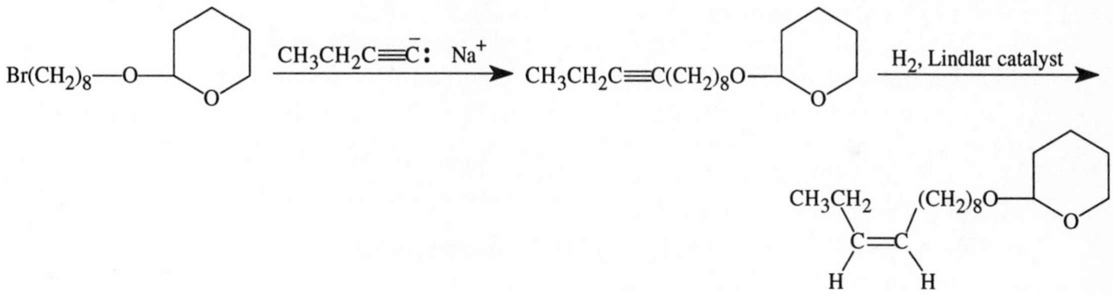

$$Br(CH_2)_8\text{—}O\text{—}\underset{O}{\overset{\bigcirc}{\diagdown}} \xrightarrow{CH_3CH_2C\equiv \overset{-}{C}:Na^+} CH_3CH_2C\equiv C(CH_2)_8O\text{—}\underset{O}{\overset{\bigcirc}{\diagdown}} \xrightarrow{H_2,\ \text{Lindlar catalyst}}$$

$$\underset{H}{\overset{CH_3CH_2}{\diagup}}C=C\underset{H}{\overset{(CH_2)_8O\text{—}\overset{\bigcirc}{\diagdown}}{\diagdown}}$$

Solutions to Additional Problems

14.26 (a)

$$\underset{Br}{\overset{}{CH_3CH_2CH_2CH_2\underset{|}{C}=CH_2}}$$

(b) $CH_3(CH_2)_4CH_3$

(c) $CH_3CH_2CH_2CH_2CH=CH_2$

(d) $CH_3CH_2CH_2CH_2CH=O$
$+$
$O=CH_2$

(e) $CH_3CH_2CH_2CH_2CH_2CH_2OH$

(f) $\underset{Br}{\overset{CH_3CH_2CH_2CH_2}{\diagdown}}C=C\underset{H}{\overset{Br}{\diagup}}$

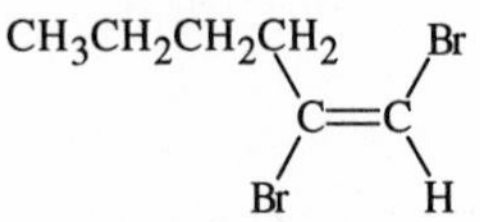

(g) $CH_3CH_2CH_2CH_2C\equiv \overset{-}{C}:Na^+$

(h) $\overset{O}{\overset{\|}{CH_3CH_2CH_2CH_2CCH_3}}$

(i) $CH_3CH_2CH_2CH_2CH=CH_2$

(j) $\overset{OH}{\overset{|}{CH_3CH_2CH_2CH_2CHCH_3}}$

14.28 In part (a), the —OH group is the principal group and receives the lower number; in part (b), the double bond receives numerical precedence when there is no other basis for deciding between the numbers for the double and triple bonds.

(a)   (c)

$$\underset{\text{4-hexyne-3-ol}}{CH_3C\equiv C\overset{\overset{\displaystyle OH}{|}}{C}HCH_2CH_3}$$ $$\underset{\text{1-buten-3-yne}}{HC\equiv CCH=CH_2}$$

14.29 (a) Since the triple bond cannot be contained within the ring, the compound can only be cyclopropylacetylene:

$$\triangleright\!\!-\!C\equiv CH \qquad \text{cyclopropylacetylene}$$

(c) $CH_3CH_2C\equiv CCH_2CH_3$ 3-hexyne

(e) The diastereomers are the *cis* and *trans* isomers of $CH_3CH=CHCH_2C\equiv CH$ (4-hexen-1-yne).

14.30 (a) Bonds with greater *s* character are shorter. (See text p. 125.) Therefore, the order of increasing C—H bond length is

$$\text{acetylene} < \text{ethylene} < \text{ethane}$$

14.31 (a) The order of increasing basicity is $F^- < CH_3CH_2O^- < HC\equiv C^-$. The reason is essentially the element effect as it operates across a row of the periodic table: the more electronegative the atom, the less basic it is. (Note that the element effect on acidity is more important than the effect of hybridization.)

14.32 (a) The alkyne should give off a gas (methane or ethane, respectively) when treated with CH_3MgBr or C_2H_5MgBr. (See Eq. 14.24 on text p. 664.) The alkene is not acidic enough to react in this way.

(c) The alkyne should react rapidly with Br_2 in an inert solvent, and should therefore rapidly decolorize such a bromine solution; the alkane will not react.

14.33 Note that in many syntheses more than one acceptable route may be possible.

(a)

$$\underset{\text{(excess)}}{HC\equiv CH} \xrightarrow{NaNH_2} HC\equiv \overset{-}{C}\!: \ Na^+ \xrightarrow{CH_3CH_2Br} HC\equiv CCH_2CH_3 \xrightarrow{NaNH_2}$$

$$Na^+ \ :\!\overset{-}{C}\equiv CCH_2CH_3 \xrightarrow{CH_3CH_2Br} CH_3CH_2C\equiv CCH_2CH_3 \xrightarrow{H_2,\ \text{Lindlar catalyst}}$$

$$\underset{H \qquad\quad H}{\overset{CH_3CH_2 \qquad CH_2CH_3}{C=C}} \xrightarrow[\text{2) } H_2O_2/NaOH]{\text{1) } BH_3/THF} \underset{}{CH_3CH_2\overset{\overset{\displaystyle OH}{|}}{C}HCH_2CH_2CH_3}$$

(c)

$$\underset{\text{prepared in part (a)}}{CH_3CH_2C\equiv CCH_2CH_3} \xrightarrow{D_2,\ Pd/C} CH_3CH_2CD_2CD_2CH_2CH_3$$

(e)

$$Na^+ \ :\!\bar{C}\!\!\equiv\!\!CH \xrightarrow{CH_3(CH_2)_6Br} CH_3(CH_2)_6C\!\!\equiv\!\!CH \xrightarrow[\text{2) } H_2O_2/NaOH]{\text{1) } \left(\overset{}{\square}\right)_2 BH} CH_3(CH_2)_7CH\!\!=\!\!O \xrightarrow[\text{2) } H_3O^+]{\text{1) } KMnO_4, \ ^-OH}$$

prepared in
part (a)

$$CH_3(CH_2)_7\overset{O}{\overset{\|}{C}}OH$$

(g)

$$Na^+ \ :\!\bar{C}\!\!\equiv\!\!CCH_2CH_3 \xrightarrow{CH_3I} CH_3C\!\!\equiv\!\!CCH_2CH_3 \xrightarrow{H_2, \text{ Lindlar catalyst}} \underset{\substack{H \qquad\quad H}}{\overset{\substack{H_3C \qquad CH_2CH_3}}{C\!\!=\!\!C}}$$

prepared in part (a)

cis-2-pentene

(i)

$$Na^+ \ :\!\bar{C}\!\!\equiv\!\!CH \xrightarrow{CH_3(CH_2)_2Br} CH_3(CH_2)_2C\!\!\equiv\!\!CH \xrightarrow{NaNH_2} CH_3(CH_2)_2C\!\!\equiv\!\!\bar{C}: \ Na^+ \xrightarrow{Br(CH_2)_2CH_3}$$

prepared in
part (a)

$$CH_3(CH_2)_2C\!\!\equiv\!\!C(CH_2)_2CH_3 \xrightarrow[\text{Lindlar catalyst}]{H_2} \underset{\substack{H \qquad\qquad H}}{\overset{\substack{CH_3(CH_2)_2 \qquad (CH_2)_2CH_3}}{C\!\!=\!\!C}} \xrightarrow[\text{2) } H_2O, NaHSO_3]{\text{1) } OsO_4}$$

$$\underset{\substack{CH_3(CH_2)_2 \qquad\quad (CH_2)_2CH_3}}{\overset{\substack{HO \qquad\qquad OH}}{H^{\cdots}C\!\!-\!\!C^{\cdots}H}}$$

meso-4,5-octanediol

14.34 This reaction consists of two successive Brønsted acid-base reactions.

$$PhC\!\!\equiv\!\!C\!\!-\!\!H \quad :\!\overset{-}{\underset{\cdot\cdot}{O}}D \ \xleftarrow{\quad\longrightarrow\quad} \ PhC\!\!\equiv\!\!\bar{C}: \quad D\!\!-\!\!\overset{\cdot\cdot}{\underset{\cdot\cdot}{O}}D \ \xleftarrow{\quad\longrightarrow\quad} \ PhC\!\!\equiv\!\!C\!\!-\!\!D \ + \ :\!\overset{-}{\underset{\cdot\cdot}{O}}D$$

$$+ \ H\!\!-\!\!\overset{\cdot\cdot}{\underset{\cdot\cdot}{O}}D$$

Notice that, because ⁻OD is a *much* weaker base than the acetylenic anion, only a *very small amount* of the acetylenic anion is present at any one time; however, because all reactions are in equilibrium, and because D$_2$O is present in large excess, deuterium is eventually incorporated into the alkyne.

14.36 (a) Use the method suggested by Problem 14.35.

$$CH_3CH_2C\!\!\equiv\!\!CH \xrightarrow[-CH_3CH_3]{C_2H_5MgI} CH_3CH_2C\!\!\equiv\!\!CMgI \xrightarrow{D_2O} CH_3CH_2C\!\!\equiv\!\!CD$$

(c)

$$CH_3CH_2C\!\!\equiv\!\!CH \xrightarrow[\text{Lindlar catalyst}]{H_2} CH_3CH_2CH\!\!=\!\!CH_2 \xrightarrow[\text{2) } H_2O_2/NaOH]{\text{1) } BH_3/THF}$$

(solution continues)

$$CH_3CH_2CH_2CH_2OH \xrightarrow[\text{2) } H_3O^+]{\text{1) } KMnO_4, \ ^-OH} CH_3CH_2CH_2CO_2H$$

(e) First prepare 1-bromobutane from 1-butyne:

$$HC{\equiv}CCH_2CH_3 \xrightarrow{H_2, \text{ Lindlar cat.}} H_2C{=}CHCH_2CH_3 \xrightarrow{HBr, \text{ peroxides}} BrCH_2CH_2CH_2CH_3$$

1-butyne   1-bromobutane

Then use 1-bromobutane to alkylate an acetylenic anion:

$$CH_3CH_2C{\equiv}CH \xrightarrow{NaNH_2} CH_3CH_2C{\equiv}\overset{-}{C}{:}\ Na^+ \xrightarrow{BrCH_2CH_2CH_2CH_3} CH_3CH_2C{\equiv}CCH_2CH_2CH_2CH_3$$

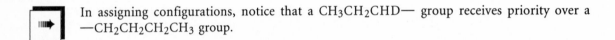

In assigning configurations, notice that a CH_3CH_2CHD— group receives priority over a —$CH_2CH_2CH_2CH_3$ group.

There is an error in the text associated with part (g). This part cannot be worked as stated with 1-butyne as the *only* source of carbon. However, with 1-butyne as *one* source of carbon, the following synthesis is possible.

(g)

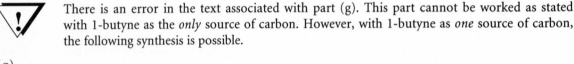

from part (e)

14.37 First look for the telltale C—H and C≡C stretching absorptions; these are present in the spectrum of compound *B* near 2100 cm^{-1} and 3300 cm^{-1}. Hence, compound *B* is 1-hexyne. Compound *A* has no trace of C≡C absorption and, more importantly, it has a 1640 cm^{-1} C=C stretching absorption and 910 cm^{-1} and 990 cm^{-1} C—H bending absorptions, all of which are consistent with a —CH=CH$_2$ group. Therefore, compound *A* is 3-methyl-1,4-pentadiene. By elimination, compound *C* must be 2-hexyne.

$$\overset{\overset{\textstyle CH_3}{|}}{H_2C{=}CHCHCH{=}CH_2} \qquad CH_3CH_2CH_2CH_2C{\equiv}CH \qquad CH_3C{\equiv}CCH_2CH_2CH_3$$

3-methyl-1,4-pentadiene   1-hexyne   2-hexyne

compound *A*   compound *B*   compound *C*

14.38 First, prepare 1-bromooctane.

$$CH_3CH_2C{\equiv}CH \xrightarrow[\text{Lindlar catalyst}]{H_2} CH_3CH_2CH{=}CH_2 \xrightarrow{HBr, \text{ peroxides}} CH_3CH_2CH_2CH_2Br \xrightarrow[\substack{\text{(for preparation}\\ \text{see Eq. 14.28c, text p. 667)}}]{Na^+ \ :\overset{-}{C}{\equiv}CH}$$

$$CH_3CH_2CH_2CH_2C{\equiv}CH \xrightarrow[\text{Lindlar catalyst}]{H_2} CH_3(CH_2)_3CH{=}CH_2 \xrightarrow{HBr, \text{ peroxides}} CH_3(CH_2)_5Br \xrightarrow{Na^+ \ :\overset{-}{C}{\equiv}CH}$$

$$CH_3(CH_2)_5C\equiv CH \xrightarrow[\text{Lindlar catalyst}]{H_2} CH_3(CH_2)_5CH=CH_2 \xrightarrow{\text{HBr, peroxides}} CH_3(CH_2)_7Br$$

1-bromooctane

As you can see from this example, the carbon chain of an alkyne can be extended two carbons at a time using the following sequence (referred to below as the *alkyne n + 2 sequence*):

$$CH_3(CH_2)_nC\equiv CH \xrightarrow[\text{Lindlar catalyst}]{H_2} CH_3(CH_2)_nCH=CH_2 \xrightarrow{\text{HBr, peroxides}} CH_3(CH_2)_{n+2}Br \xrightarrow{Na^+ \; :\bar{C}\equiv CH}$$

$$CH_3(CH_2)_{n+2}C\equiv CH$$

Consequently, 1-bromotridecane, $CH_3(CH_2)_{12}Br$, can be prepared by applying four successive alkyne $n + 2$ sequences to 1-pentyne ($n = 2$), and finishing with the hydrogenation and HBr-addition steps:

$$CH_3(CH_2)_2C\equiv CH \xrightarrow[\text{sequences}]{\text{four alkyne } n + 2} CH_3(CH_2)_{10}C\equiv CH \xrightarrow[\text{Lindlar catalyst}]{H_2} CH_3(CH_2)_{10}CH=CH_2$$

1-pentyne

$$\xrightarrow{\text{HBr, peroxides}} CH_3(CH_2)_{12}Br$$

1-bromotridecane

To complete the synthesis of muscalure, carry out two successive alkylations of acetylenic anions with the alkyl bromides prepared above and hydrogenate the resulting product.

$$HC\equiv\bar{C}: Na^+ \xrightarrow{Br(CH_2)_{12}CH_3} HC\equiv C(CH_2)_{12}CH_3 \xrightarrow{NaNH_2} \xrightarrow{CH_3(CH_2)_7Br}$$

$$CH_3(CH_2)_7C\equiv C(CH_2)_{12}CH_3 \xrightarrow[\text{Lindlar cat.}]{H_2}$$

muscalure

14.40 (a) The "anionic carbon" of ethylmagnesium bromide acts as a base towards the C—H bond of ethynylmagnesium bromide:

$$\longrightarrow CH_3CH_3 + BrMgC\equiv CMgBr$$

According to Eq. (2) in the problem, an excess of acetylene reacts with the product of Eq. (1) to regenerate ethynylmagnesium bromide. Thus, the formation of $BrMgC\equiv CMgBr$, the undesired by-product, is avoided by "recycling" it back to the desired product $HC\equiv CMgBr$.

(c) Precipitation of $BrMgC\equiv CMgBr$ would pull both reactions (1) and (2) to the right, thus depleting the desired product $HC\equiv CMgBr$ and forming more of the undesired one. Using a solvent that avoids this precipitation thus avoids these side-reactions.

14.41 (a) The IR spectrum suggests a 1-alkyne. In the NMR spectrum, the δ 3.31 singlet indicates a methoxy group, and the six-proton δ 2.41 singlet suggests two equivalent methyl groups bound to a quaternary carbon. The compound is 3-methoxy-3-methyl-1-butyne.

3-methoxy-3-methyl-1-butyne

(c) The three-proton doublet at δ 1.4 indicates the partial structure $-\overset{|}{C}H-CH_3$. The broad peak at δ 4.1 that exchanges with D_2O suggests an alcohol. The IR spectrum indicates both an alcohol and a 1-alkyne; the triple bond accounts for both degrees of unsaturation. The unknown is 3-butyn-2-ol.

$$HC\equiv C\overset{\overset{\displaystyle OH}{|}}{C}HCH_3 \quad \text{3-butyn-2-ol}$$

14.42 (a) The IR absorptions indicate a carbon-carbon triple bond (2100 cm^{-1}) and a 1-alkyne (3300 cm^{-1}). The CMR data show that there are only four nonequivalent sets of carbons. The compound is 3,3-dimethyl-1-butyne, $(CH_3)_3CC\equiv CH$.

14.43 (a) Butyllithium is a strong base; like a Grignard reagent, it converts a 1-alkyne into a lithium acetylide. Lithium acetylides, like an acetylenic Grignard reagents, reacts as if they were acetylenic anions. Consequently, the lithium acetylide, as suggested by the hint, reacts as a nucleophile, displacing chloride ion from $(CH_3)_3Si-Cl$.

$$CH_3CH_2CH_2CH_2C\equiv CH \xrightarrow{CH_3CH_2CH_2CH_2Li} CH_3CH_2CH_2CH_2C\equiv C-Li \xrightarrow{(CH_3)_3SiCl}$$
$$+ \ CH_3CH_2CH_2CH_3$$

$$CH_3CH_2CH_2CH_2C\equiv CSi(CH_3)_3 + Li^+ \ Cl^-$$

(b) The acetylenic anion, which is formed by the reaction of 1-octyne with $NaNH_2$, is alkylated by diethyl sulfate. The product is $CH_3(CH_2)_6C\equiv CCH_2CH_3$.

(c) The lithium acetylide undergoes an S_N2 reaction with the alkyl halide; the better leaving group (chloride) is expelled. The product is $F(CH_2)_5C\equiv CH + Li^+ \ Cl^-$.

(d) Bromine adds to the double bond; both bromines are subsequently eliminated in consecutive $NaNH_2$-promoted E2 reactions to give phenylacetylene.

$$PhCH=CH_2 \xrightarrow{Br_2} PhCH-CH_2-Br \xrightarrow{NaNH_2} PhC=CH_2 \xrightarrow{NaNH_2} PhC\equiv CH$$
$$\overset{\textstyle |}{Br} \qquad\qquad \overset{\textstyle |}{Br} \qquad\qquad \text{phenylacetylene}$$

(e) As suggested by the hint, chloroform and a strong base give dichloromethylene, a carbene, which adds to the triple bond to give a cyclopropene.

$$Cl_2C\colon \quad + \quad PhC\equiv CPh \longrightarrow$$

dichloromethylene
from reaction of
$HCCl_3$ and $^-OC(CH_3)_3$

14.44 (a) First, the alkyne undergoes hydration to the enol. (The mechanism of this reaction is given in Eq. 14.6a–c on text p. 655.) One tertiary hydroxy group of the enol is lost as water, and the other hydroxy group attacks the resulting carbocation *A* to close the ring. Finally, the enol is converted into the ketone. (The mechanism for conversion of an enol into a ketone is shown in Eqs. 14.6d–e on text p. 655.)

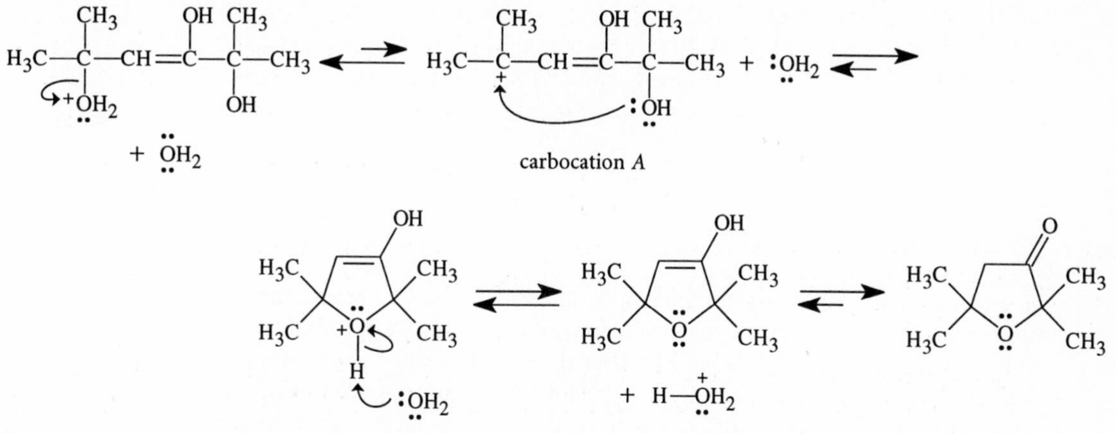

You may wonder why the enol and not the ketone itself undergoes cyclization. The reason is that the carbocation *A* derived from the enol has several resonance structures, and is therefore relatively stable and easily formed. (Can you draw these resonance structures?) The corresponding carbocation derived from the ketone is not resonance-stabilized, and is therefore much less stable. (Resonance stabilization was discussed on text pp. 20–21, and is discussed in greater detail in Sec. 15.6.) However, if you wrote a mechanism similar to the one above involving the ketone, your answer should be considered satisfactory at this point, because the issue of resonance stabilization has not been strongly emphasized.

(b) A bromohydrin is formed; however, because the bromohydrin is also an enol, it is converted into the α-bromo ketone. (See Eqs. 14.6d–e on text p. 655 for the mechanism of the enol-to-ketone conversion.)

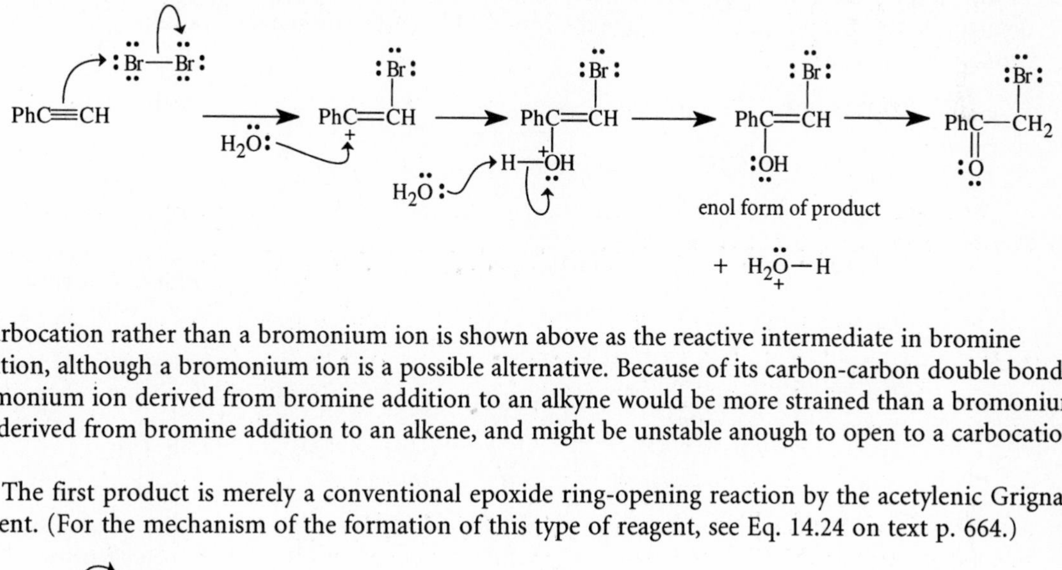

A carbocation rather than a bromonium ion is shown above as the reactive intermediate in bromine addition, although a bromonium ion is a possible alternative. Because of its carbon-carbon double bond, a bromonium ion derived from bromine addition to an alkyne would be more strained than a bromonium ion derived from bromine addition to an alkene, and might be unstable anough to open to a carbocation.

(c) The first product is merely a conventional epoxide ring-opening reaction by the acetylenic Grignard reagent. (For the mechanism of the formation of this type of reagent, see Eq. 14.24 on text p. 664.)

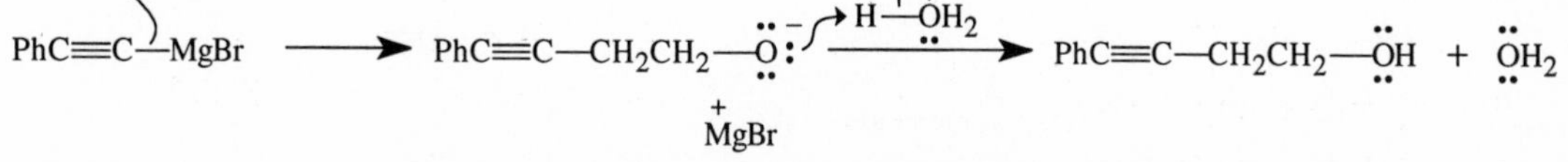

The second product results from the first product by a hydration-like mechanism in which the alcohol rather than water serves as the nucleophile.

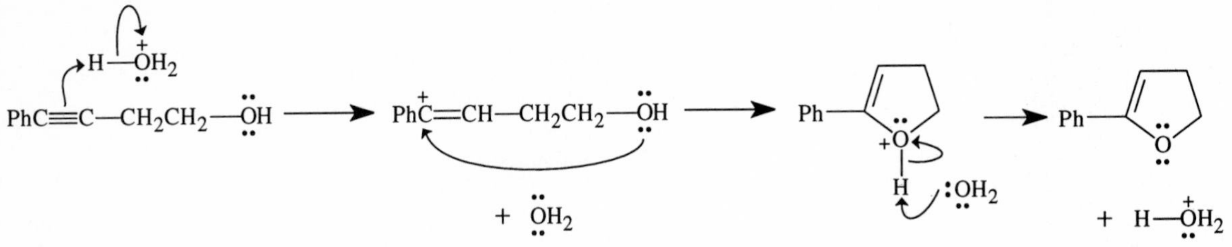

14.45 The elemental analysis of *A* gives the empirical formula C_5H_7. Because the hydrogenation product of *A* contains ten carbons, the likely molecular formula is $C_{10}H_{14}$. Compound *A* has the carbon skeleton of butylcyclohexane (a six-membered ring bearing an unbranched chain of four carbons). Compound *A* also has four degrees of unsaturation. One degree of unsaturation is accounted for by the ring. Two others are accounted for by the triple bond; it was given that compound *A* is an alkyne. Because there are no other rings, the one remaining degree of unsaturation must be a double bond. The triple bond cannot, of course, be within the ring, nor can it be at the end of the four-carbon chain because of the absence of a reaction of *A* with ethylmagnesium bromide. Hydrogenation of *A* under conditions that hydrogenate triple bonds to double bonds but leave double bonds unaffected gives an alkene which, on ozonolysis, loses *two* carbons. The fact that a *tricarboxylic* acid is formed on ozonolysis shows that the double bond of *A* is within the ring. It is now possible to write several preliminary structures for *A*:

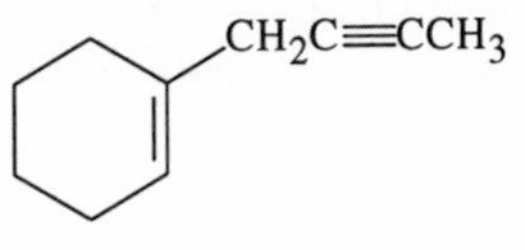

A1

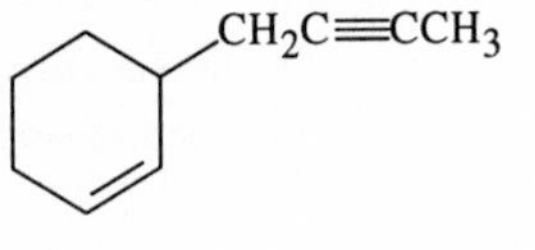

A2

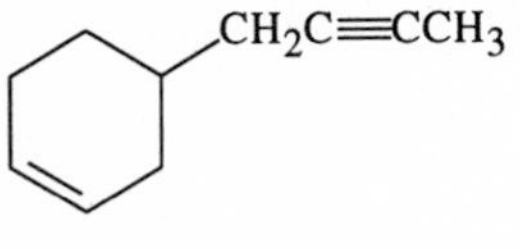

A3

Compound *A1* is ruled out because it is not chiral; recall that compound *A* is optically active. To distinguish between *A2* and *A3*, imagine the results of hydrogenation over a poisoned catalyst followed by ozonolysis:

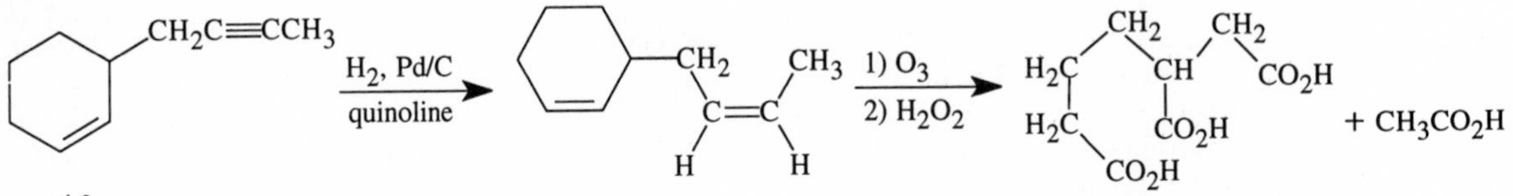

A2

chiral; would be optically active if *A2* were optically active

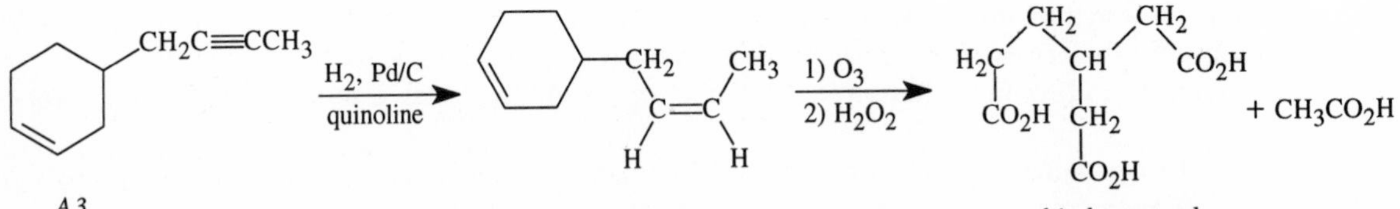

A3

achiral; cannot be optically active

The observation of optical activity in the ozonolysis product rules out *A3* and is consistent only with *A2* as the structure of compound *A*.

compound *A* = *A2*

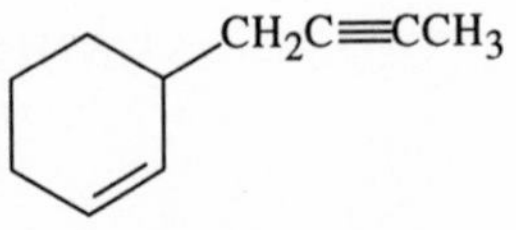

15

Dienes, Resonance, and Aromaticity

Terms

Concepts

I. *Structure and Reactivity of Dienes*

A. INTRODUCTION

1. Dienes are compounds with two carbon-carbon double bonds.
   a. Conjugated dienes have two double bonds that are separated by one single bond; the double bonds are called conjugated double bonds.
   b. Cumulenes are compounds in which one carbon participates in two carbon-carbon double bonds; the double bonds are called cumulated double bonds.
      i. Allene is the simplest cumulene.
      ii. The term allene is also sometimes used as a family name for compounds containing only two cumulated double bonds.

$$H_3C\!-\!CH\!=\!CH\!-\!CH\!=\!CH\!-\!CH_3 \qquad \text{conjugated diene}$$

$$H_3C\!-\!CH\!=\!C\!=\!CH\!-\!CH_3 \qquad \text{cumulated diene}$$

2. Conjugated dienes and allenes have unique structures and chemical properties.
3. Dienes in which the double bonds are separated by two or more single bonds have structures and chemical properties more or less like those of simple alkenes.

B. CONJUGATED DIENES

1. Conjugated dienes are more stable than their unconjugated isomers because:
   a. The carbon-carbon single bond between the two double bonds in a conjugated diene is derived from the overlap of two carbon sp^2 orbitals.
      i. An sp^2-sp^2 single bond is a stronger bond than the sp^2-sp^3 single bond of an ordinary diene and thus contributes to the stability of a conjugated diene.
      ii. The sp^2-sp^2 single bond is shorter than the sp^2-sp^3 or sp^3-sp^3 carbon-carbon single bond.
   b. There is overlap of *p* orbitals across the carbon-carbon bond connecting the two alkene units.
      i. There is π bonding between as well as within each of the two alkene units.
      ii. The additional bonding associated with this overlap provides additional stability to the molecule.

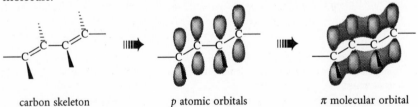

carbon skeleton   *p* atomic orbitals π molecular orbital

2. There are two stable conformations about the single bond of 1,3-butadiene:
   a. The transoid or *s-trans* conformation (the prefix *s* refers to the single bond).
   b. The cisoid or *s-cis* conformation (the prefix *s* refers to the single bond).
   c. Both forms are planar, or nearly so.
      i. The *s-trans* form of an acyclic diene is more stable than the *s-cis* form.
      ii. The internal rotation that interconverts these two forms is very rapid at room temperature.

transoid   cisoid
(*s-trans*)   (*s-cis*)

C. CUMULATED DIENES

1. The carbon skeleton of allene is linear.
   a. The central carbon of allene is *sp*-hybridized; the two remaining carbons are sp^2-hybridized and have trigonal geometry.

b. The two bonding π molecular orbitals in allenes are mutually perpendicular; these two π systems do not overlap.

c. The H—C—H plane at one end of the allene molecule is perpendicular to the H—C—H plane at the other end.

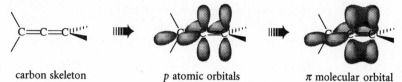

| carbon skeleton | *p* atomic orbitals | π molecular orbital |

2. Because of their geometries, some allenes are chiral even though they do not contain an asymmetric carbon atom.

   a. The two sp^2-hybridized carbons can be stereocenters.

   b. The perpendicularity of the planes of the two trigonal carbons can give rise to enantiomeric allenes.

   c. Enantiomeric allenes can be isolated since internal rotation about a double bond for all practical purposes does not occur.

3. The bond length of each cumulated double bond is somewhat less than that of an ordinary alkene double bond.

4. The cumulated arrangement is the least stable arrangement of two double bonds; allenes also have greater heats of formation than isomeric alkynes.

5. The C=C stretching absorption in the infrared spectrum of allenes is around 1950 cm^{-1}.

6. Naturally occurring allenes are relatively rare.

II. Resonance

A. Resonance Structures

1. Molecules that can be represented as resonance hybrids are said to be resonance-stabilized; that is, they are more stable than any one of their contributing structures.

   a. The additional stabilization results from additional bonding, which in turn results from the delocalization of electrons and additional orbital overlap.

   b. Resonance structures can be derived by the curved-arrow formalism.

   c. Resonance structures show the delocalization of electrons.

2. The derivation and use of resonance structures is important for understanding both molecular structure and molecular stability.

3. Resonance structures can be drawn when bonds, unshared electron pairs, or single electrons can be delocalized (moved) by the curved-arrow formalism without moving any atoms.

   a. Resonance structures are usually placed in brackets to emphasize the fact that they are being used to describe a single species.

   b. Resonance structures are used to describe single molecules; a molecule is the weighted average of its resonance structures.

   c. The resonance arrow means that two or more fictitious structures are being used to represent a compound for which a single Lewis structure is inadequate.

B. Relative Importance of Resonance Structures

1. The structure of a molecule is most closely approximated by its most important resonance structures.

2. A molecule is a weighted average of its contributing resonance structures; some resonance structures are more important than others.
   a. To evaluate the relative importance of resonance structures, compare the stabilities of the resonance structures for a given molecule by treating each structure as a separate molecule.
   b. The most stable structures are the most important ones.
3. Guidelines for important resonance structures:
   a. Identical structures are equally important descriptions of a molecule.

   identical resonance structures
   (equally important)

   b. Structures with the greater number of bonds are more important; bonding is energetically favorable.
   c. Structures that require the separation of opposite charges are less important than those that do not.

   separation of charges (less important)

   i. When a charge is separated, two opposite charges are moved away from each other.
   ii. When a charge is delocalized by resonance, charge of a given type is moved to different locations within a molecule.

charge delocalization

charge separation

   d. Structures in which charges and/or electron deficiency are assigned to atoms of appropriate electronegativity are more important.
   e. If the orbital overlap symbolized by a resonance structure is not possible, the resonance structure is not important.

highly strained resonance structure (unimportant)

4. All other things being equal, a molecule with a greater number of important resonance structures is more stable than an isomeric molecule with fewer resonance structures.

III. *Benzene*

A. INTRODUCTION

1. Benzene and its derivatives constitute the class of organic substances called aromatic compounds.
2. Benzene undergoes few of the addition reactions that are associated with either conjugated dienes or ordinary alkenes.

B. STRUCTURE OF BENZENE

1. The structure of benzene has only *one* type of carbon-carbon bond with a bond length intermediate between the lengths of single and double bonds.
   a. All atoms in benzene lie in one plane.
   b. Each carbon atom is trigonal (sp^2-hybridized).
      i. All six *p* orbitals are parallel.

ii. These *p* orbitals overlap to form a continuous bonding π molecular orbital.

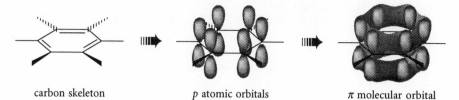

carbon skeleton   *p* atomic orbitals   π molecular orbital

2. Benzene can be depicted as the hybrid of two equally contributing resonance structures.

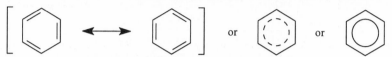

3. Benzene is more stable than a hypothetical six-carbon cyclic triene by 126–172 kJ/mol (30–41 kcal/mol); this energy difference is called the empirical resonance energy.
4. The empirical resonance energy of benzene is an estimate of how much special stability is implied by the resonance structures for benzene.

IV. *Aromaticity*

A. AROMATICITY AND THE HÜCKEL 4*n* + 2 RULE

1. The unusual stability of benzene is called aromaticity; aromaticity is a property of certain cyclic compounds.
2. To be aromatic, a compound must conform to all of the following criteria (often called collectively the Hückel 4*n* + 2 rule or simply the 4*n* + 2 rule):
   a. Aromatic compounds contain one or more rings that have a cyclic arrangement of *p* orbitals.
   b. Every atom of an aromatic ring has a *p* orbital; thus, an aromatic compound has a continuous cycle of *p* orbitals.
   c. Aromatic compounds are planar; that is, the *p* orbitals must overlap to form π molecular orbitals.
   d. The cyclic arrangement of *p* orbitals in an aromatic compound must contain 4*n* + 2 π electrons, where *n* is any positive integer (0, 1, 2, …); thus, an aromatic ring must contain 2, 6, 10, … π electrons.
3. Electron counting rules:
   a. Each atom that is part of a double bond contributes one π electron.
   b. Vinylic unshared electrons (electrons on doubly bonded atoms) do not contribute to the π-electron count.
   c. Allylic unshared electrons contribute to the π-electron count if they occupy an orbital parallel to the other *p* orbitals in the molecule.
   d. An atom with an empty *p* orbital can be part of a continuous aromatic π-electron system, but contributes no electrons.
4. Aromatic compounds are of many different types:
   a. Aromatic compounds may contain heteroatoms (aromatic heterocycles).
   b. Some ionic species are aromatic.
   c. Fused bicyclic and polycyclic compounds can be aromatic.
   d. Some remarkable organometallic compounds have aromatic character.

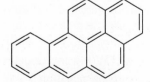

pyrrole
(an aromatic
heterocycle)

tropylium bromide
(an ionic
aromatic compound)

benzo[*a*]pyrene
(a fused polycyclic
aromatic compound)

ferrocene
(an aromatic
organometallic compund)

5. The bonding molecular orbitals of cyclic π-electron systems have particularly low energy.

B. ANTIAROMATIC COMPOUNDS

1. Compounds that contain planar, continuous rings of $4n$ π electrons are especially unstable and are said to be antiaromatic.

antiaromatic (4 π electrons)

2. The overlap of p orbitals in molecules with cyclic arrays of $4n$ π electrons is a destabilizing effect.

V. Ultraviolet Spectroscopy

A. INTRODUCTION

1. In ultraviolet-visible (UV-VIS) spectroscopy, the absorption of radiation in the ultraviolet or visible region of the spectrum is recorded as a function of wavelength.
   a. Both UV and visible spectroscopy are considered together as one type of spectroscopy, often simply called UV spectroscopy.
   b. The part of the ultraviolet spectrum of greatest interest to organic chemists is the near ultraviolet range (200–400 nm).
2. The UV spectrum of a substance is the graph of radiation absorption by a substance *vs.* the wavelength of radiation, and is determined in an instrument called a UV spectrophotometer.
   a. The wavelength λ of the ultraviolet radiation (in nanometers, nm) is plotted on the horizontal axis of the UV spectrum.
   b. The absorbance A (or optical density, O.D.) is plotted on the vertical axis of a UV spectrum; the absorbance is a measure of the amount of radiation energy absorbed.
      i. The radiation entering a sample has intensity I_0.
      ii. The light emerging from the sample has intensity I.
      iii. The absorbance A is defined as the logarithm of the ratio I_0/I.

$$A = \log \frac{I_0}{I}$$

   c. Absorption peaks in the UV spectra of compounds in solution are generally quite broad.
   d. Ultraviolet-visible spectroscopy can be used to identify organic compounds containing conjugated π-electron systems.
   e. The wavelength at the maximum of an absorption peak is called the λ_{max}; some compounds have several absorption peaks and a corresponding number of λ_{max} values.
3. UV spectroscopy is frequently used for quantitative analysis.
   a. The absorbance at a given wavelength depends on the number of molecules in the light path.
   b. The Beer-Lambert law (Beer's law) states that the absorbance A is proportional to the product of the path length (l) and the concentration (c).

$$A = \varepsilon l c$$

   where ε, the constant of proportionality, is called the molar extinction coefficient or molar absorptivity.
      i. Each absorption in a given spectrum has a unique extinction coefficient.
      ii. Each molar extinction coefficient depends on wavelength, solvent, and temperature.
      iii. The larger the ε, the greater is the light absorption at a given concentration c and path length l; thus, strong absorptions can be obtained from very dilute solutions if the absorbing molecule has a high molar extinction coefficient ε.
4. Some UV spectra are presented in abbreviated form by citing the λ_{max} values of their principal peaks, the solvent used, and the extinction coefficients.
5. UV spectroscopy is especially useful for the diagnosis of conjugated double or triple bonds.

B. PHYSICAL BASIS OF UV SPECTROSCOPY

1. Ultraviolet and/or visible radiation is absorbed by the π electrons and, in some cases, by the unshared electron pairs in organic compounds.
   a. Absorptions by compounds containing only unshared electron pairs are generally quite weak.

   b. UV and visible spectra are sometimes called electronic spectra.

2. The structural feature of a molecule responsible for its UV or visible absorption is called a chromophore.

3. UV absorptions of conjugated alkenes are due to $\pi \rightarrow \pi^*$ transitions.

   a. When a chromophore absorbs energy from light, a π electron is elevated from a bonding molecular orbital (ground state) to an antibonding π^* molecular orbital (excited state) in a process called a $\pi \rightarrow \pi^*$ transition.

   b. The energy required for $\pi \rightarrow \pi^*$ absorption must match the difference in energy between the π and π^* orbitals.

C. UV SPECTROSCOPY OF CONJUGATED ALKENES

1. When UV spectroscopy is used to determine chemical structure, the most important aspect of a spectrum is the λ_{max} values.

2. The structural features of a compound that are most important in determining the λ_{max} are

   a. the number of consecutive conjugated double or triple bonds (the longer the conjugated π-electron system, the higher the wavelength of the absorption).

      i. The λ_{max} increases with increasing number of conjugated double bonds.

      ii. Molecules that contain many conjugated double bonds generally have several absorption peaks; the λ_{max} usually quoted for such compounds is the one at highest wavelength.

      iii. One or more of the λ_{max} values of a compound that has several double bonds in conjugation will be large enough to fall within the visible region of the electromagnetic spectrum, and the compound will appear colored.

   b. the conformation of a diene unit about its central single bond.

      i. Noncyclic dienes generally assume the lower energy *s-trans*, or transoid, conformation.

      ii. Dienes that are locked into *s-cis*, or cisoid, conformations have higher values of λ_{max} and lower extinction coefficients than comparably substituted *s-trans*, or transoid, compounds.

   c. the presence of substituent groups on the double bond.

      i. Each alkyl substituent on a double bond adds +5 nm to the base λ_{max} of the conjugated system.

3. Light absorption by a pigment, rhodopsin, in the rod cells (as well as a related pigment in the cone cells) of the eye triggers the series of physiological events associated with vision.

VI. *General Concepts*

A. ALLYLIC CARBOCATIONS

1. An allylic carbocation is a carbocation adjacent to a double bond.

                                allylic carbocation

2. Allylic cations are more stable than comparably substituted nonallylic alkyl cations.

   a. The unusual stability of allylic carbocations lies in their electronic structures.

   b. The electron-deficient carbon and the carbons of the double bond are all sp^2-hybridized (each carbon has a p orbital).

      i. The overlap of these p orbitals provides additional bonding in allylic carbocations and hence, additional stability.

      ii. Both the positive charge and the double-bond character are delocalized.

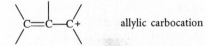

3. Allylic carbocations are represented as resonance hybrids.

4. Relative stability of carbocations:

      primary alkyl < secondary alkyl < secondary allylic ≈ tertiary alkyl < tertiary allylic

B. Kinetic and Thermodynamic Control

1. When the product distribution in a reaction differs substantially from the product distribution that would be observed if the products were at equilibrium, the reaction is said to be kinetically controlled.
2. If the products of a reaction come to equilibrium under the reaction conditions, the product distribution is said to be thermodynamically controlled.
3. When the less stable product of a reaction is the major product, then two things must be true:
    a. The less stable product must be formed more rapidly than the other products.
    b. The products must not come to equilibrium under the reaction conditions.
4. The precise reason for kinetic control varies from reaction to reaction; the relative amounts of products in a kinetically controlled reaction are determined by the relative energies of the transition states for each of the product-determining steps.
5. In a kinetically-controlled reaction, the most stable product is not always the one formed in greatest amount.

- -

⬥ Reactions

- -

I. *Addition of Hydrogen Halides to Conjugated Dienes*

A. 1,2- and 1,4-Additions

1. Conjugated dienes react with hydrogen halides to give two types of addition products:
    a. The major product results from 1,2-addition. In a 1,2-addition, the hydrogen and halogen add to adjacent carbons.
    b. The minor product results from 1,4-addition, or conjugate addition. In a 1,4-addition, the hydrogen and halogen add to carbons that have a 1,4-relationship.

1,2-addition product   1,4-addition product

2. The 1,2-addition mechanism is analogous to the addition of HX with an ordinary alkene.
3. The 1,4-addition, or conjugate-addition, mechanism involves:
    a. protonation of a double bond to give an allylic carbocation in which the positive charge in this ion is not localized, but is instead shared by two different carbons.
    b. formation of two constitutional isomers by attack of halide ion on either of the electron-deficient carbons.

1,2-addition product   1,4-addition product

4. The product distribution in hydrogen-halide addition to a conjugated diene is determined by the relative rates of the product-determining steps.

a. Formation of the 1,2-addition product is faster but reversible.

b. Formation of the 1,4-addition product is slower but virtually irreversible.

B. DIENE POLYMERS

1. The polymerization of 1,3-butadiene gives polybutadiene, an important type of synthetic rubber used in the manufacture of tires.

   a. Polybutadiene is referred to as a diene polymer because it comes from polymerization of a diene monomer.

   b. Polybutadiene has only one double bond per unit, because one double bond is lost through the addition that takes place in the polymerization process.

2. Polymerization of dienes is a free-radical reaction. Although the polymerization product is largely the result of 1,4-addition, a small amount of 1,2-addition can occur as well.

3. Styrene-butadiene rubber, SBR, is an example of a copolymer, a polymer produced by the simultaneous polymerization of two or more monomers.

II. The Diels-Alder Reaction

A. INTRODUCTION

1. Many conjugated dienes undergo addition reactions, called Diels-Alder reactions, with certain alkenes or alkynes.

   a. The conjugated diene component of this reaction is referred to simply as the diene.

   b. The alkene component of this reaction is referred to as the dienophile.

dienophile $\longrightarrow$ $H_2C{=}CH{-}CO_2CH_3$

diene $\longrightarrow$

+

$\longrightarrow$

CO_2CH_3

2. Some of the dienophiles that react most readily in the Diels-Alder reaction bear substituent groups such as esters ($-CO_2R$), nitriles ($-C{\equiv}N$), or certain other unsaturated, electronegative groups that are conjugated with the alkene double bond. (Some alkynes also serve as dienophiles.)

3. When the diene is cyclic, bicyclic products are obtained in the Diels-Alder reaction.

+ CH_3O_2C CO_2CH_3 $\longrightarrow$ CO_2CH_3 $-CO_2CH_3$ $\equiv$ CO_2CH_3 CO_2CH_3

4. The Diels-Alder reaction is

   a. a cycloaddition reaction (an addition reaction that results in the formation of a ring).

   b. a pericyclic reaction (a concerted reaction that involves a cyclic flow of electrons).

   c. a 1,4-addition, or conjugate addition, reaction (an addition that occurs across the outer carbons of the diene; 1,4-addition is a characteristic type of reaction of conjugated dienes).

   d. a stereoselective reaction.

B. EFFECT OF DIENE CONFORMATION ON THE DIELS-ALDER REACTION

1. In the Diels-Alder transition state, the diene component is in the *s-cis* conformation.

   a. Dienes "locked" into *s-trans*, or transoid, conformations are unreactive in Diels-Alder reactions.

   b. Dienes "locked" into *s-cis*, or cisoid, conformations are unusually reactive and in many cases much more reactive than corresponding noncyclic dienes.

2. The Diels-Alder reactions of a *cis*-diene are much slower than the corresponding reactions of a *trans*-diene, which does not have the destabilizing repulsion in its cisoid conformation.

cis-diene   van der Waals repulsions

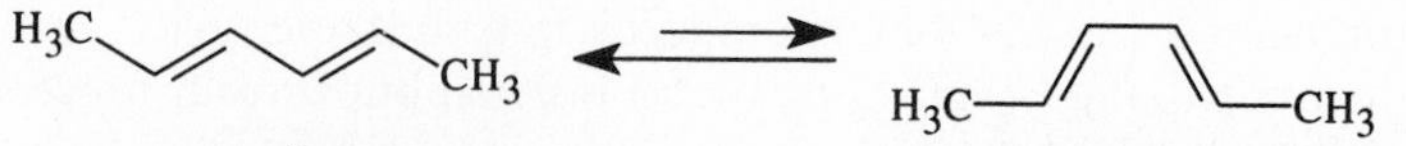

trans-diene   no methyl-methyl repulsions

    a. Severe van der Waals repulsion destabilizes the cisoid conformation of the diene.
    b. Severe van der Waals repulsion destabilizes the transition states for the Diels-Alder reactions of the cisoid conformation of the diene.

C. STEREOCHEMISTRY OF THE DIELS-ALDER REACTION

  1. When a diene and dienophile react in a Diels-Alder reaction, they approach each other in parallel planes.
    a. This type of approach allows the π-electron clouds of the two components to overlap and form the bonds of the product.
    b. Each component adds to the other at one face.
      i. The diene undergoes a *syn* addition to the dienophile.
     ii. The dienophile undergoes a 1,4-*syn* addition to the diene.
    c. Groups that are *cis* in the alkene starting material are also *cis* in the product.
    d. *Syn* addition to the diene is apparent if the terminal carbons of the diene unit are stereocenters.
  2. A *syn* addition requires that in the Diels-Alder product:
    a. the two inner substituents of the diene always have a *cis* relationship.
    b. the two outer substituents of the diene also always have a *cis* relationship.
    c. an inner substituent on one carbon of the diene is always *trans* to an outer substituent on the other.
    d. *Cis* substituents in the dienophile are *cis* in the product; *trans* substituents in the dienophile are *trans* in the product.

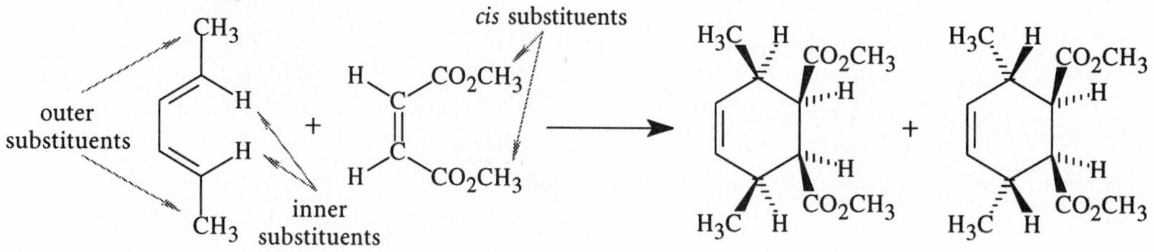

  3. For a given diene and dienophile, two diastereomeric *syn*-addition products are possible and are abbreviated with the terms *endo* and *exo*:
    a. In the *endo* product, the dienophile substituents are *cis* to the outer diene substituents.

endo addition

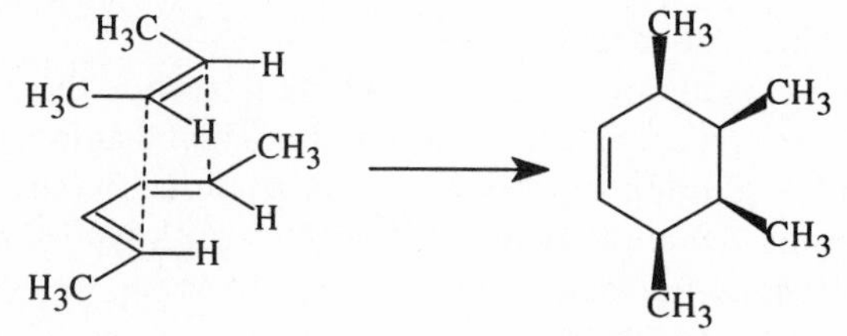

    b. In the *exo* product, the dienophile substituents are *trans* to the outer diene substituents.

exo addition

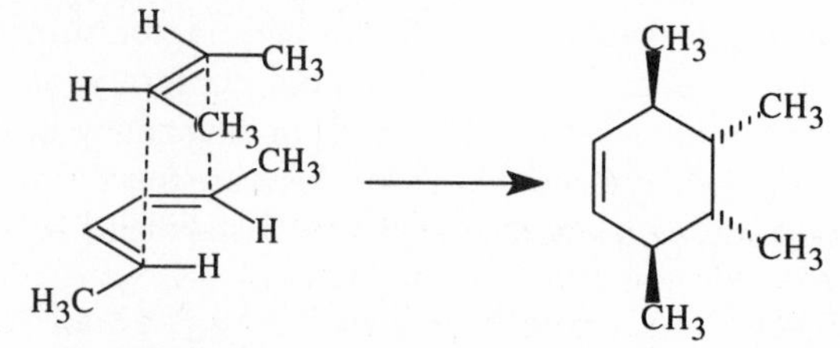

    c. In many cases, the *endo* mode of addition is favored over the *exo* mode, particularly when cyclic dienes are used.

Study Guide Links

✓ 15.1   A Terminology Review

The purpose of this Study Guide Link is to review and to distinguish between the following terms used to describe addition reactions:

1. cycloaddition reactions
2. concerted reactions
3. pericyclic reactions

These terms in some cases may seem synonymous, but they are in fact different, and have very precise meanings. A *cycloaddition* is an addition that results in ring formation. Ozone addition and the Diels-Alder reaction are two examples of cycloadditions. The term *cycloaddition* succinctly describes the *outcome* of a reaction, but says nothing about the *mechanism* of the reaction. The terms *concerted* and *pericyclic*, in contrast, are descriptions of *mechanism*. Reactions that occur by mechanisms involving a single step are referred to as *concerted reactions*. Concerted reactions that occur by a cyclic flow of electrons are called *pericyclic reactions* (pericyclic = "around the circle"). All pericyclic reactions are concerted, but some concerted reactions are not pericyclic. Thus, hydroboration of alkenes (Sec. 5.3B of the text) is a concerted pericyclic reaction (although it is not a cycloaddition because a ring is not formed). The S_N2 reaction (Sec. 9.3 of the text) is concerted, but it is not pericyclic because the flow of electrons in the mechanism is not cyclic. The Diels-Alder reaction is a concerted cycloaddition, and it is also a pericyclic reaction. (Chapter 25 is devoted to a more detailed study of pericyclic reactions.)

✓ 15.2   Resonance Structures

Remember that resonance structures have *two* major implications:

1. A *structural implication:* resonance structures show that a compound cannot be represented by a single Lewis structure.
2. *An energy implication:* molecules for which there are more important resonance structures are more stable than isomers for which there are fewer important resonance structures.

Another way to state the energy implication is to say that *a molecule represented by resonance structures is more stable than any of its fictitious resonance contributors.* Thus, the allyl cation is more stable than it would be if it were accurately described by *either one* of the following structures in brackets:

$$\left[H_2C{=}CH{-}\overset{+}{C}H_2 \longleftrightarrow H_2\overset{+}{C}{-}CH{=}CH_2 \right]$$

allyl cation

Think about this philosophical point. The term "resonance-stabilized" might sound as if the stability associated with resonance structures is somehow a *consequence of our ability to draw these structures.* The behavior and properties of a chemical substance are, of course, totally independent of how we represent the substance in our minds or on paper. Our pencil-and-paper representations of resonance structures—or any structures, for that matter—are merely symbols for bonding patterns that occur in nature. When you think about it, it is quite remarkable that such simple symbols can be used to predict chemical phenomena or to portray something that requires considerable mathematical sophistication to describe physically. Resonance is a structural representation that has been recruited to say something about bonding. The extra stability associated with resonance is a consequence of this extra bonding. *Resonance structures have energy implications only because the extra bonding they represent has such implications.* We could do without resonance structures if we were willing to sketch out the π-molecular orbitals involved every time we draw a resonance-stabilized species. However, it is easier to use the inherently simpler Lewis structures to represent the same thing.

Because resonance structures are a symbolic representation, they must not be used blindly; rather, you must always have in the back of your mind what resonance structures mean. They convey information about the overlap of orbitals; and, if such overlap is not possible, resonance structures are not appropriate, even if they may seem to be derivable on paper with the curved-arrow formalism. In order to use resonance structures really effectively you must *think about the orbital overlap involved* every time you draw resonance structures.

Solutions

Solutions to In-Text Problems

15.1   When the 1,3-butadiene molecule undergoes internal rotation about the carbon-carbon single bond such that the π systems of the two double bonds become perpendicular, overlap is lost at the cost of some stabilization. Notice, however, that the 31 kJ/mol cost of rotation about the central single bond is very modest compared with the cost of rotation about a true double bond, which is about 264 kJ/mol. In other words, the π-orbital overlap across the central single bond, while real, is relatively modest.

15.3   The σ-bond component of the double bond in a cumulene is an sp-sp^2 bond, whereas, in ordinary alkenes, it is an sp^2-sp^2 bond. (See Fig. 15.2 on text p. 683.) That is, the bond in a cumulated diene contains more s character. Bond length decreases with increasing s character in the component orbitals.

15.4   The effect of conjugation operates within both pairs; that is, the conjugated diene is more stable than the unconjugated one. But an *additional* effect governs the relative energies of the pentadienes. The double bond between carbons 3 and 4 of 1,3-pentadiene has two carbon substituents, whereas the double bond between carbons 4 and 5 of 1,4-pentadiene has only one carbon substituent. Consequently, 1,3-pentadiene is more stable than 1,4-pentadiene not only by the amount due to conjugation, but also by an additional amount due to the additional carbon substituent on the double bond. In contrast, both of the hexadienes have two carbon substituents on one double bond; hence the energy difference between them reflects *only* conjugation.

15.5   (a)

(b) Because enantiomers have specific rotations of equal magnitudes and opposite signs, the other enantiomer has a specific rotation of +30.7°.

15.6   (a) The energy of light is given by Eq. 12.3 on text p. 537 with the value of h, in kJ sec mol^{-1}, given in Eq. 12.5.

$$E = \frac{hc}{\lambda} = \frac{(3.99 \times 10^{-13}\ \text{kJ sec mol}^{-1})(3 \times 10^8\ \text{m sec}^{-1})}{(450 \times 10^{-9}\ \text{m})} = 266\ \text{kJ/mol (63.6 kcal/mol)}$$

15.7   (a) Use Eq. 15.2 in the text, with $(I_0/I) = 2$. Thus, $A = \log(2) = 0.30$.

15.8   The piece with greater absorbance transmits less of the incident radiation. Therefore, the thick piece of glass has greater absorbance.

15.9   (a) Apply Beer's law (text Eq. 15.3), with $\varepsilon = 10{,}750$ absorbance units cm^{-1} L mol^{-1}, $l = 1.0$ cm, and, from Fig. 15.3 on text p. 685, $A = 0.800$ absorbance units. The concentration of isoprene is therefore

$$\frac{0.800\ \text{absorbance units}}{(10{,}750\ \text{absorbance units cm}^{-1}\ \text{L mol}^{-1})(1\ \text{cm})} = 7.44 \times 10^{-5}\ \text{mol/L}.$$

15.10 (a) In each case, assume that the energy difference is the same as the energy of the UV radiation at the λ_{max}. (See Fig. 15.4 on text p. 687.) This energy, in turn, is equal to hc/λ. For isoprene, this energy separation is

$$E = \frac{hc}{\lambda} = \frac{(3.99 \times 10^{-13}\ \text{kJ sec mol}^{-1})(3 \times 10^8\ \text{m sec}^{-1})}{(222.5 \times 10^{-9}\ \text{m})} = 538\ \text{kJ/mol} \ (129\ \text{kcal/mol})$$

15.11 (a) The —CH_2— groups of the ring attached to the double bond are alkyl substituents. (If this point is not clear, see Study Guide Link 9.3 on p. 195 of this manual.) These two alkyl substituents contribute a total of +10 nm to the base λ_{max} of 217 nm. The predicted λ_{max} is therefore (217 + 10) nm = 227 nm.

(c) For three alkyl substituents the predicted λ_{max} is (217 + 15) nm = 232 nm. (This calculation assumes that the molecule exists predominantly in the *s-trans* conformation.)

15.12 (a) The stereochemistry of this addition, also shown below, is covered in Sec. 15.3C.

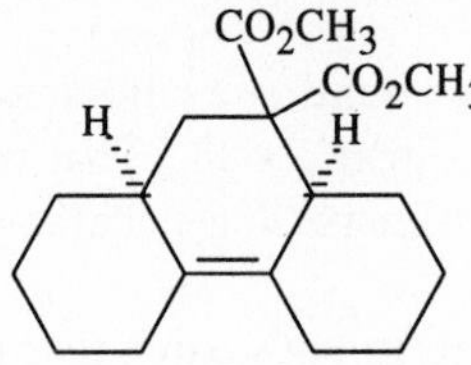

15.13 (a) An analysis like that employed in Study Problem 15.1 suggests two possibilities.

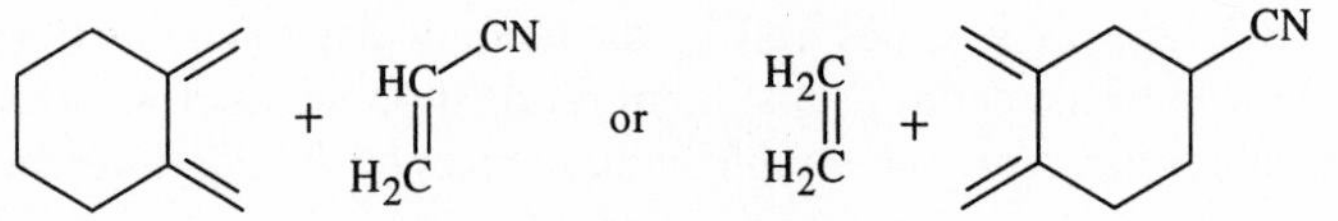

Pair *A* is preferred because, in many cases, the most reactive dienophiles are those with conjugated electronegative substituents. But if your answer was pair *B*, you have analyzed the problem correctly.

(c) This product results from the Diels-Alder reaction of two molecules of 1,3-butadiene:

15.14 (a) The dienophile can be oriented in two different ways with respect to the diene.

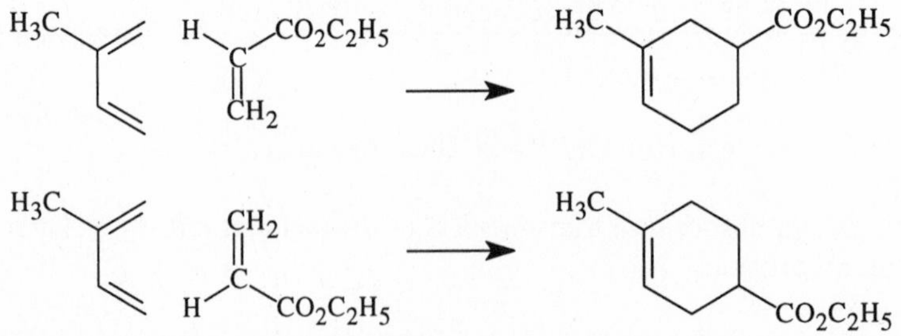

15.15 The Diels-Alder reaction requires that the diene assume an *s-cis* conformation in the transition state. In this conformation, the (2E,4Z)-diene has a methyl-hydrogen van der Waals repulsion that is absent in the (2E,4E)-diene. This repulsion raises the energy of the Diels-Alder transition state of this diene, and therefore reduces the rate of its reaction. Since the (2E,4Z)-diene reacts more slowly than the (2E,4E)-diene, it follows that the (2E,4E)-diene is consumed in the reaction, and the (2E,4Z)-diene remains unreacted.

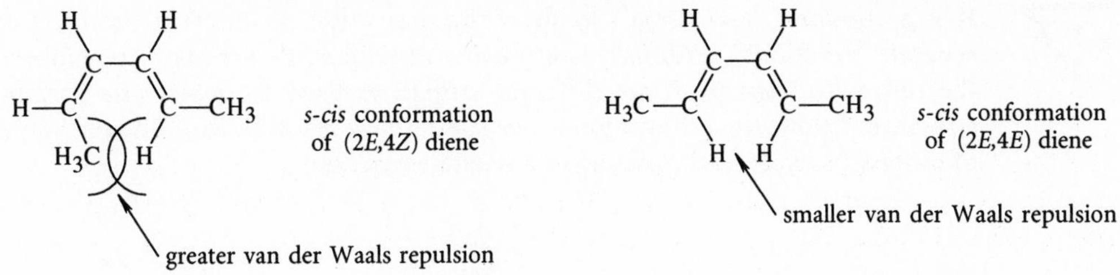

s-cis conformation of (2E,4Z) diene

greater van der Waals repulsion

s-cis conformation of (2E,4E) diene

smaller van der Waals repulsion

15.17 (a) The two acetoxy groups are *cis* in the product because they are both outer substituents in the *s-cis* conformation. (Of course, the racemate is formed; only one enantiomer is shown below.)

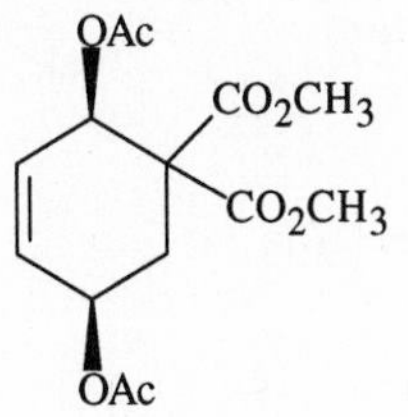

(c) The "inner" groups of the diene are "tied together" as the —CH₂— group of the five-membered ring.

CH₂ group of the diene

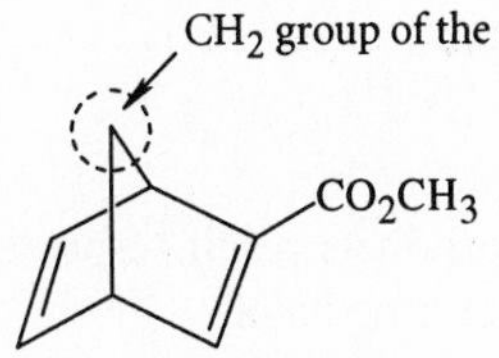

15.18 (a)

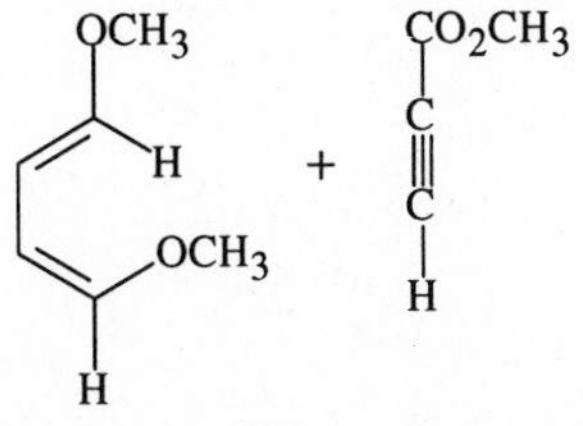

(c)

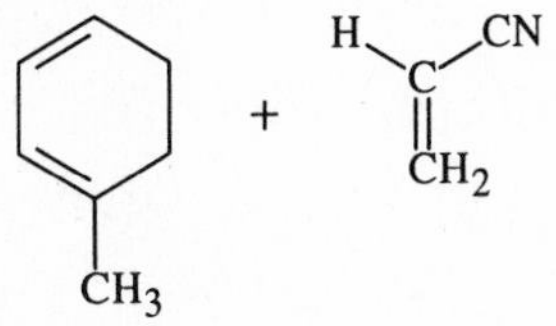

15.19 As shown by the diagram on p. 698, the *endo* diastereomer is the one in which the dienophile substituents (which, in this case, are the arms of the five-membered ring) are *cis* to the outer substituents when the diene is drawn in the *s-cis* conformation.

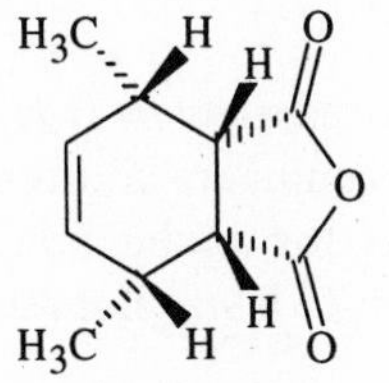

15.20 (a) The products are a mixture of constitutionally isomeric ethers because the carbocation intermediate in the S_N1 reaction has two sites of electron deficiency that can be attacked by the nucleophile ethanol. The details for the formation of 3-ethoxy-3-methylcyclohexene are shown below; you should provide the curved-arrow formalism for the formation of 3-ethoxy-1-methylcyclohexene.

(solution continues)

It is a common (and useful) to show the derivation of different products by showing the separate "reactions" of different resonance structures. Be sure you remember, however, that the resonance structures are different structures used to depict *one* species. Thus, in the structures below, the carbocation is *one* species that reacts in two different ways, *not* a mixture of different species, each reacting in a separate reaction.

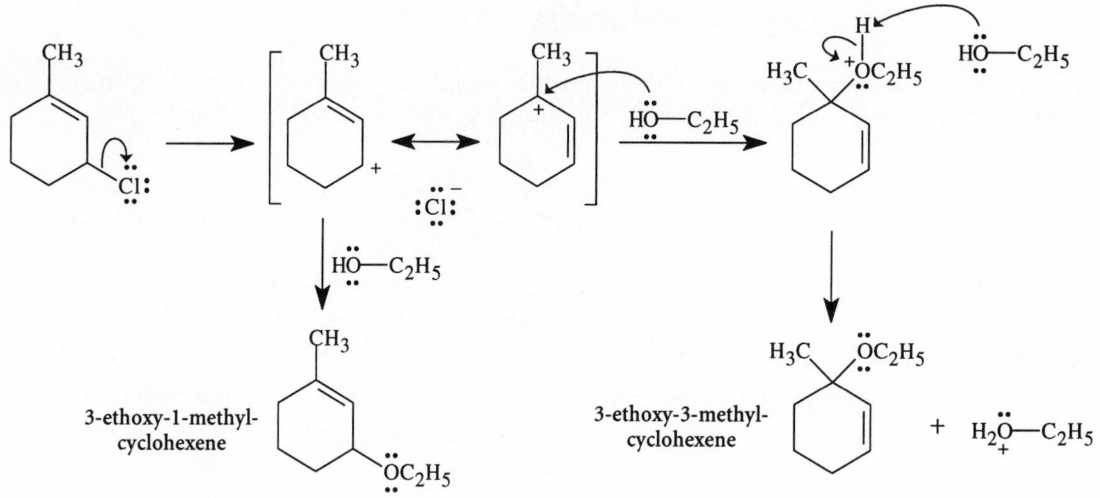

3-ethoxy-1-methyl-cyclohexene

3-ethoxy-3-methyl-cyclohexene

15.21 (a) This is an S_N1 reaction; the two products result respectively from attack of water on the two electron-deficient carbons of the carbocation intermediate.

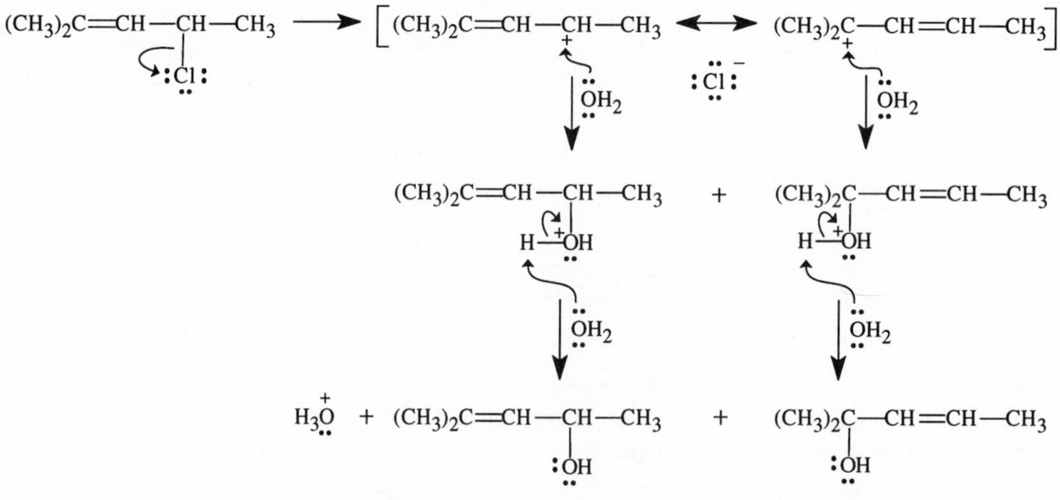

15.22 The two constitutional isomers result respectively from "normal" addition (1,2-addition, product *A*) and conjugate addition (product *B*). The conjugate-addition product *B* is the major one at equilibrium because it has an internal double bond (that is, the double bond with the greater number of alkyl branches). Although the stereochemistry of product *B* was not requested, it turns out that it is the *trans*-isomer.

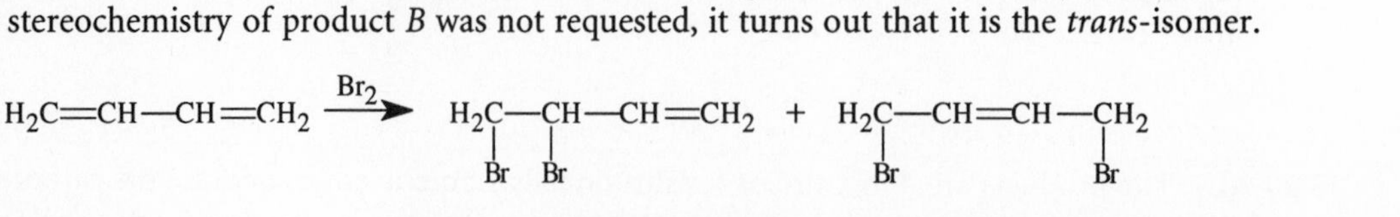

15.24 Start with a radical derived from attack of initiator on 1,3-butadiene, the alkene present in greatest amount. (1,3-Butadiene and styrene are present in a 3:1 ratio; see text p. 707.) Incorporation of the first two butadiene units is shown in Eqs. 15.27a–b on text p. 707. The mechanism for incorporation of the third butadiene unit and styrene are as follows:

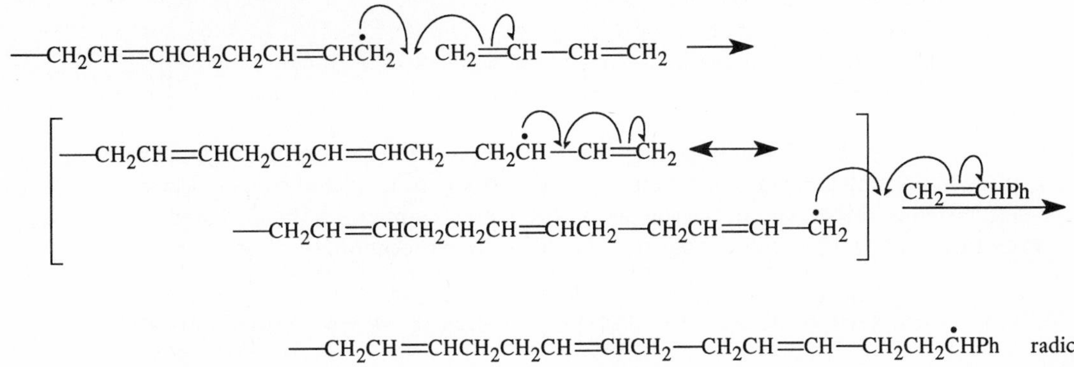

$$—CH_2CH=CHCH_2CH_2CH=CHCH_2 \quad CH_2=CH—CH=CH_2 \longrightarrow$$

$$\left[\begin{array}{l} —CH_2CH=CHCH_2CH_2CH=CHCH_2—CH_2\overset{.}{C}H—CH=CH_2 \longleftrightarrow \\ —CH_2CH=CHCH_2CH_2CH=CHCH_2—CH_2CH=CH—\overset{.}{C}H_2 \end{array} \right] \quad \xrightarrow{CH_2=CHPh}$$

$$—CH_2CH=CHCH_2CH_2CH=CHCH_2—CH_2CH=CH—CH_2CH_2\overset{.}{C}HPh \quad \text{radical } R$$

Radical *R* propagates the chain by adding to another 1,3-butadiene molecule, and so on. Notice that only conjugate addition is shown; 1,2-addition of a chain-propagating radical to 1,3-butadiene is possible at any point. (See the following problem.)

15.26 (a) The first structure is most important; the others have fewer formal bonds and involve separation of opposite charges or unpaired electrons.

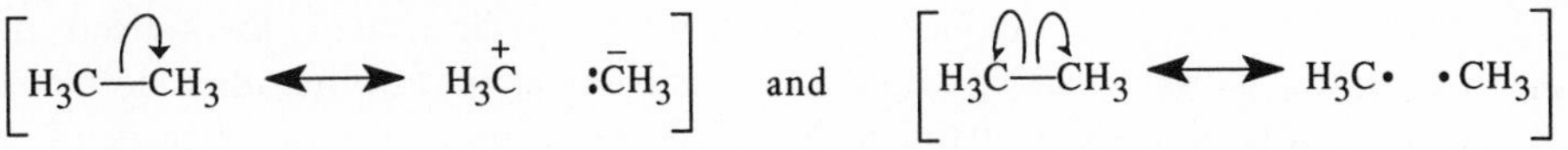

$$\left[H_3C—CH_3 \longleftrightarrow H_3\overset{+}{C} \quad :\overset{-}{C}H_3 \right] \quad \text{and} \quad \left[H_3C—CH_3 \longleftrightarrow H_3C\cdot \quad \cdot CH_3 \right]$$

(c) Both structures are important, because they have the same number of formal bonds, and show delocalization of electrons (and charge). The second structure is somewhat more important than the first, because in this structure, positive charge is on the carbon atom with the greater number of alkyl branches.

$$\left[CH_3—CH=CH—\overset{+}{C}H_2 \longleftrightarrow CH_3—\overset{+}{C}H—CH=CH_2 \right]$$

15.27 (a) The ion on the left is more stable because it has more important resonance structures. In this ion, the negative charge can be delocalized by both the double bond and the triple bond; there are a total of three resonance structures, which are shown below. In the ion on the right, the charge is delocalized only by the double bond; there are only two resonance structures.

$$\left[\overset{..}{O}=CH—\overset{-}{C}H—C≡C—CH_3 \longleftrightarrow :\overset{-}{O}—CH=CH—C≡C—CH_3 \longleftrightarrow \overset{..}{O}=CH—CH=C=\overset{-}{C}—CH_3 \right]$$

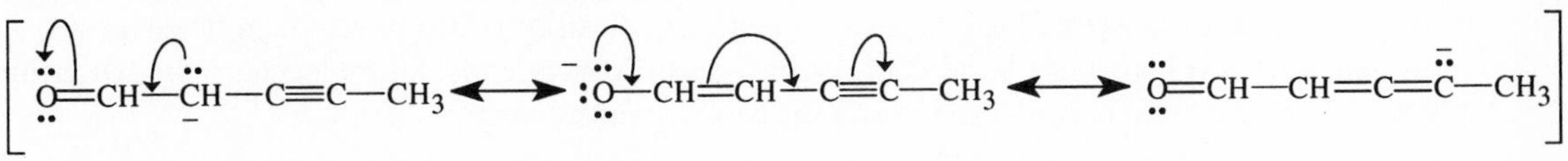

(c) The radical on the left has more important resonance structures (three) than the radical on the right. The resonance structures for the more stable radical are shown below. In the more stable radical, the unpaired electron can be delocalized by both double bonds, whereas, in the radical on the right, the unpaired electron is delocalized by only one double bond.

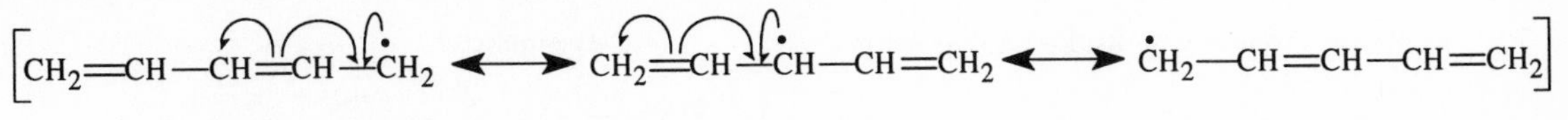

$$\left[CH_2=CH—CH=CH—\overset{.}{C}H_2 \longleftrightarrow CH_2=CH—\overset{.}{C}H—CH=CH_2 \longleftrightarrow \overset{.}{C}H_2—CH=CH—CH=CH_2 \right]$$

(d) Both ions have two resonance structures. However, the ion on the right (ion *B* below) is more stable because one of its resonance contributors is a secondary carbocation, whereas both contributors for the ion on the right (ion *A* below) are primary carbocations.

(solution continues)

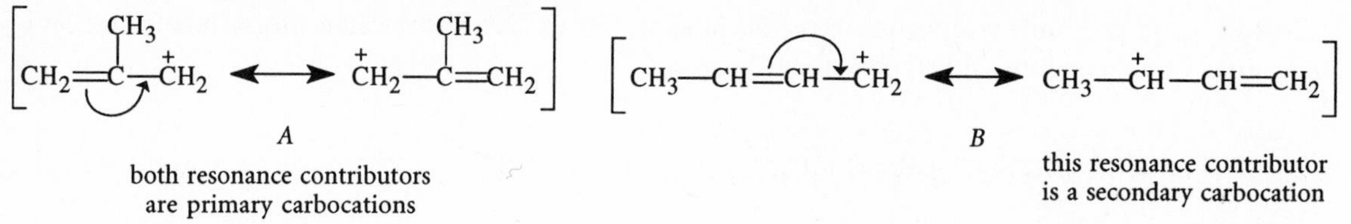

$$\left[\underset{\text{CH}_2=\underset{|}{\text{C}}-\overset{+}{\text{CH}}_2}{\overset{\text{CH}_3}{}} \longleftrightarrow \underset{\overset{+}{\text{CH}}_2-\underset{|}{\text{C}}=\text{CH}_2}{\overset{\text{CH}_3}{}} \right]$$

$$\left[\text{CH}_3-\text{CH}=\text{CH}-\overset{+}{\text{CH}}_2 \longleftrightarrow \text{CH}_3-\overset{+}{\text{CH}}-\text{CH}=\text{CH}_2 \right]$$

A

both resonance contributors
are primary carbocations

B

this resonance contributor
is a secondary carbocation

15.28 (a) Compound *A* should react much more rapidly because the carbocation intermediate (shown below) has three important resonance contributors. In particular, an unshared pair of electrons on the oxygen can be delocalized in this cation; in the solvolysis of the other compound, the unshared pairs on oxygen have no resonance interaction with the positive charge in the carbocation.

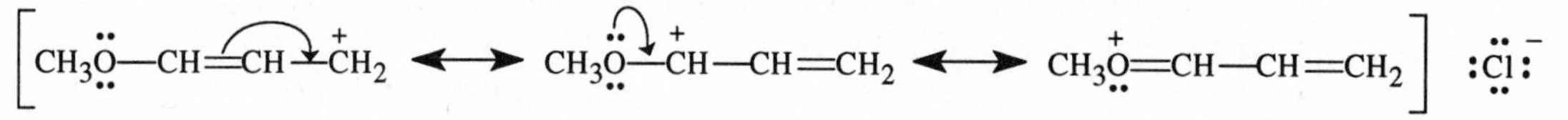

$$\left[\text{CH}_3\ddot{\text{O}}-\text{CH}=\text{CH}-\overset{+}{\text{CH}}_2 \longleftrightarrow \text{CH}_3\ddot{\text{O}}-\overset{+}{\text{CH}}-\text{CH}=\text{CH}_2 \longleftrightarrow \text{CH}_3\overset{+}{\ddot{\text{O}}}=\text{CH}-\text{CH}=\text{CH}_2 \right] \quad :\ddot{\text{Cl}}:^-$$

carbocation intermediate involved in the solvolysis of *A*

15.29 Given that furan is aromatic, it can have $4n + 2$ π-electrons (six electrons) if one of the electron pairs on oxygen is in a *p* orbital that is part of the π-electron system of the ring. The other electron pair is in an sp^2 orbital whose axis is in the plane of the ring. (The electronic structure of furan is drawn in Fig. 24.5 on text p. 1180.)

15.31 (a) This ion is not aromatic because it is not a *continuous* cycle of π-electrons.
(c) Isoxazole is aromatic. Each double bond contributes two electrons to the π-electron system. One electron pair on the oxygen is also part of the π-electron system, but the other electron pair on the oxygen is not. (See the solution to problem 15.29, above.) The electron pair on the nitrogen is vinylic and, like the electron pair in pyridine, is not part of the π-electron system.
(e) This ion is aromatic. Each double bond contributes two electrons to the π-electron system. The cationic carbon bears an empty *p* orbital that contributes no electrons but is part of the π-electron system. One electron pair on the oxygen is part of the π-electron system, and the other is not, as in Problem 15.29, or part (c) above.
(g) This compound is not aromatic because it has $4n$ rather than $4n + 2$ π-electrons. The empty *p* orbital on boron, although part of the π-electron system, contributes no electrons.

15.32 Because compounds *A* and *B* are antiaromatic, they distort as much as possible to *avoid* overlap of the *p* orbitals involved in the two double bonds. (See the discussion of cyclobutadiene on text p. 725.) The single bonds are much longer than the double bonds, and the two types of carbon-carbon bonds are not equivalent. Because the deuteriums bridge a double bond in *A* and a single bond in *B*, and because the two types of carbon-carbon bonds are distinguishable, *A* and *B* are different compounds and are therefore in principle capable of independent existence. In contrast, all carbon-carbon bonds in benzene are equivalent—each is half single bond, half double bond. Consequently, all molecules with deuteriums on adjacent positions are identical and indistinguishable. To summarize:

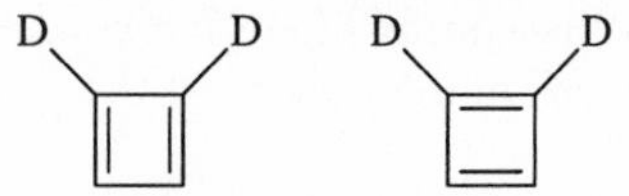

different molecules

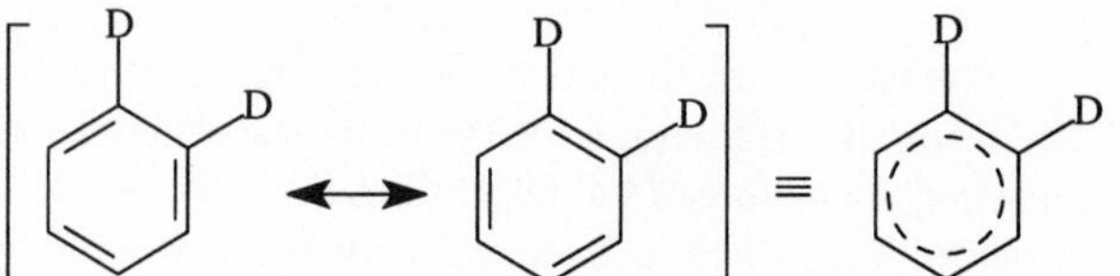

one molecule

Solutions to Additional Problems

15.34 (a) The structure on the right is somewhat more important because it is a tertiary free radical.

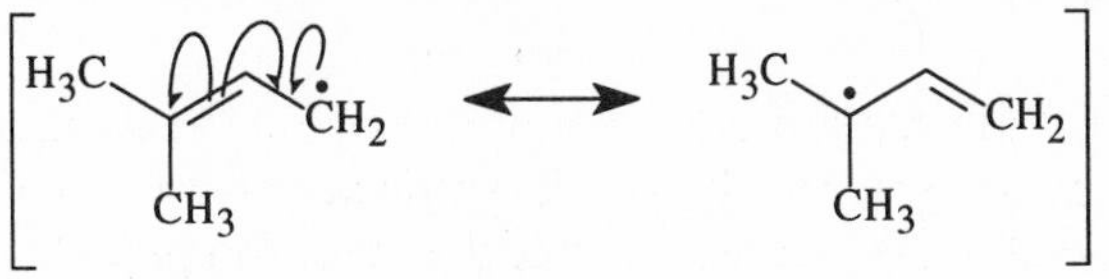

(c) The middle structure is somewhat more important because the electronegative atom (oxygen) bears the negative charge, and because the double bonds are conjugated.

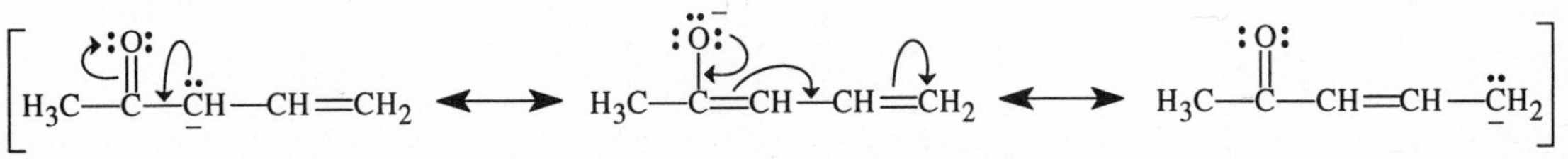

(e) The first and second structures are both secondary carbocations, but the first and third structures have conjugated double bonds; in the third structure, the double bonds have the greatest number of alkyl substituents. No structure has clearly greater importance.

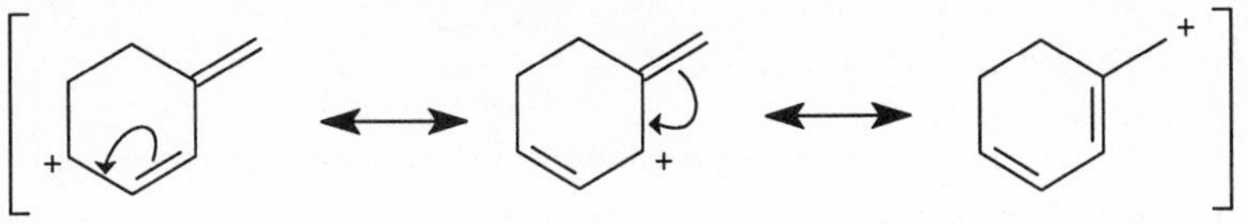

15.35 The structure of the starting material is

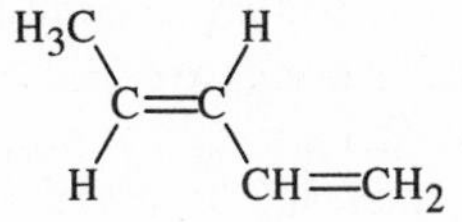

trans-1,3-pentadiene

(a)

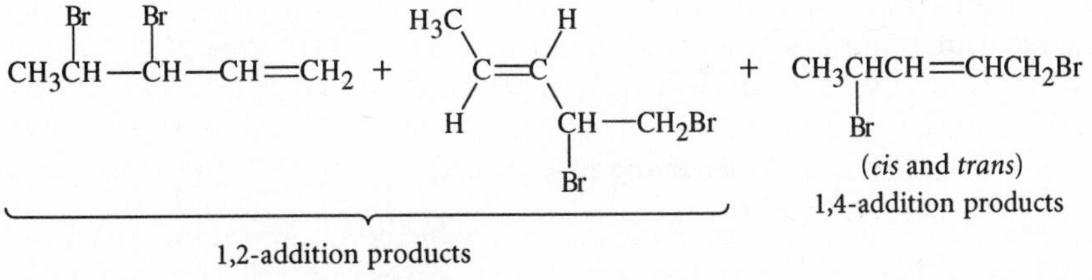

1,2-addition products

$CH_3CHCH=CHCH_2Br$
Br
(*cis* and *trans*)
1,4-addition products

(b) Within each pair, the first product is derived from the more stable carbocation intermediate (why?), and is therefore likely to be formed in greater amount.

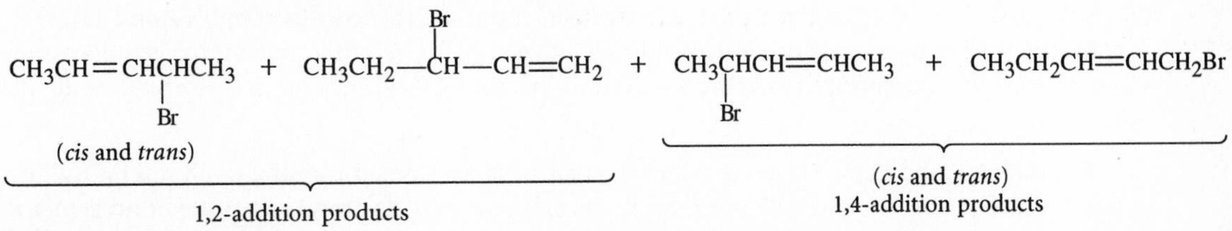

$CH_3CH=CHCHCH_3$ + $CH_3CH_2-CH-CH=CH_2$ + $CH_3CHCH=CHCH_3$ + $CH_3CH_2CH=CHCH_2Br$
Br   Br   Br

(*cis* and *trans*)

1,2-addition products

(*cis* and *trans*)
1,4-addition products

(c) $CH_3CH_2CH_2CH_2CH_3$

(d)

$$CH_3CH=CHCHCH_3 \;+\; CH_3CH_2-CH-CH=CH_2 \;+\; CH_3CHCH=CHCH_3 \;+\; CH_3CH_2CH=CHCH_2OH$$
$$\underset{OH}{|} \qquad\qquad \underset{OH}{|} \qquad\qquad\qquad \underset{OH}{|}$$

(*cis* and *trans*)

1,2-addition products

(*cis* and *trans*)
1,4-addition products

(e) No reaction occurs.

(f)

endo   *exo*

15.37 (a) 1,3-Cyclopentadiene gives 3-chlorocyclopentene by either 1,2- or 1,4-addition of HCl.

1,3-cyclopentadiene   3-chlorocyclopentene

15.38 (a) The answers to both (a) and (b) follow from the geometry of cumulenes. Each additional cumulated double bond results in a 90° twist of one end of a molecule with respect to the other. Thus, in alkenes, the atoms connected to the double bond are in the same plane; in allenes, the atoms connected to the ends of the cumulated double bonds are in *perpendicular* planes; in cumulenes with three contiguous double bonds, the atoms attached to the terminal double bonds are in the same plane, as they are in alkenes. Since the perpendicular relationship is necessary for chirality of a cumulene, allenes, as well as cumulenes with an even number of double bonds, can be chiral. Indeed, 2,3-heptadiene is a chiral allene, and therefore exists as a pair of enantiomers:

enantiomeric 2,3-heptadienes

Because 2,3,4-heptatriene has an *odd* number of cumulated double bonds, the atoms attached to the terminal double bonds are in the same plane. Hence, the twist necessary for chirality is not present, and 2,3,4-heptatriene therefore cannot be resolved into enantiomers.

15.39 (a) This compound has six π electrons in a continuous cycle, and is therefore likely to be aromatic. The electrons that contribute to the aromatic π-electron count are those of the double bond (2), the anionic carbon (2), and two of the four electrons on the oxygen (2). The other pair on oxygen, like one of the oxygen electron pairs in furan (see the solution to Problem 15.29), lies in an sp^2 orbital in the plane of the ring.

(c) Each double and triple bond contributes a pair of π electrons for a total of 18; therefore this compound is likely to be aromatic. Notice that the second pair of π electrons in a triple bond is in a π orbital that is in the plane of the ring; therefore, only the electrons in the π orbital that is in the plane of the ring contribute to the aromatic π-electron system.

15.40   The species with $4n + 2$ π electrons are likely to be planar because the planar species are aromatic. Thus, tropylium ion and cyclooctatetraenyl dianion are planar species. Oxepin, with eight π electrons, avoids planarity and thus avoids antiaromaticity.

15.42   Note the increasing heat of formation corresponds to decreasing stability.

(a) *Heats of formation:* 2 < 1 << 3. *Reasons:* Compound (2) has conjugated double bonds, and is more stable than compound (1), which does not. Because compound (3) cannot accommodate the linear geometry required by cumulated double bonds, it has great angle strain. (In fact, it is too strained to exist.)

(c) *Heats of formation:* 3 < 1 < 2. *Reasons:* Compound (3) is most stable because it is aromatic. Compound (1) is more stable than compound (2) because conjugated alkenes are more stable than isomeric alkynes.

15.43   (a) 1,4-Cyclohexadiene is not conjugated and therefore has no UV absorption in the useful region of the spectrum; 1,3-cyclohexadiene, an *s-cis* diene, has absorption at 256 nm (see text page 689). The compound with the UV spectrum is therefore the 1,3-diene.

(c) The double bonds of the second compound have two more alkyl branches than the double bonds of the first. (Note that a branch must be *on* a carbon of a double bond to affect the λ_{max}.) Hence, the λ_{max} of the second compound is greater than that of the first by about 10 nm.

(e) Although the first compound is shown in the *s-trans* conformation, it doesn't exist in that conformation because of severe van der Waals repulsions between the *tert*-butyl groups. It can avoid these repulsions by internally rotating into its *s-cis* conformation.

s-trans conformation   *s-cis* conformation

The *s-trans* conformation of the second diene, however, is more stable than the *s-cis* conformation; therefore, the second diene exists mostly in its *s-trans* conformation. Because *s-cis* dienes have considerably greater λ_{max} values than related *s-trans* dienes, the spectrum of the first diene should have a greater λ_{max} value, but a lower intensity, than the spectrum of the second diene.

15.44   First, use Beer's law to determine the concentration of the diene:

$$A = 0.356 \text{ absorbance units} = \varepsilon c l = (10{,}750) \text{ absorbance units L mol}^{-1} \text{ cm}^{-1} \, (c)(1.0 \text{ cm})$$

or
$$c = 3.31 \times 10^{-5} \text{ mol/L}$$

Since the entire sample was diluted to one liter, the entire sample contains 3.31×10^{-5} moles of isoprene. (Notice that any hydrogenation product lacks conjugated double bonds and does not have UV absorption.) The amount of isoprene present is

$$(1.00 \text{ L})(3.31 \times 10^{-5} \text{ mol/L})(68.12 \text{ g/mol}) = 2.25 \times 10^{-3} \text{ g} = 2.25 \text{ mg}$$

Because the mass of the entire sample is 75 mg, the mass percent of isoprene is

$$\frac{2.25 \text{ mg}}{75.0 \text{ mg}} \times 100\% = 3.00.$$

That is, 3% of the isolated material is unreacted isoprene.

15.45   The carbon skeleton of compound *A*, from the hydrogenation data, is that of hexane; that is, it has an unbranched chain of six carbons. The two degrees of unsaturation indicate that the compound is either a diene or an alkyne; the IR absorption and the observation of optical activity indicate that the compound is an allene. The only possible chiral allene is 2,3-hexadiene, CH_3—CH=C=CH—CH_2CH_3. Indeed, partial hydrogenation of this diene would give a mixture of *cis*-2-hexene and *cis*-3-hexene.

15.47 In order for a diene to react in a Diels-Alder reaction, it must be able to assume an *s-cis* conformation in the transition state.

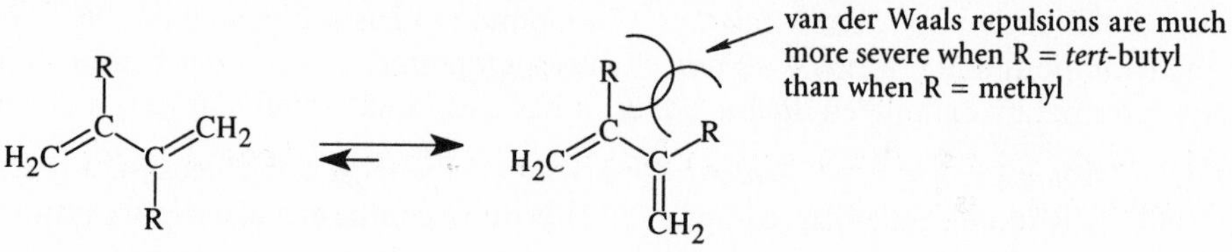

s-trans conformation   *s-cis* conformation

When R = *tert*-butyl, the van der Waals repulsions in the *s-cis* conformation are much more severe than when R = methyl. Consequently, when R = *tert*-butyl, the transition state of any Diels-Alder reaction is destabilized by these van der Waals repulsions to a greater extent than when R = methyl. Evidently, the destabilization is so great that the Diels-Alder reaction with maleic anhydride does not occur at all when R = *tert*-butyl.

15.49 The product of such a Diels-Alder reaction would be most interesting:

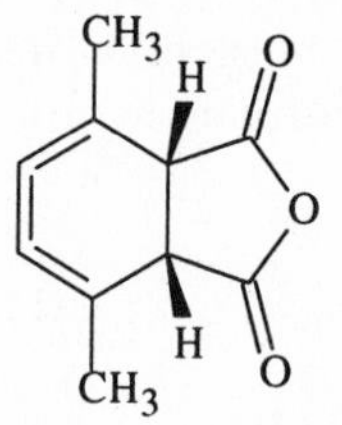

Wow! A cyclic triene containing three cumulated double bonds within the ring. Alas, such an adduct could never form, because the requirement for linear geometry in the cumulated double bonds of the product is such that the compound would have immense angle strain. A related explanation is that the *p* orbitals of maleic anhydride and those of the alkyne must overlap as shown in Figure 15.5 on text p. 695. Because the alkyne is constrained to linear geometry, the transition state for such a reaction would either be very strained (if the alkyne units were bent) or would contain extremely poor electronic overlap (if the alkyne units remained linear).

15.50 (a1) When HBr reacts with isoprene, two possible allylic carbocations could be formed:

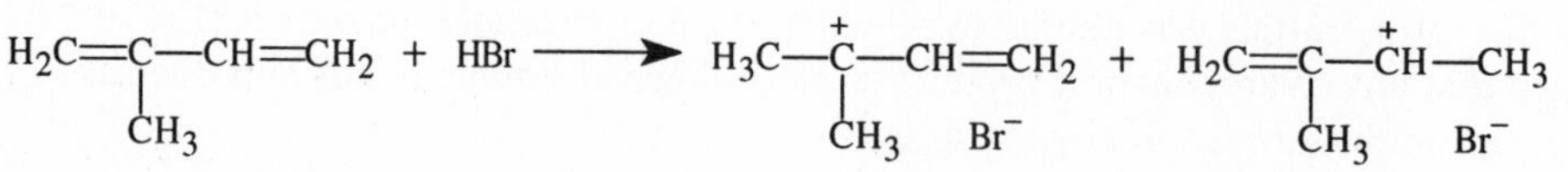

A   *B*

Carbocation *A*, a *tertiary* allylic carbocation, is more stable than cation *B*, a *secondary* allylic cation. Consequently, the products are derived from carbocation *A*:

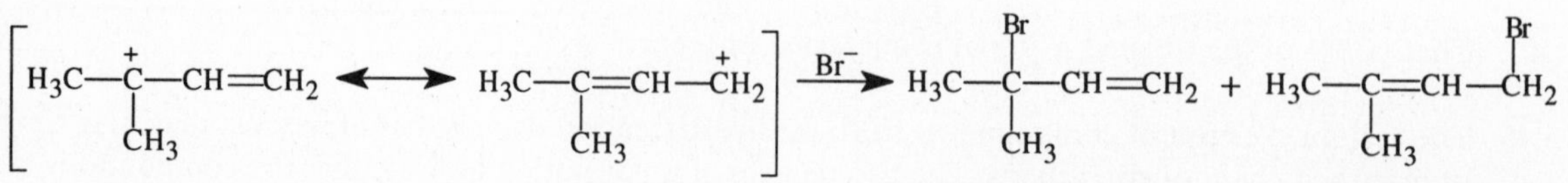

1,2-addition product   1,4-addition product

(a2)   *Trans*-1,3,5-hexatriene should give 1,2-, 1,4-, and 1,6-addition products. (The internal double bonds could have *E* or *Z* stereochemistry, although the *E* stereoisomers should predominate.)

$$H_3C-\underset{\underset{Br}{|}}{CH}-CH=CH-CH=CH_2 \ + \ H_3C-CH=CH-\underset{\underset{Br}{|}}{CH}-CH=CH_2 \ + \ H_3C-CH=CH-CH=CH-\underset{\underset{Br}{|}}{CH_2}$$

 1,2-addition product   1,4-addition product   1,6-addition product

(You should show the structure of the carbocation intermediate and how it reacts with bromide ion to give each of these products.)

(b)   In each case, the 1,2-addition product should be the major kinetically controlled product because it results from attack of bromide ion at the electron-deficient carbon closest to the site of protonation. (See p. 706 of the text.) There should be more 1,4-addition product than 1,6-addition product in the triene addition for the same reason. In the addition to isoprene, the 1,4-addition product is the more stable (thermodynamically controlled) product because it has an internal double bond, that is, the double bond with the greater number of alkyl branches. In the addition to the triene, the 1,6-addition product is the thermodynamically controlled product because it has *both* conjugated and internal double bonds.

15.52   Given that the addition occurs to the triple bond, the addition should occur regiospecifically to give the product in which the chlorine ends up on the internal carbon of the triple bond.

$$HC\equiv C-CH=CH_2 \ \xrightarrow{HCl} \ H_2C=\underset{\underset{Cl}{|}}{C}-CH=CH_2$$

chloroprene

To deduce the structure of neoprene, follow the pattern for 1,4-addition polymerization shown for polybutadiene in Eq. 15.27b on text p. 707.

$$\left[CH_2-\underset{\underset{Cl}{|}}{C}=CH-CH_2 \right]_n$$

neoprene

15.53   Assume that *A* and *B* are 1,2- and 1,4-addition products.

$$H_2C=CH-CH=CH_2 \ \xrightarrow{Cl_2} \ ClCH_2-\underset{\underset{Cl}{|}}{CH}-CH=CH_2 \ + \ ClCH_2-CH=CH-CH_2Cl$$

                                                                                                      1,4-addition product

 1,2-addition product

Which is *A* and which is *B*? And what is the stereochemistry of the 1,4-addition product? To answer these questions, note that further addition of Cl$_2$ to each gives a 1,2,3,4-tetrachlorobutane. Assume that addition of Cl$_2$ is an *anti* addition. Because the 1,4-addition product has two stereocenters, *anti* addition of Cl$_2$ will give the *meso* tetrachloride if the alkene is *trans,* and will give the racemic tetrachloride if the alkene is *cis.* (If this point is not clear, review Sec. 7.9C on text p. 313.) Because compound *B* gives a single tetrachloride *D* with *meso* stereochemistry, it follows, then, that the 1,4-addition product is compound *B*, and that it has *trans* stereochemistry. The 1,2-addition product, then, is compound *A*. Addition of Cl$_2$ to compound *A* gives both diastereomers of the tetrachloride. To summarize:

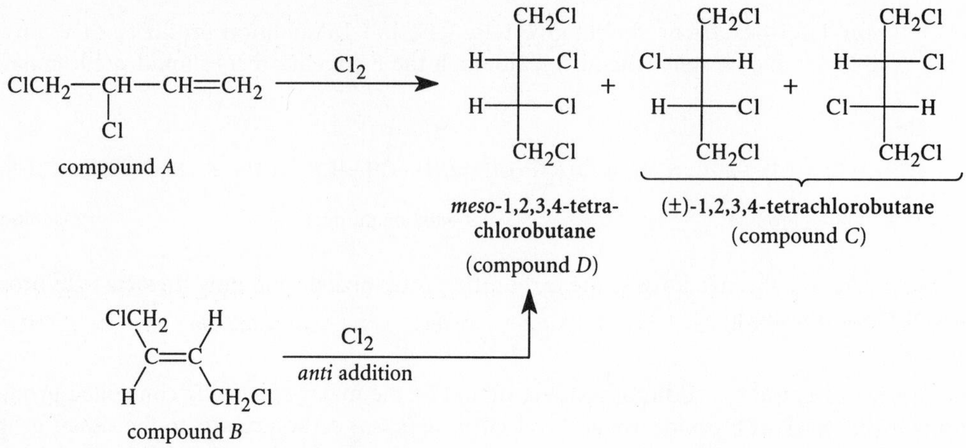

compound *A*

meso-1,2,3,4-tetra-
chlorobutane
(compound *D*)

(±)-1,2,3,4-tetrachlorobutane
(compound *C*)

anti addition

compound *B*

The addition of Cl₂ to compound *A* is not specified as *syn* or *anti* because it is not possible to determine the stereochemistry of the addition to this alkene; either stereochemistry of addition could give the same mixture of stereoisomers. (See the discussion in Sec. 7.9A, starting on text p. 311.)

15.54 (a) Reaction of KH with cyclopentadiene forms the potassium salt of the cyclopentadienyl anion (structure *X* below) plus H₂ (the gas). In this anion, all carbons, except for the isotope, are equivalent by resonance. The isotope makes no detectable difference in the relative importance of the resonance contributors; except for the position of the isotope, all resonance contributors are identical. Hence, protonation of this anion by H₂O occurs at each carbon with equal (20%) probability. Protonation at carbons 2 and 5 gives the same product; and protonation at carbons 3 and 4 gives the same product. Hence, each of these products is obtained in (2 × 20%) = 40% yield.

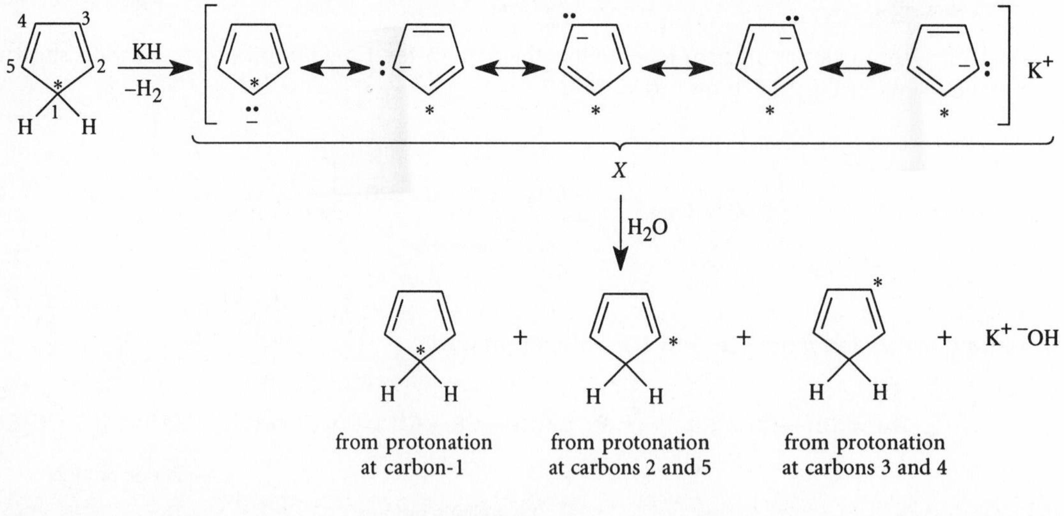

from protonation
at carbon-1

from protonation
at carbons 2 and 5

from protonation
at carbons 3 and 4

15.55 (a) The carbocation intermediate in the S$_N$1 solvolysis of compound *A* is resonance-stabilized, but the carbocation intermediate in the solvolysis of compound *B* is not. Because the relative energies of transition states reflect the relative energies of the corresponding carbocation intermediates (Hammond's postulate), the solvolysis of *A*, which gives the more stable carbocation, is faster.

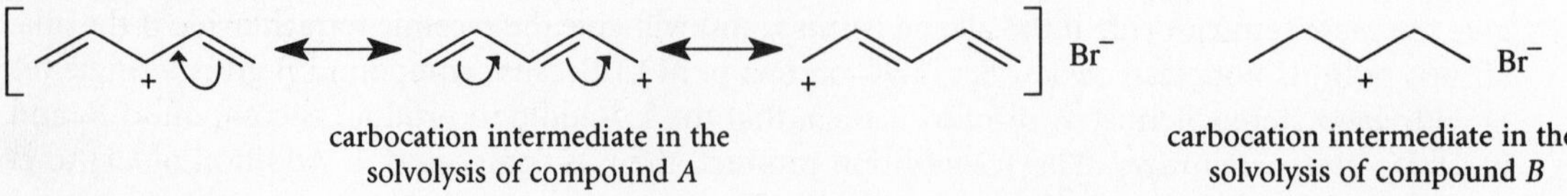

carbocation intermediate in the
solvolysis of compound *A*

carbocation intermediate in the
solvolysis of compound *B*

If compound *C* gives a doubly allylic carbocation, why should it be solvolytically inert? The answer is that the cation has a continuous cycle of π electrons. Because this cation contains $4n$ ($n = 1$) π electrons, it is *antiaromatic* and therefore very unstable. In fact, it is so unstable that it does not form.

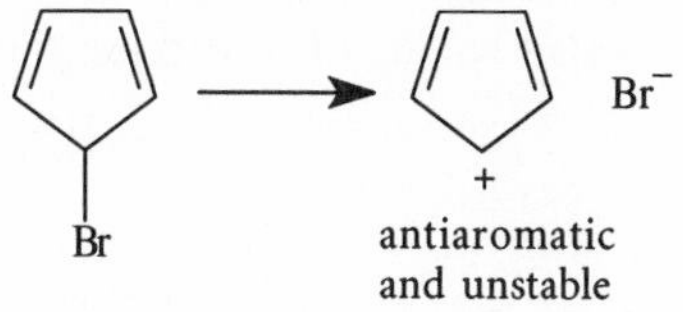

15.56 (a) This is an ordinary Diels-Alder reaction; because the diene is cyclic, a bicyclic product is formed. Two diastereomers are possible. (Each, of course, is formed as the racemate, only one enantiomer of which is shown below.)

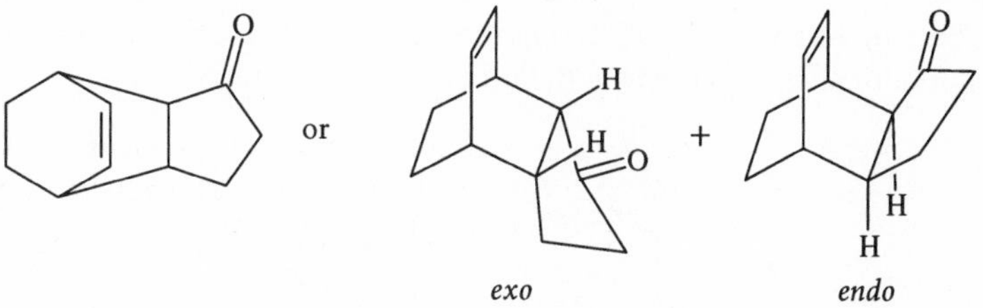

exo   *endo*

(c) The conjugated-diene part of the triene undergoes a Diels-Alder reaction with a double bond of the benzoquinone. Because there are two double bonds in the quinone, a second Diels-Alder reaction also occurs. Two constitutional isomers result from the second addition because the diene can add in two different orientations.

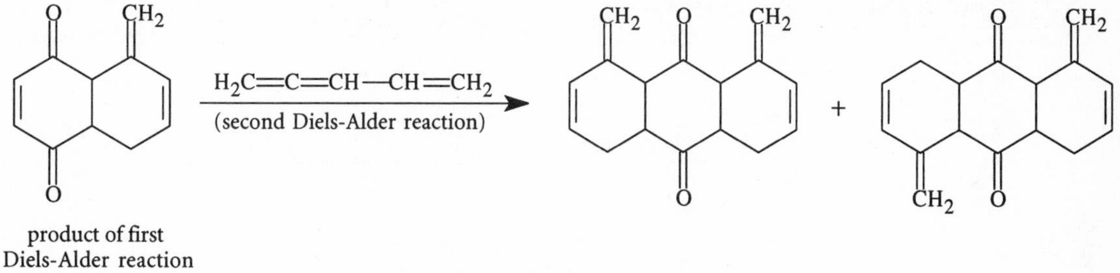

product of first
Diels-Alder reaction

Moreover, each of these constitutional isomers can exist as two diastereomers:

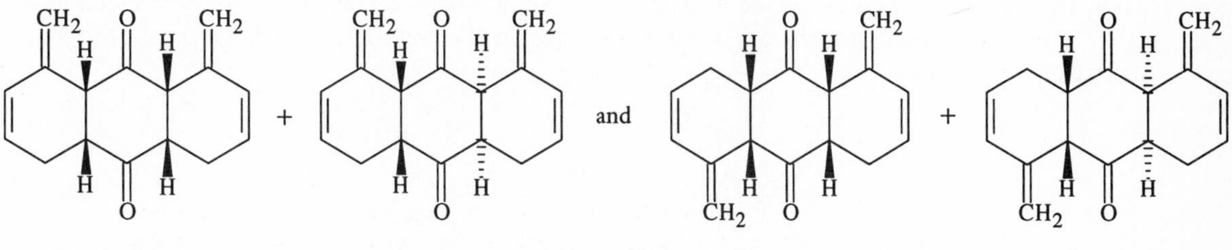

(Which of *these* can exists as enantiomers, and which are *meso*-compounds?)

(e) Because the product has the same number of carbons as the starting material, the reaction is intramolecular. In fact, it is an intramolecular Diels-Alder reaction.

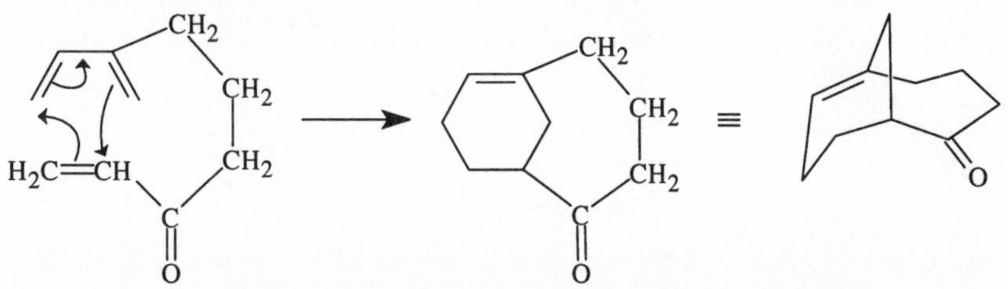

Although this product has a bridgehead double bond, it is not a severe violation of Bredt's rule (text p. 299–300) because a large ring is involved.

15.57 The product is *nickelocene*, the nickel analog of ferrocene. (See Eq. 15.44, text p. 723.)

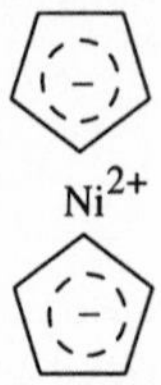

nickelocene

15.58 (a) The structure of the Diels-Alder adduct suggests the transient formation of *cyclopentyne*. This conclusion is obtained by mentally reversing the Diels-Alder reaction in such a way that diphenylisobenzofuran is obtained; cyclopentyne is the other starting material:

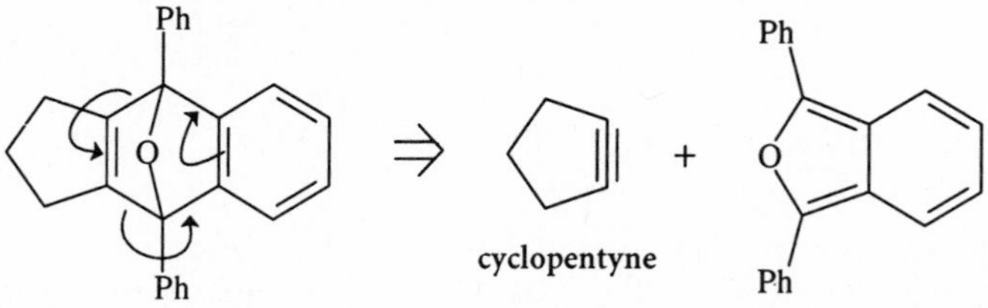

diphenylisobenzofuran

Cyclopentyne is formed by a β-elimination of the Grignard reagent formed from 1,2-dibromocyclopentene. Follow the elimination by thinking of the Grignard reagent as a carbanion:

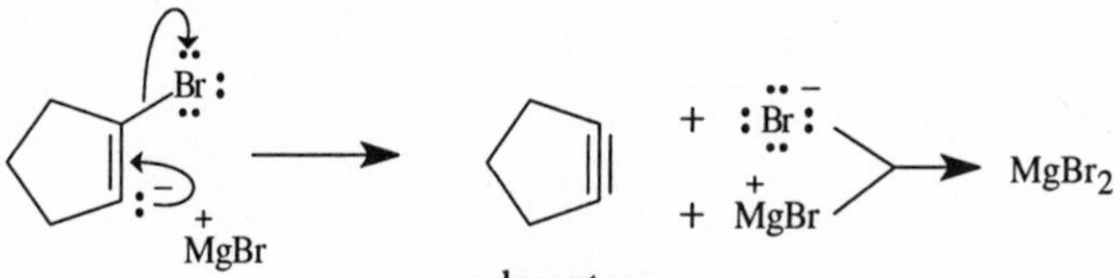

cyclopentyne
reacts with
diphenylisobenzofuran

15.59 Follow the procedure used in solving Problem 15.58(a): mentally reverse the Diels-Alder reaction so that maleic anhydride is obtained; the other fragment obtained is compound *X*, which thus has the following structure:

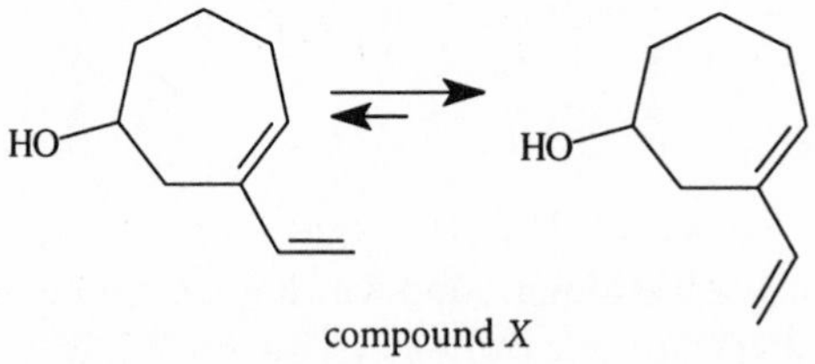

compound *X*

The formation of *X* results from ionization of the starting chloride to a tertiary allylic carbocation *A* followed by ring opening of *A* to a secondary carbocation *B*, which is captured by the nucleophile water. (Why should the tertiary allylic carbocation *A* form a secondary carbocation *B*?)

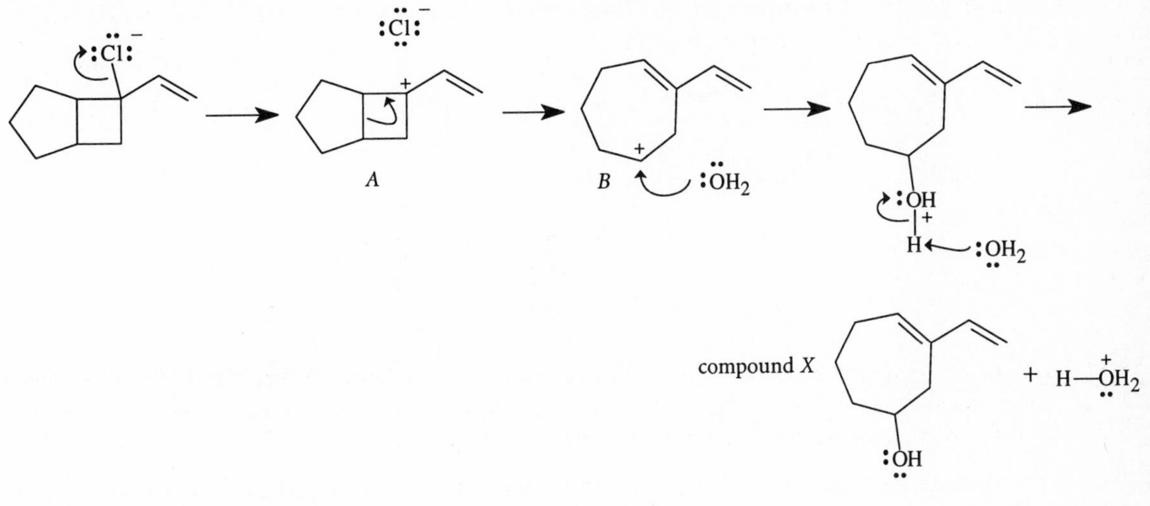

compound *X*

15.61 (a) The reaction is driven to the right (1) by the formation of a very stable product (benzene) and (2) by the formation of a volatile product (ethylene).

(b) As suggested by the first example in the problem, the final products are formed in a *reverse* Diels-Alder reaction of some intermediate *C*. Construct compound *C* by imagining a "normal" Diels-Alder of the two products. Two constitutionally isomeric candidates for *C*, labeled *C1* and *C2* in the reactions below, could result from such a process.

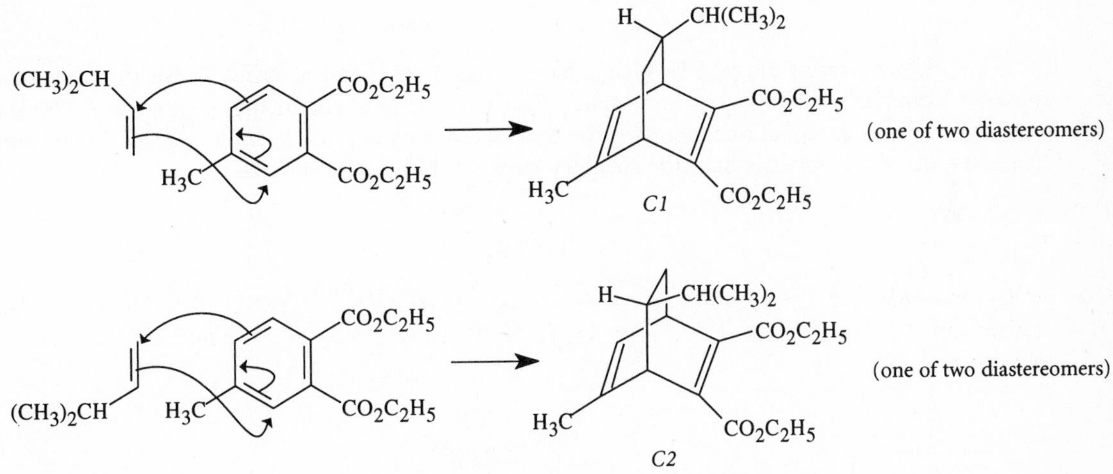

(one of two diastereomers)

(one of two diastereomers)

To determine the structure of α-phellandrene, perform a reverse Diels-Alder on *C1* and *C2* so as to form the alkyne diester. This is illustrated with intermediate *C1* in the equation on the following page. An analogous process carried out on intermediate *C2* gives a compound that would *not* undergo catalytic hydrogenation to give 1-isopropyl-4-methylcyclohexane; it would give instead 1-isopropyl-3-methylcyclohexane. (Go through the reasoning to be sure you see this point.) Consequently, the structure of α-phellandrene is as shown in the equation.

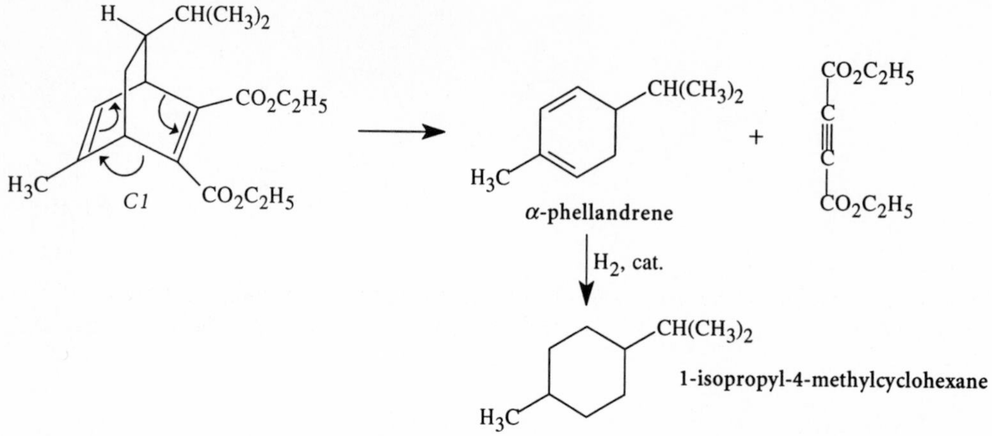

α-phellandrene

1-isopropyl-4-methylcyclohexane

15.62   (a) Protonate the exocyclic double bond in the manner shown below because this process gives the more stable carbocation—the carbocation with the greatest number of resonance structures. (The strong acid is represented by H—X.)

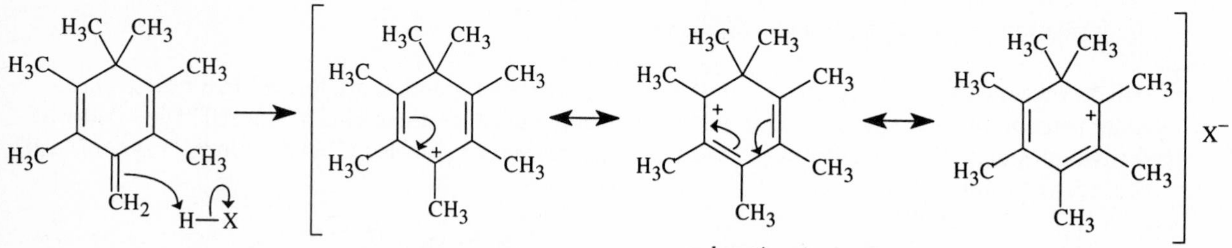

carbocation A

(b) The protons nearest the positive charge have the greatest chemical shifts. These are immediately apparent from the resonance structures above. The assignment of the methyl protons at δ 2.82 is clear from their relative integral; this methyl is the only one not equivalent to another methyl by symmetry. The two non-vinylic methyls have the smallest chemical shift. To summarize:

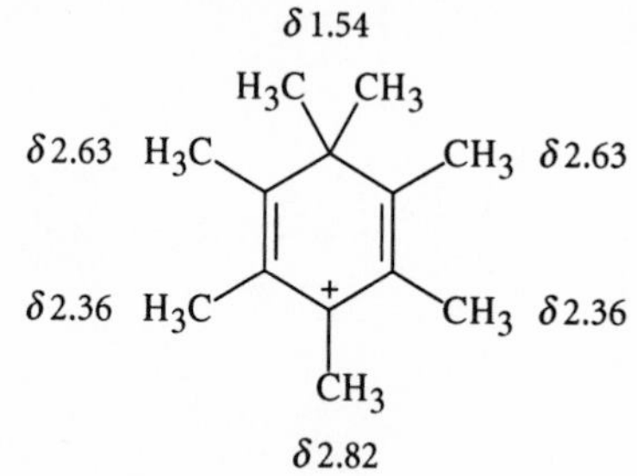

(c) The signals coalesce when the temperature is raised because a chemical process takes place that makes the methyl groups equivalent over time. This reaction is nicknamed the "methyl walk."

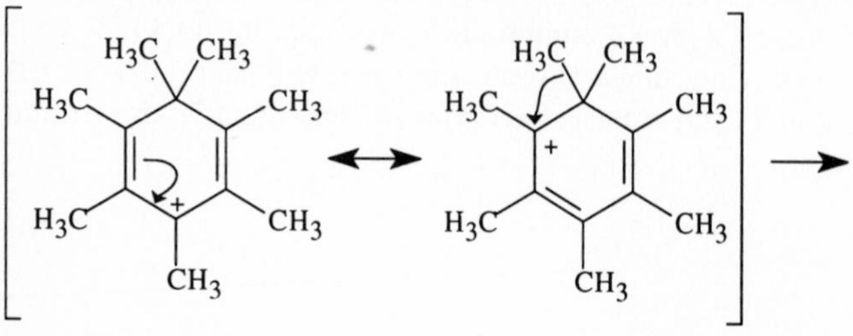

(equation continues)

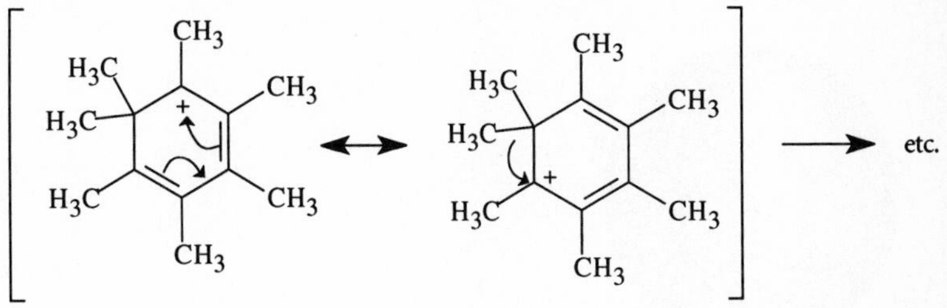

15.63 This reaction involves a rearrangement of carbocation *X1* to a more stable carbocation *X2* followed by loss of a β-proton from the rearranged ion. (Only the part of the molecule involved in the reaction is shown below.)

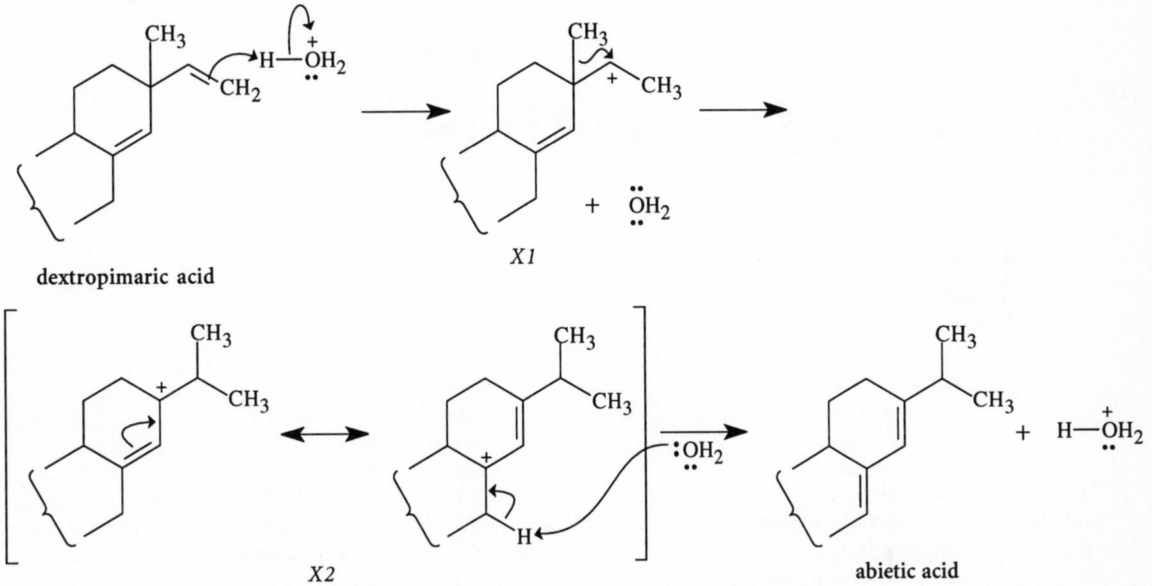

Chemistry
of Benzene and
Its Derivatives

Terms

Concepts

I. Introduction to Benzene and Its Derivatives

A. NOMENCLATURE OF BENZENE DERIVATIVES

1. The nomenclature of benzene derivatives follows the same rules used for other substituted hydrocarbons.

2. Some monosubstituted benzene derivatives have well-established common names that must be learned.

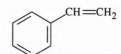
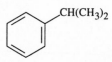

toluene   phenol   anisole   styrene   cumene

360

3. The positions of substituent groups in disubstituted benzenes can be designated in two ways:
   a. Substitutive nomenclature utilizes numerical designations in the same manner used for other compound classes.
   b. An older system employs special prefixes that can be used *only* for disubstituted benzene derivatives.
      i. *o* (*ortho*) for a 1,2 relationship.
      ii. *m* (*meta*) for a 1,3 relationship.
      iii. *p* (*para*) for a 1,4 relationship.

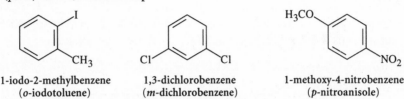

| 1-iodo-2-methylbenzene | 1,3-dichlorobenzene | 1-methoxy-4-nitrobenzene |
| (*o*-iodotoluene) | (*m*-dichlorobenzene) | (*p*-nitroanisole) |

4. Some disubstituted benzenes derivative have time-honored common names:
   a. The dimethylbenzenes are called xylenes.
   b. The methylphenols are called cresols.

o-xylene   *m*-cresol

   c. The hydroxyphenols are:
      i. catechol (1,2-disubstituted).
      ii. resorcinol (1,3-disubstituted).
      iii. hydroquinone (1,4-disubstituted).

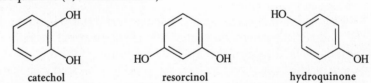

catechol   resorcinol   hydroquinone

5. When a benzene derivative contains more than two substituents on the ring, only numbers may be used to designate the positions of substituents.
   a. When none of the substituents qualify as principal groups, they are cited and numbered in alphabetical order.
   b. If a substituent is eligible for citation as a principal group, it is assumed to be at position-1 of the ring.

4-chloro-2-methylphenol

6. Sometimes it is simpler to name the benzene ring as a substituent group.
   a. When an unsubstituted benzene ring is a substituent, it is called a phenyl group, abbreviated Ph— or C_6H_5—.
   b. A benzene ring or substituted benzene ring cited as a substituent is referred to generally as an aryl group.
7. The Ph—CH_2— group is called the benzyl group.

Ph_3COH Ph—CH_2—Br

triphenylmethanol   benzyl bromide

B. PHYSICAL PROPERTIES OF BENZENE DERIVATIVES

1. Boiling points of benzene derivatives are similar to those of other hydrocarbons with similar shapes and molecular weights.
2. Melting points of *para*-disubstituted benzene derivatives are typically higher than those of the corresponding *ortho* or *meta* isomers.
3. Many *para*-substituted compounds can be separated from their *ortho* and *meta* isomers by recrystallization.
4. Benzene and other aromatic hydrocarbons are less dense than water, but more dense than alkanes and alkenes of about the same molecular mass.
5. Benzene and its hydrocarbon derivatives are insoluble in water; benzene derivatives with substituents that form hydrogen bonds to water are more soluble.

II. Spectroscopy of Benzene Derivatives

A. IR SPECTROSCOPY

1. The most useful absorptions in the infrared spectra of benzene derivatives are the carbon-carbon stretching absorptions of the ring, which occur at lower frequency than the $C{=}C$ absorption of alkenes; one is near 1600 cm^{-1} and another is near 1500 cm^{-1}.
2. Very strong absorptions below 900 cm^{-1} are characteristic of C—H bending and ring puckering.
3. Weak absorptions, called overtone and combination bands, are observed in the 1600–2000 cm^{-1} region. (These absorptions were once used to determine the substitution patterns of aromatic compounds.)

B. NMR SPECTROSCOPY

1. Low-field NMR absorptions are particularly characteristic of most benzene derivatives.
   a. Protons on aromatic rings have chemical shifts that are greater than those of typical alkenes by about 1.5–2 ppm.
   b. The NMR spectrum of benzene consists of a singlet at a chemical shift of δ 7.4.
2. When benzene is oriented relative to the applied field, a circulation of π electrons around the ring, called a ring current, is induced.
   a. The ring current induces a magnetic field that forms closed loops through the ring.
   b. The induced field opposes the applied field along the axis of the ring.
   c. The induced field augments the applied field outside of the ring and deshields the protons from the external field; these protons have a greater chemical shift than many other protons.
   d. The ring current and the large chemical shift are characteristic of compounds that are aromatic by the Hückel 4n +2 rule.
3. When the protons in a substituted benzene derivative are nonequivalent, they split each other.
   a. Their coupling constants depend on their positional relationships.
   b. The NMR spectra of many monosubstituted benzene derivatives have complex absorptions in the aromatic region.
      i. A symmetrical pattern resembling "two leaning doublets" is very typical of benzene derivatives in which two different substituents have a *para* relationship.
      ii. The chemical shifts of ring protons are affected by both the polar and resonance effects of substituent.
4. The chemical shifts of benzylic protons (δ 2–3) are slightly greater than those of allylic protons (δ 1.8–2.2).
5. The O—H absorptions of phenols (δ 5–6) are typically at lower field than that of alcohols (δ 2–3); O—H protons undergo exchange in D_2O.
6. In CMR spectra, the chemical shifts of aromatic carbons are in the carbon-carbon double bond region (δ 110–160); the exact values depend on the ring substituents that are present.
7. The chemical shifts of benzylic carbons (δ 18–30) are not appreciably different from the chemical shifts of ordinary alkyl carbons.

C. UV SPECTROSCOPY

1. Simple aromatic hydrocarbons have two absorption bands in their UV spectra:
   a. A relatively strong bond near 210 nm.
   b. A much weaker band near 260 nm.

2. Substituent groups on the ring alter both the λ_{max} values and the intensities of both peaks, particularly if:
   a. the substituent has an unshared electron pair.
   b. the substituent has *p* orbitals that can overlap with the π-electron system of the aromatic ring.
3. More extensive conjugation is associated with an increase in both λ_{max} and intensity.

III. *Introduction to Electrophilic Aromatic Substitution Reactions of Substituted Benzenes*

A. ELECTROPHILIC AROMATIC SUBSTITUTION

1. Benzene and its derivatives are aromatic compounds and thus do not undergo most of the usual addition reactions of alkenes; instead, they undergo reactions in which a ring hydrogen is substituted by another group.
2. Electrophilic aromatic substitution reactions are typical of benzene and other aromatic compounds.
   a. The reaction is a substitution because hydrogen is replaced by another group.
   b. The reaction is electrophilic because it involves the reaction of an electrophile, or Lewis acid, with the benzene π electrons.
3. All electrophilic aromatic substitution reactions have the following three mechanistic steps:
   a. Generation of an electrophile.
   b. Attack of the π electrons of the aromatic ring on the electrophile and formation of a resonance-stabilized carbocation.
   c. Loss of a proton from the carbocation intermediate to form the substituted aromatic compound.

B. DIRECTING EFFECTS OF SUBSTITUENTS

1. Further substitution on a substituted benzene ring occurs in one of two ways.
   a. If a substituted benzene undergoes further substitution at the *ortho* and *para* positions, the original substituent is called an *ortho, para*-directing group.
   b. If a substituted benzene undergoes further substitution mostly at the *meta* position, the original substituent group is called a *meta*-directing group.
2. A substituent group is either an *ortho, para*-directing group or a *meta*-directing group in all electrophilic aromatic substitution reactions.
3. These effects occur because electrophilic substitution reactions at one position of a benzene derivative are much faster than the same reactions at another position; the substitution reactions at the different ring positions are in competition.

C. ORTHO, PARA-DIRECTING GROUPS

1. Substituents containing atoms with unshared electron pairs on atoms directly attached to the benzene ring are *ortho, para*-directing substituents because the electron pairs can be involved in the resonance stabilization of the carbocation intermediate.
   a. The reaction of an electrophile at either the *ortho* or *para* positions gives a carbocation with more resonance structures; in one resonance structure, the substituent electrons are delocalized into the ring and charge is delocalized onto the substituent.
   b. The substituent electrons *cannot* be used to delocalize charge when reaction occurs at the *meta* position.
2. All alkyl groups are *ortho, para*-directing substituents.
   a. Reaction at a position that is *ortho* or *para* to an alkyl group gives a carbocation intermediate that has one tertiary carbocation resonance structure.
   b. Reaction of the electrophile *meta* to the alkyl group gives only secondary carbocation resonance structures.
3. The rate-limiting step in many electrophilic aromatic substitution reactions is formation of the carbocation intermediate.

    a. Hammond's postulate suggests that the more stable carbocation should be formed more rapidly.

    b. The products derived from the more rapidly formed carbocation (the more stable carbocation) are the ones observed.

D. The Ortho:Para Ratio

1. If a benzene derivative bears an *ortho, para*-directing group, most electrophilic aromatic substitution reactions give much more *para* than *ortho* product; the reasons for the *ortho, para* ratio vary from case to case.

    a. In some cases, predominance of the *para* isomer can be explained by van der Waals repulsions (steric effects) in the *ortho* isomer.

    b. In some cases the reasons are not well understood.

2. The *ortho* and *para* isomers obtained in many electrophilic aromatic substitution reactions have sufficiently different physical properties that they are readily separated. (See I.B.3. above.)

E. Meta-Directing Groups

1. *Meta*-directing groups are all electronegative groups that do not have an unshared electron pair on an atom adjacent to the benzene ring.

    a. Substituents that have positive charges adjacent to the aromatic ring are *meta* directors because *meta* substitution gives the carbocation intermediate in which like charges are farther apart.

    b. All *meta*-directing groups have bond dipoles that place a substantial amount of positive charge next to the benzene ring.

F. Activating and Deactivating Effects of Substituents

1. Different benzene derivatives have greatly different reactivities in electrophilic aromatic substitution reactions.

    a. If a substituted benzene derivative reacts more rapidly than benzene itself, then the substituent is said to be an activating group.

    b. If a substituted benzene derivative reacts more slowly than benzene itself, then the substituent is said to be a deactivating group.

2. A given substituent group is either activating in all electrophilic aromatic substitution reactions or deactivating in all such reactions.

    a. All *meta*-directing groups are deactivating groups.

    b. All *ortho, para*-directing groups except for the halogens are activating groups.

    c. The halogens are deactivating groups.

3. Except for the halogens, there is a correlation between the activating and directing effects of substituents.

    a. Directing effects are concerned with the relative rates of substitution at different positions of the same compound.

    b. Activating or deactivating effects are concerned with the relative rates of substitution of different compounds compared with benzene itself.

4. Two properties of substituents must be considered in order to understand activating and deactivating effects:

    a. The resonance effect of the substituent—the ability of the substituent to stabilize the carbocation intermediate in electrophilic substitution by delocalization of electrons from the substituent into the ring.

        i. The resonance effect is the same effect that is responsible for the *ortho, para*-directing effects of substituents with unshared electron pairs.

        ii. The electron-donating resonance effect of a substituent group with unshared electron pairs stabilizes the carbocation intermediate and activates further substitution, provided that the substitution occurs in the *ortho* or *para* position.

    b. The polar effect of the substituent—the tendency of the substituent group, by virtue of its electronegativity, to pull electrons away from the ring.

        i. This is the same effect discussed in connection with substituent effects on acidity.

        ii. The electron-withdrawing polar effect of an electronegative group destabilizes the carbocation intermediate and deactivates further substitution.

5. Whether a substituted derivative of benzene is activated or deactivated toward further substitution depends on the balance of the resonance and polar effects of the substituent(s).
6. Alkyl groups on a benzene ring:
   a. stabilize carbocation intermediates in electrophilic substitution.
   b. are activating groups.
7. The deactivating effects of halogen substituents reflect a different balance of resonance and polar effects.
   a. The orbitals on Cl, Br, and I and the carbon $2p$ orbitals of the benzene ring have different sizes and different numbers of nodes and thus do not overlap effectively; thus, the resonance effects of the substituents are relatively weak.
   b. *Meta* substitution in halobenzenes is deactivated even more than *ortho* or *para* substitution; this is why the halogens are *ortho, para*-directing groups.

G. USE OF ELECTROPHILIC AROMATIC SUBSTITUTION IN ORGANIC CHEMISTRY

1. Both activating/deactivating and directing effects of substituents can come into play in planning an organic synthesis that involves electrophilic substitution reactions.
2. Directing effects of the substituents must be considered carefully when designing an organic synthesis that involves electrophilic aromatic substitution reactions.
3. When an electrophilic substitution reaction is carried out on a benzene derivative with more than one substituent:
   a. The activating and directing effects are roughly the sum of the effects of the separate substituents.
   b. If one group is much more strongly activating than the other, the directing effect of the more powerful activating group generally predominates.
   c. In other cases, mixtures of isomers are typically obtained.
4. The activating or deactivating effects of substituents in an aromatic compound determine the conditions that must be used in an electrophilic substitution reaction.
   a. When a deactivating group is being introduced by an electrophilic substitution reaction, it is easy to introduce one group at a time.
   b. When an activating group is introduced by electrophilic substitution, additional substitutions can occur easily under the conditions of the first substitution and mixtures of products substituted to different extents are obtained.
   c. Deactivating substituents retard some reactions to the point that they are not useful.

···

R Reactions

···

I. Electrophilic Aromatic Substitution Reactions of Benzene

A. HALOGENATION OF BENZENE

1. Benzene reacts with bromine in the presence of a Lewis acid catalyst ($FeBr_3$) to yield a product in which one bromine is substituted for a ring hydrogen.

2. An analogous chlorination reaction using Cl_2 and $FeCl_3$ gives chlorobenzene.
3. Halogenation of benzene differs from the halogenation of alkenes by the type of product obtained and the reaction conditions.
   a. Alkenes give addition products and react spontaneously under mild conditions.
   b. Benzene gives substitution products and requires a Lewis acid as well as relatively severe conditions.
4. The mechanism of the bromination of benzene consists of the following steps:
   a. The first step is formation of a complex between Br_2 and the Lewis acid $FeBr_3$.
      i. This complexation makes one of the bromines a much better leaving group.

     ii. This complex of Br_2 and $FeBr_3$ reacts as if it contained the electron-deficient species $:\overset{\cdot\cdot}{Br}\cdot^+$.

  b. In the second step, this complex is attacked by the π electrons of the benzene ring.

     i. This step disrupts the aromatic stabilization of the benzene ring (thus the need for the harsh conditions).

     ii. The carbocation intermediate is resonance-stabilized.

  c. The reaction is completed when a bromide ion acts as a base to remove the ring proton and give the products; by losing a β-proton, the carbocation can form a stable aromatic compound.

B. NITRATION OF BENZENE

1. Benzene reacts with concentrated nitric acid, usually in the presence of a sulfuric acid catalyst, to form nitrobenzene.
2. In this reaction, called nitration, the nitro group, $-NO_2$, is introduced into the benzene ring by electrophilic substitution; the electrophile is the nitronium ion, $^+NO_2$.

$$H_2SO_4 + HNO_3 \rightleftharpoons {}^+NO_2 \ HSO_4^- + H_2O$$

C. SULFONATION OF BENZENE

1. Benzene reacts with a solution of sulfur trioxide in H_2SO_4 called fuming sulfuric acid to form benzenesulfonic acid.
2. In this reaction, called sulfonation, the sulfonic acid group, $-SO_3H$, is introduced into the benzene ring by electrophilic substitution; the electrophile is the neutral compound sulfur trioxide, SO_3.

3. Sulfonic acids such as benzenesulfonic acid are rather strong acids.
4. Sulfonation, unlike many electrophilic aromatic substitution reactions, is reversible; the sulfonic acid group is replaced by a hydrogen when sulfonic acids are heated with steam.

D. FRIEDEL-CRAFTS ACYLATION OF BENZENE

1. When benzene reacts with an acid chloride in the presence of a Lewis acid catalyst such as aluminum trichloride, $AlCl_3$, a ketone is formed.

  a. This reaction is an example of a Friedel-Crafts acylation.

  b. An acyl group, typically derived from an acid chloride, is introduced into an aromatic ring in the presence of a Lewis acid catalyst.

2. The electrophile in the Friedel-Crafts acylation reaction is a carbocation called an acylium ion.
   a. This ion is formed when the acid chloride reacts with the Lewis acid $AlCl_3$.

an acylium ion

   b. The substitution occurs when the benzene π electrons attack the acylium ion.
3. The ketone product of the Friedel-Crafts acylation reacts with the Lewis acid catalyst to form a complex that is catalytically inactive; this has two consequences:
   a. Slightly more than one equivalent of the catalyst must be used:
      i. one equivalent to react with the product.
      ii. an additional catalytic amount to ensure the presence of catalyst throughout the reaction.
   b. The complex must be destroyed before the ketone product can be isolated; this is usually accomplished by pouring the reaction mixture into ice water.
4. Because the ketone products of acylation are much less reactive than the benzene starting material, acylation occurs only once.
5. The Friedel-Crafts acylation occurs intramolecularly when acylation results in the formation of a five- or six-membered ring.

   a. The intramolecular process is much faster than attack of the acylium ion on the phenyl ring of another molecule.
   b. This type of reaction can only occur at an adjacent *ortho* position.
6. Friedel-Crafts acylation does not occur on a benzene ring substituted solely with one or more *meta*-directing groups.
7. The Friedel-Crafts acylation reaction is important for two reasons.
   a. It is an excellent method for the synthesis of aromatic ketones.
   b. It is another method for the formation of carbon-carbon bonds (see list on text page 756 or Appendix V, text page A-11).

E. FRIEDEL-CRAFTS ALKYLATION OF BENZENE

1. The reaction of an alkyl halide with benzene in the presence of a Lewis acid gives an alkylbenzene; this type of reaction is called Friedel-Crafts alkylation.

2. The electrophile in a Friedel-Crafts reaction is formed by complexation of the Lewis acid $AlCl_3$ with the halogen of an alkyl halide.
   a. Either the alkyl halide-Lewis acid complex, or the carbocation derived from it, can serve as the electrophile in a Friedel-Crafts reaction.
   b. Rearrangements of alkyl groups are observed in some Friedel-Crafts alkylations if the carbocation intermediate is prone to rearrangement.

3. A catalytic amount (much less than one equivalent) of the $AlCl_3$ catalyst can be used in this reaction.

4. The alkylbenzene products are more reactive than benzene itself.

    a. The product can undergo further alkylation, and mixtures of products alkylated to different extents are observed.

    b. A monoalkylation product can be obtained in good yield if a large excess of the starting material is used; this strategy is practical only if the starting material is cheap, and if it can be readily separated from the product.

5. Alkenes or alcohols with an acid catalyst can also be used as the alkylating agents in Friedel-Crafts alkylation reactions.

6. Friedel-Crafts alkylation is generally not useful on compounds that are more deactivated than benzene itself.

II. Addition Reactions of Benzene Derivatives

A. Hydrogenation of Benzene Derivatives

1. Because of its aromatic stability, the benzene ring is resistant to conditions used to hydrogenate ordinary double bonds.

2. Aromatic rings can be hydrogenated under extreme conditions of temperature and/or pressure.

3. Catalytic hydrogenation of benzene derivatives gives the corresponding cyclohexanes, and cannot be stopped at the cyclohexadiene or cyclohexene stages.

4. Hydrogenation of the first double bond of benzene is an endothermic (thermodynamically unfavorable) reaction and requires energy (heat or pressure) to take place.

Electrophilic Aromatic Substitution Summary

Electrophile	Reaction Name	Product
Cl—$\overset{+}{Cl}$---$\overset{-}{FeCl_3}$	chlorination	chlorobenzene
Br—$\overset{+}{Br}$---$\overset{-}{FeBr_3}$	bromination	bromobenzene
SO_3 in H_2SO_4	sulfonation	benzenesulfonic acid
$^+NO_2$	nitration	nitrobenzene
R—$C\equiv O{:}^{+}$ $^-AlCl_4$	Friedel-Crafts acylation	an aryl ketone
R—$\overset{+}{Cl}$----$\overset{-}{AlCl_3}$ or R^+ $^-AlCl_4$	Friedel-Crafts alkylation	an alkylbenzene

Study Guide Links

√ 16.1   NMR of Para-Substituted Benzene Derivatives

Be sure to realize that the NMR spectrum of a *para*-disubstituted benzene derivatives shows the "two leaning doublet" pattern when the two *para* substituents are *different*. When the substituents are the same, all ring protons are equivalent, and the resonance for the ring protons is a singlet.

The splitting pattern for *para*-disubstituted benzene derivatives is a little more subtle than might first appear, and can be rationalized as follows. Each ring proton has an *ortho* relationship to one nonequivalent proton and a *para* relationship to the other nonequivalent proton. Consider, for example, *p*-chloronitrobenzene, the structure of which is shown below, and the NMR spectrum of which is shown in Fig. 16.3 on text p. 745. In this structure, H^a is *ortho* to one H^b and *para* to the other H^b. Actually, both *ortho* and *para* splittings are present, but the *para* splittings are typically very small and not readily discernible in the spectrum. The major splitting observed is the one between adjacent protons—about 10 Hz in Fig. 16.3, a value within the range for *ortho* splitting given in Table 16.1 on text p. 744. Thus, to a useful approximation, the resonance of each proton is split into a doublet by its neighboring *ortho* proton. Since there are two chemically different types of protons—H^a and H^b—the spectrum consists of two such doublets.

two equivalent protons H^a give one resonance split into a doublet by the neighboring *ortho* proton H^b

two equivalent protons H^b give one resonance split into a doublet by the neighboring *ortho* proton H^a

√ 16.2   Lewis-Acid Assistance for Leaving Groups

In some cases a Lewis acid can used to make a molecule more reactive by complexation with the leaving group. In the case described in the text, the leaving bromine does not have to depart as *bromide*, but rather departs as $^-FeBr_4$ (*bromide complexed to FeBr$_3$*):

bromide ion as a leaving group
(very slow reaction)

$^-FeBr_4$ (bromide ion complexed with FeBr$_3$)
as a leaving group
(much faster reaction)

Why is the complex more reactive? Because when the complexed bromine, an electronegative atom, is forced to assume a positive charge, its tendency to accept a pair of electrons—that is, its tendency to act as a leaving group—is increased. Just as we think of a nucleophile as a species that provides an electronic "push," we can think of a Lewis acid catalyst as something that provides an electronic "pull." *A Lewis acid helps remove a leaving group by making it a better attractor of electrons.*

When you studied S_N1 and S_N2 reactions, you learned that bromide is a good leaving group. The fact that such a good leaving group has to be complexed to a Lewis acid in order for bromination of benzene to take place shows just how unreactive the benzene ring is.

Lewis-acid complexation can be used to accelerate other reactions. For example, alkyl halides that are normally unreactive or not very reactive in S_N1 reactions can be induced to react by addition of Ag^+ (in the form of acidic silver nitrate solution) to the reaction mixture:

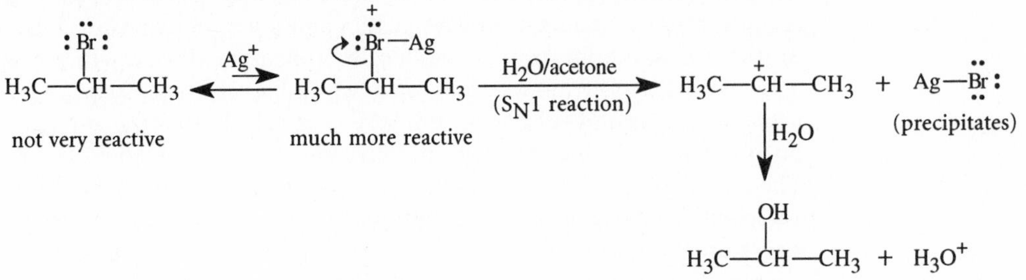

You also learned that primary alcohols react with concentrated HBr to give alkyl bromides, but that their reactions with HCl are usually very slow. The reaction of primary alcohols with HCl can be accelerated by adding the Lewis acid zinc chloride, because the Zn^{2+} ion forms a complex with the —OH group of the alcohol and thus makes it a better leaving group.

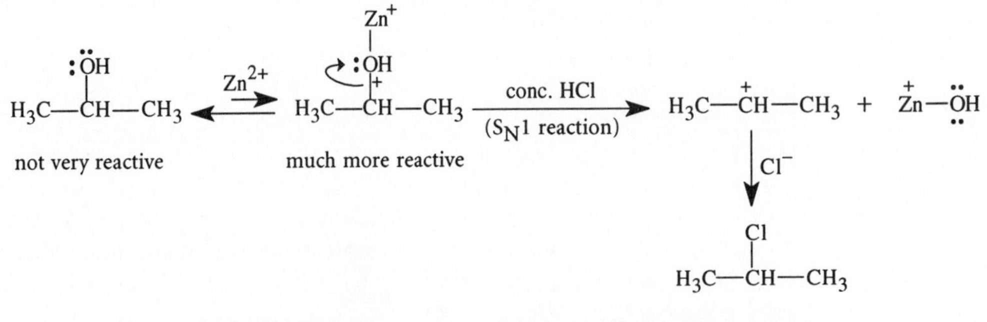

In Sec. 10.1 you learned that alcohol dehydration is acid-catalyzed because protonation of the —OH group makes this group a better leaving group. If protons were not present, the leaving group would be hydroxide ($^-$OH); protonation of the —OH group allows it to leave as the much less basic water molecule. In this case, the *proton* is the Lewis acid catalyst (and is also a Brønsted acid). In fact, it is useful to view Lewis acids conceptually as "fat protons!" Lewis acids are used instead of Brønsted acids when the latter would cause undesired side reactions.

✓16.3   Different Sources of the Same Reactive Intermediates

The example of Eq. 16.24 on text p. 759 is a further illustration of an idea that has been discussed in an earlier chapter (see Study Guide Link 11.2): *Different starting materials can serve as a source of the same reactive intermediate.*

In the Friedel-Crafts alkylation, the reactive intermediate is a carbocation. In this section you've learned that carbocation intermediates can be generated from alkyl halides and Lewis acids. But you've also learned that carbocations are formed when alkenes are subjected to strongly acidic conditions:

carbocation intermediate

The carbocation intermediates formed in such reactions can serve as electrophiles in electrophilic substitution just as carbocations generated in other ways can. Similarly, you've also learned that secondary and tertiary alcohols react with strong acids to generate carbocation intermediates (Sec. 10.1). These carbocations are transformed into alkene dehydration products if the reaction conditions are designed to remove the alkene from the reaction mixture. But if the alkene is not removed, and an aromatic hydrocarbon is present, these carbocation intermediates from alcohol dehydration can also serve as electrophiles in aromatic substitution. Problem 16.18(b) on text p. 759 is an illustration of this idea in practice.

✓16.4 Reaction Conditions and Reaction Rate

What do we mean by "harsh" or "mild" reaction conditions? In electrophilic aromatic substitution, harsh conditions include high temperature, strong Lewis acids, high concentrations of reagents, and/or the use of reagents that generate high concentrations of electrophiles. Mild conditions include lower temperature, weaker Lewis acids (or no Lewis acids), and lower concentrations of reagents.

Why is it that such conditions affect reaction rate? Remember, to say that an aromatic compound is highly activated means that it *reacts relatively rapidly* under a given set of conditions. If it is deactivated, it *reacts much more slowly* under the same conditions. Recall (Sec. 4.8A) that reaction rates increase with increasing temperature. Recall also from your study of rate laws (Sec. 9.3B) that many reaction rates increase with increasing reagent concentration. Thus, raising the temperature or the reagent concentration, or using a reagent that generates a high concentration of an electrophile, increases the reaction rate. Such strategies are necessary in order to get a highly *deactivated* (unreactive) compound to react at a convenient rate. However, use of harsh reaction conditions on a highly *activated* (reactive) compound in many cases leads to a greater degree of substitution than desired (or to other side reactions); hence, for highly activated compounds, "milder" conditions are used, such as weaker Lewis acids (or no Lewis acids), lower temperatures, or lower concentrations of reagents.

The examples in the text illustrate in a practical sense what is meant by "harsh" or "mild" conditions. It is *not* important to memorize the exact conditions for each reaction. It *is* important for you to understand why certain compounds are more activated toward substitution than others, and to understand conceptually that the success of reaction can depend on the rational choice of reaction conditions.

Solutions

Solutions to In-Text Problems

16.1 (a) *m*-chloroethylbenzene, or 1-chloro-3-ethylbenzene
(c) *p*-nitrostyrene or 1-nitro-4-vinylbenzene
(e) 2-bromo-1-chloro-5-fluoro-4-iodo-3-nitrobenzene
(f) benzylbenzene or (phenylmethyl)benzene (also often called diphenylmethane)

16.2 (a)   (c)   (e)   (g)

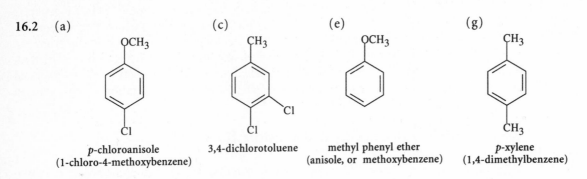

p-chloroanisole   3,4-dichlorotoluene   methyl phenyl ether   *p*-xylene
(1-chloro-4-methoxybenzene)   (anisole, or methoxybenzene)   (1,4-dimethylbenzene)

16.3 (a) The aromatic compound has NMR absorptions with greater chemical shift in each case because of the operation of the ring current. Because thiophene is aromatic, its NMR absorptions occur at greater chemical shift than those of divinyl sulfide.

16.4 (a) This hydrocarbon is a continuous cycle of 18 π electrons. Because all atoms are trigonal and all bond angles are 120°, this compound can be planar. It meets the $4n + 2$ criterion for aromaticity for $n = 4$.

(b) The ring current brings about the expected deshielding of protons *outside* the ring, but, from Figure 16.2 on text p. 742, it should have exactly the opposite effect on the protons *inside* the ring: these should be strongly shielded. The resonance at δ 9.28 is that of the outer protons, and the resonance at δ (−2.99) is that of the inner protons. The former protons are strongly deshielded, and the latter are strongly shielded. The relative integral is in agreement with this assignment.

16.5 (a) Mesitylene would give an NMR spectrum containing only two singlets in the integral ratio 3:1 for the methyl groups and ring protons, respectively. 4-Isopropyltoluene would have a more complex spectrum; even if the ring protons have identical chemical shifts, there should be four sets of signals.

16.6 (a) The signal for an exchangeable proton at δ 3.76 indicates that the compound contains an —OH group. It also contains aromatic absorptions with a splitting pattern that suggests *para* substitution. The correct structure is

$$H_3C— \quad —CH—CH_3 \quad \text{1-(4-methylphenyl)ethanol}$$
$$\overset{\displaystyle OH}{}$$

16.7 The following two methyl ethers would show five lines in their proton-decoupled CMR spectra.

1,3-dichloro-2-methoxybenzene   1,3-dichloro-5-methoxybenzene

16.9 (a) The two benzene rings are conjugated in compound *A*; because there are more π bonds in conjugation, the UV absorption occurs at longer wavelength than it does in ethylbenzene.

(b) Compound *B* is like two connected molecules of compound *C* in the same sense that compound *A* is like two connected ethylbenzene molecules. It might seem that the absorption of compound *B* should occur at longer wavelength than that of compound *C* for the reason given in part (a). That this is *not* the case means that the two rings in compound *B* are not conjugated. They are not conjugated because the rings are not coplanar; and they are not coplanar because the *ortho* methyl groups on the two rings have severe van der Waals repulsions that are relieved when the molecule adopts a nonplanar conformation:

planar conformation   nonplanar conformation

16.11 Apply the steps shown in Eqs. 16.5–16.6 on text p. 749 to the *para* position of bromobenzene.

16.13 In this case, the electrophile is the proton of H_3O^+ and the sulfonic acid group is liberated as SO_3 from the carbocation intermediate instead of a proton. (SO_3 reacts with water to give H_2SO_4; see text p. 753.)

16.15 (a)

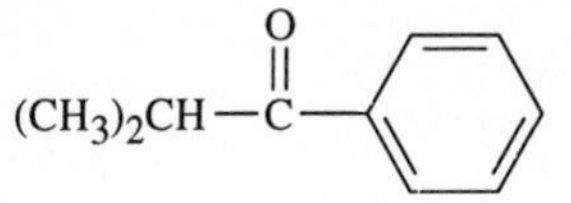

$(CH_3)_2CH-\overset{\overset{O}{\|}}{C}-\langle\text{phenyl}\rangle$ isobutyrophenone

16.16 (a)

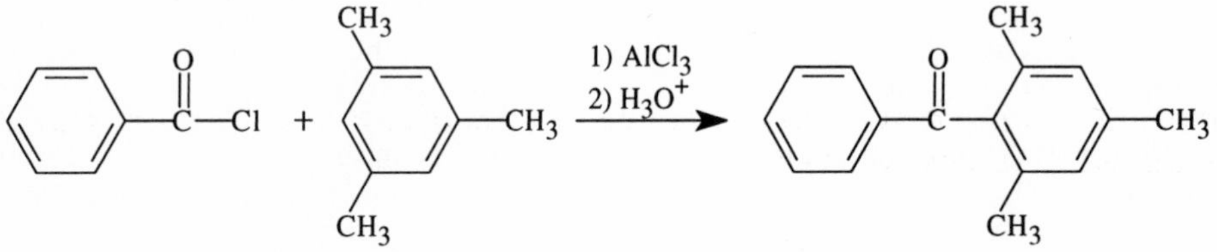

16.17 That the product contains the same number of carbons as the starting material suggests an intramolecular Friedel-Crafts reaction. The mechanism below begins with the acylium ion. (For the formation of this ion from the acid chloride, see Eq. 16.13 on text p. 754.)

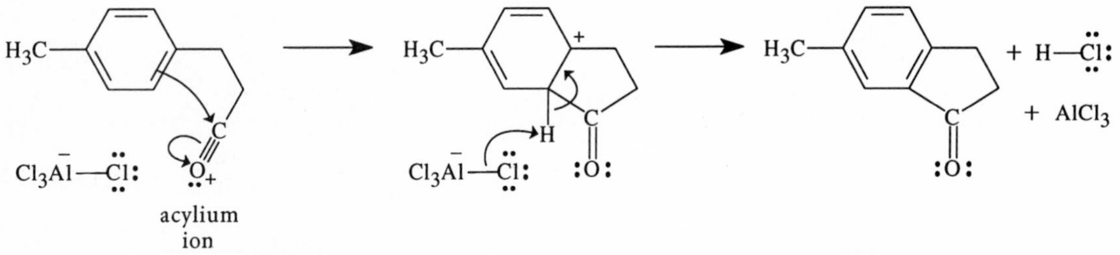

acylium
ion

> Notice that intramolecular reactions involving the formation of five- and six-membered rings are particularly rapid. This is another example of neighboring-group participation; see Sec. 11.6 of the text.

16.18 (a) As discussed in Study Guide Link 16.3, carbocations generated in a variety of ways can be used as electrophiles in Friedel-Crafts reactions. In this case, the carbocation produced by the protonation of cyclohexene is the electrophile.

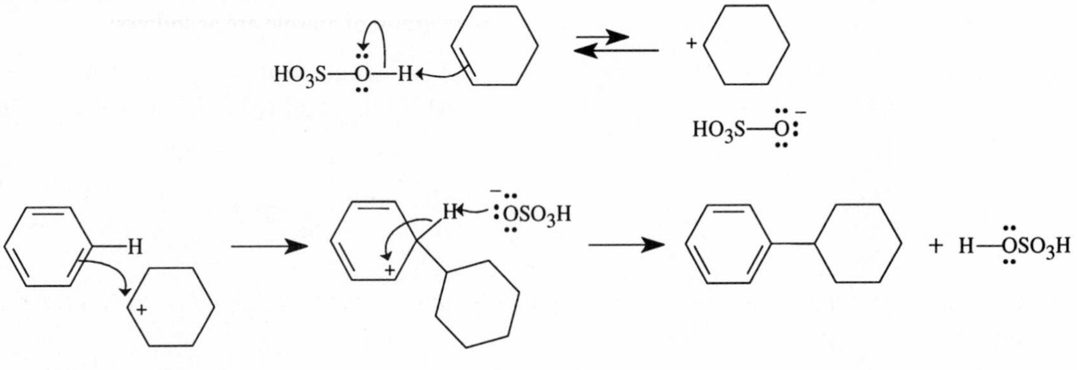

(b) The same product is obtained because the same carbocation electrophile is involved. It is generated from cyclohexanol by protonation and loss of water:

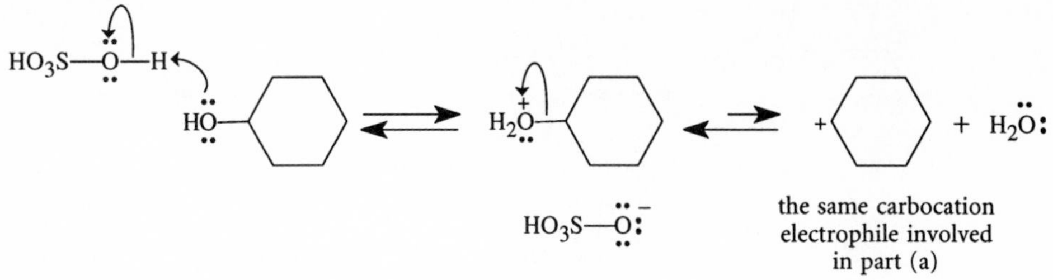

16.20 (a) The hint and the fact that the product has the same number of carbons as the starting material suggest an intramolecular Friedel-Crafts alkylation.

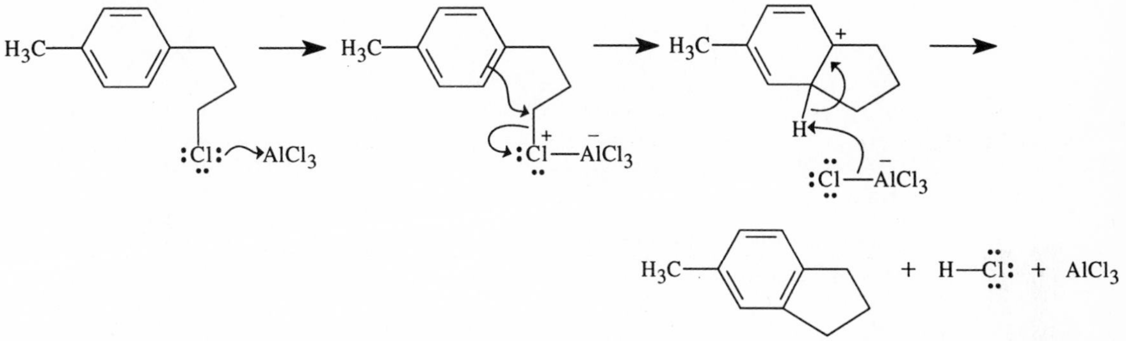

Notice that the electrophile is the complex of the Lewis acid AlCl$_3$ with the chloro group rather than a primary carbocation, which is too unstable to form.

16.21 (a) Table 16.2 indicates that the methoxy group is an *ortho, para*-directing group.

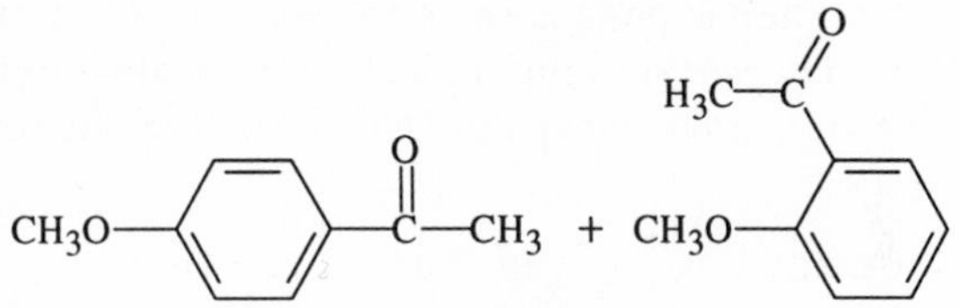

16.22 Let E$^+$ be a general electrophile. The four resonance structures of the carbocation intermediate that results from reaction of E$^+$ at the position *ortho* to the methoxy group of anisole are as follows:

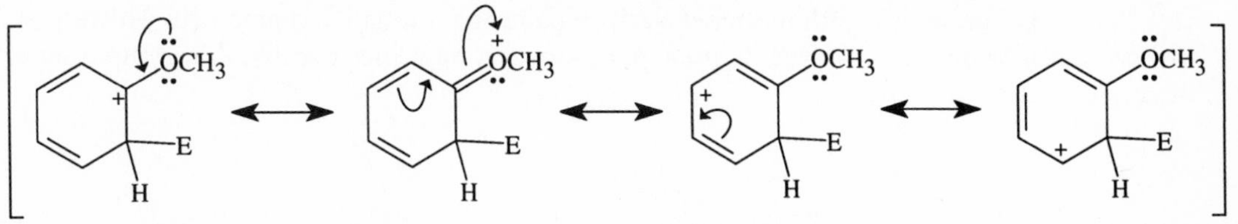

16.23 (a) Because substitution occurs *para* to the phenyl group, the phenyl group is an *ortho, para*-directing substituent.

(b) The resonance structures on the top of the following page show that the electrons of the phenyl substituent can be used to stabilize the carbocation intermediate when substitution occurs at the *para* position. The electrons of the phenyl substituent cannot be delocalized in this way when substitution occurs at the *meta* position.

16.24 Notice that, except for alkyl groups, all *ortho, para*-directing groups have electron pairs on atoms adjacent to the benzene ring.

(a) This substituent, like the methoxy group, is an *ortho, para*-directing group, a point that is confirmed by entry number 4 of Table 16.2.

(c) The carbocation intermediate involved in *meta* substitution has a greater separation between the positive charge of the carbocation and the positive charge of the substituent than does the carbocation intermediate involved in *ortho, para* substitution. (The argument is similar to that used with electrophilic substitution reactions of nitrobenzene on text p. 765.) Therefore, the substituent is a *meta*-directing group.

(e) The —O⁻ group has both unshared electron pairs and a negative charge with which to stabilize the carbocation intermediate involved in *ortho, para* substitution. Consequently, this group is an *ortho, para*-directing group.

Because of the negative charge on the oxygen, the "carbocation intermediate" is not a cation at all, but rather a neutral compound. Moreover, the —O⁻ group is so strongly activating that substitution does not stop at the monobromo derivatives but continues until all *ortho* and *para* positions have been filled. (See text p. 847.)

16.25 (a) The reaction-free energy profiles for electrophilic substitution of benzene, chlorobenzene at the *para* position, and chlorobenzene at the *meta* position are shown in Fig. SG16.1 on the following page. Notice

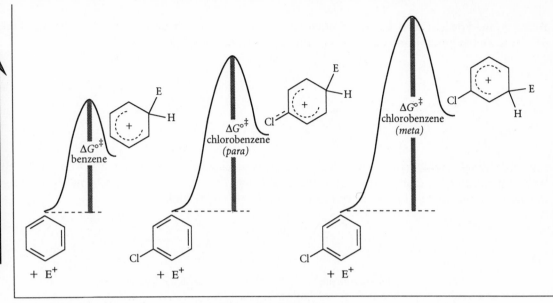

Figure SG16.1 *Reaction-free energy profiles to accompany the solution to Problem 16.25(a).*

that chlorobenzene is less reactive (that is, has a greater standard free energy of activation) than benzene because chlorine is a deactivating substituent. Notice also that *para*-substitution reactions on chlorobenzene are faster than *meta*-substitution reactions because chlorine is an *ortho, para*-directing group.

16.26 (a) Nitration of anisole should be faster. Although both the CH_3O— and the CH_3S— groups are activating, *ortho, para*-directing groups, the orbitals on sulfur containing the unshared electron pairs, like those on chlorine, are derived from quantum level 3. Consequently, the overlap of sulfur orbitals with the carbon orbitals of the ring is poorer than the overlap of the oxygen orbitals. In other words, the electron-donating resonance interaction of the sulfur is weaker than that of the oxygen. Although the electron-withdrawing (and therefore rate-retarding) polar effect of oxygen is much greater than that of sulfur, the resonance interaction of oxygen is so much more powerful that it is the dominant effect.

16.27 In both parts (a) and (c), each substituent is an *ortho, para*-directing group. There are two products in each case that satisfy the directing effects of both groups. (A third possibility is the product in which the entering group [—SO_3H in (a), —NO_2 in (c)] is *ortho* to both substituents; little of this product should be formed in either case because of the strong van der Waals repulsions that would result.)

(a)

SO_3H — Br — CH_3 + Br — HO_3S — CH_3

(c)

NO_2 — Br — I + Br — O_2N — I

16.28 (a) The order of increasingly harsh reaction conditions is *m*-xylene < benzene < *p*-dichlorobenzene. The reason is that the methyl groups of *m*-xylene are activating groups and the chlorines of *p*-dichlorobenzene are deactivating groups.

16.29 Because both groups are *meta*-directing groups, it might seem that either could be introduced first.

However, the Friedel-Crafts acylation cannot be carried out on nitrobenzene because, as discussed on text p. 774, the nitro group is too deactivating. Consequently, the acetyl group must be introduced first.

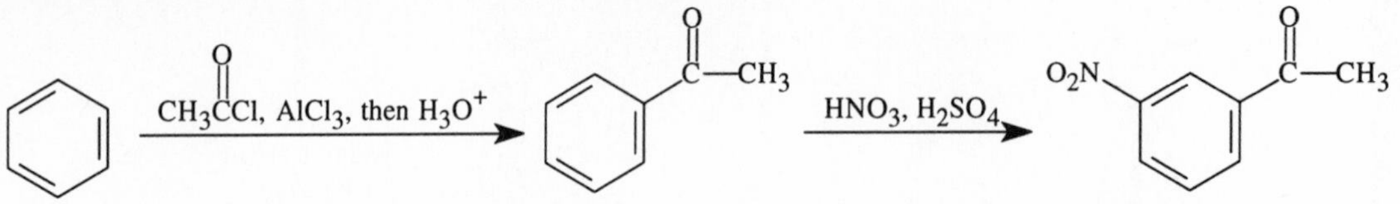

16.30 (a) Hydrogenate cyclohexylbenzene, which, in turn, is prepared as shown in Eq. 16.24 on text p. 759. Note that chlorocyclohexane and $AlCl_3$ or cyclohexanol and either H_2SO_4 or H_3PO_4 may be used as the source of the electrophile instead of cyclohexene; see Study Guide Link 16.3.

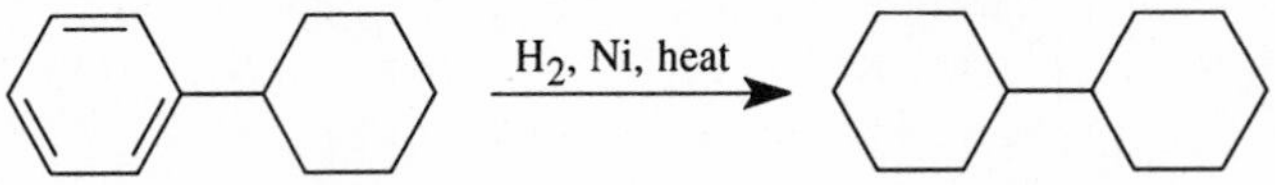

cyclohexylbenzene   cyclohexylcyclohexane

Solutions to Additional Problems
. . . .

16.31 (a) No reaction.

(b)

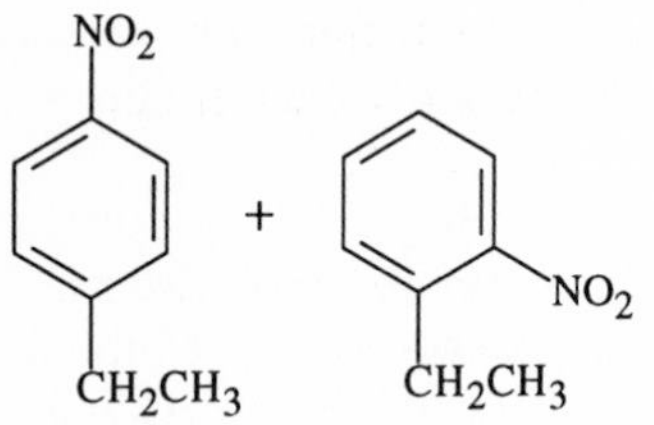

(c)

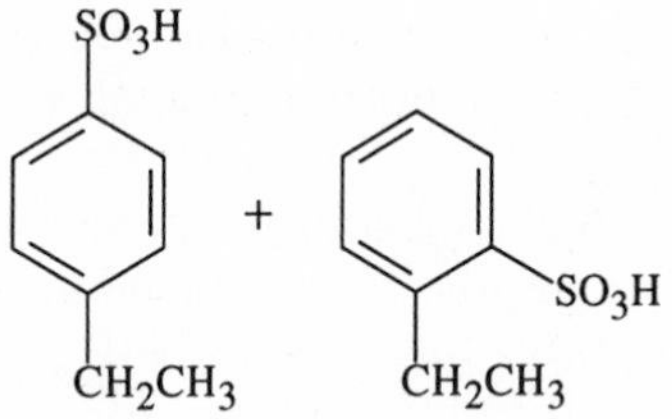

(d)

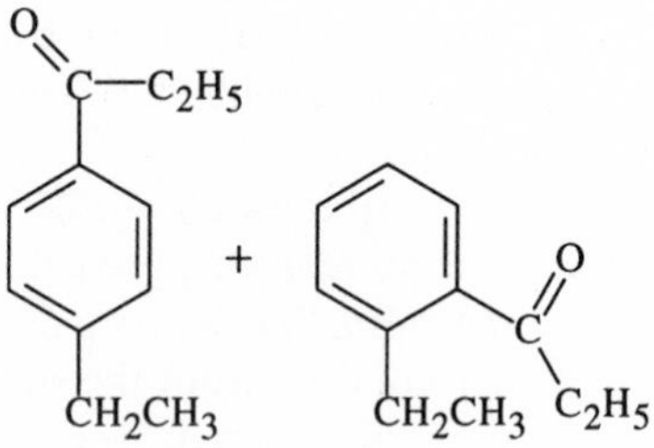

(e)

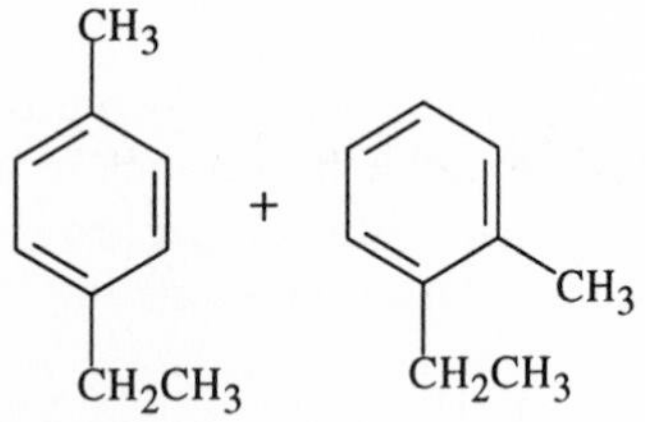

(f)

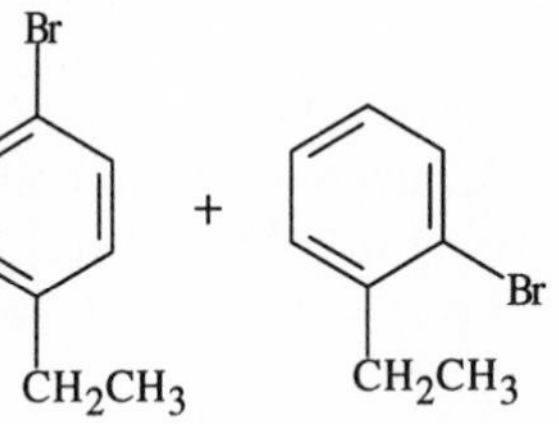

In part (e), ethylbenzene must be present in large excess in in order for the monosubstitution products to be obtained. If ethylbenzene is not present in excess, products of dimethylation, trimethylation, etc., will also be formed; see Eq. 16.22 on text p. 759 and associated discussion.

16.33 (a) Phenylacetylene will release a gas (ethane) when treated with C_2H_5MgBr and styrene will not.
(c) Benzene, with a boiling point of 80°, boils on a steam bath; ethylbenzene, with a boiling point of 136°, does not. Benzene, with a freezing point of 5.5°, freezes in an ice bath; ethylbenzene, with a melting point of −95°, does not.

16.34 A compound containing a benzene ring must have at least six carbon atoms and four degrees of unsaturation. Compound (1), with three degrees of unsaturation, cannot contain a benzene ring; compound (3) has only five carbons and therefore cannot contain a benzene ring. Compounds (2) and (4) have five and four degrees of unsaturation, respectively, and therefore could contain a benzene ring.

16.35 (a) Assume that the carbon-chlorine bond dipoles dominate the dipole moment. In the following diagram, the arrows with the small arrowheads are the C—Cl bond dipoles and the arrows with the large heads are the resultant molecular dipole moments. This diagram shows that the order of increasing dipole moment is *p*-dichlorobenzene < *m*-dichlorobenzene < *o*-dichlorobenzene.

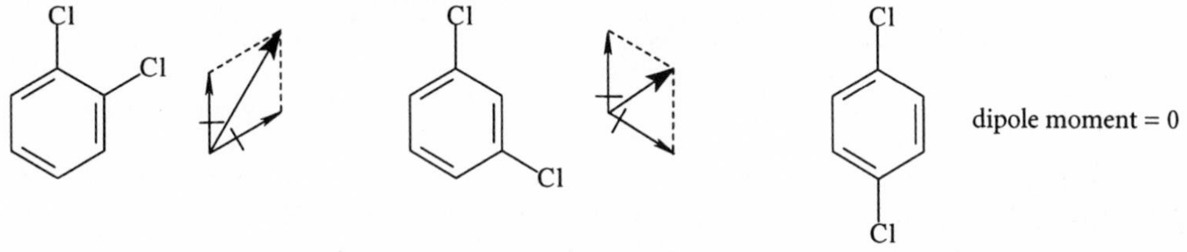

(b) Boiling points are increased by increasing attractions between molecules. To the extent that molecular attractions are dominated by attractions between molecular dipole moments, the order of increasing boiling point is predicted to be the same as the order of increasing dipole moment. (In fact, the boiling point of the *ortho* isomer (180°) is 6–7° higher than the boiling points of the *meta* and *para* isomers, which are within one degree of each other.)

16.36 Only compound *A* should have a proton NMR spectrum containing one six-proton singlet in the alkyl region (*ca.* δ 1.5). Only compound *B* should have a proton NMR spectrum that contains two closely spaced singlets in the integral ratio 2:1 in the benzylic proton region of the spectrum (about δ 2.3). Only compound *C* should have a proton NMR spectrum that contains two well-separated singlets in the integral ratio 3:1, the larger in the benzylic proton region and the smaller at δ 4.5, as well as a three-proton aromatic absorption.

16.38 (a) The cyclopentadienyl anion is aromatic; its five resonance structures show that all carbons (and hydrogens) are chemically equivalent. Hence, the proton NMR spectrum of the ion consists of one singlet.

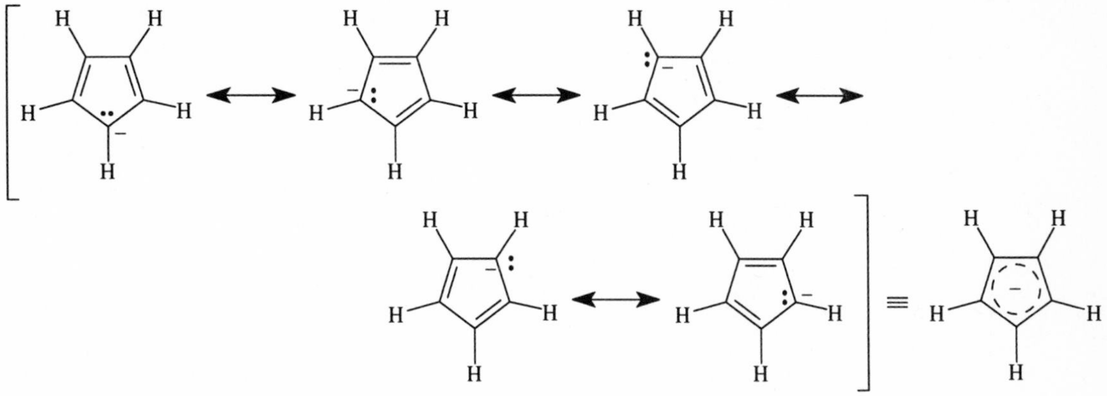

16.39 The electron density of the unshared electron pairs on the oxygen of the methoxy group is delocalized to the *ortho* and *para* positions on the ring, but not to the *meta* position:

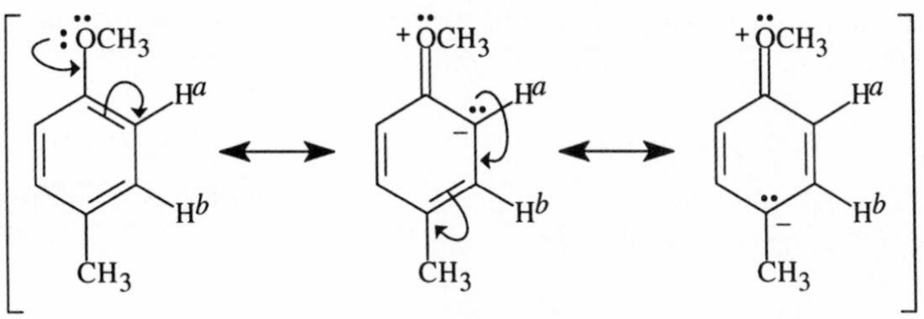

Thus, there is more electron density at the *ortho* proton H^a than at the *meta* proton H^b. Since greater electron density brings about greater shielding (see Sec. 13.2D on text p. 588), the chemical shift of proton H^a is smaller.

16.40 In each synthesis that involves substitution on a benzene derivative that contains an *ortho, para*-directing group, only the product resulting from *para* substitution is shown.

(a)

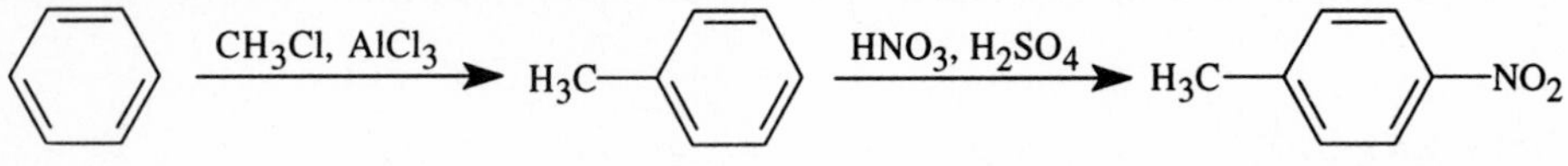

p-nitrotoluene

(c)

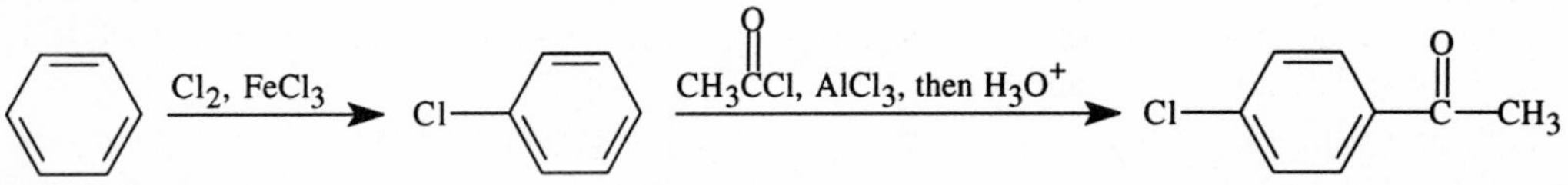

p-chloroacetophenone

(e)

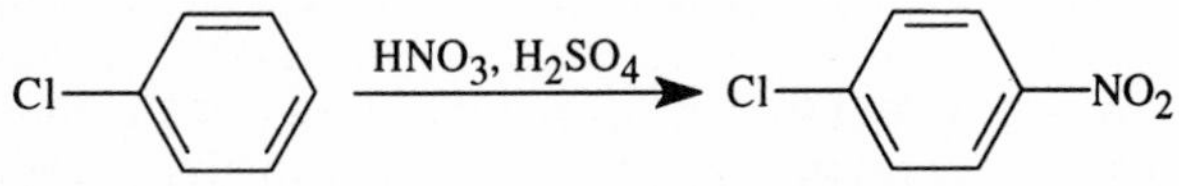

prepared in part (c)   *p*-chloronitrobenzene

(g)

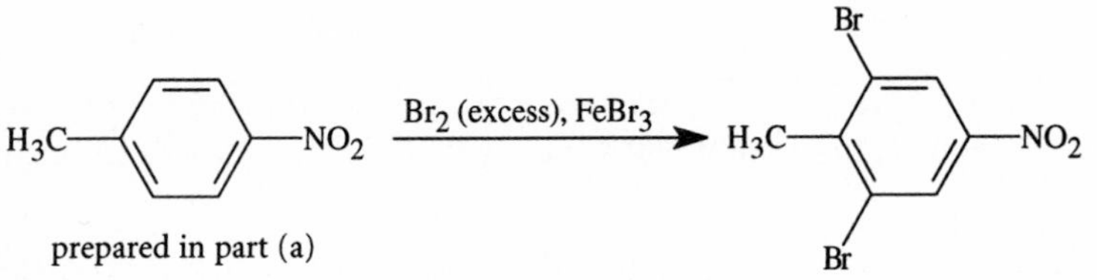

prepared in part (a)   2,6-dibromo-4-nitrotoluene

(i)

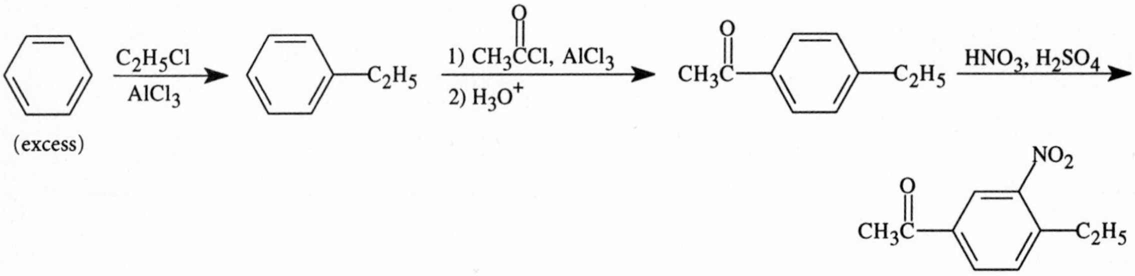

4-ethyl-3-nitroacetophenone

(j)

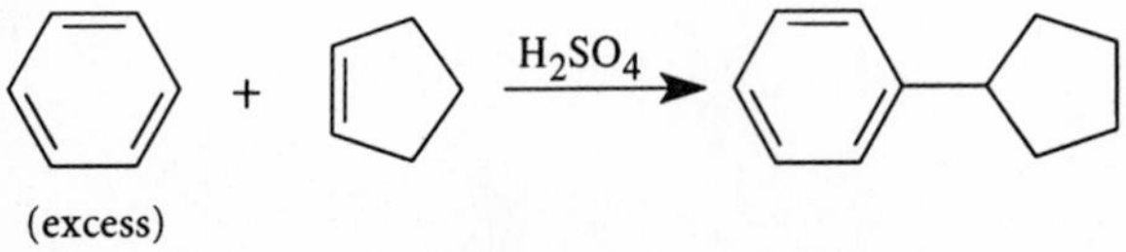

cyclopentylbenzene

Cyclopentanol may be substituted for cyclopentene in this synthesis, or chlorocyclopentane and $AlCl_3$ catalyst may be used instead of cyclopentene and H_2SO_4.

16.41 (a) In mesitylene, the three methyl groups all direct nitration to the same position; that is, they activate the same position of the ring. In 1,2,4-trimethylbenzene, only two of the three methyl groups activate the same ring position. Both trimethylbenzenes are more activated than toluene. Hence, the order is toluene < 1,2,4-trimethylbenzene < mesitylene.

(c) In *p*-chloroanisole, the chlorine and methoxy group direct to different positions; that is, the chlorine strongly deactivates the positions that the methoxy group activate, and the methoxy group deactivates the positions that are least deactivated by the chlorine. In the *meta* isomer, the directing effects of the two substituents do not conflict. In anisole, the deactivating chloro group is absent altogether. The order is *p*-chloroanisole < *m*-chloroanisole < anisole.

16.42 (a) The boron on benzeneboronic acid has an empty *p* orbital and is not capable of stabilizing a carbocation by resonance. It would be difficult to introduce electron deficiency into a compound that is already electron deficient; hence, this compound undergoes substitution much more slowly than benzene itself. Moreover, the boron-oxygen bond dipoles place substantial positive charge on boron. The boronic acid group thus has the characteristics of a *meta*-directing group, and the principal mononitration product is *m*-nitrobenzeneboronic acid (compound *A* in the structures following part (c)).

(b) For the reasons given in the solution to Problem 16.24(c), the $-\overset{+}{N}(CH_3)_3$ group is a *meta*-directing group and is therefore a *meta*-directing group. This compound reacts more slowly than benzene, and the product is the *meta*-nitro derivative *B*.

(c) One benzene ring serves as an *ortho, para*-directing substituent on the other. (The directing effect of the phenyl substituent is explained in the solution to Problem 16.23.) Because the phenyl substituent can stabilize the carbocation intermediate in nitration, it is an activating group. Hence, this compound is more reactive than benzene, and the product is the *para*-nitro derivative *C* (plus perhaps a small amount of the *ortho*-nitro isomer).

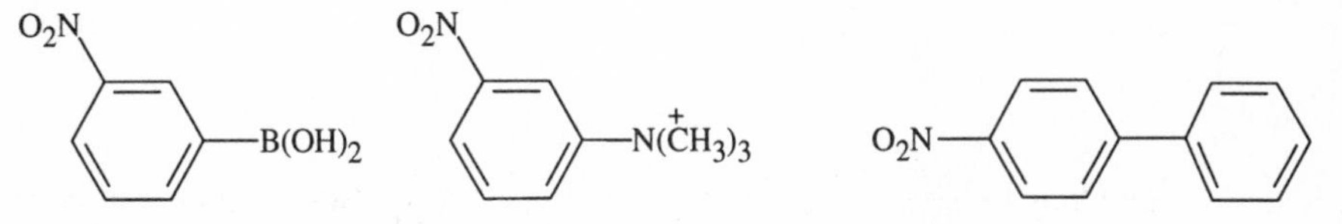

(d) Only the positions marked with an asterisk (*) are activated by *both* the phenyl and methoxy groups. (You should show that a carbocation intermediate formed in the other ring at *any* position cannot be resonance-stabilized by the unshared pair of the methoxy group.) Because of this activation, nitration at the asterisked positions is more rapid than nitration of benzene itself; the products are as follows:

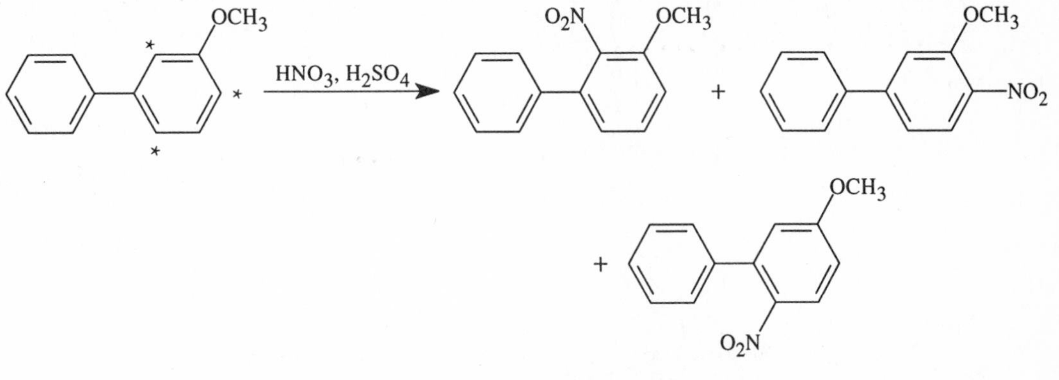

16.44 The unsaturation number of compounds *A* and *B* is 5; hence, both compounds could be aromatic. Compound *C* is a hydrocarbon; because hydrogenation of *C* gives a compound (compound *D*) with the molecular formula C_9H_{10}, then the formula of compound *C* could be C_9H_8. If so, then the reaction of compounds *A* and *B* with acid is a dehydration. If compounds *A* and *B* are alcohols, their —OH groups must be on adjacent carbons in order for them to dehydrate to the same alkene. Compound *D*, the hydrogenation product, has an unsaturation number of 5; nitration confirms that it and, by deduction, the other compounds, are aromatic. Since compound *D* can contain no ordinary double bonds (they would have been hydrogenated), it must contain a ring *in addition to* the aromatic ring. The structures of the compounds that meet all these criteria are as follows:

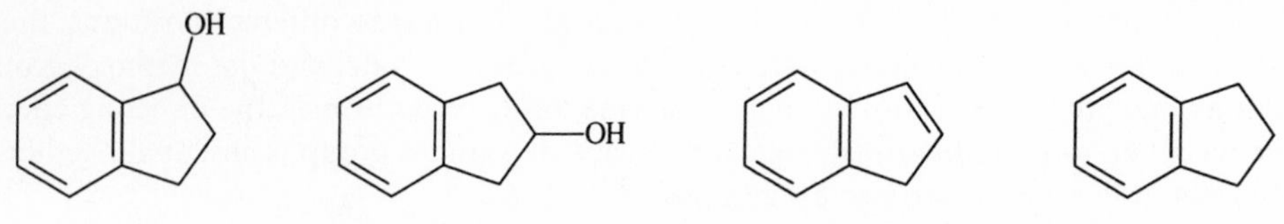

The nitration results rule out isomers such as the following:

would give four
mononitration products;
therefore cannot be D.

If compound at left cannot
be D, this cannot be A.

16.46 (a) and (b)   *Generation of the electrophile:* The electrophile is the carbocation *A* generated by protonation of the alcohol oxygen and loss of water.

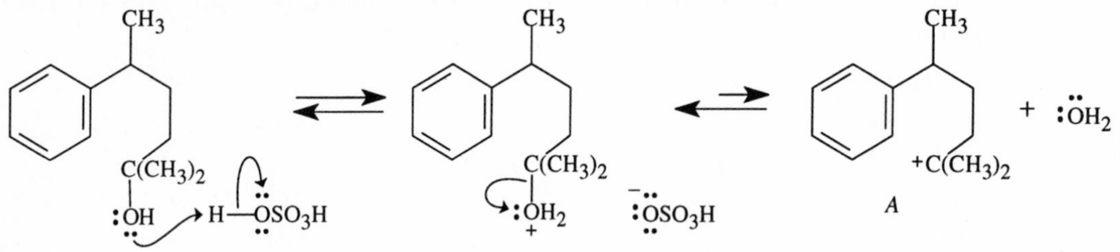

Attack of the benzene π electrons on the electrophile to generate another carbocation:

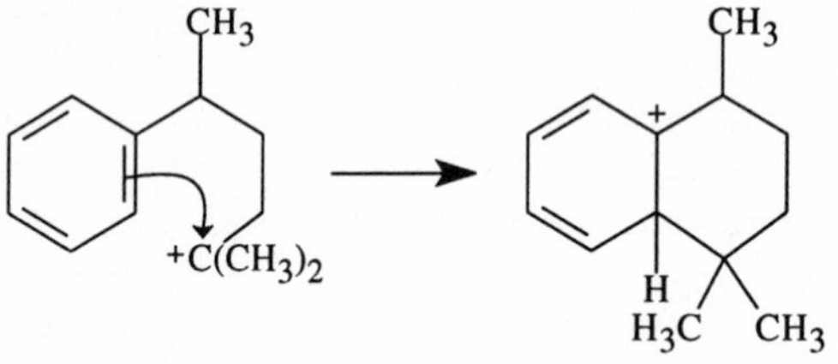

Loss of a β-proton to form the new aromatic compound:

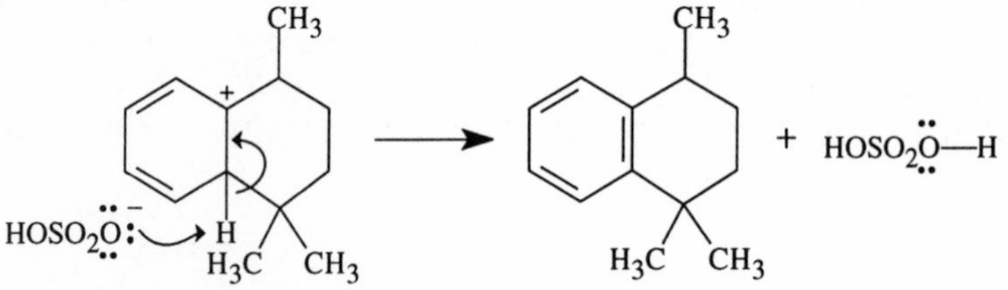

16.48 (a) The curved-arrow mechanism for reaction *(a)* is very similar to the mechanisms in the previous two problems.

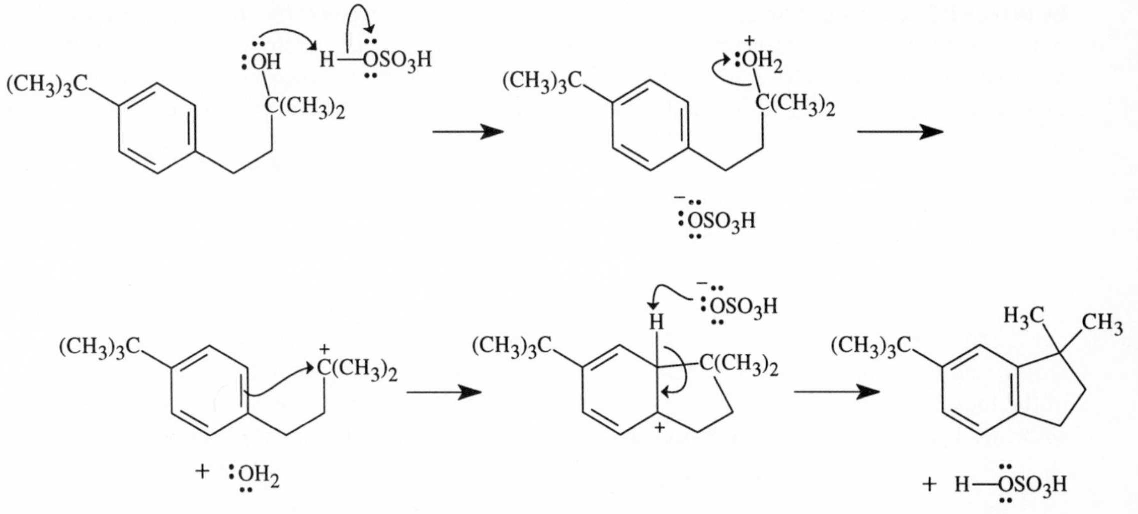

(b) The acylium ion is denied access to the positions of the ring that are *ortho* to the large *tert*-butyl group. In other words, these positions, although activated *electronically,* are deactivated *sterically.* Acylation occurs at the only remaining position to give celestolide.

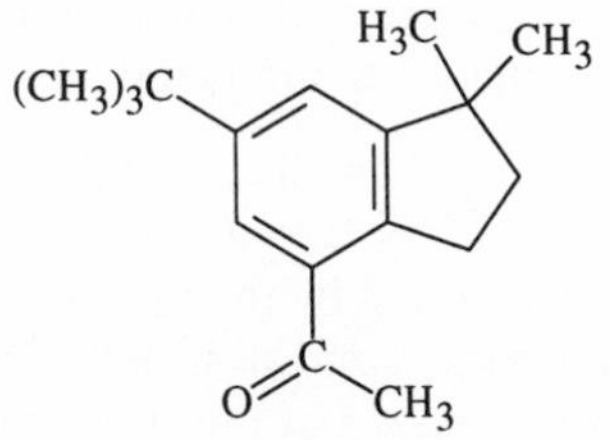

celestolide

16.49   The formula of compound *B* would be useful in solving this problem. Consider the integral in the NMR spectrum of compound *B*. If the singlet at δ 2.1 is due to a methyl group (three hydrogens), then, at 3.8 chart spaces per H, the entire spectrum integrates for ten hydrogens, and the formula of compound *B* is thus C_9H_{10}. (If the singlet at δ 2.1 were somehow due to a —CH_2— group, then the entire spectrum would integrate for 6.7 hydrogens; hence, the hypothesis of a methyl group is more reasonable.) Thus, compound *A* loses the elements of HBr when it reacts with a base and adds one molar equivalent of H_2 on catalytic hydrogenation. Therefore, compound *B* is an alkene formed by an E2 reaction from compound *A*. The NMR of compound *B* shows that it has five aromatic hydrogens; hence, the remaining unsaturation of compound *B* is due to a monosubstituted benzene ring. The resonances for the vinylic hydrogens in the NMR spectrum centered at δ 5.1 integrate for two protons. To summarize, the NMR spectrum of *B* indicates a monosubstituted benzene ring, two vinylic hydrogens, and one vinylic methyl group. The methyl group cannot be benzylic because this would require one fewer aromatic hydrogen and one more vinylic hydrogen. Considering only chemical shifts and integrals, two possibilities for compound *B* are therefore

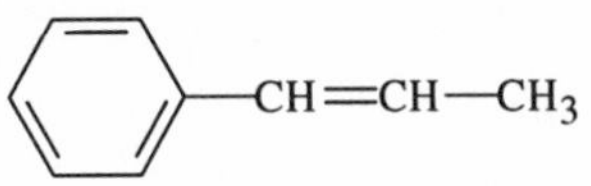

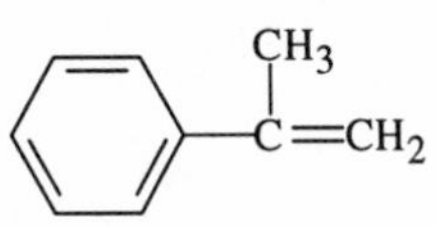

B1   *B2*

(solution continues)

Structure *B1* is ruled out by the absence of splitting in the δ 2.1 resonance; compound *B1* should show a doublet rather than a singlet for the methyl group. Structure *B2* is therefore the structure of compound *B*. Only two alkyl bromides could undergo elimination to give *B*; only the one shown below qualifies as compound *A* because only it is chiral. Compound *C* is the hydrogenation product of *B*. To summarize:

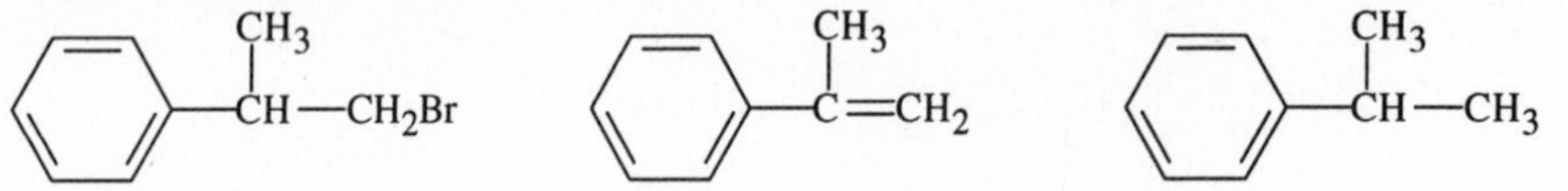

A   *B*   *C*

16.51 (a) The IR spectrum indicates the presence of an aromatic ring. The singlet at δ 3.72 suggests a methoxy group, and the apparent doublet of doublets indicates a *para*-disubstituted benzene ring. The M + 2 peak in the mass spectrum with about one-third the intensity of the M = 142 peak suggests the presence of chlorine. The compound *p*-chloroanisole (1-chloro-4-methoxybenzene) is consistent with all the data.

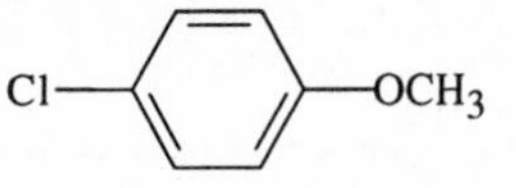

p-chloroanisole
(1-chloro-4-methoxybenzene)

When interpreting mass spectra, don't forget that you must use exact masses for isotopes. Thus, mass M = 142 corresponds to the compound containing ^{35}Cl, and the mass of the M + 2 peak corresponds to the compound containing ^{37}Cl.

(b) The IR spectrum indicates the presence of an alkene double bond; the NMR spectrum shows aromatic protons as well as alkene protons; and the UV spectrum indicates extensive conjugation. Assume that the peak at δ 3.7 is due to a methoxy group (verified by the carbon resonance at δ 54.9) and take its integral to be three protons. Then the apparent aromatic doublet of doublets at δ 6.6–7.4 integrates for four protons. These data, the splitting pattern, and the presence of four sets of carbon resonances in the aromatic region suggest a *para*-disubstituted aromatic ring. The remaining resonances integrate for three protons, and are in the alkene region. Moreover, the CMR spectrum shows two carbons in the vinylic region with a total of three attached protons. The —CH=CH$_2$ group is the only alkene functional group that has three alkene protons. Compound *B* is *p*-methoxystyrene:

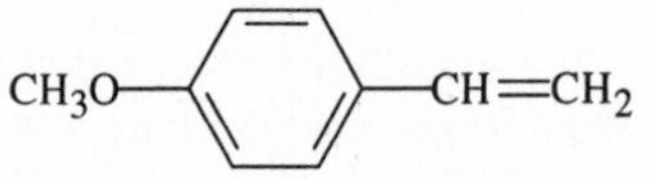

16.52 (a) *p*-Dibromobenzene can give only one mononitro derivative; hence, it must be compound *A*.
o-Dibromobenzene can give two mononitro derivatives, and is therefore compound *B*.
m-Dibromobenzene is compound *C*.

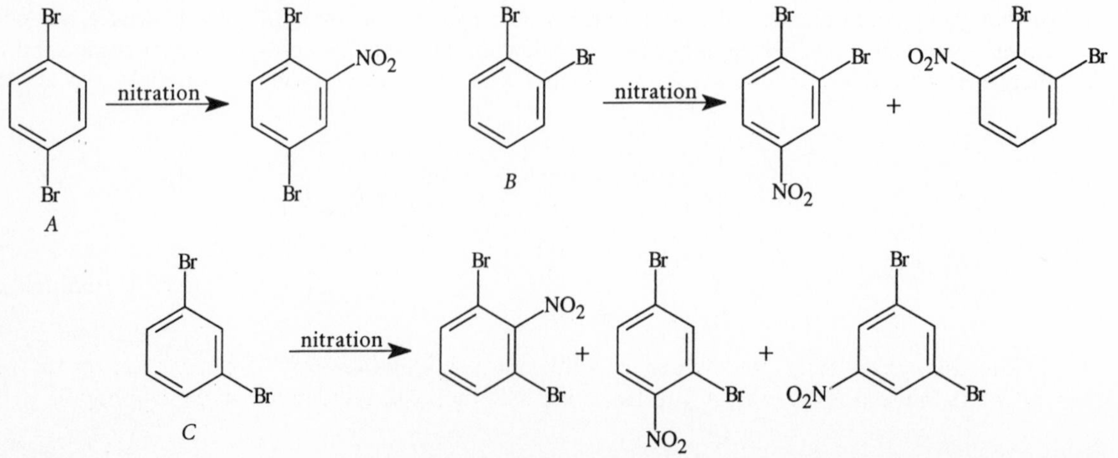

(b) Answer this question by deciding in how many different ways a single nitro group may be substituted for a hydrogen in each isomer. Because isomer separation methods (such as crystallization) based on conventional physical properties were used to differentiate isomers, enantiomeric differences between products were not evident; that is, only constitutional isomers or diastereomers are considered to be different compounds.

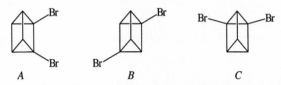

A   B   C

Compound *A* could give only one mononitro derivative; compound *B* could give two separable mononitro derivatives; and compound *C* could give three. (Can you draw the structures of these derivatives?)

(d) The three dibromobenzene isomers would be differentiated most readily by the number of resonances in their proton-decoupled CMR spectra. Compound *A*, the *para* isomer, has two resonances; compound *B*, the *ortho* isomer, has three resonances; and compound *C*, the *meta* isomer, has four resonances. The number of resonances in each case, of course, corresponds to the number of chemically nonequivalent sets of carbons.

16.53 (a) The required $\Delta H°$ is the sum of the $\Delta H°$ values in Eqs. 16.44a–c: –205 kJ/mol (–49.0 kcal/mol).
(b) The $\Delta H°$ for hydrogenation of three cyclohexenes is $3 \times (–118) = –354$ kJ/mol (–84.6 kcal/mol).
(c) The discrepancy between these two quantities is 149 kJ/mol (35.6 kcal/mol); that is, about 149 kJ/mol (36 kcal/mol) less heat is liberated when benzene is hydrogenated than would be expected from the hydrogenation of three cyclohexenes. This number is remarkably close to the empirical resonance energy of benzene calculated by comparison of its heat of formation with that of cyclooctatetraene (COT) on text page 719.

16.54 (a) Each alkyl halide group reacts with benzene in a separate Friedel-Crafts alkylation reaction.

$$\text{(benzene)}-CH_2-\text{(benzene)}-CH_2-\text{(benzene)}$$

(c) The reaction is an intramolecular Friedel-Crafts alkylation.

(structure of 4-methylchroman with O, CH₃)

(e) The cyclohexyl group, an alkyl group, directs nitration to the *ortho* and *para* positions of the benzene ring.

$$O_2N-\text{(benzene)}-\text{(cyclohexyl)} \quad + \quad \text{(NO}_2\text{ benzene)}-\text{(cyclohexyl)}$$

(mostly)

(g) The nitro group is directed by both substituents to the position *ortho* to the methoxy group, and the bromine in the second reaction is directed to the other position *ortho* to the methoxy group.

(solution continues)

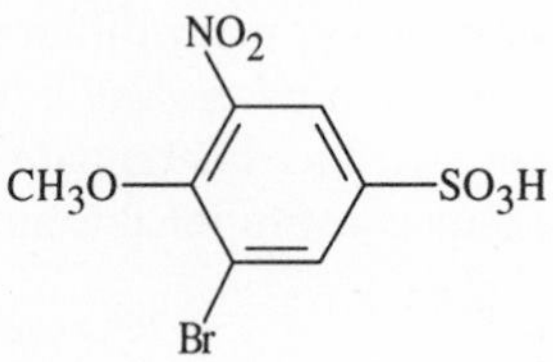

16.55 (a) The resonance structures for the carbocation intermediate in the nitration of naphthalene at carbon-1:

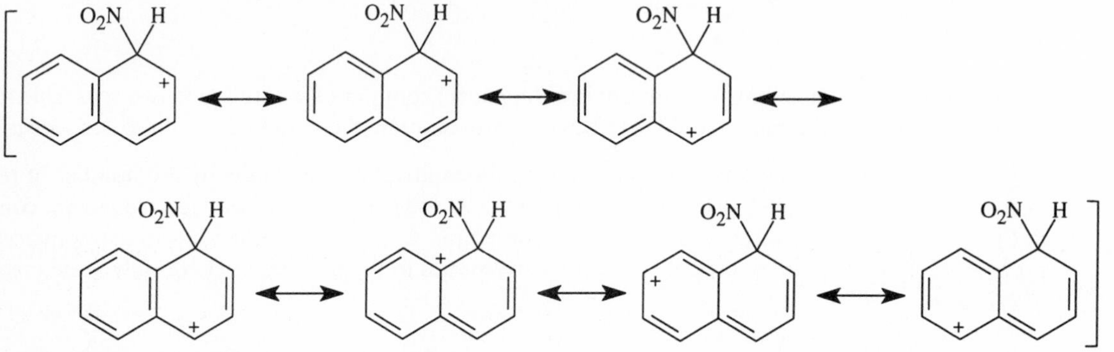

(c) If you worked part (b), you should have found that the carbocation intermediate involved in nitration at carbon-2 has six resonance structures. Naphthalene should nitrate more rapidly at carbon-1 because the carbocation intermediate in this nitration has more important resonance structures, and is thus more stable. Of two competing reactions, the one involving the more stable reactive intermediate is faster (Hammond's postulate).

16.56 (a) Because carbon-4 is *para* to the methoxy group, the carbocation intermediate involved in nitration at this carbon can be stabilized by electron delocalization from the oxygen, as it is in nitration at the *para* position of anisole. (See the colored structure in Eq. 16.27 on text p. 763.) Hence, the methoxy group activates substitution at carbon-4, and nitration at carbon-4 of 1-methoxynaphthalene is faster than nitration of naphthalene.

(c) Nitration at carbon-6, in contrast, is analogous to nitration at a *meta* position of anisole; see Eq. 16.28 on text p. 763. In this case, the oxygen electrons *cannot* be delocalized to stabilize the carbocation. The only effect of the methoxy group in this case is its rate-retarding polar effect. Therefore, nitration at carbon-6 of 1-methoxynaphthalene is slower than nitration of naphthalene.

positive charge must be delocalized to this carbon for
oxygen electrons to be involved in resonance stabilization

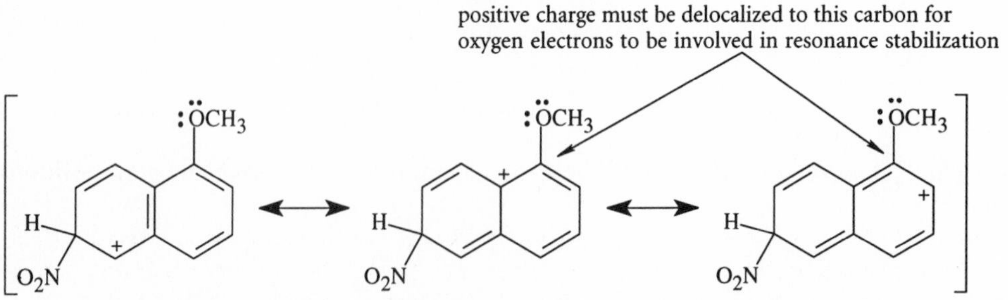

16.57 Friedel-Crafts acylation (or any other electrophilic substitution reaction) at carbon-2 of furan gives a carbocation intermediate with more resonance structures than acylation at carbon-3. Consequently, the carbocation intermediate from acylation at carbon-2 is more stable, and acylation at carbon-2 is faster.

Carbocation intermediate from acylation at carbon-2:

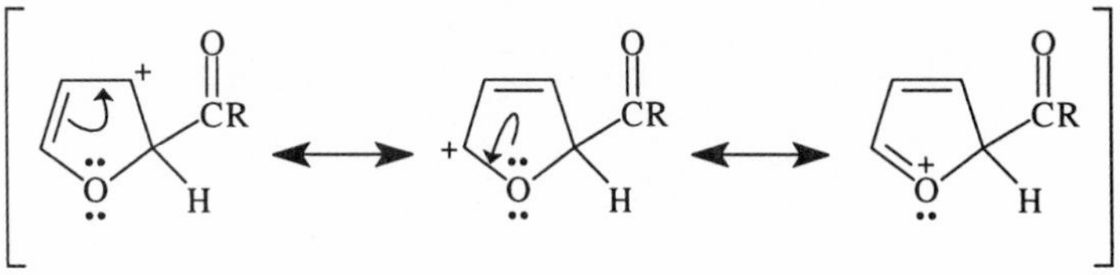

Carbocation intermediate from acylation at carbon-3:

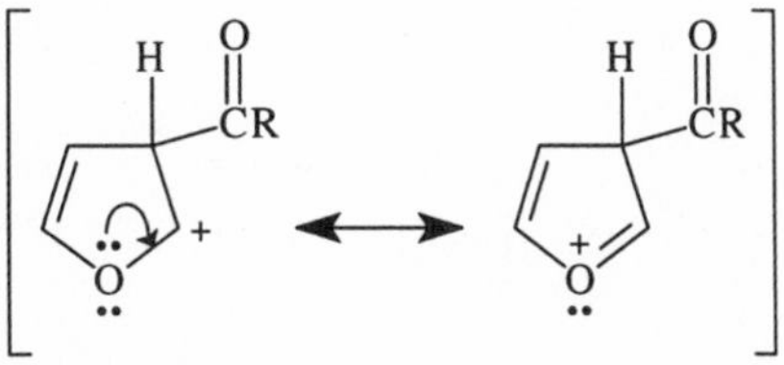

16.58 The balance between protonation on oxygen and protonation on a carbon of the ring depends on the relative basicities of an aromatic "double bond" and the oxygen of an aromatic ether. The fact that anisole protonates on oxygen shows that an ether oxygen is more basic than an aromatic "double bond."

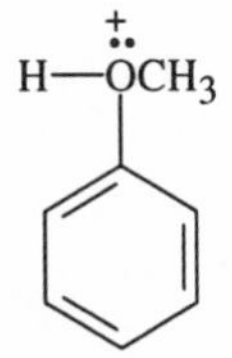

conjugate acid of anisole

The fact that 1,3,5-trimethoxybenzene protonates on carbon, then, means that the resulting conjugate-acid cation is stabilized sufficiently that the "double bonds" of the aromatic ring have enhanced basicity. The stability of this cation is due to resonance stabilization involving all three oxygens:

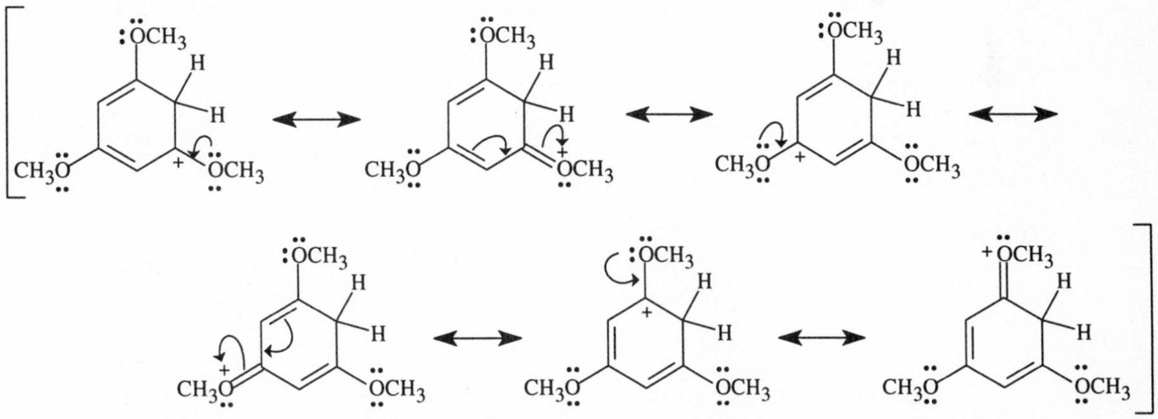

When anisole protonates on a ring carbon, the oxygen can also stabilize the resulting carbocation by analogous resonance interaction; however, in that case, the resonance effect of only one oxygen is available for such stabilization, and evidently that is not sufficient to bring about protonation on carbon.

16.59 (a) Furan derivatives, like 1,3-cyclopentadiene derivatives, are in effect conjugated dienes locked into *s-cis* conformations. Such dienes are reactive in Diels-Alder reactions. The product *A*, shown at the top of the following page, results from such a reaction.

compound *A*

(b) Evidently, the acidic conditions promote loss of the bridging ether oxygen as water. The mechanistic steps involved are protonation, carbocation formation, loss of a hydrogen, and then repetition of a similar sequence.

16.60 In this reaction a *tert*-butyl cation is lost rather than a proton from the carbocation intermediate. The electrophile, a nitronium ion $^+NO_2$, is generated by the mechanism shown in Eqs. 16.9a–b on text p. 752.

16.62 At the higher temperature, the two *ortho* methyl groups (labeled *a* in the problem) are chemically equivalent on the NMR time scale because rotation of the isopropyl group is rapid; the resonance of these groups occurs at δ 2.25. However, at the lower temperature, rotation of the isopropyl group is slow, and the two *ortho* methyl groups are no longer chemically equivalent. Hence, the resonances of the ring methyl groups occur at different chemical shifts at the lower temperature. Evidently, one methyl group is wedged between the two isopropyl methyls, and the other is in the same plane as the isopropyl C—H group, as shown by the structure in the problem.

16.63 (a) The two terminal rings of hexahelicene cannot lie in the same plane because, if they did, their hydrogens would have severe van der Waals repulsions. Consequently, the molecule is somewhat bent out of plane so that one of the terminal rings lies over the other. This causes the molecule to trace one turn of a helix, which is a chiral object.

(b) Normally the two rings of a biphenyl derivative lie in the same plane to maximize conjugation, that is, to maximize overlap of their π-electron systems. In this case, however, if the rings were to lie in the same plane, the very large sulfonic acid (— SO_3H) group on one ring would have severe van der Waals repulsions with an *ortho* hydrogen of the other ring. Consequently, the molecule adopts the conformation shown in the problem, in which the planes of the two rings are perpendicular. However, this conformation is chiral and is thus capable of showing optical activity. Evidently, rotation about the central carbon-carbon bond is slow enough that the molecule can be resolved into isolable enantiomeric conformations! When the molecule is heated, the internal rotation "reaction" (like all reactions) is accelerated, and the optically active molecule racemizes. This situation is also discussed in Study Guide Link 6.6 on page 119 of this manual; a similar situation is the subject of Problem 16.9 earlier in this chapter.

17

Allylic and Benzylic Reactivity

Terms

Concepts

I. Allylic and Benzylic Species

A. INTRODUCTION

1. An allylic group is a group on a carbon adjacent to a double bond.
2. A benzylic group is a group on a carbon adjacent to a benzene ring or substituted benzene ring.
3. In many situations allylic and benzylic groups are unusually reactive.

B. ALLYLIC AND BENZYLIC CATIONS, RADICALS, AND ANIONS

1. Allylic and benzylic carbocations are resonance-stabilized.
   a. The charge on an allylic carbocation is shared between two carbons.

$$\left[R-CH=CH-\overset{+}{C}H-R' \quad \longleftrightarrow \quad R-\overset{+}{C}H-CH=CH-R' \right]$$

an allylic cation

   b. The charge on a benzylic carbocation is shared not only by the benzylic carbon, but also by alternate carbons of the ring.

a benzylic cation

2. Allylic and benzylic radicals are resonance-stabilized; they are more stable, and thus more readily formed as reactive intermediates, than ordinary alkyl radicals.
   a. The unpaired electron of an allylic radical is shared between two carbons.

an allylic radical

   b. The unpaired electron of a benzylic radical is shared by the benzylic carbon and alternate carbons of the ring.

a benzylic radical

3. Allylic and benzylic anions are resonance-stabilized and thus more stable than their nonallylic and nonbenzylic counterparts.
   a. The polar effect of the double bond (in the allyl anion) or the phenyl ring (in the benzyl anion) also stabilizes the anion.
   b. Although allylic hydrogens and benzylic hydrogens are very weak acids, their acidities are much greater than the acidities of alkanes that do not contain allylic or benzylic hydrogens.

B.   ALLYLIC AND BENZYLIC S_N1 REACTIONS

1. Two different products are formed when a nucleophile attacks an unsymmetrical allylic cation, because there are two different electron-deficient carbons.

2. One product is formed when a nucleophile attacks an unsymmetrical benzylic cation, because the products formed by attack of the nucleophile at the other electron-deficient carbons are not aromatic and thus lack the stability associated with the aromatic ring.

3. Reactions involving benzylic or allylic carbocations as intermediates are generally considerably faster than analogous reactions involving comparably substituted nonallylic or nonbenzylic carbocations.
   a. The greater reactivities of allylic and benzylic halides in S_N1-E1 reactions are due to stabilities of the carbocation intermediates that are formed when they react; the more stable carbocation should be formed more rapidly (Hammond's postulate).
   b. *Ortho* and *para* substituent groups on the benzene ring that activate electrophilic aromatic substitution accelerate S_N1 reactions at the benzylic position; in the intermediate carbocation, additional resonance structures in which charge can be delocalized onto the substituent group itself are possible.
   c. Other reactions that involve carbocation intermediates are accelerated when the carbocations are allylic or benzylic.

C. ALLYLIC AND BENZYLIC S$_N$2 REACTIONS

1. S$_N$2 reactions of allylic and benzylic halides are relatively fast even though they do not involve reactive intermediates.
2. Allylic and benzylic S$_N$2 reactions are accelerated because the energies of their transition states are reduced by *p*-orbital overlap.
   a. In the transition state of the S$_N$2 reaction, the carbon at which substitution occurs is *sp*^{2}-hybridized.
   b. The incoming nucleophile and the departing leaving group are partially bonded to a *p* orbital on this carbon.
   c. Overlap of this *p* orbital with the *p* orbitals of an adjacent double bond or phenyl ring provides additional bonding that lowers the energy of the transition state and accelerates the reaction.

D. ALLYLIC AND BENZYLIC E2 REACTIONS

1. The acidity of the *β*-hydrogens is a structural effect in the alkyl halide that tends to promote a greater fraction of elimination.
   a. A greater ratio of elimination to substitution is observed when the *β*-hydrogens of the alkyl halide have higher than normal acidity.
   b. In the transition state of the E2 reaction, the base is removing a *β*-proton, and the transition state of the reaction has carbanion character at the *β*-carbon atom.
2. Another factor that promotes elimination is that the alkene double bond, which is partially formed in the transition state, is conjugated with the benzene ring or the double bond of the allylic group.

E. ALLYLIC AND BENZYLIC RADICALS

1. The free-radical chain reaction of bromine with a benzylic group entails the following steps:
   a. abstraction of a benzylic hydrogen by a bromine atom in preference to a nonbenzylic hydrogen.
      i. This propagation step gives rise to the selectivity for substitution of the benzylic hydrogen.
      ii. The reason for this selectivity is that the benzylic radical which is formed has greater stability than nonbenzylic radicals.
   b. reaction of the benzylic radical with a molecule of bromine to generate a molecule of product as well as another bromine atom.
2. Free-radical substitution at the allylic position of an alkene is in competition with bromine addition to the alkene double bond.
   a. Addition of bromine is the predominant reaction if:
      i. free-radical substitution is suppressed by avoiding conditions that promote free-radical reactions (light, heat, or free-radical initiators).
      ii. the reaction is carried out in solvents of even slight polarity that promote the ionic mechanism for bromine addition.
   b. When a compound with allylic hydrogens is treated with *N*-bromosuccinimide (NBS) in CCl$_4$ under free-radical conditions, allylic bromination takes place, and addition to the double bond is not observed.

G. ALLYLIC GRIGNARD REAGENTS

1. Allylic Grignard reagents undergo a rapid equilibrium in which the —MgBr group moves back and forth between the two partially negative carbons; this is an example of an allylic rearrangement.

a. An allylic rearrangement involves the simultaneous movement of an allylic group and a double bond so that one allylic isomer is converted into another.

    i. These two structures are not resonance structures; they are two distinct species in rapid equilibrium.

    ii. Because of this equilibrium, an unsymmetrical allylic Grignard reagent is a mixture of two different reagents.

b. The same mixture of reagents can be obtained from either of two allylically related alkyl halides.

c. The same mixture of products is obtained when the Grignard reagent is prepared from either allylic halide.

2. Allylic Grignard reagents resemble allylic carbanions; because the allylic carbanion is resonance-stabilized, the transition state for equilibration has relatively low energy, and consequently the equilibration occurs rapidly.

II. The Isoprene Rule; Biosynthesis of Terpenes

A. ESSENTIAL OILS AND TERPENES

1. Some pleasant-smelling substances found in nature have come to be called essential oils.

a. Terpenes (also called isoprenoids) are a class of natural products with similar atomic composition; many of these compounds are familiar natural flavorings or fragrances.

b. All terpenes consist of repeating units that have the same carbon skeleton as the five-carbon diene isoprene.

2. The basis of the terpene or isoprenoid classification is only the connectivity of the carbon skeleton.

a. The presence of double bonds and other functional groups, or the configuration of double bonds and asymmetrical carbons, have nothing to do with the terpene classification.

b. All terpenes consist of repeating units that have the same carbon skeleton as the five-carbon diene isoprene (the isoprene rule).

    i. In many terpenes, the isoprene units are connected in a "head-to-tail" arrangement.

    ii. Many examples are known in which the isoprene units have a "head-to-head" connectivity.

c. Some compounds are derived from the conventional terpene structures by skeletal rearrangements.

d. Terpene carbon skeletons contain multiples of five carbon atoms (10, 15, 20, ... 5*n*).

    i. Monoterpenes have ten carbon atoms in their carbon chains.

    ii. Sesquiterpenes have fifteen carbon atoms in their carbon chains.

    iii. Diterpenes have twenty carbon atoms in their carbon chains.

3. Criteria for terpene classification are:

a. A multiple of five carbon atoms in the main carbon chain.

b. The carbon connectivity of the isoprene carbon skeleton within each five-carbon unit.

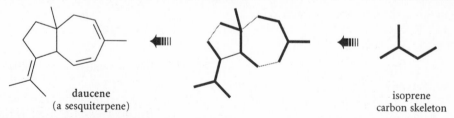

daucene
(a sesquiterpene)

isoprene
carbon skeleton

B. BIOSYNTHESIS OF TERPENES

1. Biosynthesis is the synthesis of chemical compounds by living organisms.

2. The repetitive isoprene units in all terpenes have a common origin in two simple five-carbon compounds (structures on text page 809, Eq. 17.35):

a. Isopentenyl pyrophosphate (IPP).

b. γ,γ-Dimethylallyl pyrophosphate (DMAP).

c. Alkyl pyrophosphates (R—OPP) are esters of the inorganic acid pyrophosphoric acid; —OPP is an abbreviation of the pyrophosphate group.

    i. Pyrophosphate and phosphate are "nature's leaving groups."

    ii. IPP and DMAP are readily interconverted in living systems by enzyme catalysis.

Reactions

I. Allylic and Benzylic Substitution Reactions

A. ALLYLIC AND BENZYLIC BROMINATION WITH NBS

1. The mechanism of allylic or benzylic bromination with NBS entails the following steps:
   a. The initiation step is the formation of a bromine atom by homolytic cleavage of the N—Br bond in NBS itself.

   b. The ensuing substitution reaction has three propagation steps:
      i. The bromine atom abstracts an allylic or benzylic hydrogen.

      ii. The HBr thus formed reacts with the NBS in the second propagation step (by an ionic mechanism) to produce a Br_2 molecule.

      iii. The last propagation step is the reaction of this bromine molecule with the radical formed in the first propagation step.

   c. The Br_2 concentration remains low because it can be generated no faster than an HBr molecule and an allylic radical are generated.
   d. The low solubility of NBS in CCl_4 is crucial to the success of allylic bromination with NBS.
2. Because the unpaired electron of an unsymmetrical allylic radical is shared by two chemically nonequivalent carbons, such a radical can react to give two constitutionally isomeric products.

II. *Benzylic Oxidation Reactions*

A. SIDE-CHAIN OXIDATION OF ALKYLBENZENES

1. Treatment of alkylbenzene derivatives with strong oxidizing agents under vigorous conditions converts the alkyl side chain into a carboxylic acid group.
   a. The benzene ring is left intact.
   b. The alkyl side chain, regardless of length, is converted into a carboxylic acid group.
   c. The conditions for this side-chain oxidation are generally vigorous: heat, high concentrations of oxidant, and/or long reaction times.
   d. Common oxidants are Na_2CrO_7 (sodium dichromate), CrO_3 (chromium trioxide), $KMnO_4$ (potassium permanganate), or O_2 and special catalysts.
2. Oxidation of alkyl side chains requires the presence of a benzylic hydrogen; benzene derivatives with no benzylic hydrogens are resistant to side-chain oxidation.

Study Guide Links

17.1   Addition vs. Substitution with Bromine

The purpose of this Study Guide Link is to explain further the effect of low bromine concentration in promoting free-radical substitution. The explanation of this effect lies in the rate laws for addition and substitution. The rate law for *ionic addition* (Sec. 5.1A) at low bromine concentrations is

$$\text{rate of ionic addition} = k_{\text{addition}}[\text{alkene}][\text{Br}_2]$$

Although it is not discussed in the text, addition of Br_2 can also take place by a free-radical chain mechanism. At low Br_2 concentrations, the rate of free-radical addition follows the rate law

$$\text{rate of free-radical addition} = k_{\text{fr-addition}}[\text{alkene}][\text{Br}_2]^{3/2}$$

The rather strange-looking 3/2-order in bromine, roughly speaking, is due to the involvement of a bromine atom ("half of a bromine molecule") produced in the initiation step together with the involvement of a bromine molecule in the rate-limiting propagation step—one and one-half bromine molecules total.

Finally, the rate of free-radical substitution at low Br_2 concentrations has the rate law

$$\text{rate of free-radical substitution} = k_{\text{substitution}}[\text{alkene}][\text{Br}_2]^{1/2}$$

The half-order dependence in bromine is due to the involvement of a bromine atom that is produced in the initiation step.

If you compare these rate laws, you will notice that the substitution reaction has the least dependence on bromine concentration. Hence, the rate of substitution not only *increases less rapidly* with *increasing* bromine concentration than the other reactions, but also *decreases less rapidly* with *decreasing* bromine concentration. At very low bromine concentration, then, free-radical substitution has the largest rate, and is therefore the observed reaction.

The success of *N*-bromosuccinimide (NBS) in CCl_4 as a reagent for substitution of allylic and benzylic hydrogens by bromine is due to the fact that it provides an experimentally convenient way to maintain a low bromine concentration. It wasn't planned that way; NBS in CCl_4 for many years was known as a good reagent for carrying out allylic and benzylic brominations with no competing additions. Careful research later developed the rationale presented here and in the text.

17.2   Polar Effect of Double Bonds

Why should the polar effect of a nearby double bond stabilize negative charge? Recall (Sec. 4.4, text p. 138) that the bond dipole between an sp^3-hybridized carbon atom and an sp^2-hybridized carbon is directed toward the double bond:

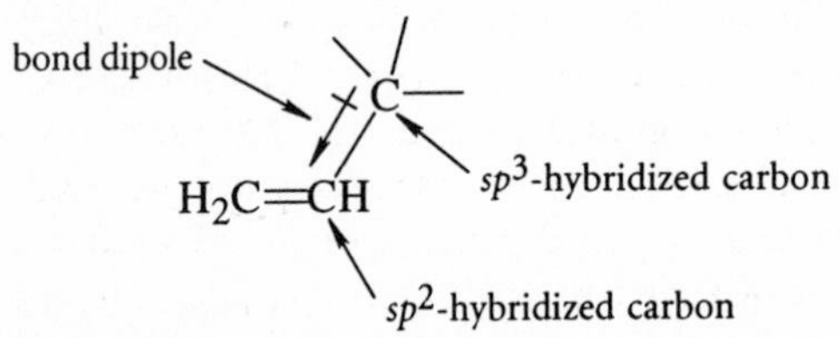

<!-- figure: bond dipole diagram -->
bond dipole
sp^3-hybridized carbon
$\text{H}_2\text{C}=\text{CH}$
sp^2-hybridized carbon

If there happens to be a negative charge on the sp^3-hybridized carbon atom, it is stabilized by interaction with the positive end of the bond dipole.

At a deeper level, we might ask why a bond dipole is oriented from an sp^3-hybridized carbon toward an sp^2-hybridized carbon, as shown above. In an sp^2-hybridized carbon atom, one of the electrons is in a p orbital, and is thus relatively far from the nucleus. (Think of this electron as being partially "pulled away" or "peeled back" from the nucleus.) As a result, the electronic shielding of the positive nucleus in an sp^2-hybridized carbon atom is reduced—that is, the nucleus is more exposed. Such a nucleus has a stronger attraction for the electrons in neighboring σ-bonds than the nucleus of an sp^3-hybridized carbon atom, in which there is no p electron. The electrons in the σ-bond are thus pulled toward the sp^2-hybridized carbon and away from the sp^3-hybridized carbon, and a bond dipole results.

✓17.3   Synthetic Equivalence

The benzylic oxidation provides an excellent way to introduce a carboxylic acid group into a benzene ring. (None of the electrophilic aromatic substitution reactions in Chapter 16 provide a way to introduce this group directly.) Suppose, for example, we wish to prepare p-nitrobenzoic acid from benzene. The desired acid can be formed by benzylic oxidation of any p-alkylnitrobenzene with benzylic hydrogens—say, p-ethylnitrobenzene:

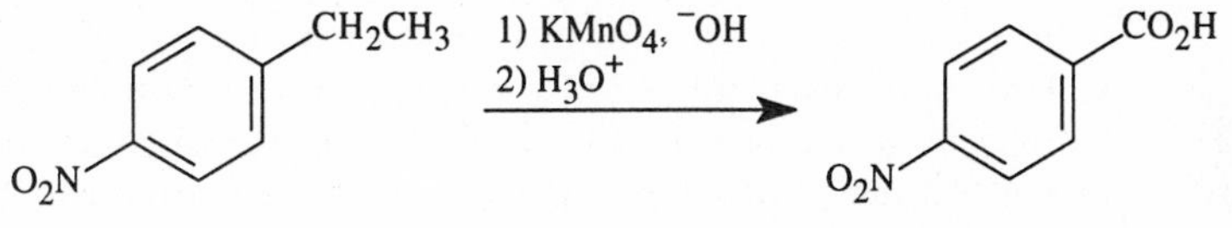

| p-ethylnitrobenzene | | p-nitrobenzoic acid |

The p-ethylnitrobenzene comes, in turn, from nitration of ethylbenzene:

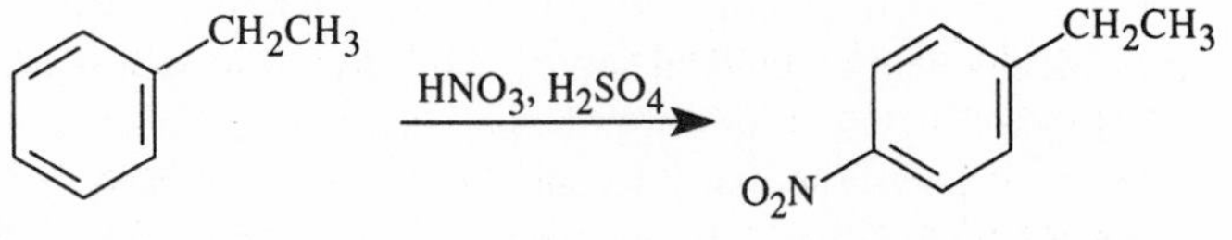

| ethylbenzene | | p-ethylnitrobenzene |

And the ethylbenzene comes from Friedel-Crafts alkylation of benzene (Eq. 16.23, text p. 759).

If we wanted to prepare m-nitrobenzoic acid, we would *first* oxidize ethylbenzene to benzoic acid, and then nitrate the benzoic acid, thus taking advantage of the *meta*-directing effect of the carboxylic acid group:

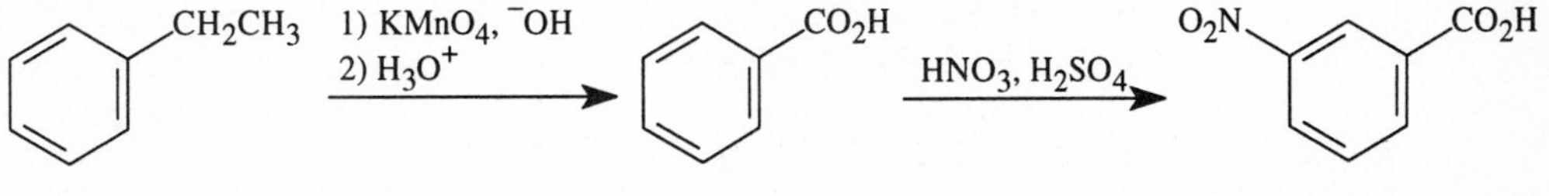

| ethylbenzene | benzoic acid | m-nitrobenzoic acid |

In these schemes, an alkyl group—specifically, the ethyl group—has been utilized as the *synthetic equivalent* of a carboxylic acid group. That is, the ethyl group is something that is *easily converted into* a carboxylic acid group. (Some people use the word *synthon* to mean the same thing as *synthetic equivalent*.) A person who is skilled in synthesis tends to see various functional groups in terms of their synthetic equivalents. For example, when such a person sees a carboxylic acid group on a benzene ring, an alkyl group pops into mind. For an aldehyde, a primary alcohol comes to mind. (What would you think of as a synthetic equivalent of a *cis*-alkene?) Of course, a group may have more than one synthetic equivalent, and the appropriate synthetic equivalency will generally depend on the exact situation at hand.

By coupling the Friedel-Crafts alkylation with benzylic oxidation, a connection has been made that was not explicitly discussed in the text. As you improve in your mastery of organic chemistry, this is the sort of connection that you should begin to make on your own. A good student will begin to think of compounds "out of context," that is, in ways not directly related to the text material. For example, a student will see a structure and ask, "How could I synthesize this?" even though its synthesis might not be the context in which the structure is encountered. Professional chemists tend to think this way. A structure or a reaction presented in one context will trigger the imagination to think of a wholly new context in which it might be useful. This is one way that new ideas are born. Awareness of this "intellectual triggering" process is one of the best reasons that scientists study the professional literature. And this works not only in chemistry; it works in fields as diverse as physics and medicine.

17.4   Essential Oils

The history of the essential oils is an important part of the early history of both chemistry and medicine. A Swiss alchemist, Theophrastus Bombastus von Hohenheim, better known as Paracelsus (*ca.* 1493–1541), believed that everything had a chemical essence. For example, he could demonstrate that the odor of spearmint could be liberated from the plant as a volatile oil on heating; this was the "essence of spearmint"—its quintessence, or "reason for being." (Paracelsus tried unsuccessfully to find the essences of rocks and other refractory objects.) He believed that a person who was ill was missing part of his or her essence—and Paracelsus sought to restore the essence with chemical cures, among them mercury and sulfur. Paracelsus was reputed to have effected some very dramatic cures. His firm belief in the chemical essence of all things was reflected in public tirades against physicians of the day, whom (he said) ". . . strut about with haughty gait, dressed in silk with rings upon their fingers displayed ostentatiously, or silver poignards fixed upon their loins and sleek gloves upon their hands . . ." while chemists "sweat whole nights and days over fiery furnaces, do not waste time with empty talk, but express delight in the laboratory." Paracelsus was driven out of Basel when he angered some local clergy during a quarrel over his fees; he died in exile.

Many (but certainly not all) of the essential oils turned out to be terpenes. And it is known today that many terpenes are neither volatile nor pleasant-smelling, and many come from animal sources rather than plant sources. Nevertheless, it was the fascination with, and curiosity about, the essential oils that ultimately led chemists to their understanding of how chemical substances are synthesized in the natural world.

✓17.5   Skeletal Structures

Skeletal structures are used in this section and with increasing frequency throughout the rest of the text. Skeletal structures are explained in Sec. 2.5 on text pp. 64–67. Be especially careful to remember that a carbon atom is located not only at *each vertex* of a skeletal structure but also at *each end of a chain*. If you have difficulty interpreting a skeletal structure, do not hesitate to draw out the structure with all the carbons and hydrogens.

Solutions

Solutions to In-Text Problems

17.1 (a) The order of increasing S_N1 reactivity is (2) < (1) < (3). The rates depend on the stabilities of the respective carbocation intermediates. One resonance structure for the carbocation intermediate in the solvolysis of compound (3) is a tertiary carbocation; the resonance structures of the carbocation intermediates derived from compounds (1) and (2) are secondary carbocations.

carbocation intermediate in the solvolysis of compound (3)

Consequently, the carbocation intermediate derived from compound (3) is more stable, relative to the starting alkyl halide, than the intermediates in the solvolysis reactions of the other compounds. The intermediate in the solvolysis of compound (2) is least stable because of the electron-withdrawing polar effect of the oxygen. Note that an oxygen in the *meta* position cannot be involved in resonance; consequently, only its rate-retarding polar effect operates.

 Note that the rates of S_N1 solvolysis reactions parallel the rates of electrophilic aromatic substitution reactions. Just as a methoxy group activates an aromatic substitution reaction at an *ortho* or *para* position, it also activates a reaction involving a benzylic cation at an *ortho* or *para* position. Just as a methoxy group *deactivates* aromatic substitution at a *meta* position, it also deactivates a reaction involving a benzylic cation at a *meta* position.

17.2 The alkyl halide is the one that reacts to give the same carbocation intermediate, that is, a carbocation with the same resonance structures:

$$(CH_3)_2\underset{\underset{Cl}{|}}{C}-CH=CH_2 \xrightarrow{-Cl^-} \left[(CH_3)_2\underset{+}{C}-CH=CH_2 \longleftrightarrow (CH_3)_2C=CH-\underset{+}{CH_2} \right]$$

the same carbocation intermediate
that is involved in text Eq. 17.7

17.4 The carbocation formed when trityl chloride ionizes, the *trityl cation* (Ph_3C^+), is stabilized by delocalization of electrons from *all three* phenyl rings. This carbocation has more resonance structures than the carbocations formed from the other alkyl halides in the table, and is thus so stable that the transition state leading to its formation also has very low energy; consequently, it is formed very rapidly.

17.5 (a) There are two chemically distinguishable allylic positions in *trans*-2-pentene. In each of the allylic radical intermediates that result from reactions at the two positions, the unpaired electron is delocalized to two different carbons. (See Study Problem 17.1.) Reaction at H^a leads to two products, and reaction at H^b leads to only one product because the resonance structures of the intermediate radical are identical (if it is assumed that the double bond retains the more stable *trans* stereochemistry).

products from abstraction of H^a

product from
abstraction of H^b

(c) Because the unpaired electron in the intermediate free radical is delocalized to two chemically nonequivalent carbons, two different products are formed.

17.6 (a) The initially formed Grignard reagent undergoes a rapid allylic rearrangement; each Grignard reagent in the equilibrium can react with D_2O.

17.7 (a) Although there are two chemically nonequivalent types of β-hydrogens in this alkyl halide, an allylic hydrogen is more acidic than a nonallylic one; hence, the product is the conjugated diene derived from elimination of the allylic hydrogen and the bromine.

1,3-cyclohexadiene

17.8 Reaction with concentrated HBr can involve an allylic carbocation intermediate; consequently, a mixture of products, *A* and *B*, could be formed:

Reaction of the tosylate with NaBr in acetone, in contrast, involves an S_N2 reaction; consequently, there is no reactive intermediate, and only product *A* is formed.

17.10 (a) The butyl group is oxidized because it has α-hydrogens; the *tert*-butyl group is not affected.

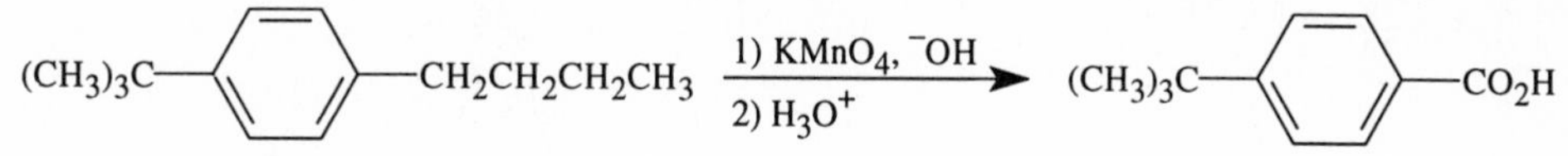

1-butyl-4-*tert*-butylbenzene   ***p*-*tert*-butylbenzoic acid**

17.11 (a) The oxidation product shows that compound *A* is an *ortho*-disubstituted dialkylbenzene. Only compound *A* = *o*-xylene is consistent with this analysis and with the formula. (The structure of *o*-xylene can be found on p. 738 of the text.)

17.12 Caryophyllene is a sesquiterpene because it contains three isoprene units. The isoprene skeletons are shown as heavy bonds.

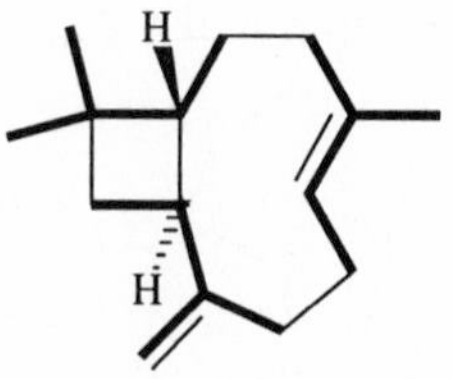

caryophyllene

17.13 (a) A biosynthetic mechanism for limonene is as follows: (B: = a base.)

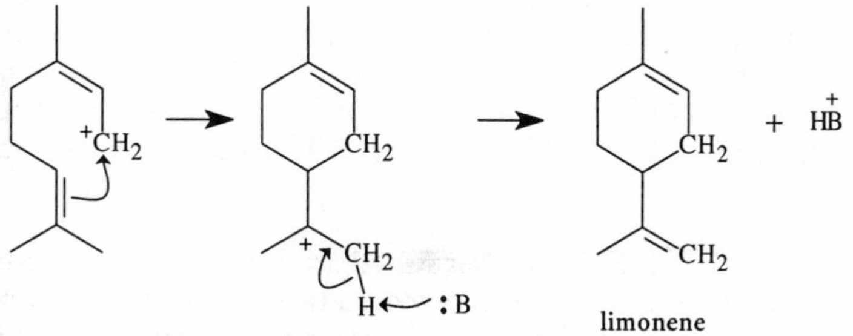

limonene

(b) Ionization of geranyl pyrophosphate is followed by attack of the pyrophosphate anion on the other electron-deficient carbon of the resonance-stabilized carbocation; rotation about a single bond is followed by ionization of pyrophosphate to give the desired carbocation.

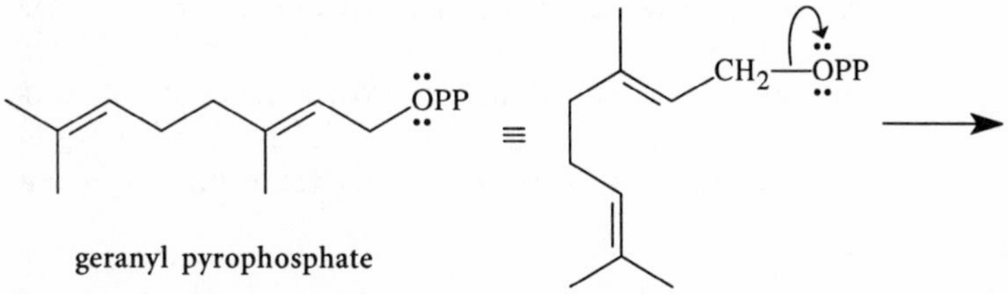

geranyl pyrophosphate

(equation continues)

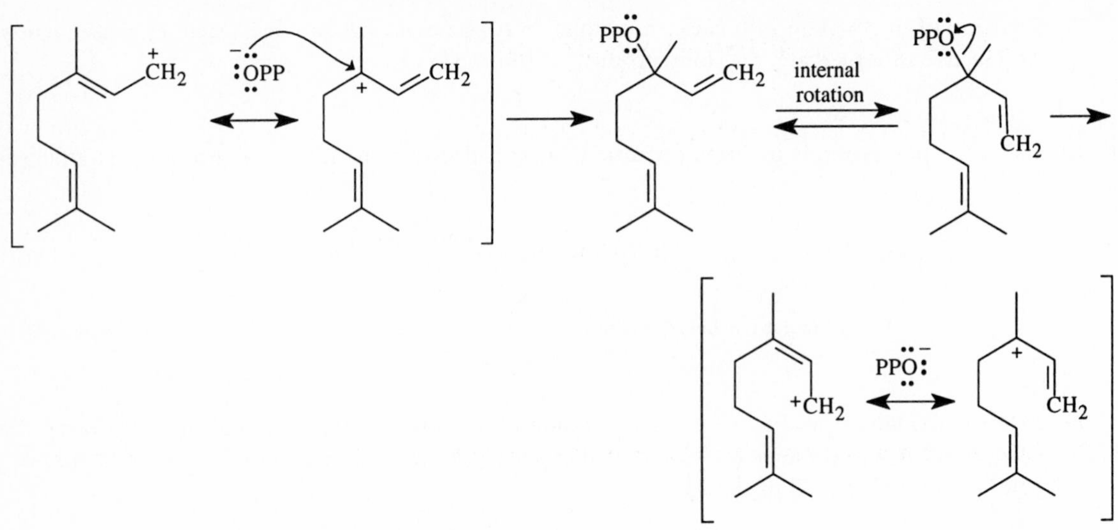

17.14 (a) To see the new bond connections that have to be made, draw α-pinene in a planar projection. The projection on the left shows the molecule as it is in the text; turning this projection clockwise 120° in the plane of the page yields the second projection; and turning this projection over yields the third projection, which can be related more easily to the carbocation in the previous problem.

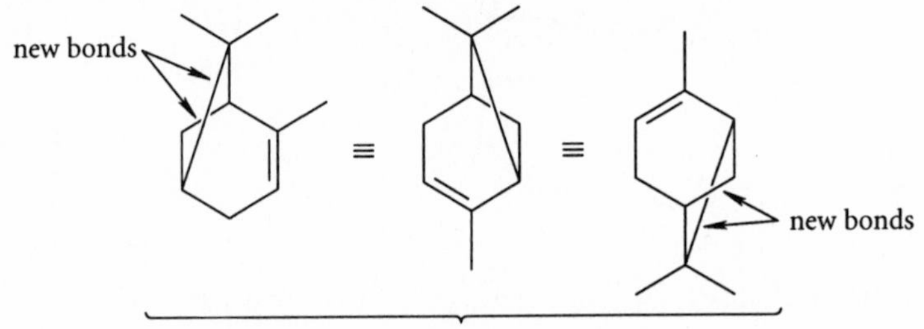

α-pinene

Here is the mechanism:

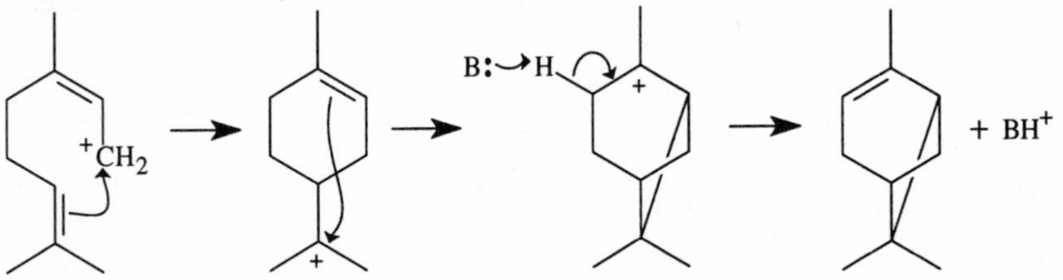

α-pinene

Solutions to Additional Problems

. . . .

17.15 (a) The allylic carbons are indicated with an asterisk (*).

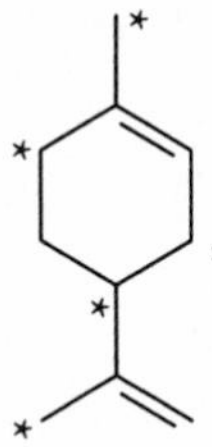

17.16 (a) The benzylic carbons are indicated with an asterisk (*).

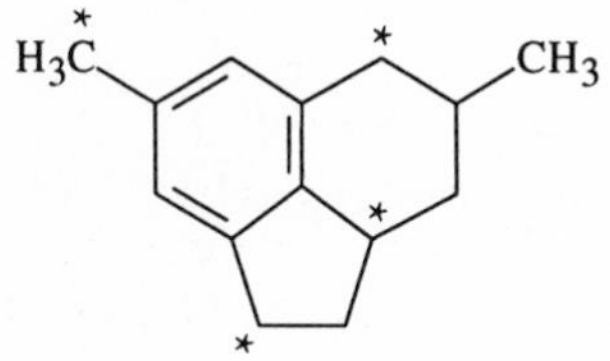

17.17 Note that all chiral compounds are obtained as racemates; the enantiomers are not shown.

(a)   (b)

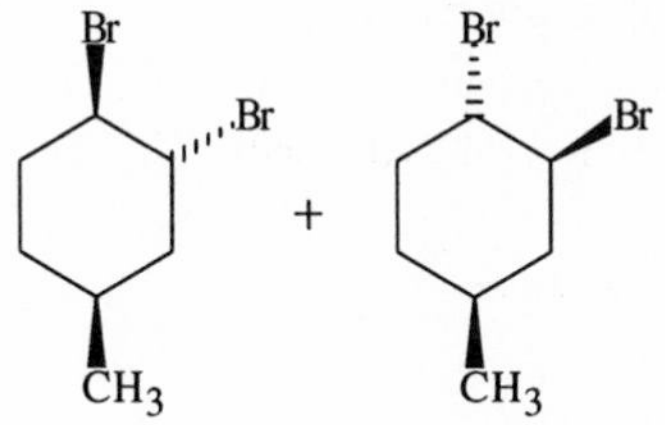

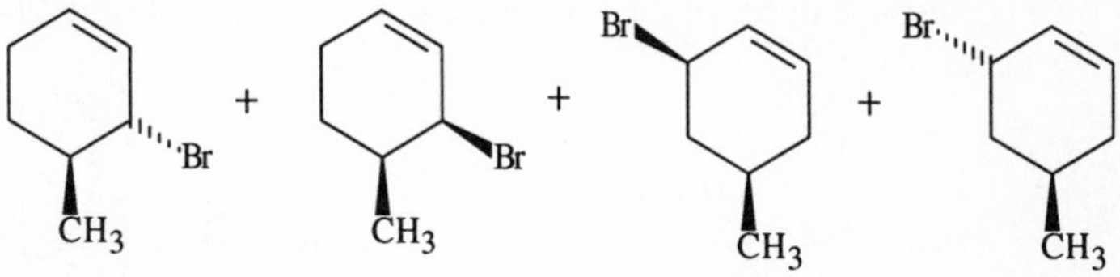

Note that the allylic isomers of the last two compounds in each part (c)–(e) are their enantiomers.

(c)

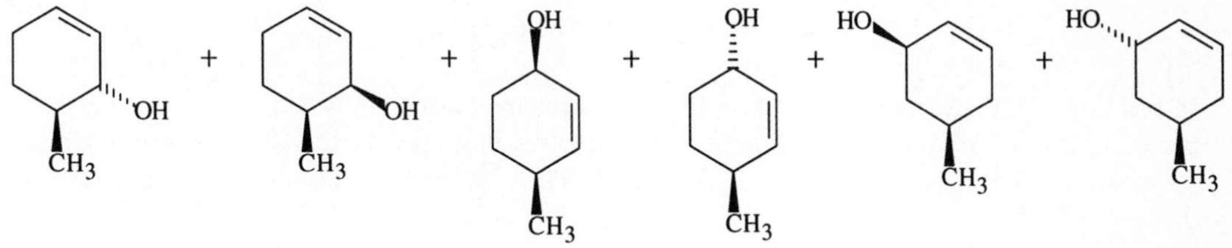

(d)

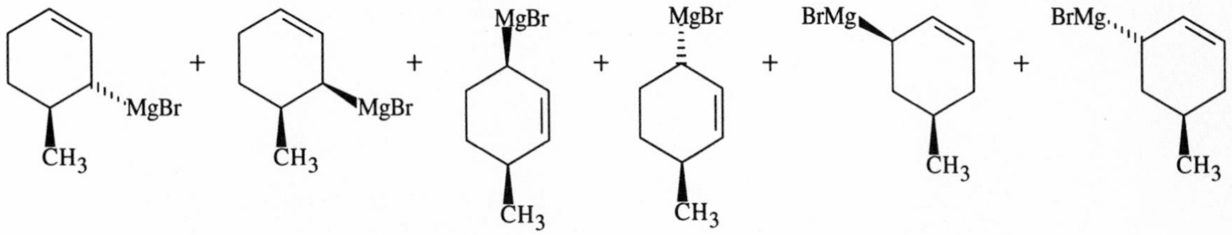

(e)

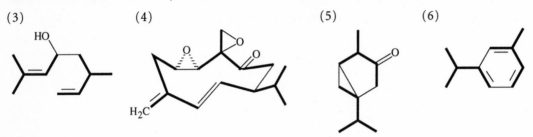

17.19 Compounds (3), (4), (5), and (6) are terpenes. (Compound (2) is believed to be formed from a terpene by a skeletal rearrangement, but its terpene origin is not obvious from its structure.) The isoprene skeletons are shown with heavy bonds.

(3) (4) (5) (6)

17.21 In reaction (1), the acidic conditions promote protonation of the alcohol and ionization to a carbocation, which is trapped by ethanol at the two nonequivalent electron-deficient carbons. (The mechanism, which is discussed in Sec. 11.1C on text pp. 500–501, is outlined below; fill in the curved-arrow formalism if necessary.)

In reaction (2), the carbon-oxygen bond is never broken; consequently, no allylic rearrangement can occur. This is a Williamson ether synthesis that involves formation of the alkoxide anion, which, in turn, is alkylated by ethyl iodide. (See Sec. 11.1A, text pp. 497-498; Problem 11.42 on text pp. 528-529 illustrates the same idea.)

17.23 (a)

$$PhCH_3 \xrightarrow[\text{peroxides}]{NBS, CCl_4} PhCH_2Br \xrightarrow{Mg, ether} PhCH_2MgBr \xrightarrow[2)\ H_3O^+]{1) \triangle} PhCH_2CH_2CH_2OH$$

3-phenyl-1-propanol

(c)

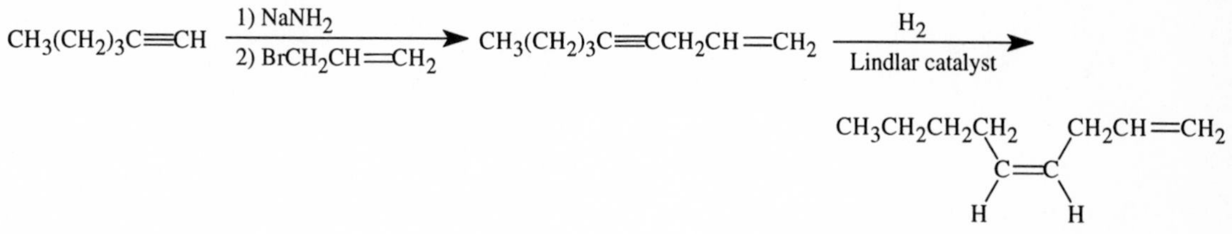

$$CH_3(CH_2)_3C{\equiv}CH \xrightarrow[\text{2) BrCH}_2\text{CH}{=}\text{CH}_2]{\text{1) NaNH}_2} CH_3(CH_2)_3C{\equiv}CCH_2CH{=}CH_2 \xrightarrow[\text{Lindlar catalyst}]{\text{H}_2}$$

(Z)-1,4-nonadiene

(e)

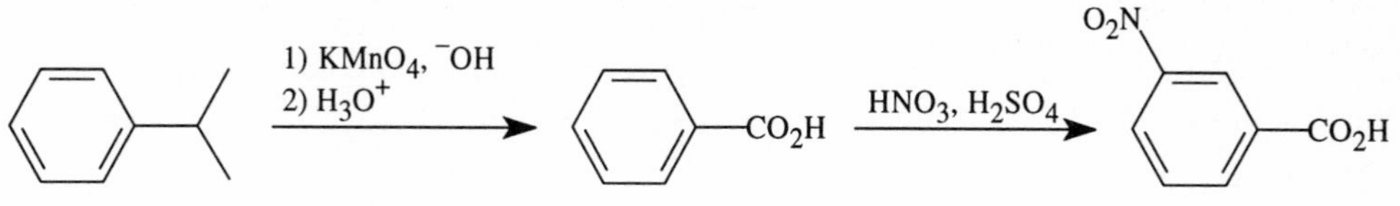

cumene

(g)

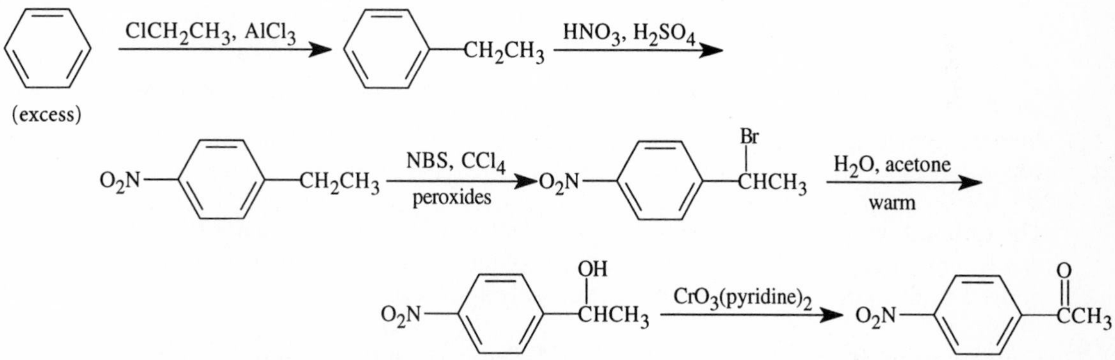

(excess)

17.24 The compounds that give the most stable carbocation intermediates are the ones that undergo the most rapid solvolysis. This problem deals with the effect of substituent on the stability of the carbocation intermediate. The key is to analyze the balance of resonance and polar substituent effects just as you would for electrophilic aromatic substitution. The order of increasing reactivity is (4) < (1) < (3) < (2). Thus, compound (2) reacts most rapidly because the carbocation intermediate is stabilized by the *electron-donating resonance effect* of the *p*-methoxy substituent:

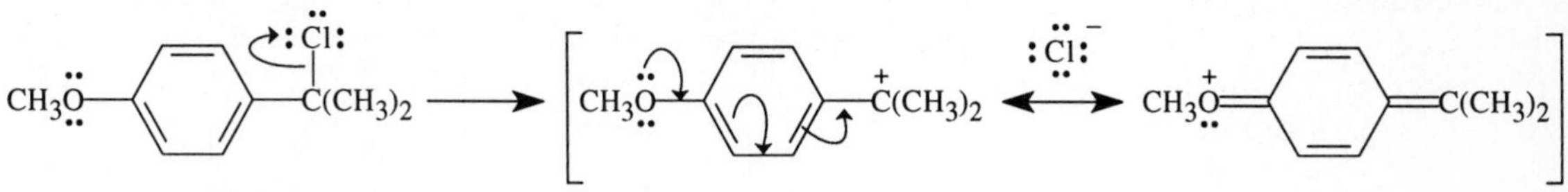

As in electrophilic substitution, the resonance effect of the *p*-methoxy group strongly outweighs its electron-withdrawing polar effect. In compound (3), there is a similar resonance effect; however, the polar effects of halogen substituents outweigh their resonance effects. Consequently, compound (3) reacts more slowly. The nitro group exerts no resonance effect in the carbocation intermediates derived from compounds (1) and (4); the question is then whether its polar effect is stonger from the *meta* or *para* position. As in electrophilic aromatic substitution, a *para*-nitro group destabilizes a carbocation intermediate more than a *meta*-nitro group because, in a *para*-nitro carbocation, positive charge is on adjacent atoms:

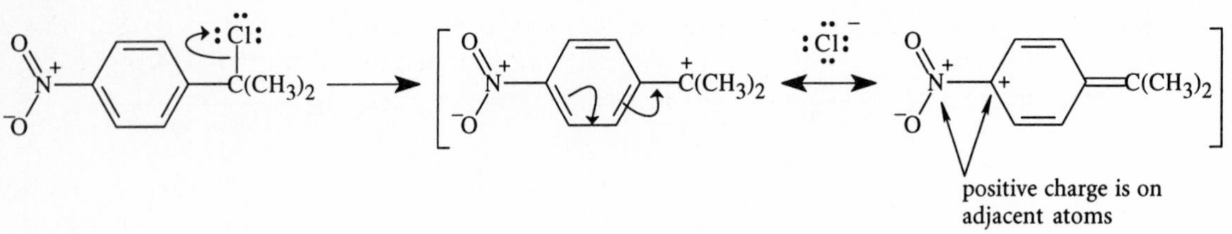

positive charge is on
adjacent atoms

In the *meta*-nitro carbocation, positive charge does *not* reside on adjacent atoms. Consequently, the *meta*-nitro carbocation is more stable than the *para*-nitro carbocation, and *m*-nitro-*tert*-cumyl chloride solvolyzes more rapidly than *p*-nitro-*tert*-cumyl chloride.

17.26   The solution to this problem, like the previous two solutions, hinges on an analysis of the relative stabilities of the carbocation intermediates involved in the S_N1 reactions of the two compounds. The carbocation intermediate in the solvolysis of compound *A* is resonance-stabilized:

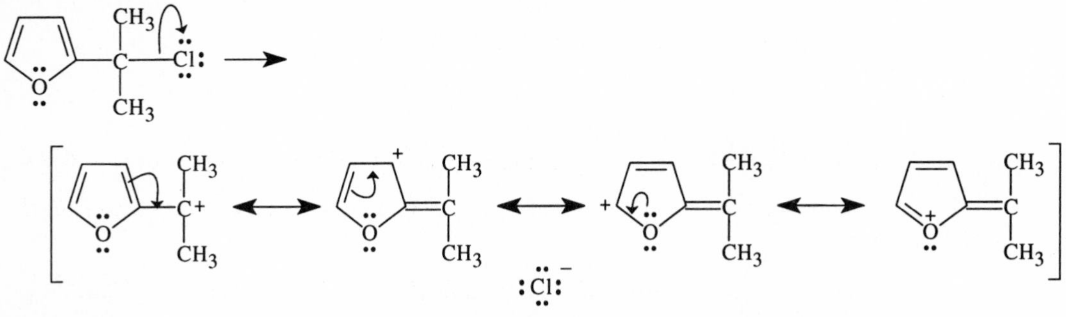

The carbocation intermediate involved in the solvolysis of compound *B* is not resonance-stabilized, and in fact is somewhat destabilized by the polar effect of the oxygen. The greater stability of the carbocation derived from compound *A* results in a greater solvolysis rate.

17.27   Compound *A* has an unsaturation number of 5. Because it ultimately affords phthalic acid, it is an *ortho*-disubstituted benzene derivative. The optical resolution and oxidation results show that compound *C* is a chiral secondary alcohol, and the α-carbon of this alcohol is the asymmetric carbon, because conversion to a ketone destroys its chirality. Compound *C*, in turn, arises from solvolysis of *B*, a product of either allylic or benzylic bromination. Because *A* contains a benzene ring, the formation of compound *B* is probably a benzylic bromination. A by-product of the formation of compound *C* is compound *D*, presumably an alkene, because it can be hydrogenated back to compound *A*. The only way to accommodate all of these data is for the additional degree of unsaturation in compound *A* to be due to a ring. Structures of all compounds that fit these data are as follows:

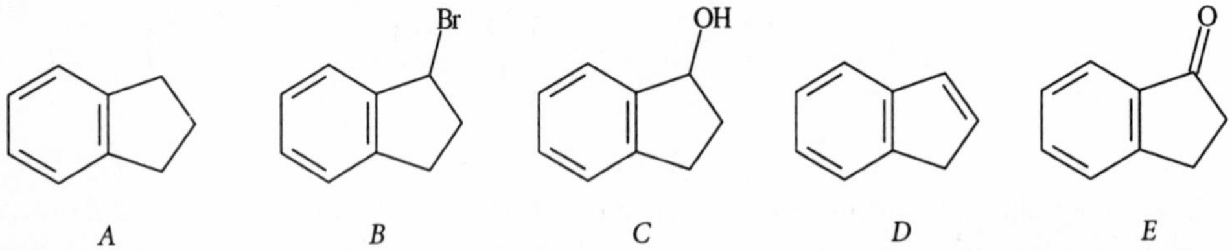

The following structure for compound *A* is ruled out because it would give a *tertiary* alkyl halide, which, in turn, would give a *tertiary* alcohol on solvolysis; a tertiary alcohol could not be oxidized to a ketone.

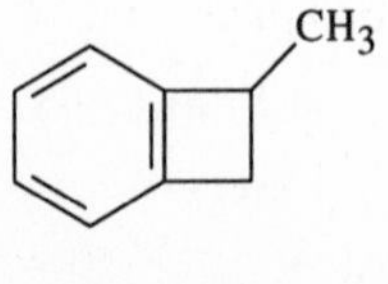

17.29   Alkyl halides with allylic or benzylic β-hydrogens undergo more rapid elimination because the allylic or benzylic hydrogens are more acidic than ordinary hydrogens (See Sec. 17.3B on text p. 802.) Because compound *A* has allylic β-hydrogens and compound *B* does not, the E2 reaction of compound *A* is faster, and this reaction gives the following product:

17.30 (a) Two allylic Grignard reagents are in equilibrium, and each reacts with ethylene oxide:

(Note that a third allylic-rearrangement product of the Grignard reagent is identical to the first.)

(c) This is an S$_N$2 reaction of the ethanethiolate anion; because there is no reactive intermediate, only one product is formed:

$$C_2H_5S—CH_2—CH=CH—CH(CH_3)_2$$

(e) The observed product is derived from substitution of the tertiary benzylic hydrogen. This substitution involves a tertiary benzylic free-radical intermediate, whereas substitution of a *p*-methyl hydrogen would involve a less stable primary benzylic free-readical intermediate.

(g) The benzylic bromination product formed in the NBS reaction undergoes an E2 reaction; note that the β-hydrogens involved are also benzylic.

benzylic bromination
product

(i) This product results from protonation of the alkene at the terminal carbon, because this gives the most stable carbocation intermediate. In this example, the 1,2-addition product not only is the more rapidly formed (kinetic) product but also is the more stable (thermodynamic) product. It is more stable thermodynamically because it is the only product that has a double bond conjugated with the benzene ring.

(j) All benzylic carbons—even the ones in the ring—are oxidized to carboxylic acid groups.

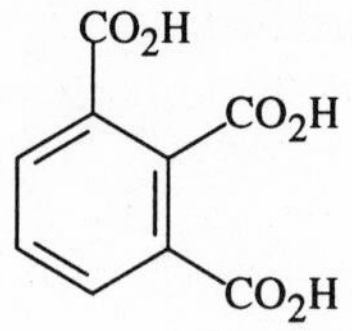

17.31 The greater λ_{max} value indicates the formation of a species with more extensive conjugation. Protonation of the hydroxy group and loss of water gives this species, the benzyl cation. Addition of $^-$OH regenerates benzyl alcohol.

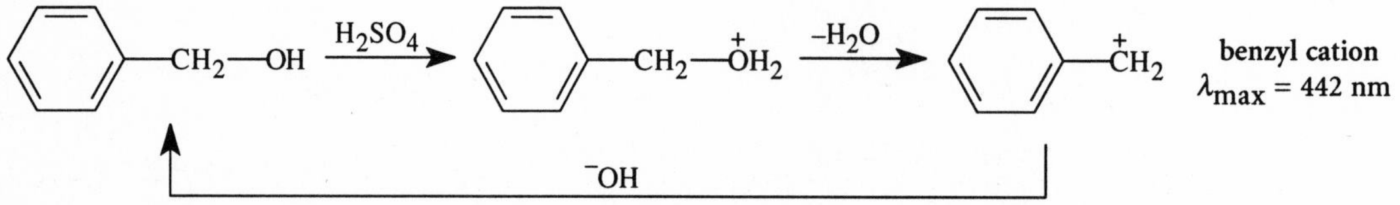

17.32 (a) First analyze the relationship of the isoprene skeletons. Then use steps like the ones shown in Eqs. 17.36–17.38 on text pp. 810–811 to assemble the parts from IPP and DMAP. Begin with the carbocation whose biosynthesis is shown in the solution to Problem 17.13(b). (B = a base.)

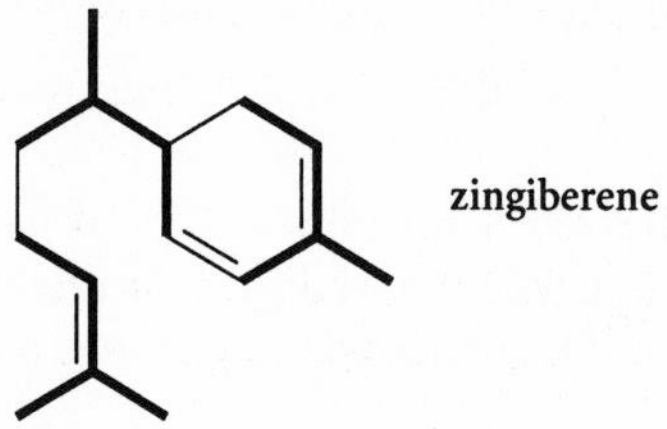

zingiberene

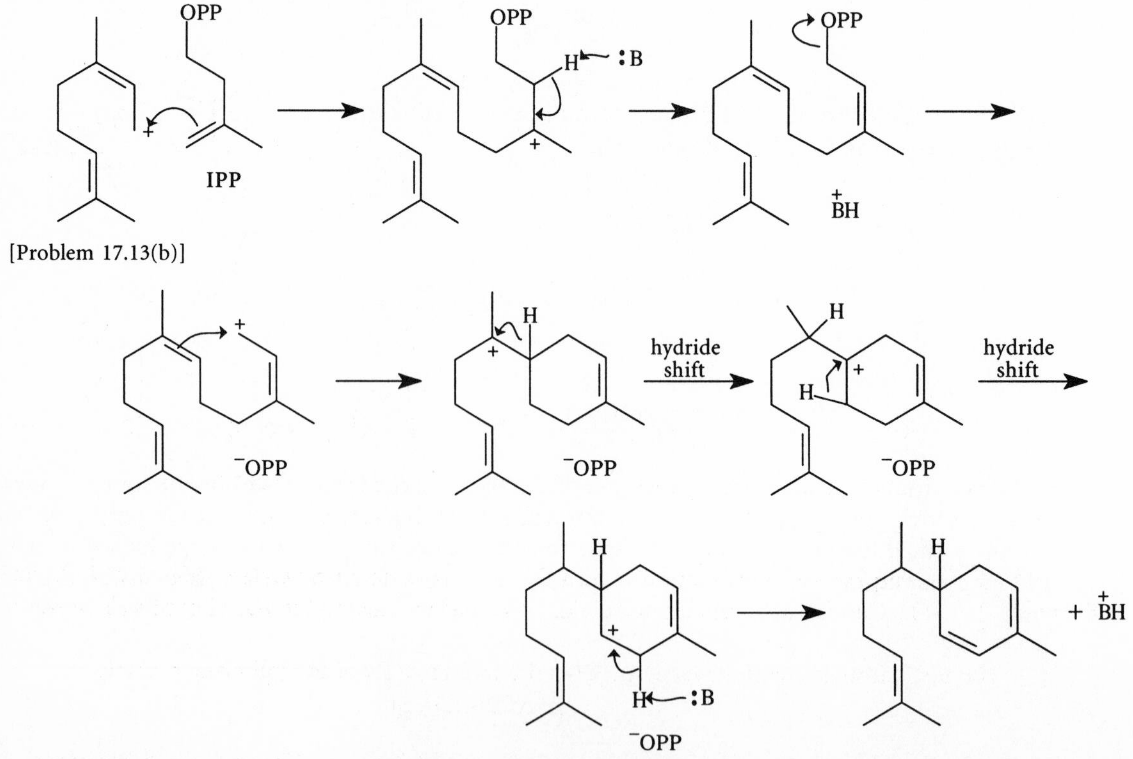

IPP

[Problem 17.13(b)]

$^+$BH

$^-$OPP

hydride shift

$^-$OPP

hydride shift

$^-$OPP

H$\longleftarrow$:B

$^-$OPP

$+\ ^+$BH

In the foregoing mechanism there are some interesting hydride shifts. One converts a tertiary carbocation into a less stable secondary carbocation, and the other interconverts two secondary carbocations. These can be thought of as equilibria that are pulled to the right by the formation of the conjugated diene product. It is also possible that the enzyme(s) that catalyze these transformations use basic groups to form covalent bonds to the secondary carbocations, so that they are more stable than "bare" cations. Whatever the reasons, such tertiary-to-secondary hydride shifts are actually rather common in terpene biosynthesis.

17.33 (a) A hydrogen on the central carbon is more acidic than an acetylenic hydrogen because the conjugate-base anion resulting from removal of the central hydrogen is both allylic and propargylic, and is therefore doubly resonance-stabilized. The conjugate-base anion is alkylated by allyl bromide.

(c) The carbocation formed by protonation of the —OH group and loss of water adds to a neighboring double bond to give a new carbocation; this carbocation adds to a neighboring double bond to give yet another carbocation; and so on, until the product is formed by attack of water on the last carbocation. The mechanism below begins with the protonated alcohol:

 As you can see, the reaction in part (c) forms a steroid ring system. (See Sec. 7.6D on text p. 300.) Conceptually similar cyclization reactions are involved in the biosynthesis of steroids.

(e) The formation of a cyclopropane suggests the intermediacy of the following carbene:

$$(CH_3)_2C=C=C:$$

This carbene is formed by loss of chloride ion from an acetylenic ion. The carbene then adds to cyclohexene to give the cyclopropane. (See text pp. 428–429.)

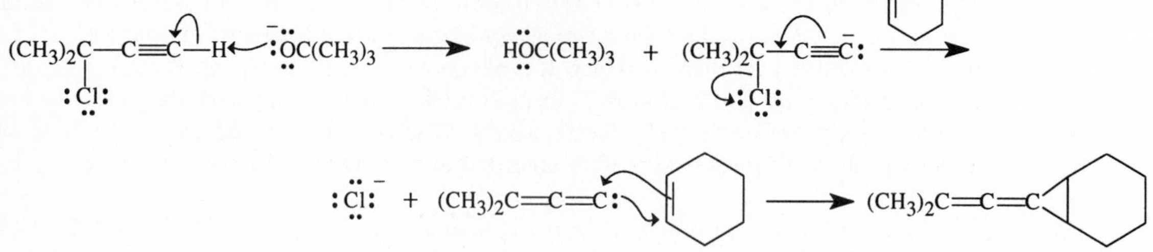

$$(CH_3)_2C-C{\equiv}C-H \;\; :\ddot{O}C(CH_3)_3 \longrightarrow H\ddot{O}C(CH_3)_3 \;+\; (CH_3)_2C-C{\equiv}C: \longrightarrow$$
$$\underset{:\ddot{C}l:}{|} \hspace{6cm} \underset{:\ddot{C}l:}{|}$$

$$:\ddot{C}l:^- \;+\; (CH_3)_2C{=}C{=}C: \hspace{2cm} \longrightarrow \;\; (CH_3)_2C{=}C{=}C$$

17.34 (a) Although the conjugate-base anion of 1,4-pentadiene is doubly allylic and resonance-stabilized, the conjugate-base anion of 1,3-cyclopentadiene is in addition aromatic. (See text p. 722 for a discussion of this case.) Consequently, much less energy is required for the ionization of 1,3-cyclopentadiene, and its pK_a is therefore much lower. (The pK_a difference between these two compounds is estimated to be about five units.)

17.35 Formation of a resonance-stabilized propargylic anion (that is, an anion that is "allylic" to a triple bond) is followed by protonation at the internal carbon. Formation of another resonance-stabilized anion followed by protonation gives the 1-alkyne, which is the most acidic species in the reaction mixture. Consequently, its conjugate-base anion is formed irreversibly, and all equilibria are thus pulled toward this anion. The net result is migration of the triple bond to the end of the carbon chain.

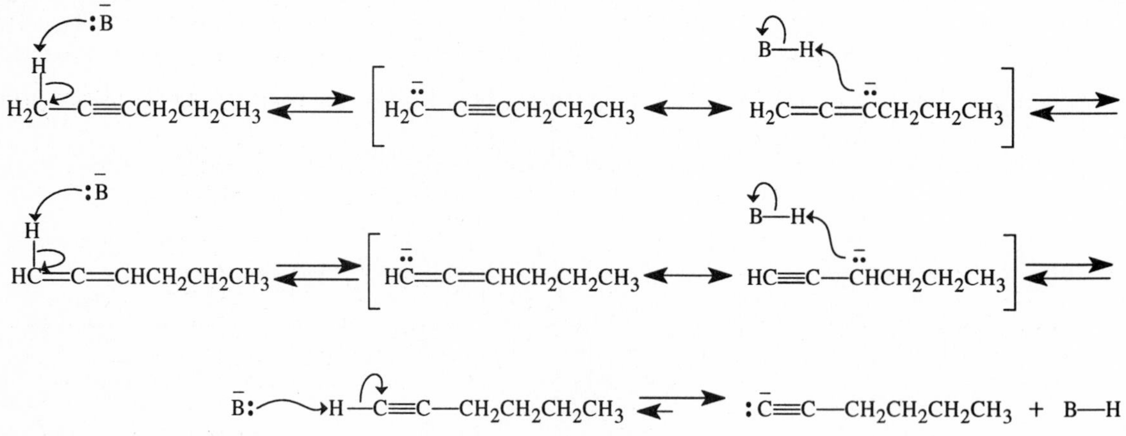

$$H_2C-C{\equiv}CCH_2CH_2CH_3 \rightleftharpoons \left[H_2\ddot{C}-C{\equiv}CCH_2CH_2CH_3 \longleftrightarrow H_2C{=}C{=}\ddot{C}CH_2CH_2CH_3 \right] \rightleftharpoons$$

$$HC{=}C{=}CHCH_2CH_2CH_3 \rightleftharpoons \left[H\ddot{C}{=}C{=}CHCH_2CH_2CH_3 \longleftrightarrow HC{\equiv}C-\ddot{C}HCH_2CH_2CH_3 \right] \rightleftharpoons$$

$$B:\; H-C{\equiv}C-CH_2CH_2CH_2CH_3 \rightleftharpoons \;:\bar{C}{\equiv}C-CH_2CH_2CH_2CH_3 \;+\; B-H$$

Addition of water to the reaction mixture gives the 1-alkyne. Thus, the reaction shown here followed by addition of water results in the migration of a triple bond from an interior position to the 1-position of a carbon chain provided there are no branches. It has been shown that the triple bond migrates with equal frequency to both ends of the chain.

You might ask why a 1-alkyne is more acidic than a propargylic hydrogen when the conjugate-base anion of the latter is resonance-stabilized, and an acetylenic anion is not. This is a reflection of the effect of hybridization on acidity discussed on text p. 664. A *doubly allylic* hydrogen is more acidic than an acetylenic hydrogen [see Problem 17.33(a)], but a singly allylic hydrogen is not.

17.37 From the structure of the Diels-Alder product *C*, compound *B* must be the conjugated diene 2-chloro-1,3-butadiene. Because compound *B* is an allylic rearrangement product of compound *A*, compound *A* must be the allene 4-chloro-1,2-butadiene. (Note that 1-chloro-1,3-butadiene would not give the same Diels-Alder product.)

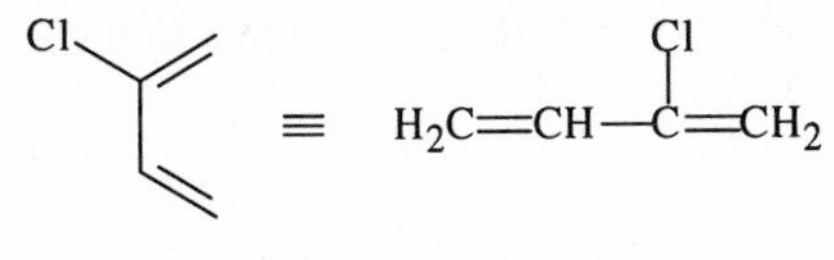
Cl
Cl
H2C=CH—C=CH2

2-chloro-1,3-butadiene
compound *B*

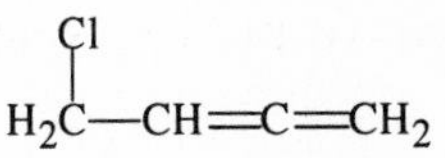
Cl
H2C—CH=C=CH2

4-chloro-1,2-butadiene
compound *A*

17.38 (a) The stability of the trityl radical is due to delocalization of the unpaired electron into all three benzene rings.

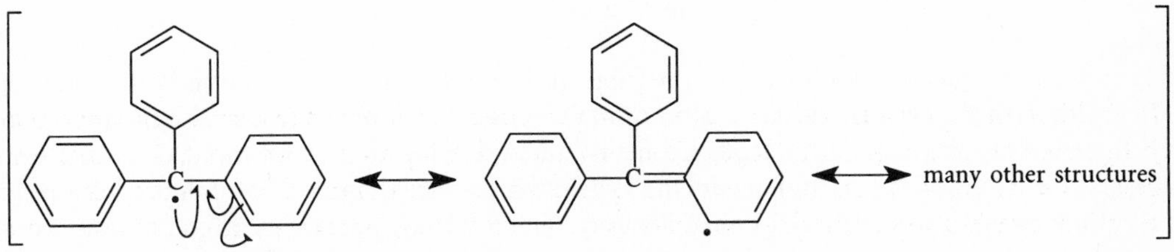
C
C
many other structures

(b) Hexaphenylethane could be formed by the recombination of two trityl radicals:

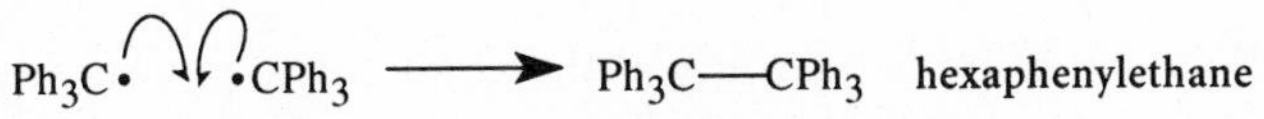
Ph3C•
•CPh3
Ph3C—CPh3
hexaphenylethane

(c) The dimer of the trityl radical is formed by the following mechanism:

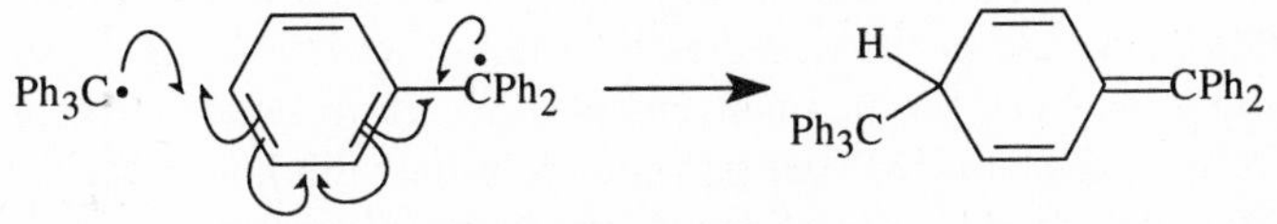
Ph3C•
CPh2
H
Ph3C
CPh2

Hexaphenylethane is destabilized by van der Waals repulsions between *gauche* pairs of phenyl groups.

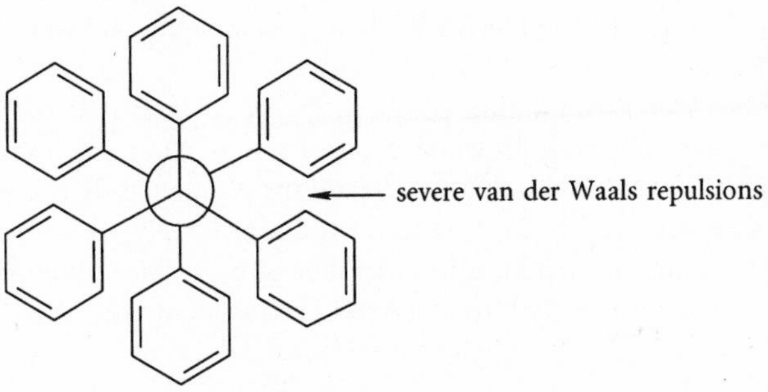
severe van der Waals repulsions

This destabilization is evidently so great, and the drive for recombination so powerful, that aromatic stability in one of the benzene rings can be sacrificed in forming the dimer shown above and in the problem.

17.39 (a) The triphenylmethyl anion (trityl anion) is stabilized by resonance interaction with all three benzene rings:

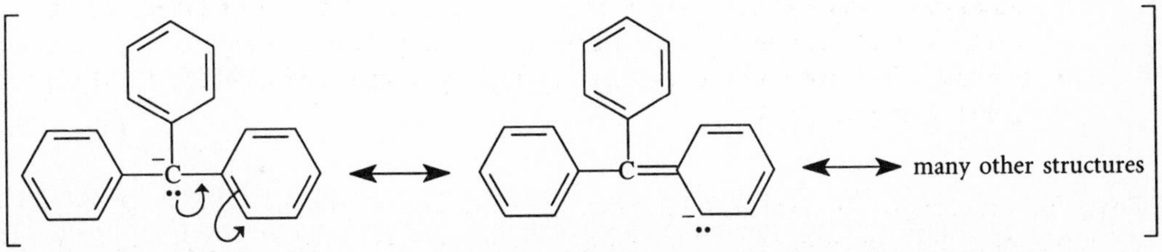
C
C
many other structures

(b) The resonance interaction shown in part (a) is optimized if the three rings are coplanar; when the rings are coplanar, there is optimum overlap between the *p* orbitals of the central carbon and those of the rings. However, when the rings are coplanar, there are significant van der Waals repulsions between hydrogens on different rings:

These van der Waals repulsions force the rings in the triphenylmethyl anion to twist significantly away from coplanarity, with a resulting cost in orbital overlap. (The molecule resembles a molecular "propeller.") In other words, in the triphenylmethyl anion, conjugation is not so effective as it would be if the rings were all coplanar. In fluoradene, the hydrogens that are the source of these repulsions are replaced by bonds that constrain the rings to coplanarity. The resulting increase in orbital overlap brings about a concomitant increase in stabilization of the anion. As a result, the pK_a of the conjugate acid is lowered significantly relative to that of triphenylmethane.

17.40   Section 7.9C shows that *anti* stereochemistry is one of the major pieces of evidence that a bromonium ion is involved in bromine addition to alkenes; similar conclusions hold for chlorine-addition reactions. The text also shows that a mechanism involving carbocation intermediates predicts mixed *syn* and *anti* addition (Eqs. 7.43a–b, text p. 316). This problem deals with the *competition* between the two mechanisms: one involving a cyclic *chloronium ion* intermediate and the other involving a carbocation intermediate. The loss of stereoselectivity is evidence for the involvement of a carbocation intermediate. When the carbocation intermediate is resonance-stabilized (last two entries in the table), it is stable enough to compete energetically with the chloronium ion as a reactive intermediate. When R = *p*-methoxyphenyl, the carbocation intermediate is further resonance-stabilized by the *para* substituent. (See, for example, Eq. 17.6 on text p. 792.) In this case, the carbocation intermediate is stable enough, and the carbocation mechanism thus important enough, that nearly complete loss of stereoselectivity is observed.

17.41   Compound *A* is a benzylic bromination product, and it appears that meperidine and compound *B* are S_N1 products, each derived respectively from reaction of one of the nucleophiles present with a carbocation intermediate derived from *A*. Recall that S_N1 reactions are generally accompanied by E1 reactions; hence, it is reasonable to suppose that MPTP is such a product, namely, an alkene. The molecular formula of MPTP, $C_{12}H_{15}N$, indicates an unsaturation number of 6. Because five unsaturations are accounted for by the two rings in the starting material, the formula is consistent with the hypothesis that MPTP is an alkene.

17.42   (a) Fact (1) indicates that the transition state involves a molecule of $(C_2H_5)_2NH$ and a molecule of alkyl halide. Fact (2) indicates that the indicated starting materials undergo the reaction shown; had this not been established, one could have postulated that one of the starting alkyl halides is converted into the other by the reaction conditions, and that the observed product could have originated from only one of the two starting compounds. Fact (3) establishes that the observed product is the actual product of the reaction, and is not derived from a subsequent reaction of its allylic isomer.

Now to the mechanisms. In the reaction of the second alkyl chloride an allylic rearrangement has occurred. Such allylic rearrangements generally suggest the involvement of carbocation intermediates, which have two resonance forms, and can be attacked at two different carbons. However, the observation of second-order kinetics rules out an S_N1 reaction. Hence, the reaction is evidently a direct substitution at the *allylic carbon*, thus:

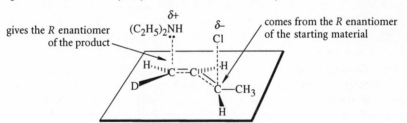

This is the so-called S_N2' mechanism. The other isomer reacts by the conventional S_N2 mechanism, in which the nucleophile attacks at the carbon bearing the halide.

(b) The large size of the nucleophile dictates that its attack occurs at the unbranched carbon. In one case, this is the allylic carbon; in the other, it is the α-carbon.

17.43 Imagine that, in the transition state of the S_N2' reaction, the three carbons that either start or finish the reaction as part of a double bond, as well as the atoms attached to these carbons, lie in a common plane with the nucleophile and the leaving group either above or below this plane. Placing all groups with their stereochemical configurations corresponding to those of the reactant and the first product shows that approach of the nucleophile and departure of the leaving group occur from the *same face* of the plane. Hence, the major product is formed by a *syn* substitution, that is, a *frontside* substitution.

gives the R enantiomer of the product — $(C_2H_5)_2NH$ ... comes from the R enantiomer of the starting material

Note that the product is a *trans*-alkene with *R* stereochemistry at the asymmetric carbon. The minor product, a *cis*-alkene with *S* stereochemistry, also comes from a *syn* substitution in which attack of the nucleophile and loss of the leaving group both occur from the lower face of the same plane. In order for this to happen, the transition state must adopt a conformation that leads to the *cis*-alkene. (You should demonstrate this to yourself with a diagram like the one above.) Because *trans*-alkenes are more stable than *cis*-alkenes, the "transoid" transition state—the one shown above—is more stable and accounts for 95% of the reaction. However, the point is that *both* products result from *syn* substitution. You should show that if the nucleophile and leaving group were situated on opposite faces, the *R* enantiomer of the product would have a *cis* double bond, and that the *S* enantiomer of the product would have a *trans* double bond, a result not consistent with the observations given in the problem. The *syn* stereochemistry observed in the S_N2' reaction contrasts with the stereochemistry of the S_N2 reaction, which is an *anti* substitution; that is, attack of the nucleophile and loss of the leaving group occur from opposite sides of the molecule. (See Fig. 9.2 on text p. 397; note the relationship of the nucleophile and the leaving group with respect to the plane shown in color.)

18

Chemistry of Aryl Halides, Vinylic Halides, and Phenols

Terms

Concepts

I. Aryl Halides and Their Unique Reactivity Under S_N1-E1, S_N2, and E2 Conditions

A. INTRODUCTION

1. Aryl halides are compounds in which a halogen is bound to the carbon of a benzene ring (or other aromatic ring)—benzylic groups are on the carbon *adjacent* to an aromatic ring.
2. Vinylic halides are compounds in which a halogen is bound to a carbon of a double bond—allylic groups are on the carbon *adjacent* to the double bond.
3. The reactivity of an aryl or vinylic halide is quite different from that of an ordinary alkyl halide.

B. LACK OF REACTIVITY IN S_N1 AND E1 REACTIONS

1. Vinylic and aryl halides are virtually inert to the conditions that promote the S_N1 or E1 reactions of alkyl halides.
2. To undergo S_N1 or E1 reactions, vinylic halides must ionize to form vinylic cations.
   a. A vinylic cation is a carbocation in which the electron-deficient carbon is part of a carbon-carbon double bond.
      i. The geometry at a vinylic cation carbon is linear.

     ii. The electron-deficient carbon is *sp*-hybridized.

     iii. The vacant *p* orbital is not conjugated with the π-electron system of the double bond.

   b. Vinylic cations are considerably less stable than alkyl carbocations because of their *sp* hybridization and because of the electron-withdrawing polar effect of the double bond.

   c. Carbon-halogen bonds are stronger in vinylic halides than they are in alkyl halides.

     i. A vinylic carbon-halogen bond has more *s* character.

     ii. It takes more energy to break the carbon-halogen bond of a vinylic halide; this energy is reflected in a smaller rate of ionization.

3. To undergo S_N1 or E1 reactions, aryl halides must ionize to aryl cations.

   a. An aryl cation is a carbocation in which the electron-deficient carbon is part of an aromatic ring.

     i. The aryl cation prefers a linear geometry which is virtually impossible to achieve because it would introduce too much strain in the six-membered ring.

     ii. The vacant orbital cannot become a *p* orbital and must remain an sp^2 orbital.

   b. Because an aryl cation is forced to assume a nonoptimal geometry and hybridization, it has a very high energy.

   c. The electron-withdrawing polar effect of the ring double bonds also destabilize an aryl cation.

C. LACK OF REACTIVITY IN S_N2 REACTIONS

1. Vinylic and aryl halides are inert under S_N2 conditions.

2. Two factors within the transition state retard the S_N2 reactions of vinylic halides to such an extent that S_N2 reactions do not occur.

   a. Hybridization:

     i. In reaching the transition state, the carbon in the carbon-halogen bond must be rehybridized from sp^2 to *sp*.

     ii. The relatively high energy of the transition state caused by *sp* hybridization reduces the rate of S_N2 reactions of vinylic halides.

   b. van der Waals repulsions (steric effects):

     i. The attacking nucleophile must approach the vinylic halide at the backside of the halogen-bearing carbon and in the plane of the alkene.

     ii. The resulting van der Waals repulsions raise the energy of the transition state and lower the reaction rate.

3. S_N2 reactions of aryl halides have the same problems as those of vinylic halides; in addition:

   a. Backside attack on the carbon-halogen bond would place the nucleophile on a path that goes through the plane of the benzene ring.

   b. Since the carbon that is attacked undergoes a stereochemical inversion, the reaction would necessarily yield a twisted and highly strained benzene derivative.

D. ELIMINATION REACTIONS OF VINYLIC HALIDES

1. Base-promoted β-elimination reactions of vinylic halides do occur and can be useful in the synthesis of alkynes.

2. Vinylic eliminations require rather harsh conditions, and some of the more useful examples of this reaction involve elimination of β-hydrogens that have enhanced acidity.

II. *Nucleophilic Substitution Reactions of Aryl Halides*

A. NUCLEOPHILIC AROMATIC SUBSTITUTION

1. Aryl halides that have one or more strongly electronegative groups (especially nitro groups) *ortho* or *para* to the halogen undergo nucleophilic aromatic substitution reactions under relatively mild conditions.

2. A nucleophilic aromatic substitution reaction is a substitution that occurs at a carbon of an aromatic ring by a nucleophilic mechanism.

   a. It involves a nucleophile and a leaving group.

   b. It obeys a second-order rate law.

   c. It does *not* involve a one-step backside attack.

   d. It is faster when there are electron-withdrawing groups such as nitro groups *ortho* and/or *para* to the halogen leaving group that can stabilize the anionic intermediate by resonance.

  e. The effect of the halogen on the rate of this type of reaction is quite different from that in the S_N1 or S_N2 reaction of alkyl halides.

2. The nucleophilic aromatic substitution mechanism involves:

  a. Attack of a nucleophile (Lewis base) on the π-electron system above (or below) the plane of the aromatic ring to yield a resonance-stabilized anion called a Meisenheimer complex.

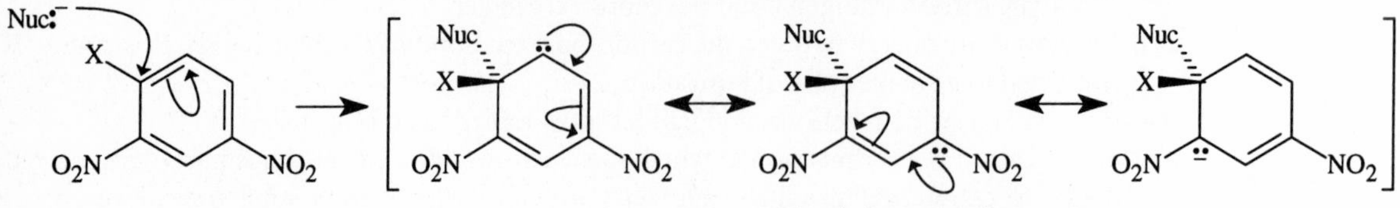

Meisenheimer complex

   i. The negative charge is delocalized throughout the π-electron system of the ring.

   ii. Formation of this anion is the rate-limiting step in many nucleophilic aromatic substitution reactions.

   iii. The negative charge in this complex is also delocalized into an *ortho* or *para* nitro group if present.

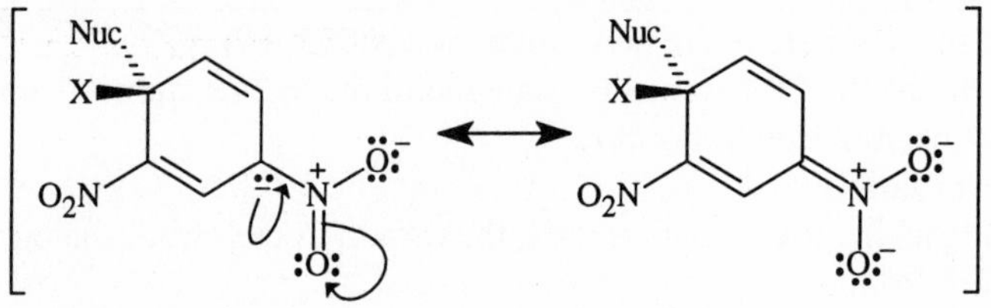

  b. The Meisenheimer complex breaks down to products by loss of the halide ion.

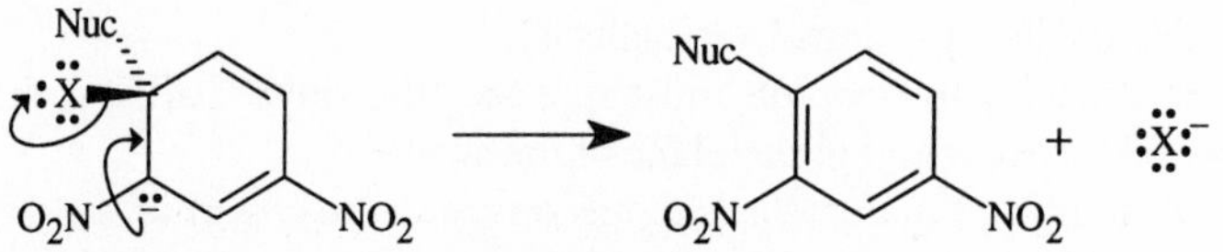

3. *Ortho* and *para* nitro groups accelerate the reaction because:

  a. the transition state resembles the Meisenheimer complex

  b. the nitro groups stabilize this complex by resonance.

4. In nucleophilic aromatic substitution reactions, aryl fluorides are most reactive.

  a. The reactivity of aryl halides is

$$Ar{-}F \gg Ar{-}Cl \approx Ar{-}Br \approx Ar{-}I$$

  b. Fluorine stabilizes the negative charge by its electron-withdrawing polar effect.

  c. Because the loss of halide is not rate-limiting, the basicity of the halide, or equivalently, the strength of the carbon-halogen bond, is not important in determining the reaction rate.

5. The nucleophilic aromatic substitution reaction is an overall frontside substitution; it requires no inversion of configuration.

B. Substitution by Elimination-Addition: Benzyne

1. β-Elimination of an aryl halide gives a very unstable "alkyne" called benzyne.

  a. One of the two π bonds in the triple bond of benzyne is perpendicular to the π-electron system of the aromatic ring.

  b. The orbitals from which benzyne is formed are more like sp^2 orbitals than p orbitals.

  c. Benzyne is highly strained.

  d. Benzyne is a reactive intermediate in certain reactions of aryl halides.

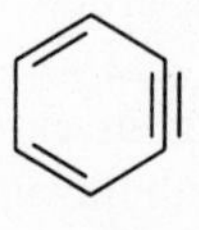

benzyne

2. The mechanism of this reaction involves:
   a. Formation of an anion at the *ortho* position; this step requires a strong base because benzene derivatives are only weakly acidic.
   b. Expulsion of a halide ion completes the β-elimination reaction and forms benzyne.

   c. Attack by amide ion ($^-NH_2$) can occur at either carbon of the triple bond to give a new anion. If the benzyne intermediate is unsymmetrical, two different anions can form.

   d. Protonation of the two anions gives a mixture of products.

   e. The overall substitution is really an elimination-addition process.
3. Substitution at a site different from the one occupied by the leaving group is called *cine* substitution.

III. Phenols

A. INTRODUCTION
1. Phenols are compounds in which a hydroxy (—OH) group is bound to a benzene ring.
2. Phenols, like alcohols, can ionize.
3. Phenols are much more acidic than alcohols.
   a. The enhanced acidity of phenols is due to stabilization of its conjugate-base anion.
   b. The conjugate base of a phenol is named as a phenoxide ion (common nomenclature) or as a phenolate ion (substitutive nomenclature).
   c. Phenolate anions are stabilized by resonance.
   d. The polar effect of the benzene ring stabilizes the negative charge.
   e. Since pK_a is directly proportional to free energy of ionization, phenols have lower pK_a values, and are thus more acidic, than alcohols.
4. Phenols are completely converted into their conjugate-base anions by NaOH solution.
   a. Alkali metal salts of phenols have considerable solubility in water because they are ionic compounds.
   b. Acidification of an aqueous phenolate solution gives the phenol; if it is water-insoluble, it separates from the solution because, after acidification, it is no longer ionized.
   c. The acidities of water-insoluble phenols can sometimes be used to separate them from mixtures with other organic compounds.

B. OXIDATION OF PHENOLS TO QUINONES

1. Phenols undergo oxidation to quinones; quinones are compounds having two oxygens doubly bonded to a cyclohexadiene.
   a. If the quinone oxygens have a 1,2 (*ortho*) relationship, the quinone is called an *ortho*-quinone.
   b. If the quinone oxygens have a 1,4 (*para*) relationship, the quinone is called a *para*-quinone.

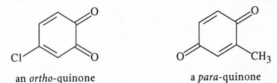

an *ortho*-quinone   a *para*-quinone

   c. *Ortho*-quinones are considerably less stable than the isomeric *para*-quinones because the C=O bond dipoles are nearly aligned, and therefore have a repulsive, destabilizing interaction.
2. Hydroquinone and other phenols are sometimes used to inhibit free-radical reactions that result in the oxidation of other compounds.
   a. Many free radicals abstract a hydrogen from hydroquinone to form a very stable radical called a semiquinone; the semiquinone radical is resonance-stabilized.
   b. A second free radical can react with the semiquinone to complete its oxidation to quinone.
   c. Hydroquinone terminates free-radical chain reactions by intercepting free radical intermediates R· and reducing them to R—H.
3. Oxidation of phenols is a key element in the operation of some food preservatives and photographic developers.

C. LACK OF REACTIVITY OF THE ARYL-OXYGEN BOND

1. Phenols and phenyl ethers do not react under conditions used for S_N2 reactions of alcohols and their derivatives for the same reasons that aryl halides do not react (see Sec. 18.1).
2. Phenols also do not react under conditions used for S_N1 or E1 reactions of alcohols and their derivatives for the same reasons that aryl halides do not react (see Sec. 18.3).

Reactions

I. Reactions of Aryl and Vinyl Halides

A. ELIMINATION REACTIONS OF VICINAL HALIDES

1. Base-promoted β-elimination reactions of vinyl halides do occur and can be useful in the synthesis of alkynes.

$$Ph-CH=\overset{\overset{\displaystyle Br}{|}}{C}-Ph \xrightarrow[\text{(CH}_3)_3\text{COH}]{\text{(CH}_3)_3\text{CO}^-\text{ K}^+} Ph-C\equiv C-Ph + K^+Br^-$$

B. ARYL AND VINYLIC GRIGNARD REAGENTS

1. Preparation of Grignard and organolithium reagents from aryl and vinylic halides is analogous to the corresponding reactions of alkyl halides.
2. Arylmagnesium bromides or iodides can be prepared from the corresponding bromo- or iodobenzenes in either tetrahydrofuran (THF) or diethyl ether solvent; formation of arylmagnesium chlorides from chlorobenzenes requires the use of THF.
3. The preparation of vinylic Grignard reagents also requires THF as a solvent.

C. ELIMINATION REACTIONS OF ARYL HALIDES

1. Strong bases can promote β-elimination in aryl halides to give benzyne intermediates, which are rapidly consumed by addition reactions with the bases; a mixture of products is usually observed.

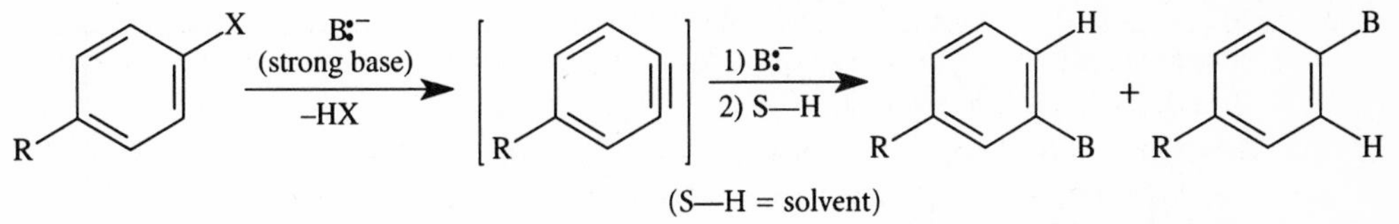

(S—H = solvent)

II. Reactions of Phenols and Phenolates

A. USE OF PHENOXIDES AS NUCLEOPHILES

1. Phenoxides can be used as nucleophiles.
2. Aryl ethers can be prepared by the reaction of a phenoxide anion and an alkyl halide (another example of the Williamson ether synthesis).

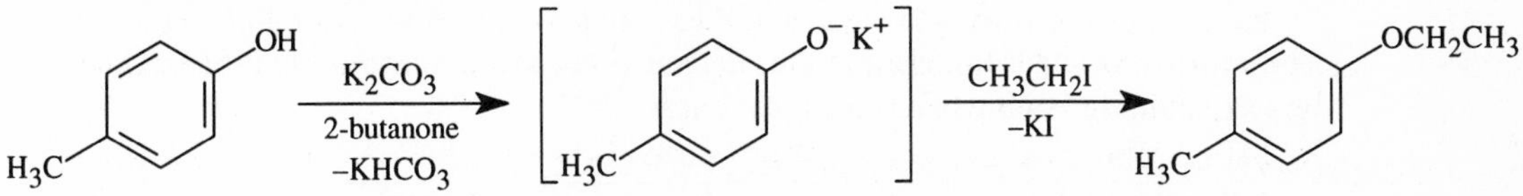

B. ELECTROPHILIC AROMATIC SUBSTITUTION REACTIONS OF PHENOLS

1. Phenols undergo electrophilic aromatic substitution reactions.
2. Phenols react rapidly with bromine.
   a. Phenol can be halogenated once under mild conditions that do not affect benzene itself.

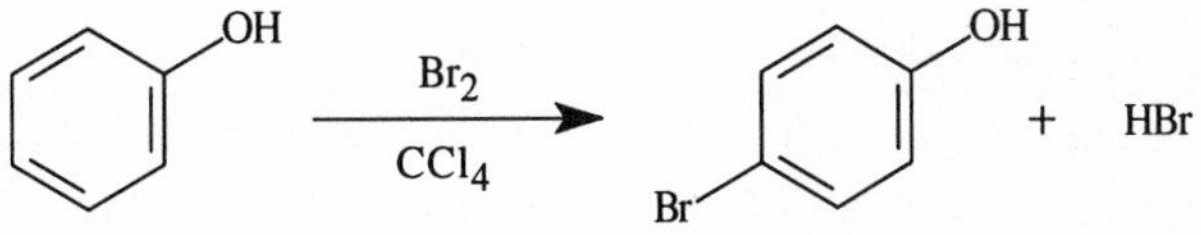

   b. Phenol reacts with Br_2 in H_2O (bromine water) to give 2,4,6-tribromophenol.
      i. Bromine reacts with water to give protonated hypobromous acid, a more potent electrophile than bromine itself.
      ii. In aqueous solutions near neutrality, phenol partially ionizes to its conjugate-base phenoxide anion; this anion is very reactive and brominates instantly, thereby pulling the phenol-phenolate equilibrium to the right.
      iii. Phenoxide ion is much more reactive than phenol because the reactive intermediate is not a carbocation, but is instead a more stable neutral molecule.

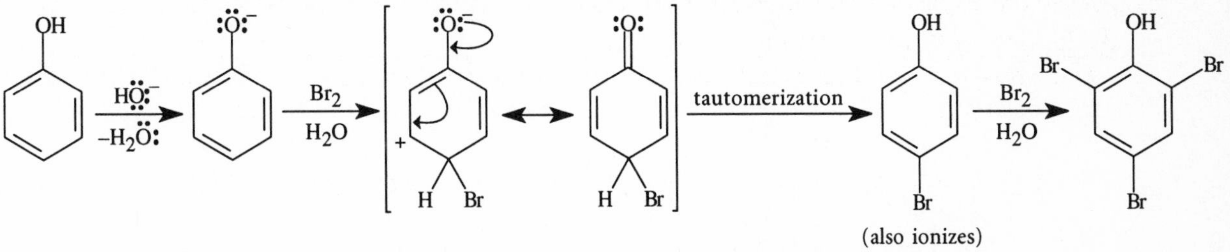

(also ionizes)

   c. In strongly acidic solution, in which formation of the phenolate anion is suppressed, bromination can be stopped at the 2,4-dibromophenol stage.
3. Phenol is very reactive in other electrophilic substitution reactions.
   a. Phenol can be nitrated once under mild conditions.
   b. Direct nitration is not the preferred method for synthesis of di- and trinitrophenol, because the concentrated HNO_3 required for multiple nitrations is an oxidizing agent.
   c. The great reactivity of phenol in electrophilic aromatic substitution does not extend to the Friedel-Crafts acylation reaction.
      i. Friedel-Crafts acylation of phenol occurs slowly, but may be carried out successfully at elevated temperatures; the ring is acylated only once.
      ii. Phenols are less reactive because they react rapidly with the $AlCl_3$ catalyst.

C. Oxidation of Phenols to Quinones

1. *p*-Hydroxyphenols (hydroquinones), *o*-hydroxyphenols (catechols), and phenols with an unsubstituted position *ortho* or *para* to the hydroxy group are oxidized to quinones.

2. The oxidation of hydroquinone and its derivatives to the corresponding *p*-benzoquinone can also be carried out reversibly in an electrochemical cell.

D. Industrial Preparation and Use of Phenol

1. Phenol and acetone are prepared from a single starting material, cumene, which comes from benzene and propene, two compounds obtained from petroleum.
2. The production of phenol and acetone is a two-step process.
   a. Cumene undergoes an autoxidation with molecular oxygen to form cumene hydroperoxide.

   b. The cumene hydroperoxide undergoes an acid-catalyzed rearrangement that yields both acetone and phenol.

Organohalide Reaction Mechanism Summary

Type of Organohalide	Reaction Mechanism	Comments
methyl	S_N2	usually a very rapid reaction
primary	S_N2	inversion of configuration
	E2	observed with strong, highly branched bases only
secondary	S_N1 and S_N2	S_N1 slower than S_N1 of tertiary halides; S_N2 slower than S_N2 of primary halides
	E1 and E2	highly branched products observed except in E2 reactions with highly branched bases
tertiary	S_N1	predominant racemization
	E1 and E2	Saytzeff products observed except in E2 reaction with hindered bases
allylic	S_N1 and S_N2	products of allylic rearrangement observed in S_N1 reactions
benzylic	S_N1 and S_N2	rearrangement does not occur
vinylic	S_N1 or S_N2	not observed
	β-elimination	requires strong base
aromatic	S_N1 or S_N2	not observed
	nucleophilic aromatic substitution	requires electron-withdrawing groups, typically nitro groups, in *ortho* and/or *para* positions
	elimination-addition (benzyne)	requires strong base; both direct and *cine* substitution occur
	electrophilic aromatic substitution	proton is substituted by electrophile in *ortho* or *para* positions; deactivated towards substitution relative to benzene

Study Guide Links

18.1   Vinylic Cations

Despite their instability, vinylic carbocations are reactive intermediates in some reactions, but typically in these situations the vinylic cation is stabilized by some factor. For example, in the acid-catalyzed hydration of phenylacetylene, the carbocation intermediate is not only vinylic, but also benzylic.

$$Ph-C\equiv CH \longrightarrow Ph-\overset{+}{C}=CH_2 \longrightarrow Ph-C=CH_2 \longrightarrow$$

phenylacetylene   a vinylic carbocation
                         (also benzylic)

$$Ph-C=CH_2 \ + \ H_3\overset{\cdot\cdot}{O}^+$$

an enol; spontaneously
converted into a ketone
(Sec. 14.5A)

Notice also that it takes less energy to form a vinylic carbocation from an alkyne than from a vinylic halide, because the alkyne is already *sp* hybridized; an unfavorable hybridization change does not have to occur in order to form the carbocation.

✓18.2   Contrast of Aromatic Substitution Reactions

The name "aromatic substitution" may suggest some connection to the reactions you studied in Chapter 16, such as nitration, halogenation, sulfonation, etc. Other than the fact that a group is replaced on an aromatic ring, *nucleophilic aromatic substitution and electrophilic aromatic substitution have little in common.* In *electrophilic* aromatic substitution, an electrophile, or Lewis acid, is attacked by the benzene π electrons, and a *proton* is lost from the benzene ring. In *nucleophilic* aromatic substitution, a *nucleophile,* or Lewis base, attacks the benzene π electrons, and a *halide* is lost from the benzene ring.

Electrophilic aromatic substitution involves introduction of *electron deficiency* (and positive charge) into the benzene ring. Nitro groups and other electron-attracting groups *destabilize* this charge and thus retard the reaction. Nucleophilic aromatic substitution involves introduction of *more electrons* (and negative charge) into the benzene ring. Nitro groups and other electron-attracting groups *stabilize* this charge and thus accelerate this type of reaction.

Notice that *substitution,* like *addition,* is a type of reaction that can occur by a variety of mechanisms. The word *substitution* by itself conveys no information about mechanism. Thus, a substitution can occur by a concerted (S$_N$2) mechanism, an ionization (S$_N$1) mechanism, a two-step nucleophilic mechanism (as in nucleophilic aromatic substitution), a two-step electrophilic mechanism (as in electrophilic aromatic substitution), a free-radical mechanism, or (as shown in Section 18.4B of the text) an elimination-addition mechanism. The mechanism that operates depends on the reactants and the reaction conditions. Students sometimes forget this point and assume that the word *substitution* means "S$_N$2 reaction," because that is the first substitution reaction they learn.

18.3   Resonance Effects on Phenol Acidity

There is an important subtlety in the resonance effect on phenol acidity. *Un-ionized* phenols are also stabilized by resonance. However, the resonance structures for an un-ionized phenol involve *separation of charge,* but the structures of a phenoxide ion involve *dispersal of charge.*

structures for un-ionized phenol involve *separation* of charge

structures for phenolate ion involve *dispersal* of charge

Because structures that disperse charge are more important than those that separate charge (Guideline 3, Sec. 15.6B, text p. 710), a phenoxide ion is stabilized *more* by resonance than its conjugate-acid phenol. If the two were equally stabilized by resonance, the resonance effect would cancel when the free energies of phenol and phenoxide are compared, and it would have no effect on the pK_a.

Notice how this situation differs from the ionization of a benzylic hydrogen in toluene: There is no resonance stabilization in un-ionized toluene involving the benzylic carbon to offset the resonance stabilization of the benzylic anion. *Only the anion is stabilized by resonance involving the benzylic carbon.*

no resonance interaction of this carbon with the ring

stabilized by resonance

Consequently, the pK_a of toluene should be affected more by resonance than that of phenol. As shown in the text, a phenol is about seven pK_a units more acidic than an alcohol; however, toluene (p$K_a \approx 41$), is about 14–19 pK_a units more acidic than an alkane (pK_a 55–60).

18.4   Fries Rearrangement

There's more to the Friedel-Crafts reactions of phenols than meets the eye. It turns out that phenols react with acid chlorides to give phenyl esters; this is a reaction you'll study in Chapter 21.

phenol   a phenyl ester

(Acid chlorides are *electrophiles* in this reaction just as they are in the Friedel-Crafts reaction. The phenol oxygen, like the aromatic ring, is a source of electrons because of its unshared pairs.) In the presence of AlCl₃—the catalyst used in Friedel-Crafts reactions—phenyl esters rearrange to ketones:

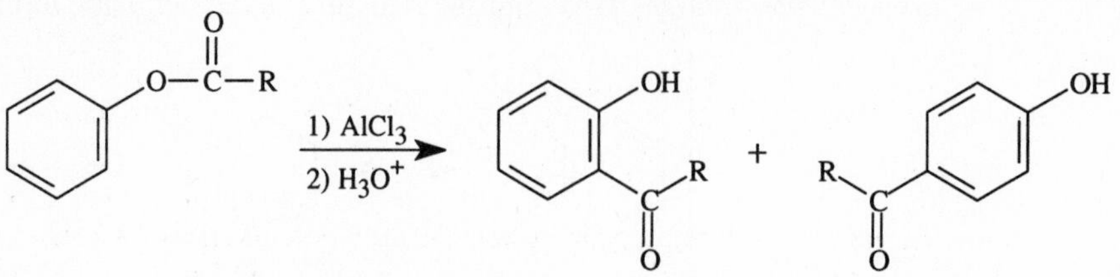

This reaction, known as the *Fries rearrangement,* can be clearly demonstrated because phenyl esters, which can be prepared as pure compounds in the absence of AlCl₃, can be shown to rearrange when treated with AlCl₃. The products of the Fries rearrangement, as you can see, are exactly what would be expected from a direct Friedel-Crafts reaction. Hence, the "Friedel-Crafts reaction" of a phenol almost surely involves ester formation followed by the Fries rearrangement, or a *combination* of the Fries rearrangement and a direct Friedel-Crafts reaction. In any event, the *end result* is as if a Friedel-Crafts reaction had taken place directly.

It is interesting that the "Friedel-Crafts reactions" of phenols in many cases give significant amounts of *ortho*-substituted products. These may be separated from the *para* isomers by application of the principles of hydrogen bonding. The *ortho* isomers contain *intramolecular* hydrogen bonds:

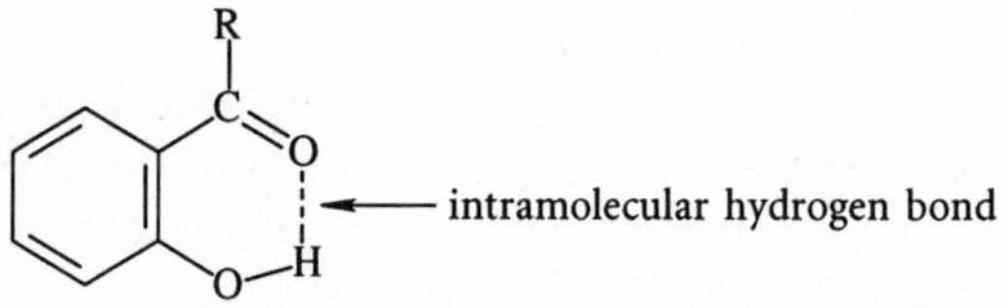

Such intramolecular hydrogen bonding is not possible for the isomeric *para*-substituted phenol. Because the hydrogen-bonding capability of the phenolic —OH group is satisfied within the *ortho*-substituted molecule, this —OH group forms hydrogen bonds less effectively to acceptor atoms of *other molecules*—for example, solvents or chromatography supports such as silica gel—than the —OH group of the *para*-substituted isomer. Hence, the *ortho*-substituted isomer is generally much less soluble in solvents that accept hydrogen bonds, and moves through a silica gel chromatography column more rapidly, than the *para*-substituted isomer. Thus, the great difference in hydrogen-bonding properties of the two isomers makes separation a relatively simple matter. Exactly the same principles can be utilized in designing separations of *o*- and *p*-nitrophenol.

✓18.5   The Cumene Hydroperoxide Rearrangement

The formation of phenol and acetone (Eq. 18.53b) is an acid-catalyzed rearrangement of cumene hydroperoxide. This Study Guide Link discusses the mechanism of this reaction.

In the first step of the mechanism, an oxygen of the hydroperoxide is protonated:

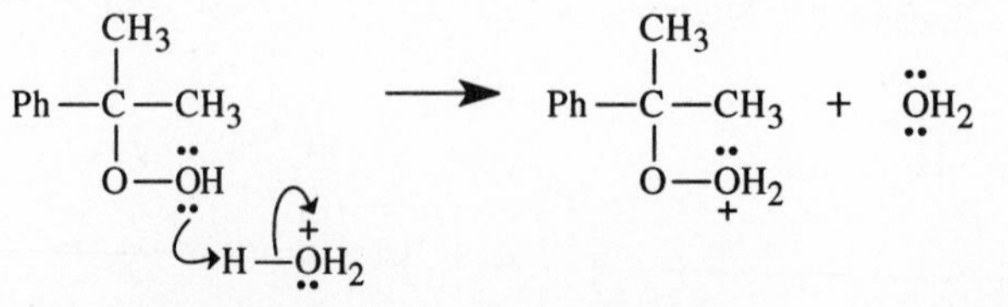

Loss of water gives an *electron-deficient oxygen,* the oxygen analog of a carbocation. Rearrangement occurs to give a much more stable resonance-stabilized carbocation. (The electron-deficient oxygen structure is so unstable that loss of water and the rearrangement undoubtedly take place in a concerted manner.)

Attack of water on the carbocation gives a type of alcohol known as a hemiacetal.

a hemiacetal

You should now be able to complete the acid-catalyzed conversion of the hemiacetal into phenol and acetone. Start by protonating the oxygen of the PhO— group. What has to happen next to form phenol? Finish the mechanism to give acetone.

Solutions

Solutions to In-Text Problems

18.1   (a)   *p*-Bromotoluene reacts most slowly (that is, not at all) because it is an aryl halide; aryl halides do not undergo S_N2 reactions. Benzyl bromide is a benzylic halide and therefore reacts most rapidly in an S_N2 reaction (see text p. 803).

Br—⟨benzene ring⟩—CH₃   <   ⟨benzene ring⟩—CH₂CH₂CH₂Br   <   ⟨benzene ring⟩—CH₂Br

p-bromotoluene   (3-bromopropyl)benzene   benzyl bromide

18.2   Assume that E2 reactions of *vinylic* halides follow the same trends as E2 reactions of *alkyl* halides. Therefore, because an *anti* elimination is faster than the corresponding *syn* elimination, the reaction of compound *B* is faster than that of compound *A*; and because bromide is a better leaving group than chloride, the reaction of compound *A* is faster than that of compound *C*. Because eliminations involving benzylic *β*-hydrogens are faster than other eliminations, compounds *A–C*, in which the *β*-hydrogens are benzylic, react more rapidly than compound *D*, in which the *β*-hydrogens are not benzylic. Hence, the desired order is *D* < *C* < *A* < *B*.

18.3   (a)   The order of increasing S_N1-E1 reactivity is (1) < (2) < (3). The benzylic halide (3) reacts fastest because its ionization gives a resonance-stabilized benzylic carbocation; and the vinylic halide (1) reacts slowest because vinylic halides are nearly inert in S_N1 reactions for the reasons given in the text.

18.4   (a)   The nitrogen of the amine acts as a nucleophile in a nucleophilic aromatic substitution reaction.

O_2N—⟨benzene ring, NO₂ at ortho⟩—$\overset{+}{N}H_2C_2H_5$ Cl⁻   ⇌ O_2N—⟨benzene ring, NO₂ at ortho⟩—$\overset{..}{N}HC_2H_5$ + HCl

> The nitro groups reduce the basicity of the product to the point that it is deprotonated under the reaction conditions. (Can you give a reason why?) Because you have not yet studied the basicity of aromatic amines, however, either answer is correct.

(c)   No reaction, because the compound contains no substituent that activates nucleophilic aromatic substitution.

18.5   (a)   The first compound should react more rapidly because the Meisenheimer complex is stabilized by the resonance interaction of both substituents; the complex in the case of the second compound is stabilized by resonance involving only a nitro substituent.

(solution continues)

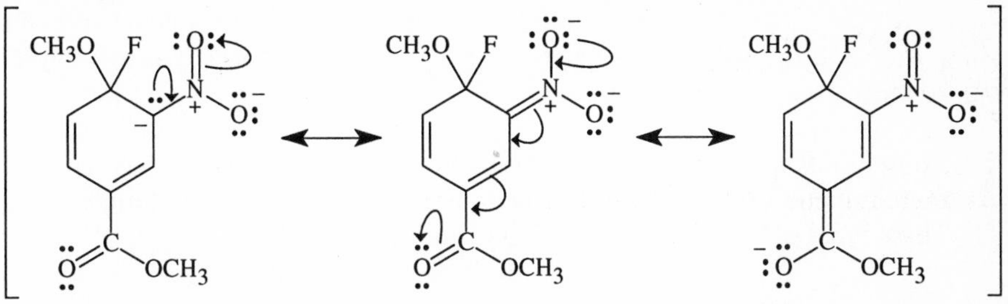

18.6 (a) Because this compound has two chemically nonequivalent β-hydrogens, two possible benzyne intermediates, *A* and *B*, can be formed. Each can give two products, one of which is common to both intermediates.

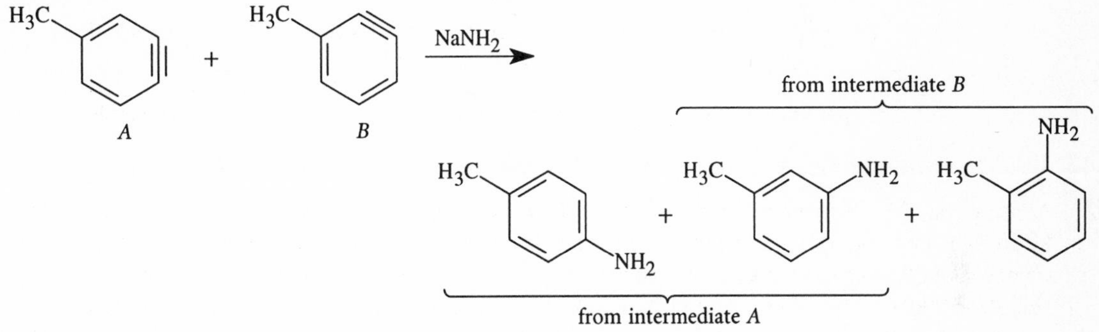

(c) Because there is no β-hydrogen, there is no reaction.

18.7 (a)

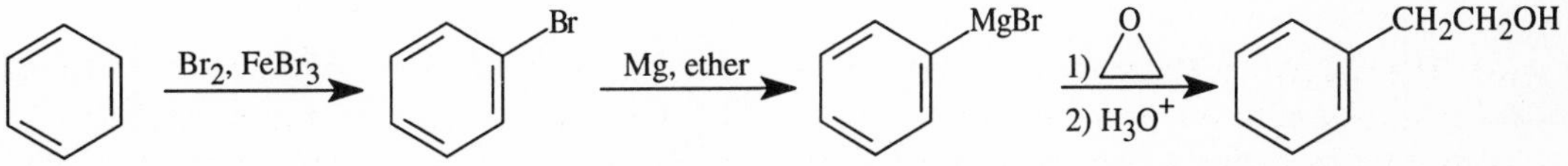

18.8 (a) 2,4-Dinitrophenol is more acidic because its conjugate-base anion has more important resonance structures. In the 2,4-dinitrophenoxide anion, both nitro substituents are involved in resonance stabilization; in 2,5-dinitrophenoxide anion, only the 2-nitro group is involved in resonance stabilization. (The 5-nitro group is not involved for the reasons shown in the discussion of Eq. 18.31 in the text.)

(c) The *para*-isomer of the conjugate-base phenoxide anion has more important resonance structures. The key resonance structure is the one below, in which the substituent group participates in the resonance stabilization.

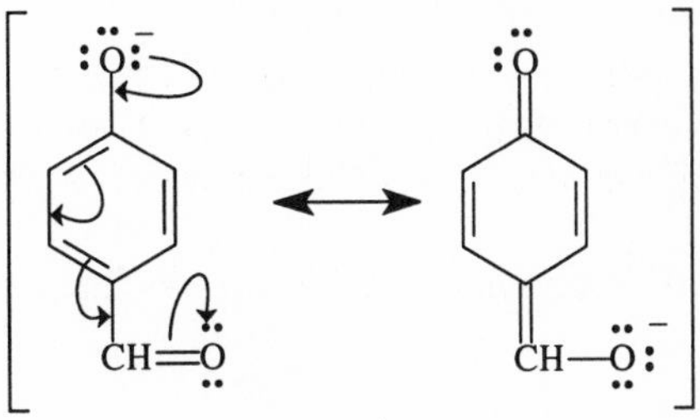

18.9 (a)

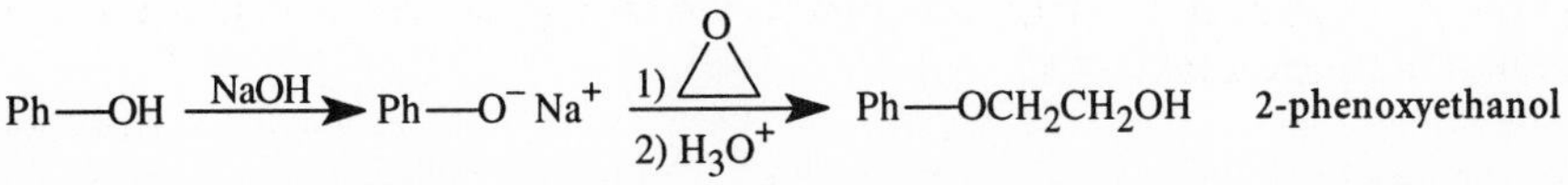

18.10 The conjugate-base anion of the compound shown in the problem has a net charge of zero, whereas the compound shown in the problem has a net charge of +1. Neutral species are less soluble in water than compounds with a net charge.

> ⇨ Compounds that contain both a positive and a negative charge, such as the conjugate base of the compound shown in the problem, are called *zwitterions* ("double-ions"), or *inner salts*. These types of compounds are discussed in Chapter 26.

18.11 (a) 9,10-Phenanthraquinone is an *ortho*-quinone.

9,10-phenanthraquinone

18.12 (a) The *para*-quinone is formed because it is more stable than the alternative *ortho*-quinone.

18.13 (a) Let the long alkyl chain in vitamin E (structure on text p. 844) be abbreviated as —alkyl. The most stable radical is formed by abstraction of the phenolic hydrogen atom.

18.14 (a)   (c)

18.15 The electrophile is the *tert*-butyl cation, which is formed by protonation of the alcohol and loss of water.

The carbocation intermediate in the electrophilic aromatic substitution reaction is, of course, resonance-stabilized; it is left to you to draw the important resonance structures, including one involving the hydroxy group.

18.16 (a) Diphenyl ether does not cleave with hot, concentrated HBr, because such a cleavage would require either an S_N1 reaction or an S_N2 reaction at a phenyl-oxygen bond; as this section of the text shows, such reactions do not occur. In contrast, *tert*-butyl phenyl ether cleaves by an S_N1 mechanism involving protonation of the oxygen and loss of phenol to form a *tert*-butyl cation, which reacts with bromide ion to form *tert*-butyl bromide.

tert-butyl phenyl ether

18.17 The mechanism is exactly like the one shown in the solution to Problem 18.15, except that a different carbocation electrophile is involved.

Solutions to Additional Problems

18.18 (a) No reaction   (b) No reaction

(c)

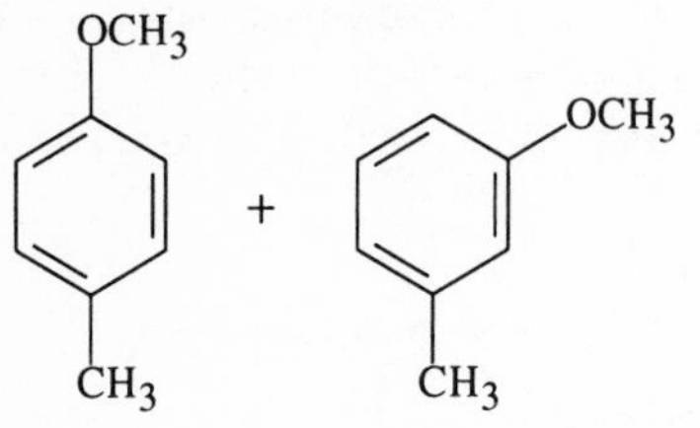

$+$

(d) 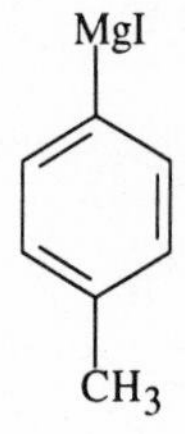   (e) 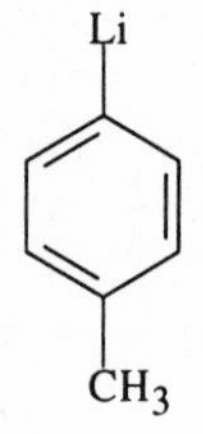   (f)

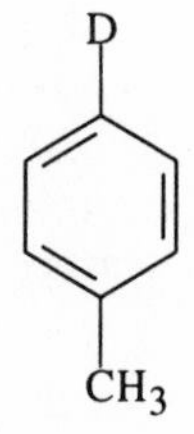

The two products in part (c) arise from a benzyne mechanism.

18.19 (a)   (b)

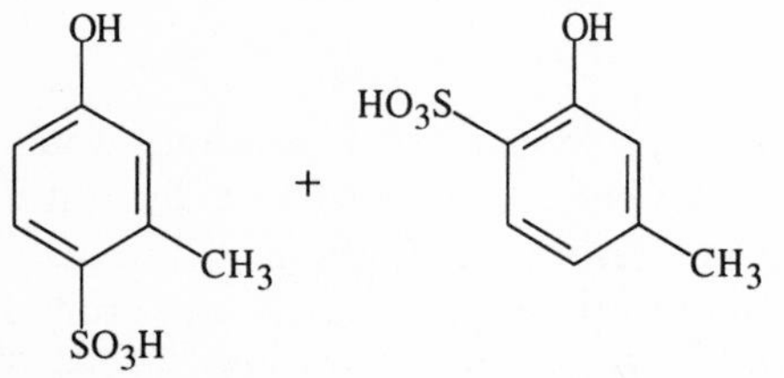

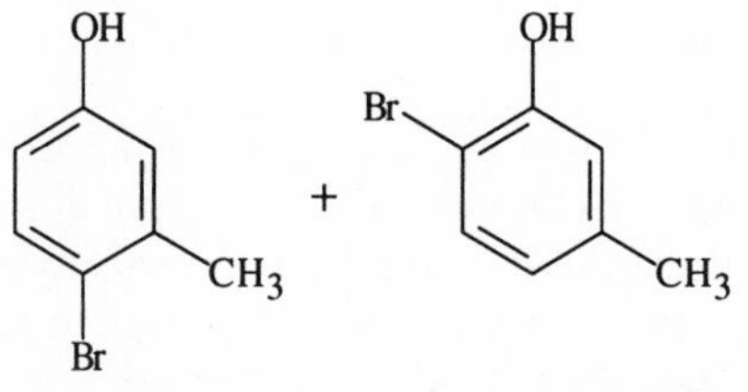

$+$

In part (c), both benzylic bromination and ring bromination take place; see Eq. 18.40 on text p. 846.

(c)   (d) No reaction.   (e)

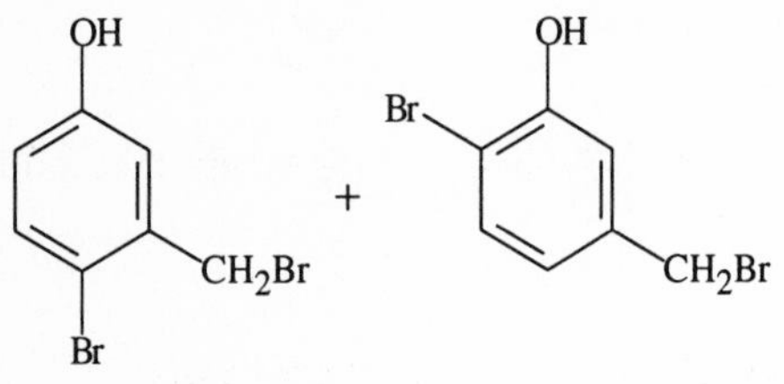

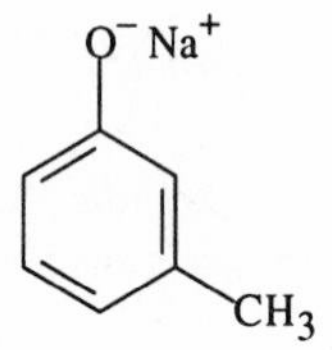

$+$

(f) 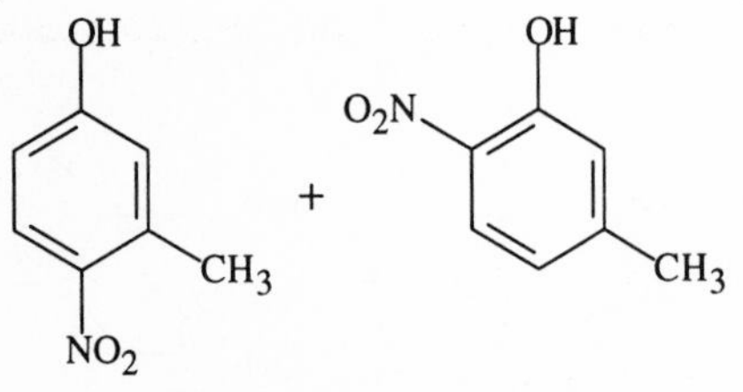   (g) 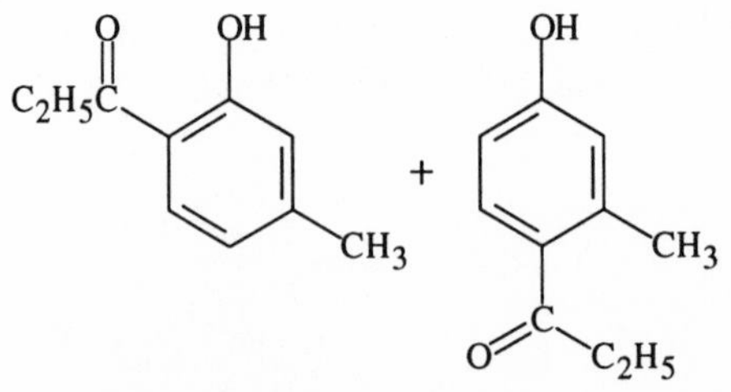   (h)

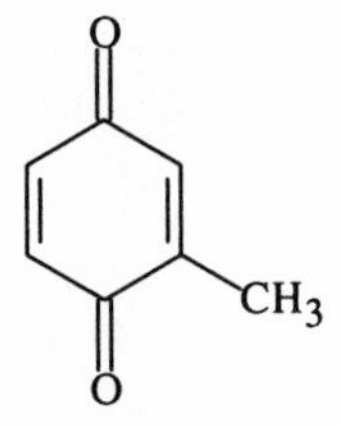

$+$

Because HNO_3 is an oxidizing agent, the products of (f) may also include some of product (h) as well as some nitrated quinones. Note that oxidation of the phenol to a quinone is much faster, and can be accomplished with much milder conditions, than oxidation of the methyl group to a carboxylic acid. Side-chain (benzylic) oxidation requires harsh conditions; see, for example, Eq. 17.32–17.33 on text p. 804.

18.20 (a) Mercaptans are more acidic than phenols; and aryl mercaptans (thiophenols) are more acidic than alkyl mercaptans for the same reason that phenols are more acidic than alcohols. The acidity order is cyclohexanol < cyclohexyl mercaptan < benzenethiol

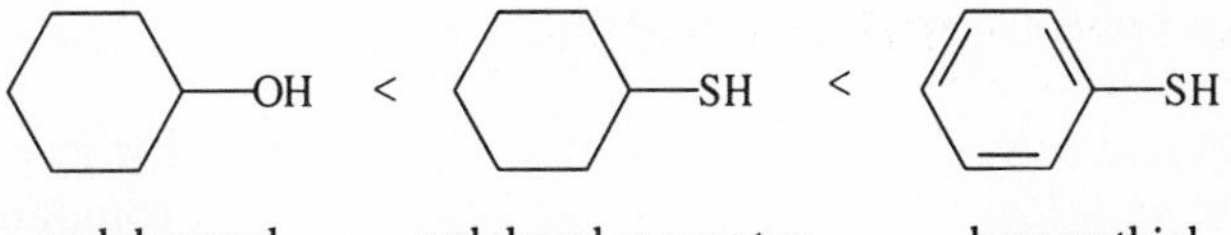

(c) The *p*-nitro substituent stabilizes the *p*-nitrophenolate ion by both its resonance effect (see text Eq. 18.30 on p. 838) and its polar effect; the *p*-chloro substituent stabilizes *p*-chlorophenolate ion only by its polar effect. The third compound is nitric acid, a strong acid. Its conjugate-base anion, the nitrate ion, is stabilized both by the polar effect of the nitro group, which is much closer to the anionic oxygen than it is in the *p*-nitrophenolate ion, and by a resonance effect:

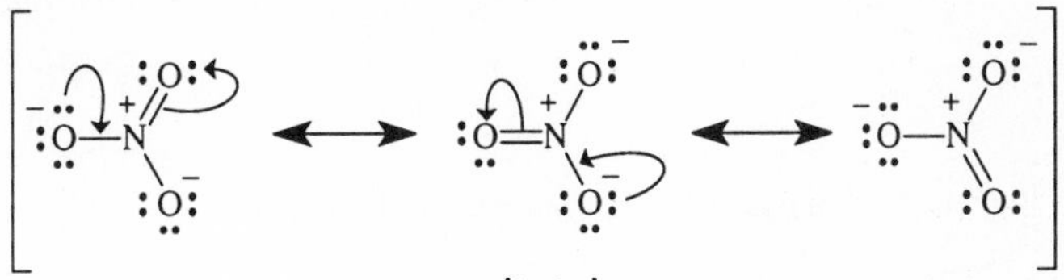

nitrate ion

The acidity order is

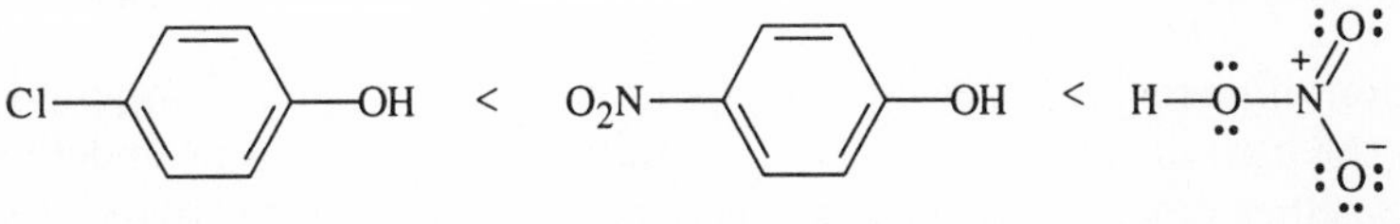

p-chlorophenol < *p*-nitrophenol < nitric acid

(e) The effect of large *ortho* substituents on acidity is much like that of α-branching in an alcohol: acidity is reduced because solvation of the conjugate-base anion is less effective (see the discussion at the top of text p. 366). This effect is more pronounced for the large *tert*-butyl groups than it is for the chlorines or the methyl group. The *ortho*-chloro groups stabilize the conjugate-base of 2,6-dichlorophenol by their polar effect. The acidity order is

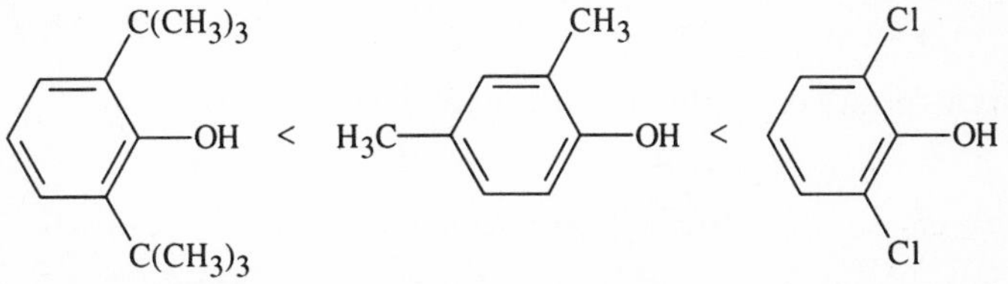

2,6-di(*tert*-butyl)phenol < 2,4-dimethylphenol < 2,6-dichlorophenol

(f) The *para*-aryl substituents all stabilize the respective conjugate-base alkoxide ions by both a resonance and a polar effect. The nitro group of the *meta*-nitrophenyl substituent exerts an additional stabilizing polar effect; and the nitro group of the *para*-nitrophenyl substituent stabilizes the conjugate-base phenoxide ion by both a polar effect and a resonance effect:

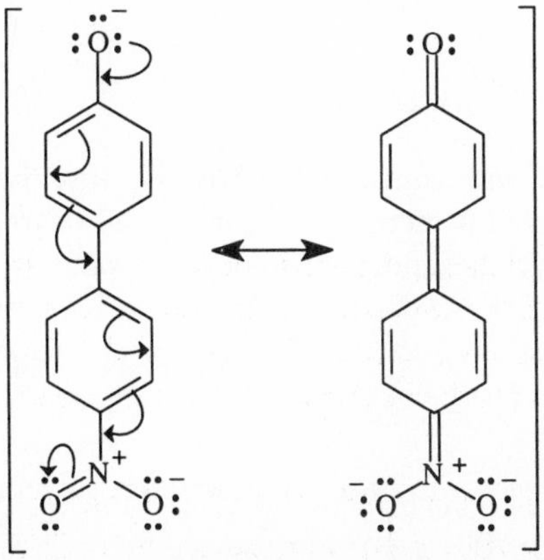

(solution continues)

The acidity order is (3) < (1) < (4) < (2).

18.21 The enol *A* is more acidic (by about five pK_a units). The reason is that its conjugate-base anion is stabilized by resonance as shown below, whereas the conjugate-base anion of compound *B* is not resonance-stabilized.

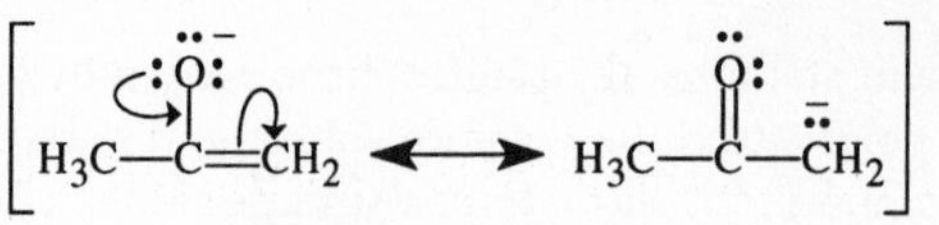

conjugate base of compound *A*

18.22 (a)   (b)   (c)

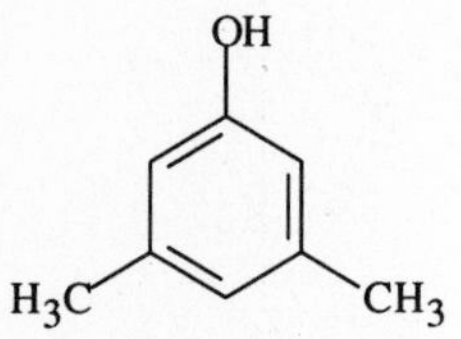

3,5-dimethylphenol
(compound *A*)

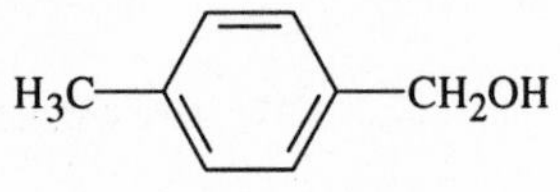

p-methylbenzyl alcohol
(compound *B*)

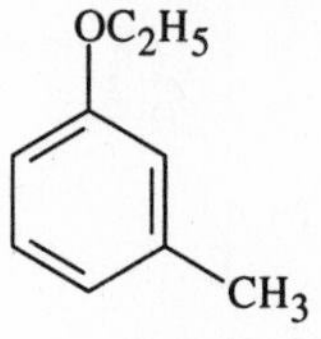

m-ethoxytoluene
(compound *C*)

Hydrogenation of phenol *A* yields the corresponding cyclohexanol. Compound *B* reacts readily with HBr to give the benzylic bromide; the corresponding Grignard reagent undergoes protonolysis to give *p*-xylene. The ether group in compound *C* is cleaved by HBr to give C_2H_5Br (which is volatile under the high-temperature reaction conditions) and *m*-cresol.

18.24 (a) Cyclohexanol is ionized only to a very small extent in aqueous NaOH solution; phenol ionizes almost completely.

(b) Both compounds are converted into their tosylate derivatives.

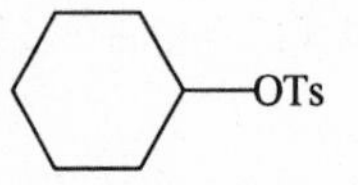

cyclohexyl tosylate

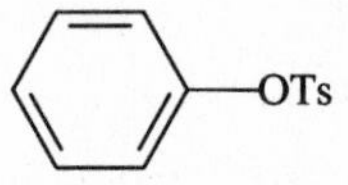

phenyl tosylate

(c) Both compounds react rapidly with NaH to give H_2 and their conjugate bases: sodium cyclohexanolate in the case of cyclohexanol, and sodium phenolate (sodium phenoxide) in the case of phenol.

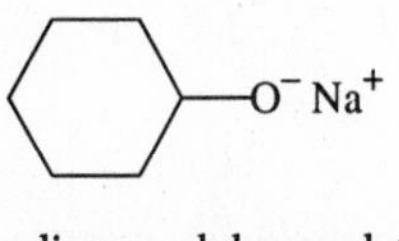

sodium cyclohexanolate

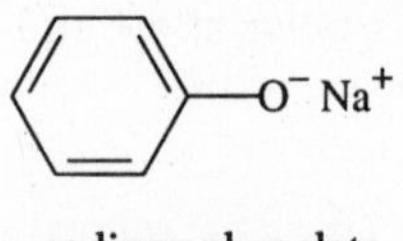

sodium phenolate

(d) Only cyclohexanol reacts with concentrated HBr; because the C—O bond in phenol cannot be broken by S_N1 or S_N2 reactions, phenol is inert, except for a small degree of protonation of the phenol oxygen. The product in the case of cyclohexanol is bromocyclohexane.

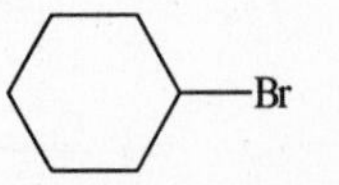

bromocyclohexane

(e) Cyclohexanol does not react with bromine in an inert solvent; phenol is brominated on the ring, as shown in Eq. 18.40 on text p. 846.

(f) Cyclohexanol is oxidized to cyclohexanone, whereas phenol is oxidized to 1,4-benzoquinone.

cyclohexanone   1,4-benzoquinone

(g) Cyclohexanol undergoes dehydration to cyclohexene; phenol undergoes sulfonation, an electrophilic aromatic substitution reaction.

cyclohexene   ring sulfonation products of phenol

18.26 (a) The second compound reacts in a nucleophilic aromatic substitution reaction. The first compound does not react in this way because it is not activated toward nucleophilic aromatic substitution; the nitro group is not attached to the ring.

In the interest of total accuracy we point out that the first compound actually does react, but in a different way. A hydrogen of the $-CH_2-$ group is acidic enough to ionize in alcoholic KOH; can you see why?

(b) The last compound, diphenyl ether, cannot be prepared by a Williamson ether synthesis, because such a synthesis would require an S_N2 reaction to occur at the carbon of a benzene ring. Notice that the first ether, tetrahydrofuran, can be prepared by an *intramolecular* Williamson synthesis from 4-bromo-1-butanol, $Br-(CH_2)_4-OH$. The second ether, ethoxybenzene, can be prepared by a reaction of sodium phenolate, $PhO^- Na^+$, with ethyl iodide.

(c) The first compound undergoes solvolysis by an S_N1 mechanism to give a mixture of an alcohol, an ether, an alkene, and HBr (thus the acidic solution). The second compound is inert, because aryl halides do not undergo solvolysis under neutral conditions.

(d) Only benzyl chloride, $PhCH_2Cl$, reacts, because it is an *alkyl* chloride. Note that *p*-chlorotoluene is an *aryl* chloride; aryl chlorides require THF as a solvent for conversion into Grignard reagents.

(e) Diphenyl ether will not cleave with HI because such a cleavage would require an S_N1 or S_N2 reaction at the carbon of a benzene ring, and such reactions do not occur. Phenyl cyclohexyl ether, however, would cleave to phenol and iodocyclohexane:

cyclohexyl phenyl ether   diphenyl ether
(inert to HI)

(f) Because of the symmetry of the benzyne intermediate *A*, the first compound would give only a single product; the last compound does not react because it has no β-hydrogen. Only the second compound gives two products.

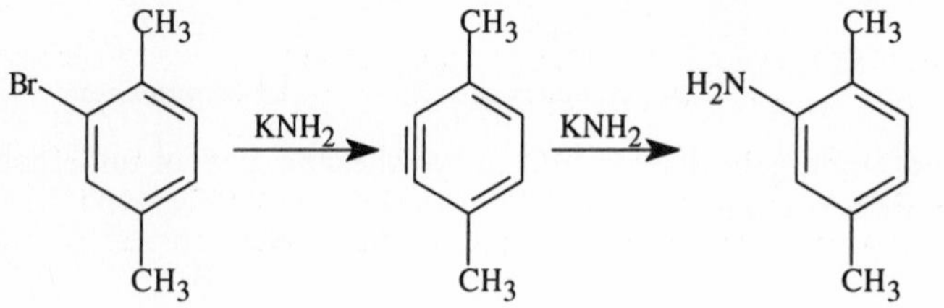

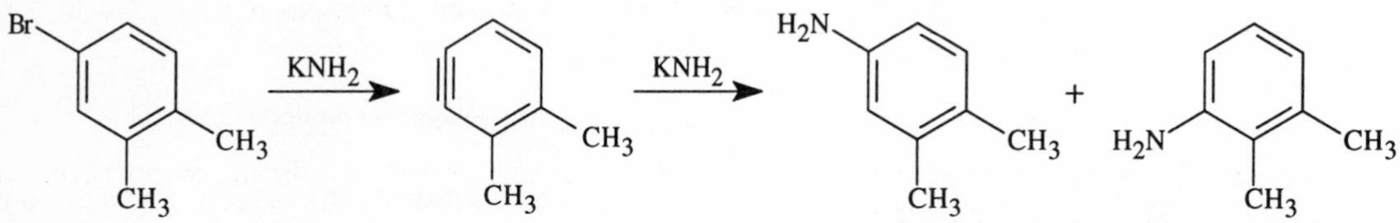

18.27 Potassium benzenethiolate, a base, reacts with the tertiary alkyl halide in an E2 reaction:

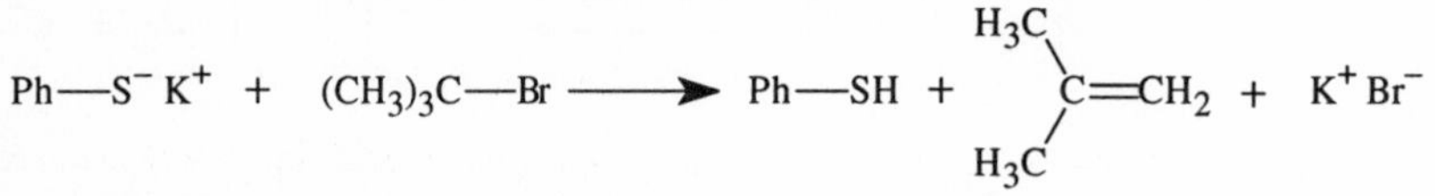

18.29 The two principles involved in the solution to this problem are, first, that phenoxide ions are more soluble in water than are their conjugate-acid phenols; and second, that an acid is ionized when the pH of the solution is well above the pK_a of the acid. The second point can be demonstrated simply with the concentration expression for the dissociation constant of an acid HA to its conjugate base A^- (Eq. 3.14, text p. 98):

$$K_a = \frac{[H_3O^+][A^-]}{[HA]}$$

Taking logarithms, and letting $-\log [H_3O^+] = pH$,

$$pH - pK_a = \log \frac{[A^-]}{[HA]}$$

$$\text{or } 10^{(pH - pK_a)} = \frac{[A^-]}{[HA]}$$

This formula shows that in order for the ratio $[A^-]/[HA]$ to be large, the pH of the *solution* must be much greater (typically two or more units greater, or *more basic*) than the pK_a of the *acid*.

Condition (1) corresponds to a pH of about 1; neither phenol is ionized, and therefore neither is soluble in the buffer. Under condition (2), some fraction of 2,4-dinitrophenol is ionized, but under condition (3), which is a pH value almost three units higher than the pK_a of the phenol, 2,4-dinitrophenol is >99% ionized. Because this pH value is more than three units *below* the pK_a of *p*-cresol, this phenol is *not* ionized at pH = 7. Therefore, 2,4-dinitrophenol should be relatively soluble at this pH because it is ionized, and *p*-cresol remains insoluble. This is the best condition to separate the two phenols by extraction. Condition (4) corresponds to a pH of about 13; both phenols are ionized at this pH, and both should be relatively soluble (as their conjugate-base phenoxides).

18.30 (a) The reaction of phenol as a base with the acid H_2SO_4:

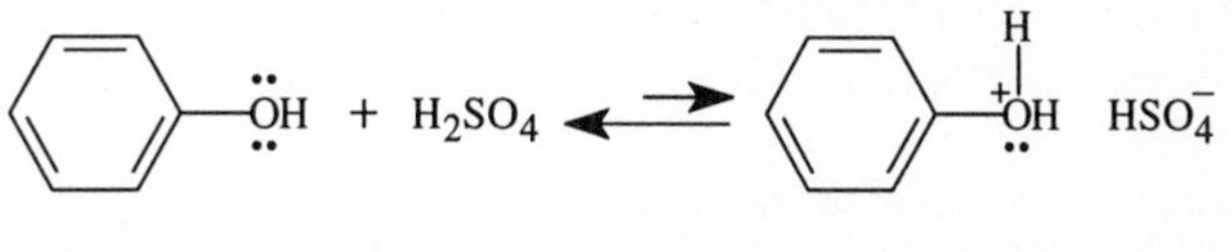

(b) First, consider polar effects. The phenyl group has an *electron-withdrawing* polar effect; this destabilizes the positively-charged conjugate acid of phenol. Such an effect is absent in the conjugate acid of the alcohol cyclohexanol because its ring has no double bonds. Therefore, the polar effect reduces the basicity of the phenol relative to the alcohol. Second, consider the resonance effect of the benzene ring. The source of this effect is the delocalization of an oxygen unshared electron pair into the ring:

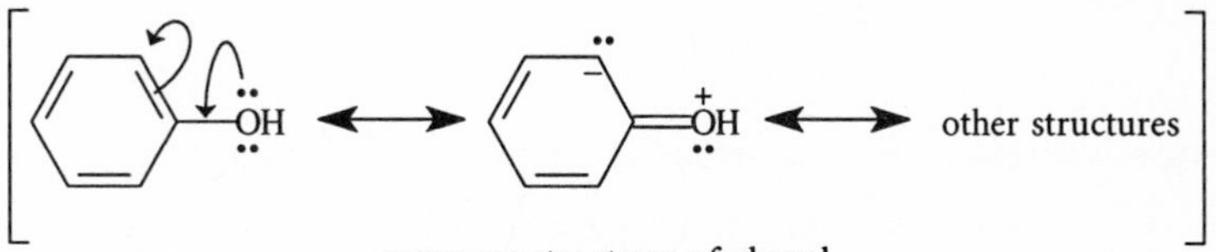

resonance structures of phenol

Such an effect is absent in the conjugate acid of the phenol because it would place *two* positive charges on the oxygen, an electronegative atom.

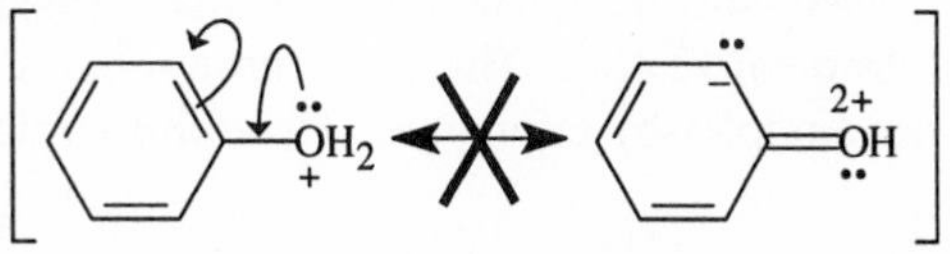

conjugate acid of phenol

Therefore, resonance stabilizes phenol relative to its conjugate acid and therefore reduces its basicity. (Another way to think of this situation is to imagine that delocalization of the oxygen unshared electron pair makes it "less available" for protonation.) Such a resonance effect is not present in the alcohol. The conclusion is that a phenol is much less basic than an alcohol because of both the polar effect and the resonance effect of the ring. Therefore, the alcohol is the stronger base.

18.31 Because NaOH can ionize a phenol, particularly a relatively acidic phenol such as *p*-nitrophenol, which has a pK_a of 7.21, the trace *B* (and the corresponding yellow color) is due to the conjugate base of the phenol. The UV absorption reflects conjugation of the extra unshared pair with the ring and its delocalization into the *p*-nitro group, as shown in Eq. 18.30, text p. 838. Addition of acid lowers the pH so that the un-ionized phenol is regenerated, and the UV spectrum of the phenoxide disappears. To summarize:

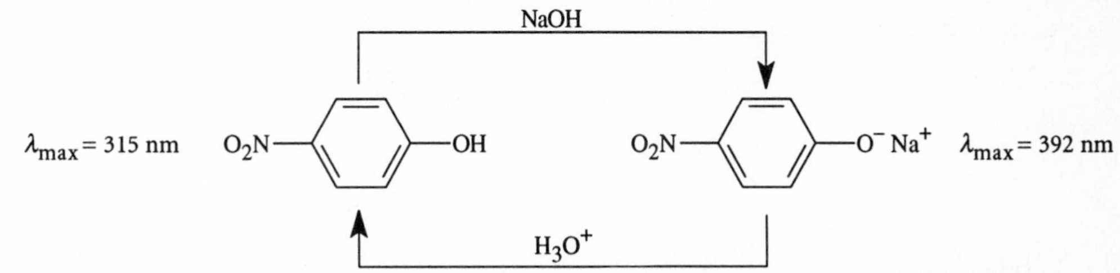

18.33. Ionization of the benzylic bromide gives a carbocation that can lose a proton to give the bracketed intermediate.

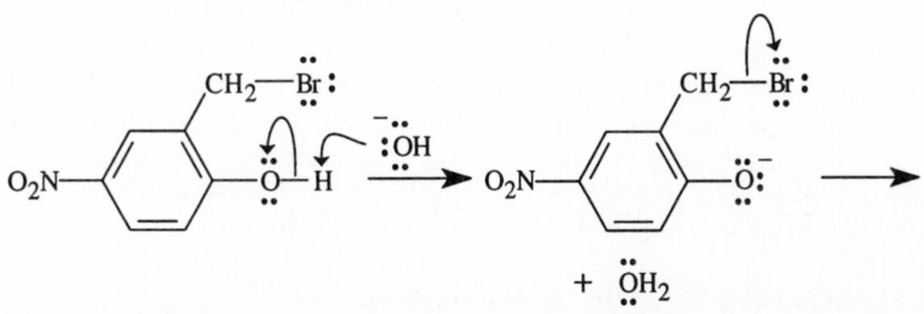

(equation continues)

18.35 (a) Fluoride ion is displaced in a nucleophilic aromatic substitution reaction by ethanethiolate ion.

(c) The amine acts as a nucleophile in a nucleophilic aromatic substitution reaction. (See the note following the solution to Problem 18.4(a).) The cyano group, like a nitro group, can stabilize the intermediate Meisenheimer complex by resonance. (Draw this complex and its resonance structures.)

(e) The products result from a benzyne mechanism.

(g) Bromine is introduced at the ring position that is activated by two hydroxy groups.

(i) Oxidation occurs to give the quinone.

1,4-naphthoquinone

(k) The methanesulfonate derivative of 2,4-dinitrophenol undergoes a nucleophilic substitution reaction in which the methanesulfonate group is displaced by methoxide.

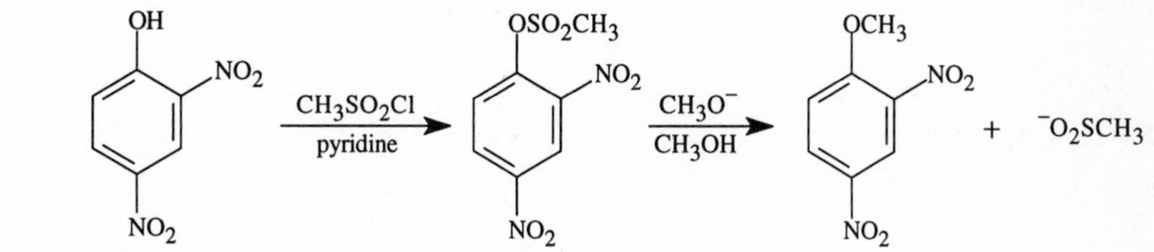

(m) The methyl ethers are cleaved but the phenyl ethers are not. (See the solution to Problem 18.26(e).)

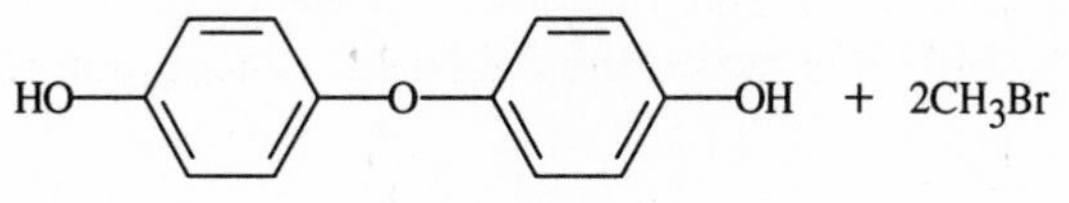

18.36 (a) The $\Delta H°$ of hydrogenation of 2-butyne is $(-7.1 - 146) = -153$ kJ/mol (-36.6 kcal/mol). The $\Delta H°$ of hydrogenation of 2-pentyne is $(-28 - 129) = -157$ kJ/mol (-37.5 kcal/mol). These data show that hydrogenation of a typical alkyne to a *cis*-alkene liberates about 155 kJ/mol (37 kcal/mol) of energy.

(b) The $\Delta H°$ for hydrogenation of the "alkyne" benzyne to the corresponding "*cis*-alkene," benzene, is $(83 - 494) = -411$ kJ/mol (-98 kcal/mol). The difference between this figure and the typical $\Delta H°$ for alkyne hydrogenation, -155 kJ/mol (-37 kcal/mol), can be attributed to the amount by which benzyne is unstable. This difference is 256 kJ/mol (61 kcal/mol). Benzyne is less stable than a typical alkyne by about 256 kJ/mol (61 kcal/mol)!

18.37 (a)

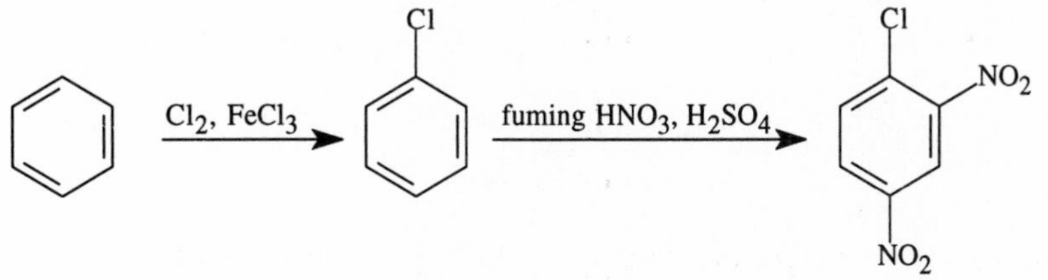

(c)

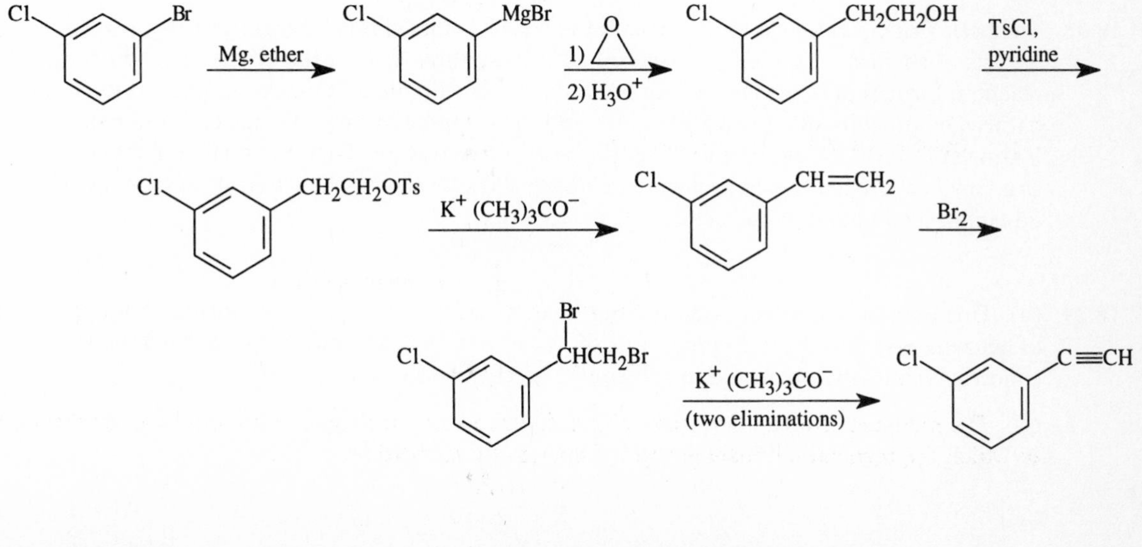

(e)

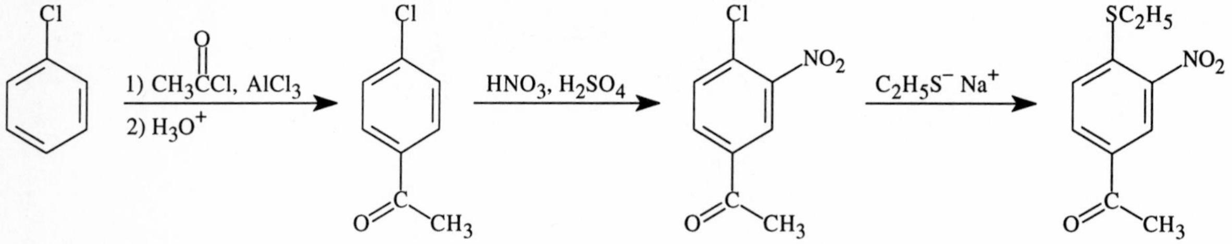

(g) (See Eq. 18.50, text p. 849.) Other sources of the *tert*-butyl cation electrophile could also be used, such as 2-methylpropene and acid. *Tert*-butyl chloride and AlCl₃ would probably not work well because of the tendency of phenols to form complexes with AlCl₃.

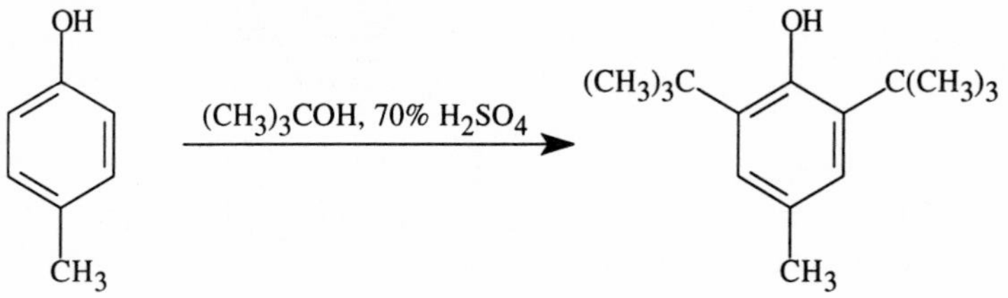

(i)

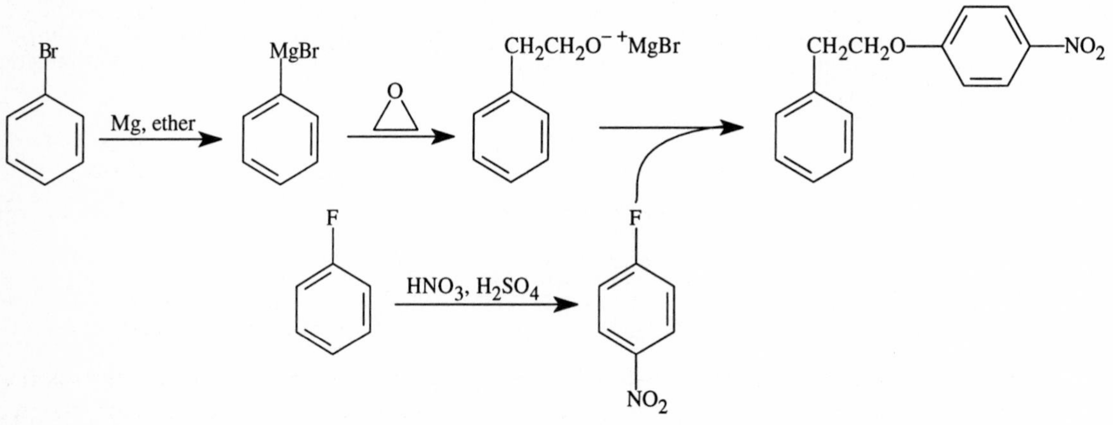

18.38 The basic principle needed to understand the results is that elimination is most rapid when it occurs with *anti* stereochemistry. In the first reaction, *anti* elimination leads to the observed product. In the second reaction, formation of the alkyne requires a slower *syn* elimination; hence, another process can compete, namely, elimination of a methyl hydrogen and the bromine to form the allene. Elimination of the methyl hydrogen to form the allene is a slower process for two reasons. First, the methyl hydrogen is not as acidic as a vinylic hydrogen. Second, allenes are not so stable as alkynes. The transition state for elimination is destabilized by both of these factors.

18.39 (a) This is an overall substitution reaction that occurs by a benzyne mechanism. (Although the formation of benzyne may be a two-step process, as shown in Eqs. 18.22a–b on text pp. 833–834, it is shown as a concerted elimination here and in subsequent mechanisms to save space.)

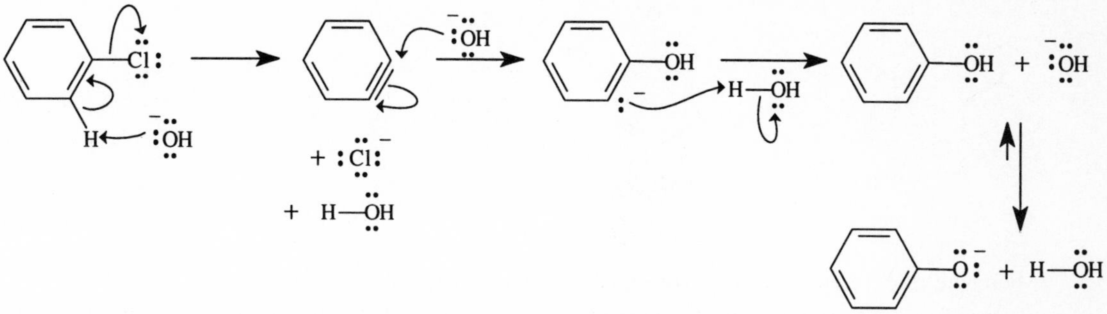

(b) Diphenyl ether is formed by attack of the phenoxide product on the benzyne intermediate.

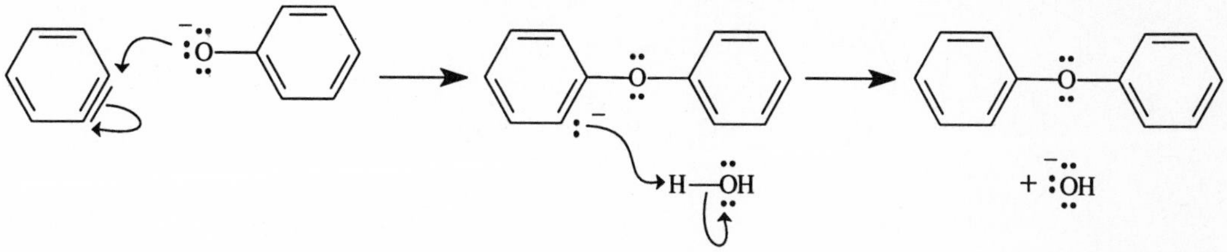

18.41 In base, chloroacetic acid and 2,4,5-trichlorophenol are ionized. Attack of the conjugate base of 2,4,5-trichlorophenol on chloroacetate ion in an S_N2 reaction gives the conjugate base of 2,4,5-T; un-ionized 2,4,5-T is formed on acidification.

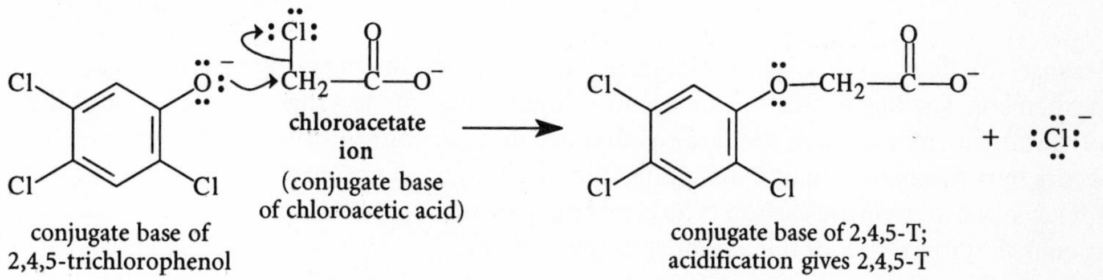

The formation of dioxin is accounted for by a benzyne mechanism similar to that in the solution to Problem 18.39, except that in this case there are two consecutive benzyne reactions, the second of which is intramolecular. The first benzyne intermediate *A* is formed by hydroxide-promoted elimination of HCl from the conjugate base of 2,4,5-trichlorophenol.

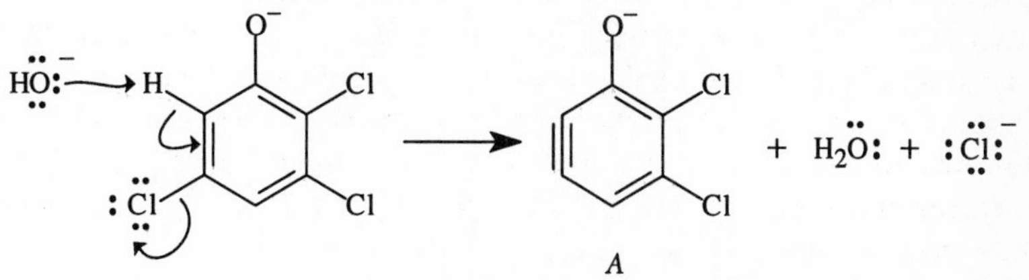

2,4,5-Trichlorophenolate ion attacks this benzyne intermediate to form ether *B*. Although the reaction of the benzyne intermediate may be a two-step process, as shown in Eqs. 18.22c–d on text p. 834, it is shown below and in subsequent mechanisms as a concerted process to save space.)

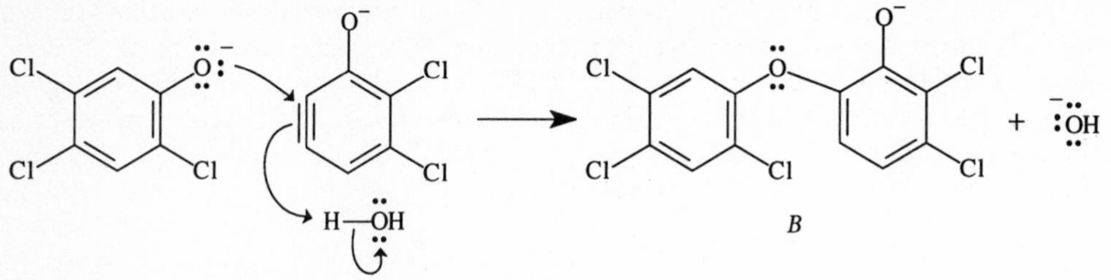

A benzyne intermediate derived from ether *B* then undergoes an intramolecular reaction to form dioxin.

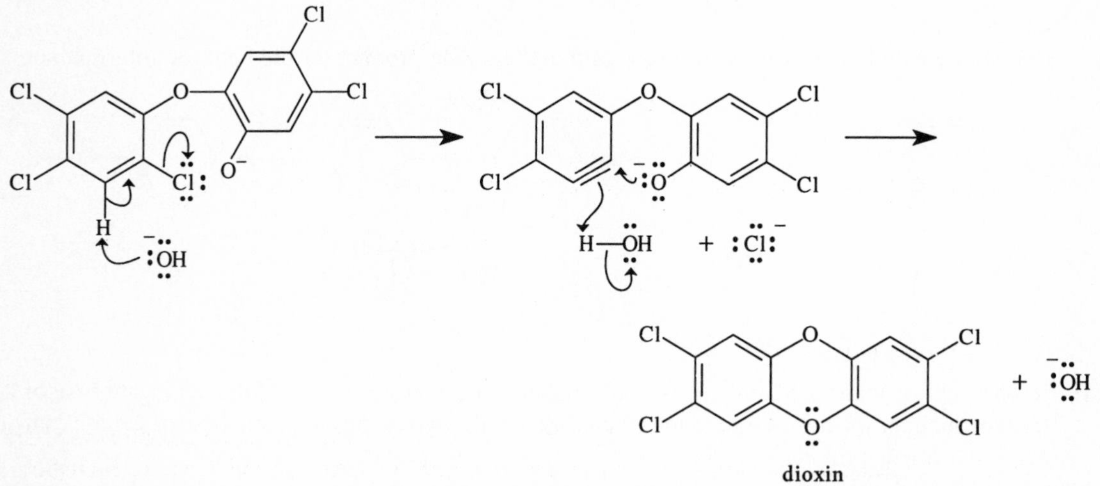

dioxin

18.42 Because vinylic halides do not undergo S_N2 or S_N1 reactions, this substitution must occur by an unusual mechanism. The key to this solution is to recognize that the leaving-group effect and the effect of a *para*-nitro substituent are much the same as they are in nucleophilic aromatic substitution. This reaction, in fact, occurs by a mechanism that is much like the mechanism of nucleophilic aromatic substitution, except that it takes place at a vinylic carbon. The *para*-nitro group accelerates the reaction by affording additional resonance stabilization of the anionic intermediate *A*.

anionic intermediate *A*

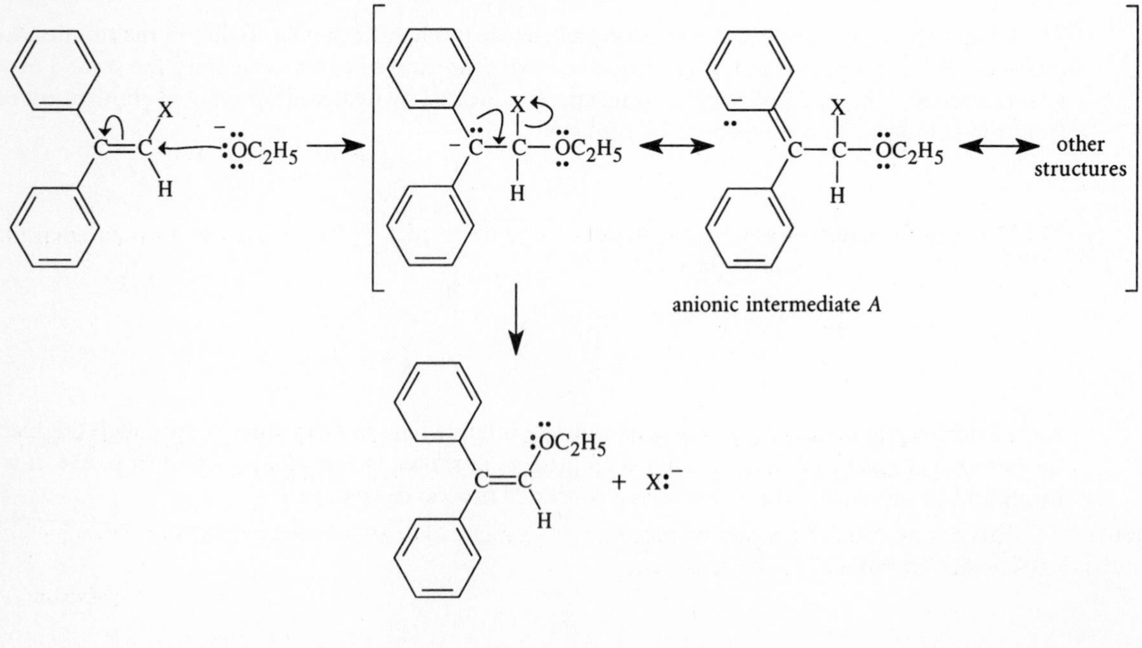

18.43 The spectrum is consistent with the formation of an anionic intermediate—a stable Meisenheimer complex. (The dotted lines symbolize resonance delocalization of the negative charge.) If you draw out the resonance structures for this ion, you will see that charge is delocalized to the carbons bearing the nitro groups, and it can also be delocalized into the nitro groups. However, charge is *not* delocalized to the other carbons. Hence, the protons on these carbons do not show the smaller chemical shift that would be expected if there were high electron density on these carbons.

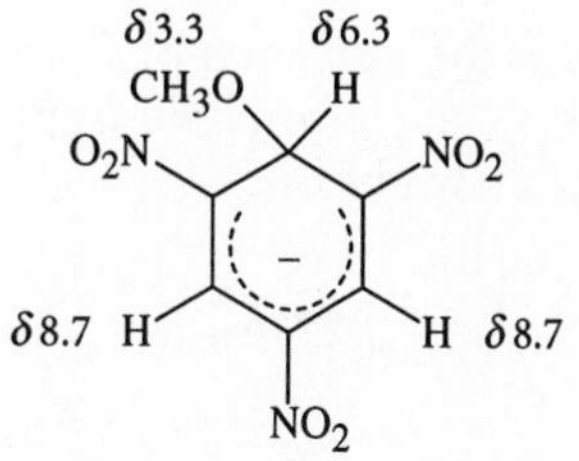

18.44 (a) In the elimination, a proton could be removed from either β-carbon atom:

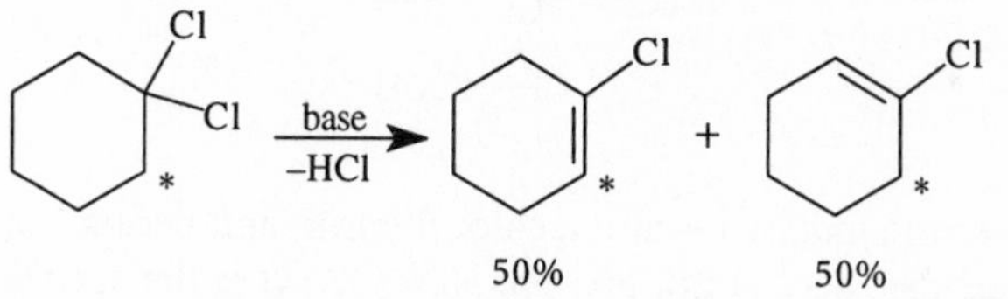

(b) A *cyclohexyne* intermediate—the cyclohexane analog of benzyne—is formed from each of the labeled 1-chlorocyclohexenes, and each cyclohexyne can react with a "phenyl anion" (that is, phenyllithium) at two different carbon atoms with equal probability. This leads to the labeling pattern shown in the problem.

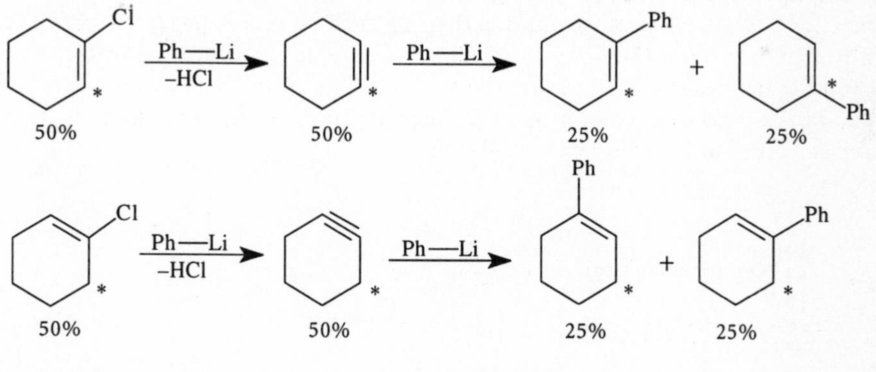

This labeling experiment was used to gather evidence for or against a cyclic allene intermediate, 1,2,-cyclohexadiene:

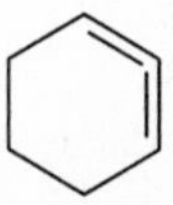

1,2-cyclohexadiene

The labeling pattern of the products *clearly excludes* this intermediate. Can you see why? What labeling pattern would be predicted if this were the intermediate?

18.45 (a) This is a nucleophilic aromatic substitution reaction. Notice that the cyano group ($:N{\equiv}C{-}$) can stabilize the intermediate *A* by resonance.

(solution continues)

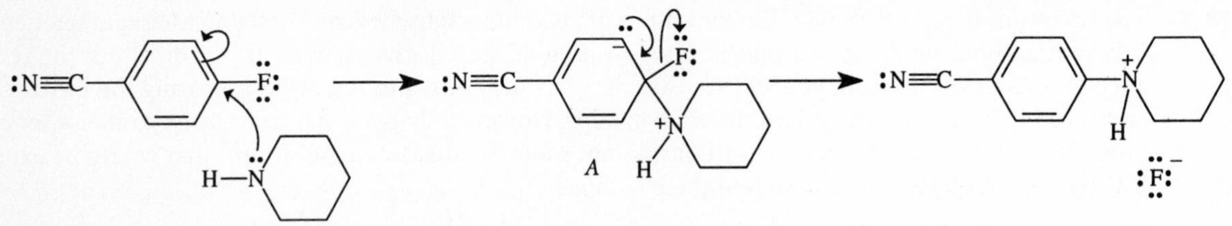

(c) The Lewis acid $ZnCl_2$ promotes the formation of a complex *B* or a benzyl carbocation *C*, either of which can serve as the electrophile in an electrophilic aromatic substitution reaction. (The carbocation *C* is the electrophile in the mechanism shown below.)

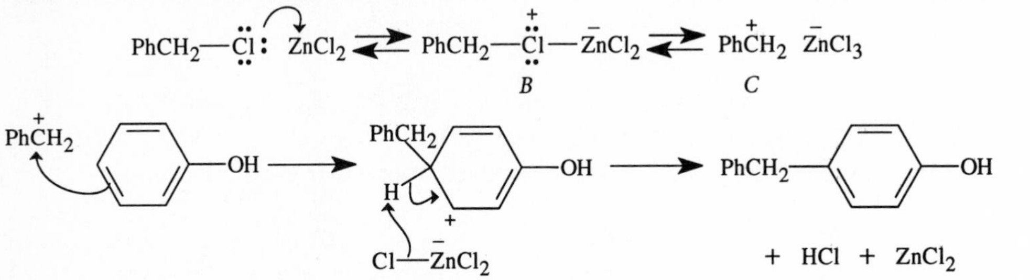

18.46 Because the glycol starting material is also a chlorohydrin, and because the phenoxide salt is a base, epoxide formation takes place. (See Sec. 11.2B, text pp. 505–506.) It is the epoxide that actually alkylates the phenol. The hydroxide ion produced in the last step maintains the phenol in its ionized form.

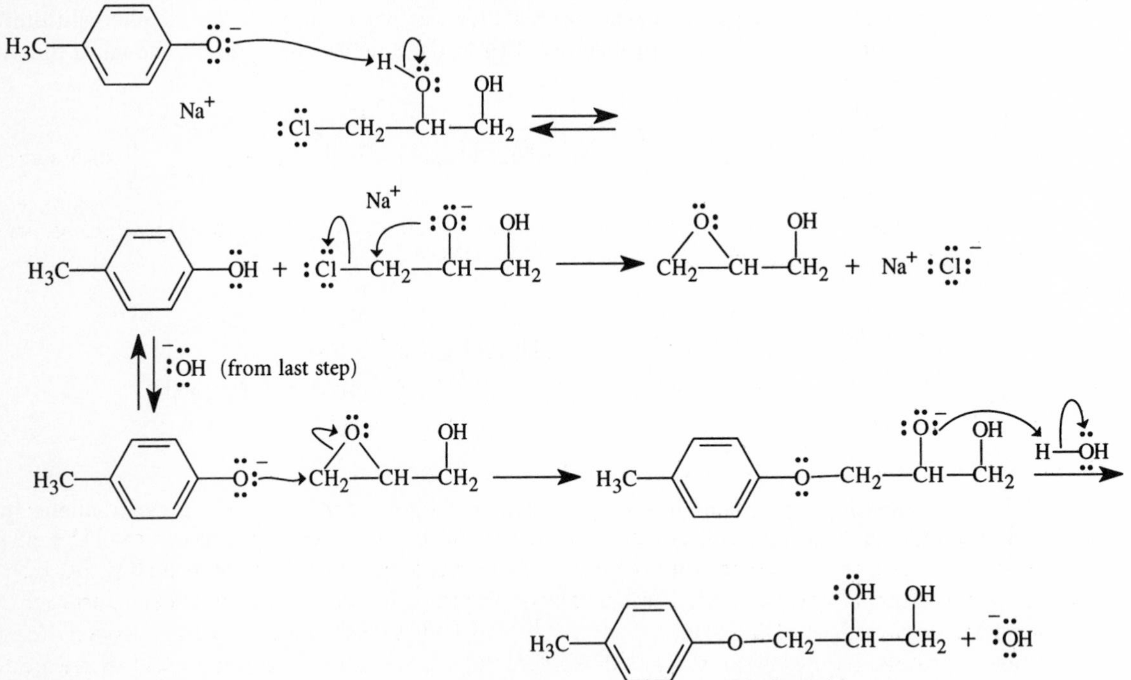

18.47 In 1-chloro-4-nitrobenzene, the dipole contributions of the carbon-chlorine bond and the nitro group are oriented in opposite directions; consequently, the net dipole moment is smaller than it is in nitrobenzene, in which only a dipole contribution of the nitro group is present.

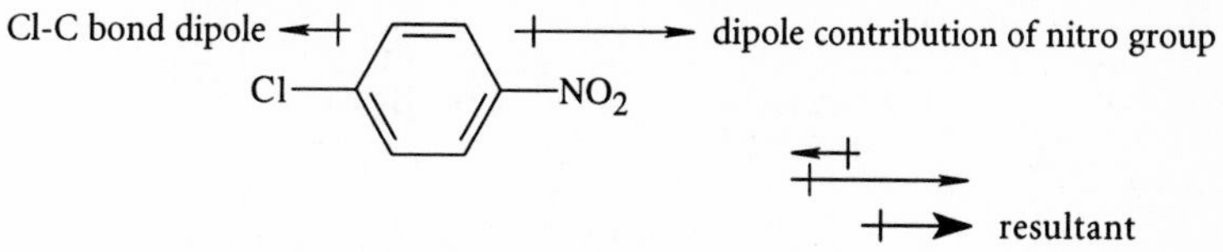

Cl-C bond dipole ←+   +→ dipole contribution of nitro group

Cl—⟨ ⟩—NO₂

+
→ resultant

The analysis of *p*-nitroanisole is similar, except that there is an additional effect: the very powerful electron-donating resonance effect of the methoxy group. Delocalization of an unshared electron pair from the oxygen of the methoxy group into the nitro group creates separation of charge that results in a large dipole moment. A similar resonance effect is present in 1-chloro-4-nitrobenzene, but, because the resonance effect of a chloro group is much weaker than that of a methoxy group (Fig. 16.8, text p. 770), it is not sufficient to offset the opposing contribution of the chlorine-carbon bond dipole.

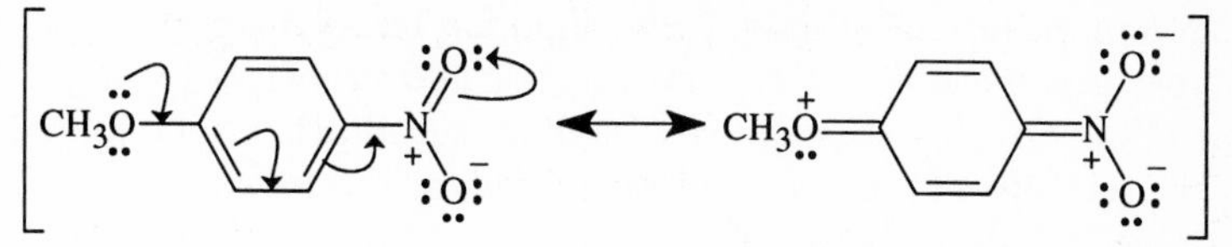

This resonance structure has
a large separation of charge, and
hence a significant contribution
to the dipole moment.

18.48 Deduce the structure of the "very interesting intermediate" by mentally imagining a "reverse Diels-Alder" reaction of triptycene that yields anthracene and the intermediate, which is benzyne:

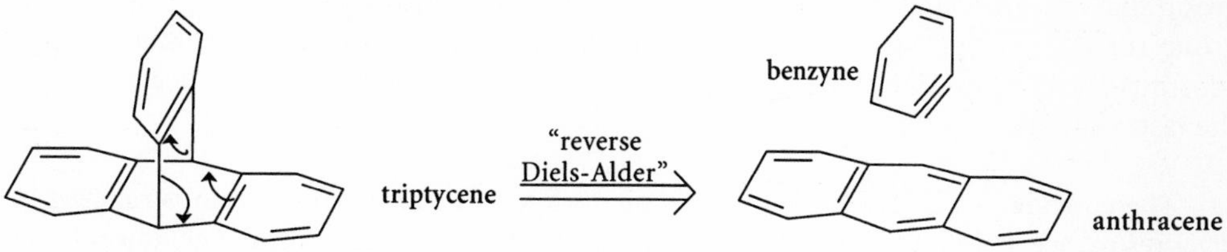

benzyne

triptycene   "reverse
Diels-Alder"→   anthracene

The Grignard reagent has carbanion character, and this "carbanion" is a strong base. Elimination of the weaker base fluoride gives benzyne:

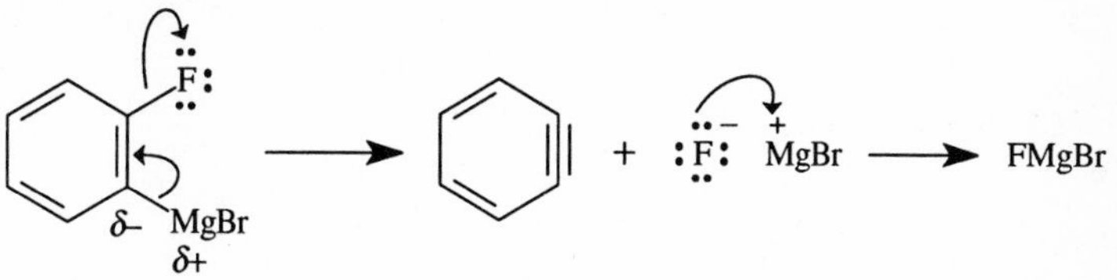

δ− MgBr
δ+   →   +   :F⁻   ⁺MgBr   →   FMgBr

18.49 The six-proton singlet in the NMR spectrum indicates that the two methyl groups *ortho* to the oxygen have been retained; this resonance cannot be due to one *ortho* methyl and one *para* methyl, because these would not be chemically equivalent. The two-proton singlet at δ 5.49 indicates two alkene protons, and the two-proton singlet at δ 6.76 is accounted for by the two ring protons, which evidently remain. The total number of protons accounted for by the NMR spectrum are two fewer than are present in the starting material. The conditions are much like those for oxidation of a phenol to a quinone. The structure below for compound *A* fits the data; the singlets are all broad because of very slight splitting over more than three bonds.

(solution continues)

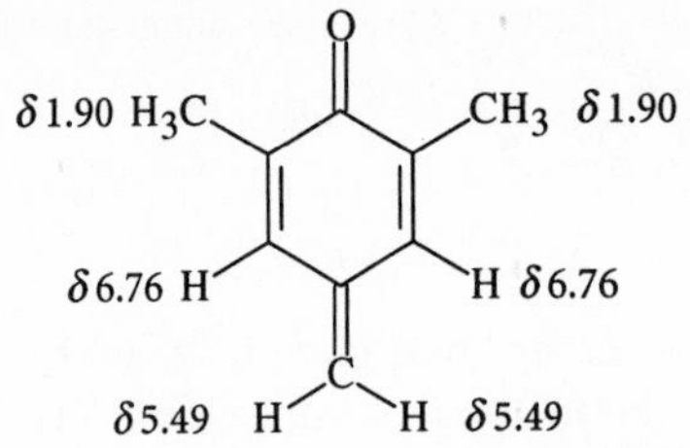

Compound *A*

18.50 The two reactions are intramolecular Friedel-Crafts acylations. The first equivalent of $AlBr_3$, a strong Lewis acid, forms a complex with the acid chloride group to generate the electrophile, an acylium ion, as discussed in the text on pp. 754–755, and is also consumed by formation of a complex with the product. As the lower reaction in the problem shows, the acylium ion reacts at a position *para* to a methoxy group, because a methoxy group is both *ortho, para*-directing and strongly activating. The question, then, is why reaction occurs at the ring that does *not* have methoxy substituents when more Lewis acid is present. The answer is that the second and third equivalents of Lewis acid form complexes with the methoxy groups:

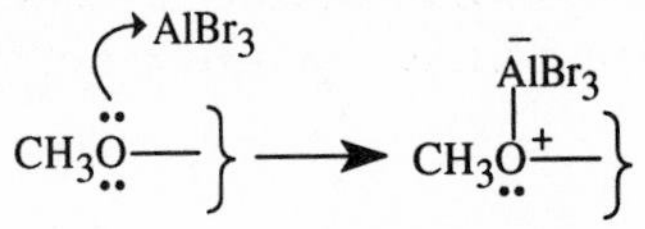

A complexed methoxy group cannot donate an unshared electron pair by resonance, because its unshared pair is already "donated" to the aluminum. Consequently, the methoxy group, through this complexation, is transformed into an electronegative group that is incapable of a resonance effect. An electronegative group that cannot donate electrons by resonance is a *deactivating group*, and a ring containing such a group (or two such groups) by definition undergoes substitution more slowly than a ring without such substituents. In other words, acylation of the ring that does not contain such deactivating substituents is the faster acylation.

18.51 (a) The epoxide slowly and spontaneously opens for three reasons: (1) strain is relieved in the epoxide; (2) a resonance-stabilized carbocation is formed; and (3) the heterolysis of a chemical bond (that is, its dissociation into fragments of opposite charge) is favored by the polar, protic, donor solvent that can solvate both the carbocation and the oxygen anion. The mechanism, then, is as follows:

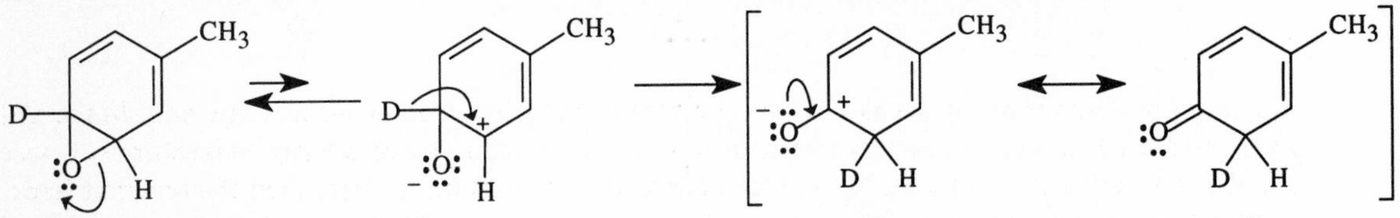

The driving force for this rearrangement is formation of the neutral ketone *K*. The "enol" of ketone *K* is the product phenol; this enol forms because of its aromaticity. In the formation of this product either the H or the D must be lost; loss of the H gives the product that contains deuterium; loss of the deuterium gives the product that contains hydrogen. Because of the primary deuterium isotope effect (see text pp. 405–406), a carbon-deuterium bond is broken more slowly than a carbon-hydrogen bond. (The product distribution suggests that the isotope effect is 75/25, or 3.) A mechanism for conversion of *K* into the deuterium-containing product is as follows:

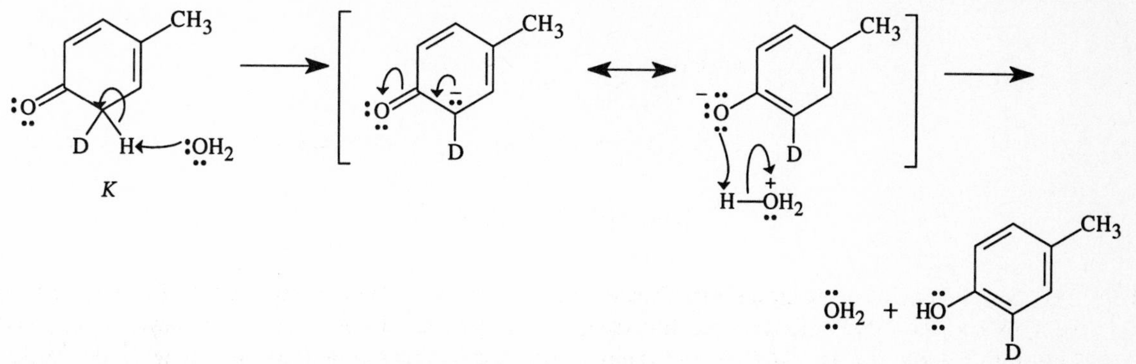

Some of the deuterium is lost from intermediate *K* (by the same mechanism) because the primary isotope effect is not very large. (Only an "infinite" isotope effect would result in the retention of all deuterium.)

(b) The position of the methyl group relative to the oxygen of the product is determined by which bond of the epoxide breaks in the first step of the mechanism. Breaking different bonds leads to carbocations of different stabilities. The carbocation intermediate involved in the formation of *p*-cresol has the following resonance structures:

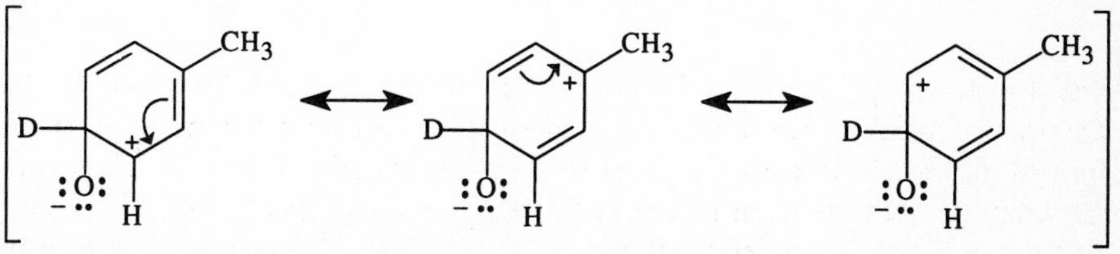

Notice that the middle structure is a *tertiary* carbocation. The carbocation intermediate involved in the formation of *m*-cresol has the following resonance structures:

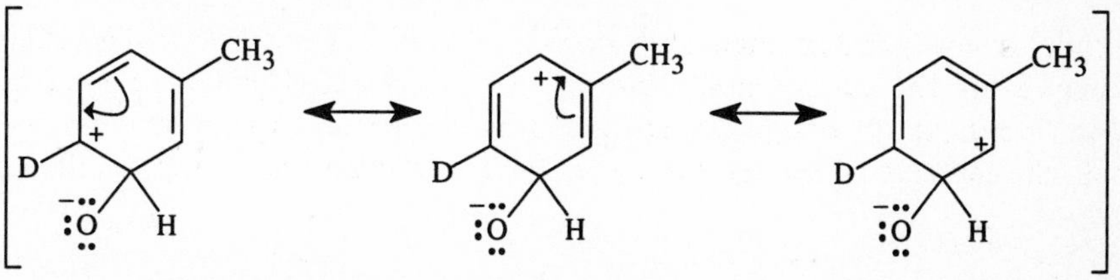

(Convince yourself that application of a mechanism to this carbocation similar to the one shown above gives *m*-cresol.) All of the resonance structures of this carbocation are *secondary* carbocations. Consequently, the carbocation involved in the formation of *p*-cresol is more stable than the one that leads to *m*-cresol. As is usually the case, the pathway involving the more stable intermediate—the pathway that gives *p*-cresol—is faster. This is why *p*-cresol is the observed product.